£75-

VOLUME 22, WINTER THEMATIC ISSUE　　　　　　　　1987

Geological Journal

African Geology Reviews

Published by John Wiley & Sons Ltd
on behalf of the
Liverpool Geological Society and the
Manchester Geological Association
ISSN 0072–1050
ISBN 0 471 91595 5

GEOLOGICAL JOURNAL (Geol. j.)

WINTER THEMATIC ISSUE
CONTENTS

VOLUME 22, TI   WINTER 1987

Foreword: P. Bowden and J. Kinnaird 1

Section 1: Studies on the Archaean Rocks of Early Precambrian Times: P. Bowden and J. Kinnaird 3

The 3.5 Ga Barberton Greenstone Succession, South Africa: Implications for Modelling the Evolution of the Archaean Crust: I. A. Paris 5

Mineralization in the Migori Greenstone Belt, Macalder, Western Kenya: J. S. Ogola 25

Stable Isotopes in the Carbonates and Kerogen from the Archaean Stromatolites in Zimbabwe: P. I. Abell and J. McClory 45

Polymetamorphic, Al, Mg-rich Granulites with Orthopyroxene-sillimanite and Sapphirine Parageneses in Archaean Rocks from the Hogger, Algeria: J. R. Kienast and K. Ouzegane 57

Section 2: Proterozoic Orogenies, Magmatism, and Related Metallogeny: P. Bowden and J. Kinnaird 81

The Kibaran Cycle in Angola—a Discussion: H. Carvalho, J. P. Crasto, Z. C. G. Silva and Y. Vialette 85

Metallogeny of the Northeastern Kibaran Belt, Central Africa: W. Pohl 103

Implications of a Palynological Study in the Upper Precambrian from Eastern Kasai and Northwestern Shaba, Zaire: D. Baudet .. 121

Precambrian Gabbro–anorthosite Complexes, Tete Province, Mozambique: M. W. C. Barr and M. A. Brown 139

Contents continued on page (iv)

Advertising: For details contact—
U.K. and Europe: Michael J. Levermore, Advertisement Manager, John Wiley & Sons Ltd., Baffins Lane, Chichester, Sussex PO19 1UD, England (Telephone 0243 770350).
North America: Arthur C. Lipner, National Advertising Sales Representative, Weston Media Associates, 184 Judd Road, Easton, CT 06612, U.S.A. (Telephone 203 261 2500).

Indexed or abstracted by 'Current Contents', 'Science Citation Index' and other leading Indexing/Abstracting Services.

GELJA8 22(TI WINTER) 1–578 (1987)
ISSN 0072–1050
ISBN 0 471 91595 5

Late Proterozoic Tectonic Terranes in the Arabian–Nubian Shield and their Characteristic Mineralization: J. R. Vail 161

Ductile Shear Zones in the Northern Red Sea Hills, Sudan and their Implication for Crustal Collision: D. C. Almond and F. Ahmed 175

The Subduction- and Collision-related Pan-African Composite Batholith of the Adrar des Iforas (Mali): a Review: J. P. Liégeois, J. M. Bertrand and R. Black 185

Geochemistry of the Tioueine Pan-African Granite Complex (Hoggar, Algeria): A. Azzouni-Sekkal and J. Boissonnas 213

Reconnaissance Geological Mapping and Mineral Exploration in Northern Sudan using Satellite Remote Sensing: P. S. Griffiths, P. A. S. Curtis, S. E. A. Fadul and P. D. Scholes 225

Section 3: Rare-metal Pegmatites and their Mineralization: P. Bowden and J. Kinnaird 251

Mineralized Pegmatites in Africa: O. von Knorring and E. Condliffe . . 253

Nigerian Rare-metal Pegmatites and their Lithological Framework: G. Matheis 271

Section 4: Phanerozoic Anorogenic Magmatism: Plate Tectonic Implications and Mineralization: P. Bowden and J. Kinnaird 293

African Anorogenic Alkaline Magmatism and Mineralization— a Discussion with Reference to the Niger–Nigerian Province: J. Kinnaird and P. Bowden 297

Palaeozoic Drift of Gondwana: Palaeomagnetic and Stratigraphic Constraints: F. B. Van Houten and R. B. Hargraves 341

The Goedynamic Significance of Alkaline Magmatism in the Western Mediterranean Compared with West Africa: B. Bonin, B. Platevoet and Y. Vialette 361

Geochemical Characteristics of the Nigerian Anorogenic Province: R. A. Batchelor 389

Accessory Mineralogy of the Ririwai Biotite Granite, Nigeria, and its Albitized and Greisenized Facies: R. A. Ixer, J. R. Ashworth and C. M. Pointer 403

Section 5: Continental Rifting and Continental Margins: Implications from Structural and Sedimentological Studies: P. Bowden and J. Kinnaird 429

Post-Hercynian Tectonics in Northern and Western Africa: R. Guiraud, Y. Bellion, J. Benkhelil and C. Moreau 433

CONTENTS

Cretaceous Deformation, Magmatism, and Metamorphism in the Lower Benue Trough, Nigeria: J. Benkhelil 467

A Model for Rift Development in Eastern Africa: J. Chorowicz, J. Le Fournier and G. Vidal 495

Petrology and Geodynamic Significance of the Tertiary Alkaline Lavas from the Kahuzi–Biega Region, Western Rift, Kivu, Zaire: R. T. Lubala, A. B. Kampunzu, J. P-H. Caron and P. J. Vellutini 515

African Transform Continental Margins: Examples from Guinea, the Ivory Coast and Mozambique: J. Mascle, D. Mougenot, E. Blarez, M. Marinho and P. Virlogeux 537

Quaternary Continental Margin Sedimentation off the Southeast Coast of South Africa: F. Westall 563

The editors would like to thank the following for their helpful and constructive improvements to papers included in this volume:

J-P. Bassot, Clermont-Ferrand, France; R. A. Batchelor, St Andrews, Scotland; J. D. Bennett, Keyworth, England; B. Bonin, Orsay, France; L. Burnol, BRGM, France; R. Caby, Montpellier, France; J. Chorowicz, Paris, France; M. P. Coward, London, England; C. H. Donaldson, St Andrews, Scotland; S. A. Dury, Milton Keynes, England; K. J. Dorning, Sheffield, England; J. G. Fitton, Edinburgh, Scotland; I. G. Gass, Milton Keynes, England; I. Gilmour, Milton Keynes, England; B. D. Hackman, Nairobi, Kenya; P. J. Hamilton, East Kilbride, Scotland; C. J. Hawksworth, Milton Keynes, England; R. A. Ixer, Birmingham, England; J. Klerkx, Tervuren, Belgium; A. R. MacGregor, St Andrews, Scotland; J. McManus, Dundee, Scotland; G. Matheis, Berlin, Germany; P. Moseley, Keyworth, England; J. S. Ogola, Nairobi, Kenya; I. A. Paris, Christchurch, New Zealand; W. S. Pitcher, Liverpool, England; R. J. Reavy, Liverpool, England; R. Sacchi, Torino, Italy; S. Saradeth, Munich, Germany; R. A. Scrutton, Edinburgh, Scotland; B. W. Sellwood, Reading, England; J. Sougy, Marseille, France; F. B. Van Houten, Princeton, USA; R. T. Watkins, Cape Town, South Africa; J. A. Weir, St Andrews, Scotland; P. Wright, Bristol, England; T. Young, Sheffield, England.

The editors would also like to thank all the reviewers who wished to remain anonymous but nevertheless aided the contributors in the preparation of the final manuscripts. Misha Sanderson assisted the editors in checking the references and Kit Finlay valiantly retyped the index. May and Kate Bowden provided meals and coffee, and support with the proof reading. Finally special mention is due to Helen Bailey, Nicky Smith, and Karen Hawes at John Wiley for helping this compilation to reach fruition.

Foreword

The papers in this volume represent some of the best presentations at the 13th Colloquium of African Geology, held at the University of St Andrews in September 1985. For simplicity the papers are grouped into five sections based partly on a chronostratigraphic timescale starting with the Archaean and ending with modern sedimentological studies in coastal basins.

There are however important themes linking all the papers irrespective of their time dependence. In particular the studies on rifting, sedimentary basins, and mineralization are developed as major reviews in this volume. In our selection of papers for inclusion we have been fortunate to have the support of specialist reviewers who have willingly provided critical comments to which the authors have responded. We have also been aided by the loyal assistance of Francoise Blackbourn who valiantly acted as secretary and translator up to November 1986.

Several authors were unable to return their revised manuscripts in time for complete translation from French into English. Because of our intention to have a carefully reviewed volume with the majority of papers in English there has inevitably been some delay in publication. We hope however that readers will find the contributions to be of lasting value, and a permanent record of the successful Colloquium of African Geology held at the University of St Andrews in September 1985.

Peter Bowden
Judith Kinnaird
Department of Geology,
University of St Andrews,
St Andrews, Scotland.
April 1987

© 1987 by John Wiley & Sons, Ltd.

Section 1
Studies on the Archaean rocks of early Precambrian times

1. Introduction

The papers dealing with the Archaean provide different insights into the problems of interpretation of dynamic processes, the formation of the Archaean crust, and the influence of greenstone belts as potentially metalliferous horizons.

The paper by Isobel Paris examines the Barberton greenstone succession and concludes that there are major problems with the classic Barberton greenstone belt model. A revised stratigraphy suggests that Archaean oceanic crust was overlain by submarine fan deposits and alluvial fan sediments. This sedimentation occurred contemporaneously with a collision event, which resulted in the obduction of the Barberton succession. The important implication is that Archaean greenstone horizons may be interpreted as ancient ophiolites.

Greenstone belts are also potentially metalliferous, particularly if hydrothermal fluids from late granitic intrusives can redeposit and concentrate economic ore minerals. Jason Ogola has studied the Macalder copper sulphide deposit, located to the western end of the Migori greenstone belt in western Kenya at the northern margin of an intensely dislocated metamorphosed Archaean volcanogenic sedimentary sequence. It is an ore deposit interpreted as having a metasomatic origin with hydrothermal fluids interacting with an Archaean banded iron formation. Copper, zinc, gold, and silver have all been mined from the Macalder ore body.

For some workers the banded iron formations represent the period in Earth's history when the atmosphere began to build up in oxygen and blue-green algae commenced to flourish. One then finds stromatolite colonies as important markers. A detailed isotopic analysis of a half-metre core from the Cheshire Formation stromatolites of the Belingwe greenstone belt, Zimbabwe, is presented by Abell and McCory, including carbon isotopic ratio measurements on the kerogen fraction, and carbon and oxygen isotope ratio measurements on the limestone.

Metamorphic overprinting of Archaean rocks by later African orogenies is considered by Kienast and Ouzegane. In particular they are concerned with the mineral parageneses in an orthopyroxene–sillimanite association from Al–Mg rich granulites within Archaean aluminous metapelites at In Hihaou (NW Hoggar). Sapphirine is characteristically present as a reaction product of a high pressure mineral association related to an important metamorphic event (M1). Other reaction products include cordierite and a second generation of orthopyroxene, confirming the presence of a second P–T event (M2). The authors consider that the second event may be correlated with a younger orogeny possibly linking it to the Eburnean.

© 1987 by John Wiley & Sons, Ltd.

The 3.5 Ga Barberton Greenstone succession, South Africa: implications for modelling the evolution of the Archaean crust

I. A. Paris

Department of Geology, University of Canterbury, Christchurch, New Zealand

---

Recent structural and sedimentological analyses of the Barberton greenstone succession have revealed the existence of major problems with the classic Barberton greenstone belt model that was developed in the 1960s. In the light of recent results, a revised stratigraphy is discussed. The sequence is shown to have undergone at least four distinct deformation events, of which the first two involved important tectonic repetition and thickening of the original succession. The sequence is interpreted as Archaean oceanic crust overlain by submarine fan deposits and alluvial fan sediments. This sedimentation which terminates the stratigraphic succession occurred contemporaneously with a collison event, which resulted in the obduction of the Barberton succession.

KEY WORDS   Archaean evolution   Barberton Greenstones   Ophiolites

1. Introduction

The 3.5 Ga Barberton greenstone succession is located in the eastern Transvaal (South Africa) and in Swaziland (Figure 1). It is formed by the non-granitoid component of the Barberton granite/greenstone terrain (*cf* Vearncombe *et al.* 1986) and comprises volcanic and sedimentary rocks that have been metamorphosed to lower greenschist facies. The greenstone succession, the base of which has been dated at 3.5 Ga (Sm/Nd, Hamilton *et al.* 1979) is surrounded by granitoids varying in age between 3.5 and 2.5 Ga (Rb/Sr whole rock, Anhaeusser and Robb, 1981; Barton *et al.* 1980, 1983). The oldest surrounding granitoids are coeval or younger than the supracrustals and both intrusive and tectonic contacts have been observed (Anhaeusser 1983).

An important amount of research was carried out in the Barberton greenstone/granite terrain in the 1960s, principally by M. J. Viljoen and C. R. Anhaeusser. From their results, these authors formulated a model of evolution for the Barberton greenstone succession which has been widely used for the study of Archaean successions elsewhere in the world (Condie 1981; Tankard *et al.* 1982; amongst

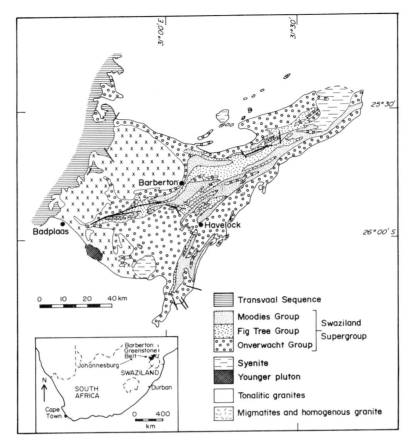

Figure 1. Simplified geological map of the Barberton greenstone belt (modified after SACS 1980).

others). This model relies on the existence of conditions particular to the Archaean such as primary peridotitic/komatiitic magma, absence of sialic crust and plates, absence of tangential forces during deformation (Viljoen and Viljoen 1969, 1970; Anhaeusser 1978). The stratigraphy of the terrain was then defined in view of this model by the South African Committee for Stratigraphy (SACS 1980). For an extensive review of the Barberton model, the reader is referred to Anhaeusser (1978, 1983); Viljoen and Viljoen (1969, 1970) and Tankard et al. (1982) as only a brief outline of this model is presented below, prior to discussion of more recent work.

2. The Barberton model

In the Barberton greenstone belt, the supracrustal rocks form a 24 km thick stratigraphic pile, the Swaziland Supergroup, which comprises the Onverwacht, the Fig Tree and the Moodies Group (Table 1, Figure 1).

Table 1. The stratigraphy of the Barberton greenstone belt, as defined by SACS (1980)

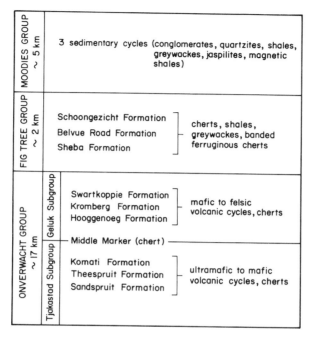

2a. The Onverwacht Group

The Onverwacht Group forms the base of the sequence and is about 17 km thick (SACS 1980). It is subdivided into six formations (Table 1). The lower three, which form the Tjakastad Subgroup, consist of cycles of mafic to ultramafic lava commonly capped by chemical chert deposits (Table 1). Serpentinized ultramafic sills are also present within the Tjakastad Subgroup.

The upper three formations of the Onverwacht Group, which form the Geluk Subgroup (Table 1), comprise cycles of lava varying from mafic at the base to intermediate and acid at the top. They are also generally separated by a chert horizon, the Middle Marker, which as it separates an ultramafic to mafic from a mafic to felsic unit, is believed to represent a major crustal break (Table 1).

2b. The Fig Tree Group

The Fig Tree Group is a 2 km thick argillaceous sedimentary sequence, divided into three formations, comprising banded iron formations, ferruginous cherts and shales, greywackes and sandy shales (Table 1). In the northern part of the belt, the top of the Fig Tree Group is marked by a trachytic horizon. In the southern part of the belt, a southern facies comprising banded iron formations and conglomerates is developed. The Fig Tree Group is overlain by the Moodies Group.

2c. The Moodies Group

The Moodies Group is an arenaceous sedimentary sequence about 5 km thick, formed by three successive sedimentary cycles. It comprises banded iron formations, jaspilite and shales, greywackes, quartzites and conglomerates (Table 1).

The 24 km sequence described above developed as an evolving volcanic complex on primitive Archaean simatic crust, to be subsequently overlain by a prograding submarine fan system and marginal marine sediments. During the formation of the greenstone succession, the nature of the earth's volcanicity varied from ultramafic and mafic to mafic and felsic, this fundamental break being marked by the Middle Marker. During the formation of the supracrustals, intrusion of adjacent granitoids led to deformation and to the emplacement of a sialic crust. Deformation took place through folding, faulting and warping due to downsagging of the greenstone terrain and upwelling of the granitoids. The degree of metamorphism does not exceed the lower greenschist facies away from the granitoid margins, where it can locally reach the amphibolite facies. Three successive cycles of granite emplacement have been distinguished (Anhaeusser and Robb 1981).

Several aspects of the above model were soon criticized: Burke *et al.* (1976) questioned the existence of a 24 km thick conformable supracrustal pile devoid of structural breaks; Williams and Furnell (1979) reported that the southern part of the belt and more particularly the lower Onverwacht Group, was more deformed than suggested in the Viljoen's model; Fripp *et al.* (1980) and De Wit (1982) pointed out that the polyphase deformation which has been observed in the northern part of the belt by Ramsay (1963) was not taken into account. With these problems in mind, an important amount of work was carried out over the last few years by De Wit, Lamb and the present author. Although it takes into account the results of coworkers as well as recently published results, this paper mostly summarises the results of the author.

3. Stratigraphic and sedimentological results

Detailed mapping carried out by Williams and Furnell (1979), De Wit (1983), Lamb (1984a), and Paris (1985a, 1986) has shown that the southern part of the greenstone succession underwent polyphase deformation and intense metasomatism. Structural breaks have been observed throughout the supracrustal pile of the Barberton greenstone succession, which are not taken into account in the presently accepted stratigraphy. As a result this stratigraphy is a very impractical tool as so-called 'stratigraphic' entities commonly refer to more than one tectonic block. Furthermore, tectonic blocks which contain similar rock types or sequences are given different stratigraphical values in different parts of the belt and vice versa.

An attempt has been made to reconstruct the original supracrustal succession for the southern part of the belt (Table 2), using the study of lithofacies assemblages, lateral and horizontal facies variations and taking into account the tectonic breaks. This revised stratigraphy comprises the Onverwacht Group, the Diepgezet Group and the Malolotsha Group and is presented below (Table 2; Lamb and Paris in press).

3a. The Onverwacht Group

The Onverwacht Group, as defined by SACS (1980) is a lithotectonic complex (De Wit *et al.* 1983; Paris 1985a) of which the original stratigraphy has not yet been entirely reconstructed (work under progress, De Wit personal communication). However, it can be broadly described as being composed of a 2 to 3 km thick ultramafic to mafic igneous unit overlain by a thin (a few hundred metres) volcaniclastic layer (De Wit *et al.* 1983, 1986; Paris 1985a).

Table 2. Revised stratigraphy of the southern part of the Barberton greenstone belt (after De Wit 1985; Lamb and Paris under review)

MALOLOTSHA GROUP ~ 2 km	F 5	quartz-arenite, siltstone	CONTINENTAL ALLUVIAL FAN
	F 4	conglomerate with a matrix of both chert and single crystal quartz grains, chert-quartz arenite, conformable to unconformable	
DIEPGEZET GROUP ~ 2 km	F 3	chert-arenite, conglomerate with a matrix formed by chert grains	OCEANIC PROGRADING SUBMARINE FAN
	F 2	ferruginous and tuffaceous siltstone, ferruginous chert-arenite	
	F 1	jaspilites, ferruginous chert, ferruginous tuff, shale and siltstone, conformable (?)	
ONVERWACHT GROUP ~ 3 km		volcaniclastic unit (distal and proximal turbidites facies and subaerial facies), mafic and ultramafic unit	OPHIOLITE ARCHEAN OCEANIC CRUST
	Unconformity ? tectonic contact ?		
granitoid			SIALIC CRUST

(Predominantly silicified — F1 through upper Onverwacht)

Granitoid gneisses, similar to those of the Ancient Gneiss Complex of Swaziland and unconformably overlain by a diamictite, have been observed within one of the basal Onverwacht tectonic slivers (De Wit et al. 1983). Such relationships suggest the existence of a pre-greenstone sialic crust and may indicate the presence, at least locally, of a basement to the greenstone succession (De Wit et al. 1983). Such an hypothesis was previously advocated by Hunter (1974) who considered the Ancient Gneiss Complex to be an extensively reworked remnant of pre-existing gneissic crust.

The upper part of (<1 km) of the Onverwacht Group has been totally to partially silicified, with both igneous and sedimentary rocks affected by the silicification (De Wit et al. 1982; Paris et al. 1985). A graduation can be observed from ultramafic and mafic rocks to partially silicified rocks to totally silicified rocks. Such variations were previously interpreted as the result of cyclic volcanism (Viljoen and Viljoen 1969, 1970). However, textures of both mafic and ultramafic volcanics can be recognized in the silicified igneous rocks, thus indicating their original composition. Such textures include spinifex textures, pseudomorphs after olivine, hopper and skeletal crystals and ophitic textures (Paris et al. 1985). In the totally silicified rocks, the crystals showing these textures have been entirely replaced by microquartz. In addition, partially silicified igneous rocks are commonly strongly carbonated (Figure 2).

The silicification of the volcaniclastic sediments of the Onverwacht Group has led to excellent preservation of sedimentary grains and structures, including graded bedding, cross laminae, and ripple laminae (Figure 3).

No evidence of compaction has been observed in either the igneous or the sedimentary rocks and the preservation of delicate primary textures such as

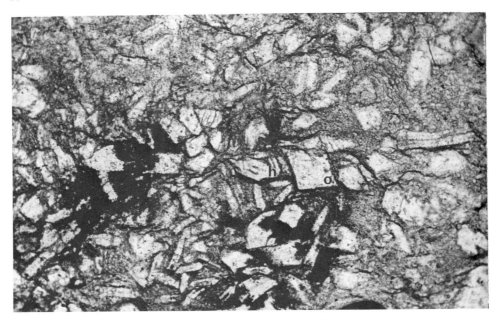

Figure 2. Photomicrograph of pillow lava showing silicified euhedral crystals, presumed to be olivine (O) and silicified hopper crystals (h) in a sericite and iron oxide groundmass. The silicified crystals are composed of microquartz. Josefsdal Farm. Field of view = 3.5 mm. Plane polarized light.

spinifex textures and sedimentary structures and grains (Figures 2, 3) suggests that very fine intergranular silica replacement occurred on a molecular scale. In places, hydraulic fracturing is associated with the silicification (Paris *et al.* 1985). Zones of banded black and white cherts formed entirely after colloform silica can be seen to pass along strike into silicified sedimentary rocks and are interpreted as representing a more advanced stage of silicification (Paris 1985a). The silicification of the Onverwacht Group has been interpreted as the result of synsedimentary hydrothermalism (De Wit *et al.* 1982; Paris *et al.* 1985).

The recognition of marker horizons, such as units of lapilli and/or accretionary lapilli (Figure 4) allows broad correlations from one tectonic unit to the next (Heinrichs 1984; Paris 1985a and in preparation). Sedimentological studies of the silicified volcaniclastic rocks have shown that three different facies can be distinguished; namely a distal turbidite facies, a proximal turbidite facies (Stanistreet *et al.* 1981; Heinrichs 1984; Paris 1985a, in preparation), and a subaerial facies (Byerly *et al.* 1986; Lowe and Knauth 1977, 1978; Paris 1985a, and in preparation). In all three facies, similar volcanogenic products can be recognized. The Onverwacht volcaniclastic sediments have been interpreted as resulting from a phase of phreatomagmatic explosion (Heinrichs 1984; Paris 1985a, in preparation). The characteristics of the volcanogenic products indicate that the volcanic centres were subaerial and that deposition took place both subaerially on the volcanic slopes and as a result of submarine pyroclastic flow (Heinrichs 1984; Paris 1985a).

The Onverwacht Group is then overlain, apparently conformably, by a × 3 km thick upward coarsening sedimentary sequence informally labelled the Diepgezet Group (Lamb 1984a; Lamb and Paris under review). The contacts between the

Figure 3. Silicified volcaniclastic sediments of the Onverwacht Group. Note the preservation of delicate sedimentary structures such as trough cross-laminae (arrow), Dunbar Farm.

Onverwacht Group and the Diepgezet Group are commonly masked by silicification and could contain an obscured but nevertheless important break: for example, mud pool structures interpreted as subaerial deposits (De Wit *et al.* 1982) have been observed at the base of the Diepgezet Group, within rocks which otherwise have deep water affinities, overlying the turbiditic facies of the

Figure 4. Accretionary lapilli of the proximal turbiditic facies. Onverwacht Group. Sample from a Bouma unit of a turbidite cycle, Dunbar Farm.

Onverwacht sedimentary rocks. Furthermore, the presence of such mud pool structures seems to be linked to the formation of D1 tectonic stacks which probably formed while basal Diepgezet sediments were being deposited in other parts of the basin(s?).

3b. The Diepgezet Group

The Diepgezet Group comprises three different facies, F1, F2, F3, which are interbedded in an upward coarsening manner (Figure 5). The three different facies of the Diepgezet Group have been defined as follows (Paris 1985a; Lamb 1984a; Lamb and Paris under review): F1, jaspillite, ferruginous chert, ferruginous shale and ferruginous tuff; F2, ferruginous and tuffaceous siltstone and ferruginous chert–arenite; F3, massive chert–arenite, pebbly chert–arenite, matrix supported granule conglomerate and clast supported conglomerate. The matrix of the Diepgezet arenites and conglomerates is formed by chert grains and the conglomerate clasts are derived from the Onverwacht silicified rocks as well as from the F1 facies.

The Diepgezet Group comprises successively from bottom to top, hemipelagic sediments, fine to medium grained turbidity current deposits and 'coarse grained turbidites' with a gradational transition between the different types (Figure 5).

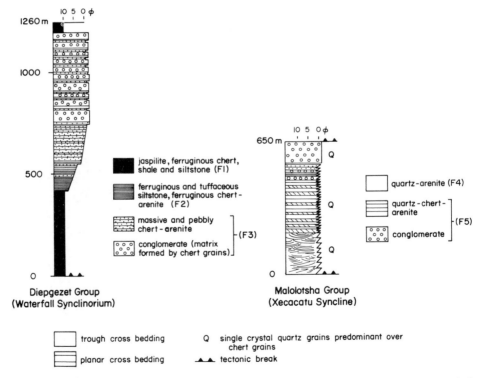

Figure 5. Measured sections of the post-Onverwacht sedimentary sequences. The section of the Diepgezet Group is from the Waterfall Synclinorium and the one of the Malolotsha Group from the Xecacatu Syncline (see Figure 9 for location).

When incorporated within the D1 tectonic slides, the basal Diepgezet sediments are generally silicified.

The Diepgezet Group has been interpreted as a prograding submarine fan (Paris 1985a). This submarine fan is in turn overlain by the Malolotsha Group (Lamb 1984a; Lamb and Paris under review).

3c. The Malolotsha Group

The Malolotsha Group is a coarse clastic sequence with a maximum observed thickness of 1.8 km. The nature of the contact between the Diepgezet and the Malolotsha Groups varies, from conformable contacts passing along strike into unconformities (Lamb 1984a, b).

The Malolotsha Group consists two separate facies (F4 and F5) which pass into one another laterally and/or vertically (Figure 5): an immature quartz-chert–arenite, conglomerate facies and a supermature quartz-arenite, siltstone facies. However, sedimentary rocks of the Malolotsha Group are further characterized by a predominance of detrital quartz grains over detrital chert grains in the matrix. In the conglomerates, the clasts are derived from the underlying Diepgezet and Onverwacht Groups, and in the eastern part of the belt, granitoid clasts, considered to be derived from the Ancient Gneiss Complex have been observed (Lamb 1984a).

The unconformities developed between the Diepgezet and the Malolotsha Groups can cut down to the Onverwacht Group (Lamb 1984a; Paris 1985a). Internal unconformities are also present within the Malolotsha Group itself (Lamb 1984a, b). Conformable sequences between the two groups can be observed along strike from the unconformities and in such cases, the passage between the two groups is gradational, with sediments of mixed characteristics (for example, conglomerates with a matrix of both chert and quartz grains in approximately equal proportions). This type of succession can be problematical when trying to attribute thin fault bonded sequences to one group or the other. In order to avoid biased interpretation, it was decided that such ambiguous sequences would be classified as intermediate between the Diepgezet and Malolotsha Groups (Lamb 1984a; Paris 1985a; Lamb and Paris under review).

In summary, the original stratigraphy of the Baberton granite/greenstone terrain can be resumed as follows (Table 2): a locally preserved sialic crust, which is mostly strongly reworked, appears to predate a mafic/ultramafic pile. It must be stressed that the nature of the sialic/mafic–ultramafic pile contact(s?) is very poorly understood. The mafic/ultramafic pile is overlain by a thin volcaniclastic layer deposited during subaerial volcanic activity, with which it forms the Onverwacht Group.

Such a succession is very similar to that found on present day oceanic crust, and this led to the comparison of the Onverwacht Group with an ophiolite (De Wit and Stern 1980; De Wit et al. 1986). The study of oxygen isotope profiles in the Onverwacht mafic and ultramafic rocks support this comparison as such profiles are indistinguishable from those obtained in phanerozoic ophiolites (Smith et al. 1984; Hoffman et al. 1986).

This Archaean oceanic crust is then overlain by a prograding submarine fan, the Diepgezet Group, which in turn passes conformably or unconformably into an alluvial fan, the Malolotsha Group. The entire succession probably did not exceed 8 km in thickness, which is in accordance with the geophysical data available for the belt (Darracott 1975).

4. Structural results

Structural mapping carried out in the southern part of the belt by De Wit, Lamb and the present author has shown that at least four phases of deformation have affected the 8 km thick supracrustal succession described above. D1 and D2 successively dismembered and repeated the stratigraphic pile, leading to the formation of a 24 km thick tectonic complex, previously interpreted as a straight-forward stratigraphy (Viljoen and Viljoen 1969; SACS 1980). D3 and D4 further refolded the lithotectonic complex developed during D1 and D2.

4a. The D1 event

D1 is the earliest phase of deformation recorded in the Barberton greenstone belt. It is manifested by the duplication of a thin and well defined portion of the stratigraphy: the silicified upper part of the Onverwacht Group, *viz.* the top of the volcanic pile and the overlying volcaniclastic sediments, and the silicified base of the Diepgezet Group, *viz.* sediments of the F1 facies (Table 2).

Individual D1 tectonic stacks can vary from 1 m to 50 m in thickness and contain part of the whole of the following succession: silicified mafic and ultramafic lavas, silicified volcaniclastic sedimentary rocks, silicified ferruginous shales. The D1 tectonic units are generally underlain by 1 to 30 m thick zones of fuchsitic and carbonated flaser banded gneisses (see De Wit 1982) which are a characteristic feature of this phase of deformation (Figure 6). The flaser banded gneisses are formed by fuchsite, chlorite, sericite and serpentinite layers in which protomylon-itic fabrics can be identified (Paris 1985a). These rocks are also criss-crossed by anastomosing and folded extension veins of calcite and quartz (De Wit 1986). The fabric of the flaser banded gneisses is generally parallel to bedding and is folded by D2 together with the cherts, indicating their pre-D2 nature (Figure 6).

In some cases, intense silicification of the cherts deformed by D1 makes it imposible to determine whether they belong to the Onverwacht or to the Diepgezet group (Figure 6). However, D1 tectonic units are generally easy to identify as they are underlain by brightly coloured fuchsitic flaser banded rocks and commonly contain sedimentary markers such as accretionary lapilli and/or lapilli (Figures 3, 4 and 6). So far, D1 tectonic units with such silicified markers and with turbiditic cycles have been recognized at the following levels of the Onverwacht Tectonic Complex (Table 1, Figure 7): in the 'Theespruit level'; the 'Komati level', within the Middle Marker; the 'Kromberg level' and the 'Zwartkoppie level' (Figure 7). In the latter two 'levels', up to four successive D1 stacks, all containing turbiditic sequences and accretionary lapilli, have been observed (Figure 6 locations A and B). In such piles, individual D1 stacks all face the same way, being separated by lenses or layers of flaser banded gneisses (Figure 6 locations A, B. C. D, E). Most of the D1 stacks have been refolded during D2 and the plunge and younging directions of these D2 folds clearly indicate that some of the D1 stacks were already downward facing prior to D2 (Figure 8). The emplacement of inverted nappes during D1, probably through recumbent folding and gravity gliding, has been well demonstrated in this part of the belt (De Wit 1982).

D1 flaser banded gneisses can be traced down and along strike into hydrotherm-ally altered and hydrothermally fractured igneous rocks, below a strongly silicified zone under high pressure. As D1 deformation only involved a thin column of rocks, high fluid pressure could not have resulted from the overburden pressure but more probably as a consequence of the hydrothermal activity which caused the silicification of the Onverwacht and basal Diepgezet rocks (De Wit *et al.* 1982;

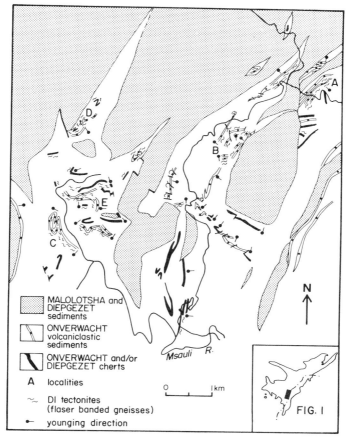

Figure 6. Simplified geological map showing the D1 deformation (insert shows the location of the area). Note that the cherts and D1 are fuchsitic tectonites are folded by D2.

Paris 1985a, b). The following sequence of events is thought to have led to the formation of the D1 imbricates (De Wit *et al.* 1982; Paris 1985a, b):

— During hydrothermal activity, silicification led to lithification of recently deposited sediments and to the formation of a cap rock.
— Hydraulic fractures developed through this cap rock were eventually sealed by the precipitation of further hydrothermal fluids.
— With continual input of hydrothermal fluids, a zone of high pressure was formed under the sealed cap rocks and gliding was initiated.
— Further gravity gliding led to the development of recumbent folds.

4b. The D2 event

D2 is a long lived, polyphase and probably diachronous event which deformed the entire supracrustal sequence of the Barberton greenstone succession. It was a phase of strong N–S compression in the northern and western part of the belt (Ramsay 1963; Fripp *et al.* 1980; De Wit *et al.* 1983); and was predominantly

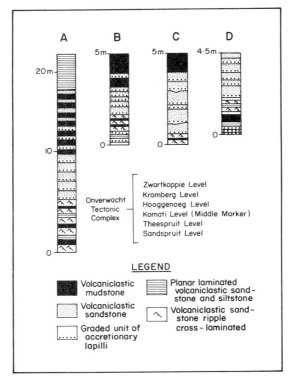

Figure 7. D1 tectonic units which contain similar sedimentary sequences, within the Onverwacht Tectonic Complex. A: Theespruit level (after De Wit et al. 1983); B: Komati level (Middle Marker); C: Kromberg level (Paris 1985a); D: Zwartkoppie level (Paris 1985a). A, B, C and D belong to the proximal turbiditic facies of the Onverwacht sedimentary rocks. For ease of reference in this paper, a level is a tectonostratigraphic unit of equivalent geographical distribution to the formations defined in the currently accepted stratigraphy of SACS (1980).

marked by NW–SE compression in the southeastern part of the belt (Lamb 1984a, b; Paris 1985a, 1986).

The D2 compression led to the development of tight to isoclinal folds with near vertical axial planes and shallow plunging fold axes. These folds have E–W to NE–SW trending axes. South dipping thrust slices have been observed in the northern part of the belt which probably resulted from the northerly tectonic transport of the entire greenstone sequence (Fripp et al. 1980). D2 has led to a minimum of 70 per cent shortening (De Wit et al. 1983; Lamb 1984a; Paris 1985a).

In the southern part of the belt, D2 is marked by a NE–SW oriented, SE dipping tectonic fan (the D2 Tectonic Fan, Figure 9) in which tectonic slices of Onverwacht material are interleaved with younger sedimentary sequences of the Diepgezet and Malolotsha Groups (Figure 9). The D2 fan is about 10 km wide and 50 km long and the relationships which can be observed within the fan are illustrated in Figure 9. Along the AB traverse for example (Figure 10) a wedge of Onverwacht material, characterized by the presence of cherts with accretionary lapilli, is present between the Waterfall Synclinorium and the Xecacatu Syncline (Figures 9 and 10). This wedge peters out at the level of the traverse and reappears in two separate lenses, to the north and along strike (Figure 9), indicating the

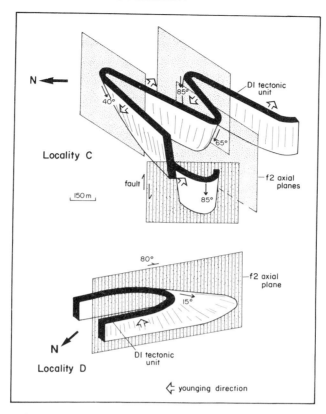

Figure 8. D1 tectonic units folded by D2. Younging directions as well as the plunges of the folds indicate that the units were inverted to overturned prior to the D2 event (see Figure 6, locality C and D for location).

presence of a tectonic contact between the Diepgezet and the Malolotsha sediments. The eastern limb of the Xecacatu Syncline is cross cut and overlain by another Onverwacht unit which is overturned, forming the western limb of another anticlinal D2 fold (Figures 9 and 10).

Although the contacts between the different units are generally poorly exposed, lenses of crush breccia, pseudotachylite, mylonite and ferruginous and talcose shear zones can be observed in places. Such contacts are interpreted as thrust zones (Paris 1985a, b, 1986).

The various units described above are all folded independently of one another by D2. The D2 fold axes are broadly parallel to the underlying thrust and can be cross cut by the overlying one, suggesting that the D2 folds were contemporaneous with thrust emplacement.

Similar relationships to the ones presented above have been noted throughout the southern part of the belt (De Wit 1983; Lamb 1984a, b; Paris 1985a, 1986). Within the D2 Tectonic Fan, the thrusts generally dip steeply (60° to 80°) to the southeast; however, this could have resulted from the later D3 shortening of the structures. Some shallow dipping contacts have nevertheless been observed, as with the T-junction thrust (Figure 10). This thrust brings steeply dipping Onver-

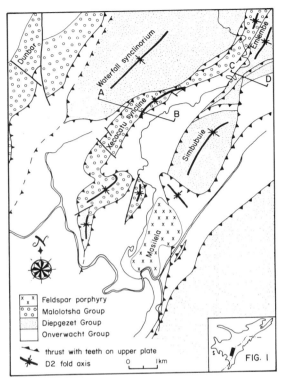

Figure 9. Simplified geological map showing the type of relationship which can be observed within the D2 Tectonic Fan (insert shows the location of the area). The buckling of the D2 Tectonic Fan is due to the D4 phase of deformation.

wacht and Diepgezet material over shallow dipping Diepgezet ferruginous shales (Figure 10). The contact itself is marked by lenses of pseudotachylite and chert breccia (Paris 1985a). To the east, the Diepgezet sediments located above the T-junction thrust are themselves cross cut by a 50 m thick shear zone which dips 70° SE and underlies a downward facing pile of D1 imbricates of Onverwacht Group material (Figures 9 and 10).

The D2 folds are generally tight to isoclinal in the Diepgezet Group and more open in the Malolotsha Group (Figure 10). In the southeastern part of the belt, the NE–SW trending D2 folds were developed during the deposition of the Malolotsha group sediments. In fact, the deposition of the Malolotsha sediments is intimately related to the D2 folding (Lamb 1984a, b). These sediments unconformably overlie folded Diepgezet group sediments and are derived from Diepgezet Group material and from a quartz-rich (presumably granitic) source. Folded unconformities also occur within the Malolotsha Group itself (for example, in the Masali and the Ngwenya synclines, Lamb 1984a, b; Lamb and Paris under review). Where these particular settings have been observed, angular unconformities are very noticeable in the limbs of the fold, while in the core, apparently conformable sequences are preserved.

The link between D2 folding and deposition of the Malolotsha sediments suggests that these sediments were deposited in tectonically ponded basins, folding

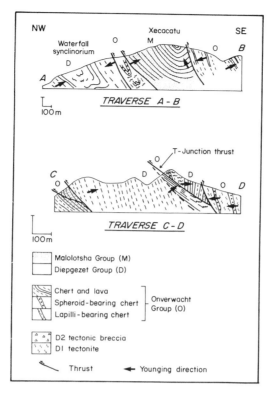

Figure 10. Two traverses across the D2 fan (see Figure 9 for location).

taking place during propagation of a basal thrust and sedimentation during periods of tectonic quiescence. Detrital material was derived both from the folded sequence and from an emerging quartz-rich source perhaps located at the back of the moving complex (Lamb 1984a). The relationships observed in the D2 synsedimentary folds appear to be similar to those described in the Himalayas by Burbank and Renolds (1984). If such an hypothesis is correct, the entire greenstone succession would then be allochthonous, as suggested by Coward (1976) and Fripp et al. (1980), the sole thrust being locally exposed in the northern part of the belt.

Lamb (1984a) argued that the E–W trending D2 folds predated the NE–SW trending folds and accordingly, subdivided D2 into D2a and D2b. His interpretation was based on the fact that the E–W trending folds are cross cut by Malolotsha sediments which, to the east, have been folded along NE–SW axes. However, it is felt that the two trends could be broadly contemporaneous as NE–SW trending folds are also cross cut by unconformities (Lamb 1984a; Paris 1985b).

Within the D2 fan, possible ramp structures are developed along NW–SE axis, suggesting movement to NW (for example south of Simbubule, Figure 9). Wider ramps could also be present in western part of the belt where the D2 compression was N–S (De Wit personal communication).

The development of the D2 thrust was probably induced by locking of the D2 folds. The presence of sheared serpentinite and talc-schists at the base of many of the D2 thrust sheets suggests that decollement was initiated along the serpentinized bodies as is often the case in more recent orogenic belts. Movement along a level of serpentinite would have led to D2 folding, which in turn was partly responsible for the deposition of the Malolotsha sediments. The thrust then cut up stratigraphy and across structures, reactivating zones of weakness such as the base of the Diepgezet Group (Paris 1985a, 1986) and planes of unconformities (De Wit 1982; Lamb 1984a).

Geometrical relationships as well as analysis of microstructures in the southern part of the belt indicate that movement was predominantly to the NW during D2 (Lamb 1984a, b; Paris 1985a).

D2 structures are cross cut by undeformed intrusive bodies such as the Masilela porphyry (Figure 9), dated at circum 2.7 Ga (Rb/Sr whole rock, Barton et al. 1983). Thus a minimum age of circa 3 Ga can be placed on the D2 event.

Lamb (1984a, b) suggested that the D2 shortening was coeval with uplift and lateral spreading of the adjacent gneissic terranes documented by Jackson and Robertson (1983). The quartz rich nature of the Malolotsha sediments further indicates the emergence of granitic terrains during D2.

Although more data is still needed in order to fully understand the nature of the D2 event, it is suggested that this widespread and long lived deformational event corresponded to an Archaean collision which led to the obduction and preservation of the greenstone succession. Such an interpretation is based on the following observations:

— There is a dominant sense of movement of the D2 thrusts to the north in the northern and western part of the succession (Table 3; Fripp et al. 1980) and to the NW in the southeastern part (Table 3; Lamb 1984a; Paris 1985a, 1986).
— D2 folds were at least partially, associated with the deposition of alluvial sediments derived from a granitic source and from the underlying deformed sequence, which suggests that continental crust was being reworked and that the closure of the oceanic basin had started during the transition from Diepgezet to Malolotsha sediments.
— The link between D2 and the formation of the Malolotsha Group suggests deposition in a tectonically active regime.
— The Malolotsha sediments have the characteristics of a molasse deposite (coarse grained and conglomeratic alluvial fan).
— Granitic terrain was being eroded during the formation of the Malolotsha Group.
— D2 was a long lived and regionally widespread event.
— During D2, there was a change from oceanic (Diepgezet Group) to continental (Malolotsha Group) sedimentation.

4c. The D3 and D4 events

D3 was a phase of E–W shortening which caused steepening of the D2 structures and, in the eastern part of the belt only, led to the formation of N–S trending isoclinal folds with an associated vertical axial planar cleavage (Table 3 ; Lamb 1984a). Lamb (1984a) interpreted D3 as syntectonic with the emplacement of the Mpuzuli batholith (circum 3.0 Ga). A similar, although E–W trending phase of deformation has been observed in the northern and southwestern part of the belt, close to the granite/greenstone margins and was also related to granitoid emplacement (Table 3; De Wit 1983; Fripp et al. 1980). The last deformational

Table 3. Résumé of the deformational events in the Barberton succession (D2 and D3 correlations are provisional). Northern part of the belt after Ramsay (1963) and Fripp et al. (1980); southwestern part after De Wit (1982), De Wit et al. (1983); southeastern after Lamb (1984a, b) and Paris (1985a, 1986).

Event	Type of structures developed	Part of the stratigraphy involved	Tectonic trend and transport direction		
			Northern	South Eastern	South Western
			Part of the Succession		
D1 repetition of thin silicified units	recumbent folds, tectonic slides	O and base of D			E - W ?
D2 development of a south-dipping tectonic fan	isoclinal and open folds, thrust sheets	O + D + M (+ G in the Northern part of the belt)	ENE/WSW → N ★	EW → N NE-SW → NW ★	ESE - WNW → N ★
D3 refolding and steepening of D2 fan	isoclinal folds	O + D + M + G only visible around the margins of the belt	E - W ★	N - S ★	E - W
D4 buckling	open folds, faults	O + D + M		E - W ENE - WSW ★	

O = Onverwacht Group
D = Diepgezet Group
M = Malolotsha Group
G = granitoids
→ transport direction
★ associated with a well developed cleavage

event which can be observed in the southern part of the Barberton granite/greenstone succession, D4, is marked by the buckling of the D2 tectonic complex and by the formation of large scale open folds along trends varying from E–W to NW–SE (Table 3; Lamb 1984a; Paris 1985a).

The data presented above show that the supracrustal terrains of the Barberton granite/greenstone terrain have been strongly affected by polyphase deformation (Table 3). The first two phases of deformation are the most important, as far as the understanding of the belt is concerned, as they both strongly disrupted the original stratigraphic pile and affected sedimentation. D1, which only affected a restricted part of the stratigraphy (Table 3), was an early and probably intraoceanic phase of gravity gliding which was associated with active hydrothermal centres. The D2 event was a complex phase of thrusting and folding which was probably triggered by movement in the adjacent granitic terrain and could mark the closing and obduction of the greenstone terrain. The last two phases appear to have developed more locally and steepened as well as refolded the D2 tectonostratigraphic complex, D3 by E–W shortening and D4 by NE–SW shortening (Table 3).

5. Conclusions

More research is still under progress in an attempt to fully understand the development of the Barberton granite/greenstone terrain and correlations between the northern and the southern part should eventually be established. However, from the results obtained so far, it appears that there are many similarities between the formation and emplacement of Archaean greenstone successions and more recent ophiolite belts. Volcanic, sedimentological and structural processes which acted in the greenstone successions be compared to those found in modern orogenic belts. The Barberton greenstone succession is thus considered to represent an Archaean ophiolite sequence overlain by submarine and continental sediments, which was originally formed along an oceanic spreading centre and emplaced through obduction during a collision phase.

Acknowledgements. This paper draws considerably upon the data of other workers and an attempt has been made to acknowledge this as fully as possible. In particular I thank Maarten de Wit and Simon Lamb for many useful discussions. The author's work was conducted while at the University of the Witwatersrand (South Africa) with financial support from Elf Aquitaine. I am also grateful to Professor R. Black and Professor J. Lameyre (Laboratoire de Pétrologie, Paris VI) for providing office space and to J. Dyin (Laboratoire de Tectonique, Paris VII) for providing access to drafting equipment and useful advice. L. Leonard (University of Canterbury, Christchurch) drafted the final diagrams and W. Nuthall (University of Canterbury, Christchurch) helped with the typing. The original draft has been improved thanks to the comments of Professor J. Sougy and Professor J. S. Ogola.

References

Anhaeusser, C.R. 1978. The geological evolution of the primitive earth: evidence from the Barberton Mountain Land. In Tarling, D.H. (Ed.), *Evolution of the Earth's Crust*, Academic Press, London, 71–106.

—— **1983.** Structural elements of Archaean granite–greenstone terrane as exemplified by the Barberton greenstone belt, southern Africa. *Economic Geology Research Unit University of Witwatersrand, Johannesburg, Information circular,* **162**.

—— **and Robb, L.J. 1981.** Magmatic cycles and the evolution of the Archaean granitic crust in the Eastern Transvaal and Swaziland. *Special Publication of the Geological Society of Australia,* **7**, 457–467.

Barton, J.M. Hunter, D.R. Jackson, M.P.A. and Wilson, A.C. 1980. Rb–Sr age and source of the bimodal gneiss suite of the Ancient Gneiss Complex, Swaziland. *Nature,* **283**, 756–758.

——, **Robb, L.J., Anhaeusser, C.R. and Van Nierop, D.A. 1983.** Geochronologic and Sr isotopic studies of certain units on the Barberton granite–greenstone terrane, South Africa. *Geological Society of South Africa Special Publication,* **9**, 63–72.

Burbank, D.W. and Reynolds, R.G.H. 1984. Sequential late Cenozoic structural disruption of the northern Himalayan foredeep. *Nature,* **311**, 114–118.

Burke, K., Dewey, J.F. and Kidd, W.S.F. 1976. Dominance of horizontal movements, arc and microcontinental collisons during the later permobile regime. In Windley, B.F. (Ed.), *The Early History of the Earth,* Wiley, Chichester, 113–129.

Byerly, G.R., Lowe, D.R. and Walsh, M.M. 1986. Stromatolites from the 3,300–3,500 Myr Swaziland Supergroup, Barberton Mountain Land, South Africa. *Nature,* **319**, 489–491.

Condie, K.C. 1981. Archean greenstone belts. *Developments in Precambrian Geology,* **3**, Elsevier, Amsterdam.

Coward, M.P. 1976. Archaean deformation patterns in southern Africa. *Philosophical Transactions of the Royal Society of London, Series A,* **283**, 313–331.

Darracott, B.W. 1975. The interpretation of the gravity anomaly over the Barberton Mountain Land,

South Africa. *Transactions of the Geological Society of South Africa*, **78**, 123–128.
De Wit, M.J. 1982. Gliding and overthrust nappe tectonics in the Barberton greenstone belt. *Journal of Structural Geology*, **4**, 117–136.
—— **1983.** Preliminary geologic map of part of the southern sector of the Barberton Greenstone Belt. *Geological Society of South Africa Special Publication*, **9**, 185–187.
—— **1985.** What the oldest rocks say. Workshop on *The Earth as a planet*, LPI and GSA, Orlando, Florida, Oct 27, Abstract.
—— **1986.** Extensional tectonic during the emplacement of the mafic–ultramafic rocks of the Barberton greenstone belt. *The Tectonic Evolution of Greenstone Belts*. Lunar and Planetary Institute workshop, Houston, Texas, Abstract., 25–26.
—— **and Stern C. 1980.** A 3500 Ma ophiolite complex from the Barberton greenstone belt, South Africa: Archaean oceanic crust and its geotectonic implications. *Extended Abstract 2nd International Archaean Symposium*, Perth, 83–85.
——, **Hart, R., Martin, A., and Abbott, P. 1982.** Archaean abiogenic and probable biogenic structures associated with hydrothermal vent systems and regional metasomatism, with implications for greenstone belt studies. *Economic Geology*, **77**, 1783–1802.
——, **Fripp, R.E.P., and Stanistreet, I.G. 1983.** On some field observations within the Barberton greenstone belt along part of its southern margin. *Geological Society of South Africa Special Publication*, **9**, 21–29.
——, **Hart, R. and Hart R. 1986.** A Mid-Archean ophiolite complex, Barberton Mountain Land. *The Tectonic Evolution of the Greenstone Belts*, Lunar and Planetary Institute workshop, Houston, Texas, Abstract, 27–28.
Fripp, R.E.P., Van Nierop, D.A., Callow, M.J., Lilly, P.A. and Du Plessis, L.V. 1980. Deformation in part of the Archaean Kaapvaal Craton, South Africa. *Precambrian Research*, **13**, 241–251.
Hamilton, P.J., Evensen, N.M., O'Nions, R.K., Smith, H.S. and Erlank, A.J. 1979. Sm–Nd dating of Onverwacht Group volcanics, southern Africa. *Nature*, 279–300.
Heinrichs, T. 1984. The Umsoli chert, turbidite testament for a major phreato plinian event at the Onverwacht/Fig Tree transition (Swaziland Supergroup, Archaean, South Africa). *Precambrian Research*, **24**, 237–293.
Hoffman, S.E., Wilson, M. and Stakes, D.S. 1986. Inferred oxygen isotope profile of Archaean oceanic crust, Onverwacht Group, South Africa. *Nature*, **321**, 55–58.
Hunter, D.R. 1974. Crustal development in the Kaapvaal Craton, I. The Archaean. *Precambrian Research*, **1**, 259–294.
Jackson, M.P.A. and Roberston, D.I. 1983. Regional implications of early Precambrian strains in the Onverwacht Group adjacent to the Lochiel Granite, north west Swaziland. *Geological Society of South Africa Special Publication*, **9**, 45–63.
Lamb, S.H., 1984a. Structures and sedimentology on the eastern margin of the Archaean Barberton greenstone belt, northwest Swaziland. *Unpublished PhD Thesis*, University of Cambridge.
—— **1984b.** Structures of the Archaean Barberton greenstone belt, northwest Swaziland. In Kröner, A. and Greiling, R. (Eds), *Precambrian Illustrated*, Nägle v. Obermiller, Stuttgart, 19–39.
—— **and Paris, I.A. under review.** Proposals for revision of the stratigraphy in the 3.5 Ga old Barberton greenstone belt, South Africa.
Lowe, D.R. and Knauth, L.P. 1977. Sedimentology of the Onverwacht Group (3.4 billion years), Transvaal, South Africa, and its bearing on the characteristics and the evolution of the early earth. *Journal of Geology*, **85**, 699–723.
—— **and Knauth, L.P. 1978.** The oldest marine carbonate ooids reinterpreted as volcanic accretionary lapilli. Onverwacht Group, South Africa. *Journal of Sedimentary Petrology*, **48**, 709–722.
Paris, I.A. 1985a. The geology of the farms Josefsdal, Dunbar and Diepgezet in the southern part of the Barberton greenstone belt. *Unpublished PhD Thesis*. University of Witwatersrand.
—— **1985b.** Métasomatisme et déformation dans la ceinture archéenne de Barberton, Afrique du Sud. In *Evolution Géologique de l'Afrique*, CIFEG, Publication occasionnelle, 42–46.
—— **1986.** Polyphase thrust tectonics in the Barberton greenstone belt. *The Tectonic Evolution of Greenstone Belts*. Lunar and Planetary Institute workshop, Houston, Texas, Abstract, 89–90.
—— **in preparation.** Depositional environment of the Onverwacht sedimentary rocks. Barberton greenstone belt, South Africa.
——, **Stanistreet, I.G. and Hughes, M.J. 1985.** Cherts of the Barberton greenstone belt interpreted as products of submarine exhalative activity. *Journal of Geology*, **93**, 111–129.
Ramsay, J.G. 1963. Structural investigations in the Barberton greenstone belt, Eastern Transvaal. *Transactions of the Geological Society of South Africa*, **66**, 354–401.
South African Committee for Stratigraphy (SACS) 1980. Stratigraphy of South Africa Part I. Lithostratigraphy of the Republics of Bophutaswana, Transkei and Venda. *Handbook of the Geological Survey of South Africa*, **8**.
Smith, H.S., O'Neil, J.R. and Erlank, A.J. 1984. Oxygen isotope compositions of minerals and rocks and chemical alteration patterns in pillow lavas from the Barberton greenstone belt, South Africa. In Kröner, A., Hanson, G.N. and Goodwin, A.M. (Eds), *Archean Geochemistry*, Springer–Verlag, Berlin, 115–137.
Stanistreet, I.G., De Wit, M.J. and Fripp, R.E.P. 1981. Do graded units of accretionary spheroids in the Barberton greenstone belt indicate an Archean deep water environment? *Nature*, **293**, 280–284.
Tankard, A.J., Jackson, M.P.A., Eriksson, K.A., Hobday, D.K. and Winter, W.E.L. 1982. *Crustal evolution of Southern Africa*, Springer–Verlag, New York.

Vearncombe, J.R., Barton, J.M., Jr., Van Reenen, D.D. and Phillips, G.N. 1986. Greenstone belts: their components and structure. *The Tectonic Evolution of Greenstone Belts*, Lunar and Planetary Institute workshop, Houston, Texas, Abstract, 19–25.

Viljoen, M.J. and Viljoen, R.P. 1969. In Haughton, S.H. (Ed.), *Upper Mantle Project. Geological Society of South African Publication*, **2**, 2–297.

—— **and Viljoen, R. P. 1970.** Archean vulcanicity and continental evolution in the Barberton region, Transvaal. In Clifford, T. N. and Gass, I.G. (Eds), *African Magmatism and Tectonics*, Oliver and Boyd, Edinburgh, 27pp.

Williams, D.A.C. and Furnell, R.G. 1979. Reassessment of part of the Barberton type area. *Precambrian Research*, **9**, 325–347.

Mineralization in the Migori greenstone belt, Macalder, Western Kenya

J.S. Ogola
Department of Geology, University of Nairobi, Kenya

Macalder copper sulphide deposit is located to the western end of the Migori greenstone belt in Kenya at the northern margin of an intensely dislocated metamorphosed Archaean volcanogenic sedimentary sequence known as the Nyanzian system. It is a copper sulphide deposit interpreted as having a metasomatic origin with hydrothermal solutions interacting with and modifying a banded iron formation. Copper, zinc, gold, and silver have all been sporadically mined from the ore bodies in the period from 1935 to 1966.

The ore body occurs as sheets, veins, and lenses located at the contact between metabasalts, banded iron formation and greywackes, and also at fold hinges and along faults and fissures.

The process of ore formation occurred in two distinct stages:
1. The deposition of magnetite occurred during the metamorphism of Nyanzian volcanosedimentary sequences. Parts of this sedimentary succession were iron-rich giving rise to horizons of banded iron formation.
2. At a later stage, sulphide ore formation developed as sheets, veins, and lenses at the stratigraphic boundaries between banded iron formation, sandstone, and metabasalts. The magnetite in the banded iron formation is corroded and represents remobilization of iron-rich horizons by circulating hydrothermal fluids.

The sulphide ores are characterized by four main mineral assemblages: quartz, sphalerite–pyrite, pyrrhotite–chalcopyrite, and carbonate. Each mineral assemblage is characterized by its own paragenetic association. Mineralization occurred in various stages at temperatures between 245–271°C. The presence of native bismuth further confirms the relatively low temperatures of ore formation.

KEY WORDS   Mineralization   Migori greenstones   Kenya   Sulphide ore deposits

1. Introduction

Macalder mine is located near the western margin of the Migori greenstone belt, to the south of Wiman gulf in western Kenya and about 18 km from the Kenyan–Tanzanian Border (Figure 1). Systematic study of this belt and in particular of the Macalder area has been carried out by many researchers, among them Hitchen (1936), Pulfrey (1938), Shackleton (1946), Sanders (1964), and Hutchinson(1981). Shackleton (1946) was the first to compile a geological map of

0072–1050/87/TI0025–20$10.00
© 1987 by John Wiley & Sons, Ltd.

the Migori greenstone belt (1:50,000 and 1:25,000). Sanders (1964) mapped the Macalder area (1:15,000) and deposit (1:500).

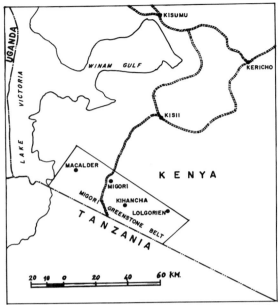

Figure 1. Location map of the Migori Greenstone Belt.

Mining at Macalder was carried out from 1935 to 1949 and then from 1955 to 1966. The deposit was worked to the 8th level at a depth of 260 m below the surface. About one million tons of primary massive sulphide ore, grading 2·3 per cent Cu, 2·9 per cent Zn, 0·8 per cent Pb, 3·7 g/t Au, and 53·7 g/t Ag was recovered. By 1966 total copper production was 20,000 tons. Gold and silver were initially obtained from about 220,000 tons of auriferous gossan ore which yielded about 950 kg of gold and 1113 kg of silver (Homa Bay Rept. No 32 1944). The closure of the mine in 1966 was mainly a result of the decrease of mineral values and the change of ore bodies downwards from sheet-like to lenses and veinlets.

The main purpose of the present work was to study the conditions under which sulphide mineralization developed in the Macalder deposit. The following techniques were used:
1. Petrological study of the rocks of the Macalder area.
2. Interpretation of the structural geology of the Macalder deposit and the morphology of the ore body.
3. A reflected-light study of the ore–mineral assemblages.
4. Decrepitation and fluid inclusion study of the thermal conditions under which sulphide ore formation took place.

2. Geological succession in the Migori region

The Migori greenstone belt comprises a metamorphosed Archaean volcanogenic–sedimentary succession belonging to the Nyanzian system. The presence of the Migori greenstone rocks in the Macalder area has been confirmed and

their petrological characteristics studied. The following stratigraphic succession is established (Figure 2):
1. Upper group—phyllite, andesite, and andesitic–dacitic porphyrites and their tuffs, porphyritic andesites.
2. Middle group—sandstones (greywacke), banded iron formation.
3. Lower group—metabasalt and pillow lavas.

2a. Lower Group

Metabasalt and pillow lavas are traceable within the tectonic zone, which occupies the central part of the area (Figure 2). These rocks are well exposed on the banks of the Migori and Kuja rivers and in the open-cast section of the mine. The width of metabasalt and pillow lavas decreases from 550 m in the northwestern part to 50 m in the southeastern margin of the area. The metabasalt contains microlites of plagioclase which in places has been replaced by albite, and seriticized. The rock exhibits aphyric and relict textures. Alteration resulted in foliation of the rock and banding is predominant in zones of maximum alteration.

Pillow lavas are common around the deposit and also along the bank of the Migori river, east of the mine. These rocks contain pillow-shaped masses that range in size from a few centimetres to a metre or more. Individual pillows usually fit closely together. In thin section, pillow lavas comprise of radially-spaced fibrous crystals of plagioclase and monoclinic pyroxene. Rock texture is variolitic.

2b. Middle Group

Banded iron formation and sandstone (greywacke) constitute the Middle Group. Around the deposit the banded iron formation extends 1000 m from northwest to the southeast with a width varying between 1 m and 8 m. The formation is composed of interbedded bands of ferruginous chert, chlorite stringers and actinolite. Individual bands of chert, oxides and hydroxides of iron and chlorite stringers vary in width from 10 to 25 cm, from 2 to 5 cm and from 0·2 to 2 cm respectively. Microfissures and minor folds are well developed in the banded iron formation. In thin sections, phenocrysts of actinolite up to 0·3 mm are seen in quartz grains.

Greywacke generally overlies the banded iron formation; in places, however, it lies directly on top of the metabasalt. The rock is light green when fresh and blue to brownish-grey when weathered, cleaved, and indistinctly bedded. This sedimentary formation is exposed on either side of the tectonic zone. It varies in thickness from 15 m to 200 m. The formation is composed of medium to fine grained sandstone (greywacke), containing argillaceous materials with quartz grains and fragments of carbonate. In thin section, quartz and angular oligoclase are set in a matrix of chlorite and sericite with accessory rutile and calcite.

2c. Upper Group

The Upper group consists of phyllites and dominantly calc-alkaline volcanic rocks. The occurrence of phyllite is restricted mainly to the central part of the area. Good exposures of these rocks are traceable along the banks of the Migori Kuja rivers. The width of their outcrop ranges from 10 m to 750 m. Phyllite overlies metabasalt, banded iron formation, and sandstone. Rarely these rocks form lenses in the andesitic rocks. Sanders (1964) noted, that these rocks consist of shales, slates and slaty tuffs. However, the present investigation concludes that, these rocks are phyllites and have been metamorphosed to greenschist facies.

In thin section they show well orientated aggregates of sericite with quartz grains, grains, minor chlorite and organic materials which are evenly distributed in the matrix.

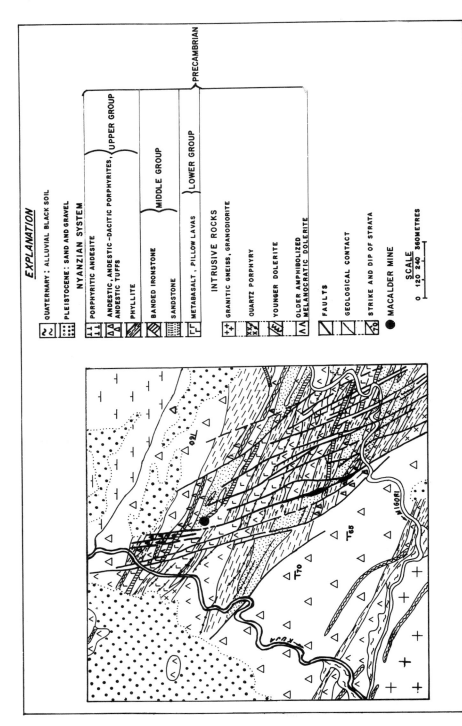

Figure 2. Geological map of the area surrounding Macalder Mine (compiled by Sanders, 1964) modified by Ogola, J.S. – 1984.

Andesitic rocks are widespread on the southern and northern parts of the area. They conform to the general trend of the Nyanzian rocks in the area, striking to the northwest. Those occurring on the northern part dip northwards and those on the south, southwards at 65–80°. Good exposures are traceable along the river banks. In thin section, flow textures can be observed. The andesitic tuffs include andestic agglomerates and crystal tuffs as well as fine-grained andesitic ash. The agglomerates are well exposed on the southern part of the mine, above the bank of the Migori river.

Two types of andesitic rocks were established; andesitic porphyrites, which occupy the southern portion of the mine and andesitic–dacitic porphyrites, which have good exposures in the area, north of Macalder mine (Figure 2).

The andesitic porphyrites have a microporphyritic texture (Figure 3). Phenocrysts consist of tabular plagioclase (albite), ranging in size from 0.1×0.2 mm to 0.3×0.9 mm. Thin sections show that plagioclase phenocrysts have been intensively sericitized, in places, and in some places totally replaced. Sericitization was initiated along grain margins. Minor aggregates of monoclinic pyroxene measure 0·1–0·4 mm. The phenocrysts are set in a matrix of plagioclase microlites, calcite, and chlorite.

Thin sections of andesitic–dacitic porphyrites show phenocrysts of plagioclase, measuring 0·23 mm, 10-15 per cent anhedral quartz, 0·28 mm, epidote 0·20 mm and actinolite 0·31 mm. Plagioclase has been intensively either to sericite and serpentine, or towards the base to chlorite and calcite. Field investigations, microscopic and chemical analysis of these rocks support the petrological identification as andesitic–dacitic porphyrites (Table 1).

Porphyritic andesite occupies the northeastern part of the area and overlies andesitic–dacitic porphyrite. This rock contains euhedral phenocrysts of zoned oligoclase with minor augite, marginally altered to actinolite, and hornblende.

Table 1. Major element composition of representative rocks from the Macalder area

	1	2	3	4	5	6	7	8	9	10	11
SiO_2	50·02	48·60	60·76	60·72	64·44	62·34	46·20	46·96	46·10	46·00	72·20
TiO_2	1·48	0·91	0·67	0·52	0·46	0·50	1·12	0·92	0·82	1·75	0·26
Al_2O_3	15·98	14·32	16·13	15·79	15·53	16·08	15·90	15·75	17·70	13·45	13·46
Fe_2O_3	11·62	11·90	5·70	5·06	4·83	5·52	2·80	11·97	2·80	4·70	3·00
FeO	7·24	8·72	3·29	3·22	2·39	3·53	9·70	8·16	8·26	13·58	2·15
MnO	0·19	0·22	0·07	0·06	0·07	0·07	0·11	0·16	0·06	0·20	0·04
MgO	7·00	7·89	3·52	2·56	1·53	1·97	8·10	7·76	3·10	5·76	0·55
CaO	4·07	6·26	2·54	3·56	3·79	3·23	10·80	11·78	11·72	9·05	1·35
Na_2O	2·75	2·80	4·25	5·12	4·37	5·00	1·94	1·55	1·30	2·00	3·67
K_2O	1·29	0·02	2·19	1·92	2·57	3·40	0·26	0·12	0·34	0·61	4·83
P_2O_5	0·73	0·40	0·46	0·34	0·17	—	0·20	0·06	0·07	0·30	0·07
CO_2	0·20	—	0·87	0·77	0·43	0·20	—	0·20	—	—	0·20
H_2O	—	—	1·27	1·05	1·80	—	3·00	2·44	2·44	2·40	—

All Values in per cent
1 Metabasalt from the first level of the Macalder mine.
2 Pillow lavas from Migori river, east of the Macalder mine.
3,4 Andesitic porphyrite from the southern part of Macalder area.
5,6 Andesitic–dacitic porphyrite from the northern part of Macalder area.
7–10 Amphibolized melanocratic dolerite from the tectonic zone of Macalder area.
11 Migori granite from the southwestern margin of the Macalder area.

3. Intrusive Rocks

The intrusive rocks range from granite and quartz porphyry to dolerites. There are two types of dolerite: dykes of younger dolerite and sills of older amphibolized melanocratic dolerite which are the oldest intrusives in the area (Figure 2). Shackleton (1946) considered the younger dolerites to be post-Kavirondian, while the older dolerites are Nyanzian in age and are closely associated with the Nyanzian basic volcanic group. Amphibolized melanocratic dolerites are restricted to the tectonic zone and form sheet-like masses 30–650 m thick at the contacts of metabasalt, sandstone, and phyllite. The sheets generally conform to the regional strike. Thin sections show that the amphibolized melanocratic dolerite has a well defined poikilophitic texture with 80–90 modal % of light green hornblende replacing phenocrysts of pyroxene. Tabular idiomorphic phenocrysts of labrodite form 5–20 modal % of the rock.

Granites and granodiorites of the Migori massif are exposed in the southwest of the area. The massif is northwestern trending and forms part of the Tanganyikan craton. Thin sections show two varieties of granitic rocks; fine-grained granodiorite and medium to coarse-grained granite. Granodiorite occurs in the endocontact zone of the granite massif with the Nyanzian rocks, whereas medium to coarse-grained granite occupies the exocontact of the massif. The granodiorite contains phenocrysts of plagioclase (30–50 per cent), microcline (20–25 per cent), quartz (20 per cent), hornblende (15 per cent), and biotite (5 per cent). The granite is composed of 35 per cent microcline, 25 per cent quartz, 25–30 per cent plagioclase, 10 per cent biotite, and ore minerals 1 per cent. The granite phenocrysts which range in size from 0·01 to 0·4 mm are evenly distributed.

Hornblende from the Migori granite gave a K–Ar age of 2530 ±50 Ma (Ogola 1984). This age is in close agreement with the ages of 2595 Ma and 2510 Ma given by Cahen and Snelling. (1966) on hornblende and biotite respectively from the post-Kavirondian Maragoli granite. Consequently, this age is adopted as the age of the Migori granite and as the younger limit for the age of the Nyanzian rocks.

There is no direct evidence on the relationship between the amphibolized melanocratic dolerite and the Migori granites. However, since the melanocratic dolerite intrudes rocks of the upper group, whereas the melanocratic dolerites occur at the contacts of metabasalt, sandstone of the Lower and Middle Groups, it is concluded, that amphibolized melanocratic dolerites are older than the Migori granites.

Dykes of quartz porphyry are widespread in the tectonic zone. They conform to the regional strike, although in places, they cross–cut the Nyanzian rocks obliquely (Figure 2). Being resistant to weathering, these dykes tend to outcrop at the crests of ridges and spurs. They extend several kilometres along strike, are generally 3–6 m and rarely 10 m thick. The quartz porphyry dykes can be seen to cross-cut the melanocratic dolerite sheets in the mine workings. However, their relationship with ore bodies remains unclear. The porphyry contains phenocrysts of quartz set in a quartz and orthoclase matrix with secondary chlorite and sericite.

Dykes of younger dolerite cross-cut practically all rocks formations in the area (Figure 2). The dykes trend parallel to the general strike of the Nyanzian rocks, with individual dykes extending 3 km or more and ranging in width from 15 to 50 m. Two varieties of dykes were noted; strongly altered dolerite, commonly occurring in amphibolized melanocratic dolerite, and a less altered type. The strongly altered dolerite was sheared and dislocated together with amphibolized melanocratic dolerite. It consists of phenocrysts of monoclinic pyroxene — 0·05 mm in size which are marginally replaced by actinolite and, idiomorphic crystals of zoned labradorite — 0.03 mm in size — which are altered to oligoclase or albite and are often completely chloritized. Ore minerals represent 10–15 modal % of the rock, with ilmenite and titanomagnetite as the major minerals. Less altered dykes of younger dolerite consists of idomorphic plagioclase and monoclinic pyroxene, ranging in size from 0.6 to 1.5 mm. The plagioclase is altered by saussuritization which is more intense at the centre of the grains.

4. Metamorphism and structure of the Migori region

The Nyanzian rocks of the Migori greenstone belt were subjected mainly to regional metamorphism up to greenschist facies. Amphibole-rich rocks containing hornblende and biotite are restricted to the contact zone of the Migori granite. The presence of a tectonic zone of highly dislocated, metamorphosed volcanogenic–sedimentary rocks is a major feature of the area. The zone conforms to the regional strike of the Nyanzian rocks (Figure 2). The stresses which accrued in this zone resulted in the formation of overturned anticlines and synclines which plunge to the southwest. The zone stretches beyond the Macalder area for several kilometres and is approximately 2700 m in width.

Five ages of faulting were identified in the Migori region:
1. The oldest faults are those whose fault planes are occupied by sills of amphibolized melanocratic dolerites, with their bedding planes parallel to the general strike of the Nyanzian rocks. The formation of these faults resulted from the shearing of the volcanosedimentary succession.
2. These were succeeded by the formation of a deep-seated fault on the southern margin of the Migori greenstone belt which acted as a channel for the granite magma forming the Migori granite. This intrusion reflects the post-Nyanzian orogenic phase which was responsible for shearing and deformation of the Nyanzian rocks at their contact with the granites.
3. The next episode of faulting was associated with quartz-porphyry dyke emplacement. These faults are predominant in the tectonic zone and conform to the regional strike.
4. Later faulting in the area involved those faults which were responsible for the localization of younger dolerite. They were developed in all types of rocks, including those of the Upper group and the granites. They are north-westerly trending, although temporarily changing direction to the north–northwest on the southern part.
5. The youngest group of faults in the area strike in a north–north westerly direction and they define the northern and southern boundaries of the tectonic zone. They cut the Precambrian sequence dislocating the succession into tectonic blocks with lateral displacements. The amplitude of displacement of the individual blocks varies from 15 m in the south to 830 m in the north.

4a. Structural control of the mineralization in the deposit

The Macalder deposit is located on the northern margin of the intensively dislocated and metamorphosed volcanogenic–sedimentary rocks within the tectonic zone (Figure 2). Sulphide bodies occur as sheets, veins, and lenses (Figure 3) which were emplaced at the contact of metabasalts, banded iron formation and sandstone and also at the fold hinges and along the fault planes within these rocks.

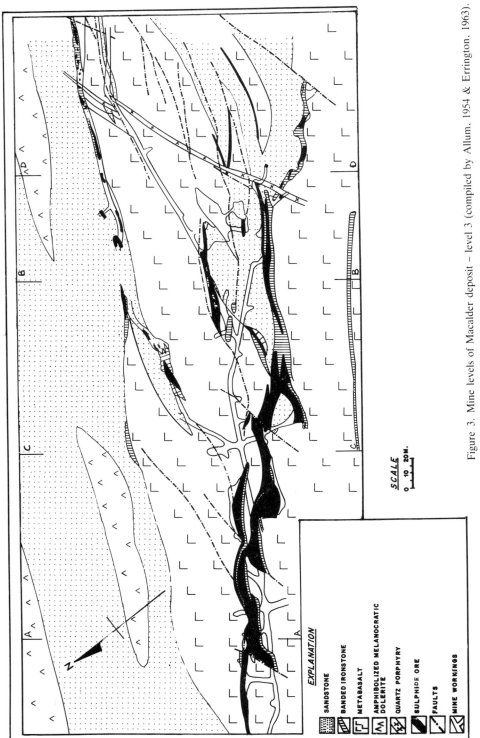

Figure 3. Mine levels of Macalder deposit – level 3 (compiled by Allum, 1954 & Errington, 1963).

The deposit occurs within a zone of faulting and isoclinal folding and the location of the ore bodies was determined by the nature and position of the sheared zones, faults and rock fissures which created passageways for the hydrothermal ore fluids.

Within the fold belt of the Macalder deposit, metabasalt forms the core of the anticline, whereas sandstone occupies the trough of the syncline. Two elongated wedges of sandstone forming synclines pinch downwards and terminate at a depth of about 73 m on the western margin and 260 m in the eastern end of the deposit (Figure 4).

Two types of faults were intersected underground: reverse and wrench faults. Reverse faults are common, particularly in the western half of the deposit. They trend east or northeast and dip northwards at 65–70°. These faults displace the footwall contact between the ore and metabasalt southwards in a series of steps (Figure 3). The amplitude of displacement of the host rocks is 100–150 m. Sanders (1964) suggested, that these faults are thrusts with upthrow on the northern side. He took note of the drag in metabasalt, orientation of slickensides and displacement of banded ironstone across the shear zone. These faults were responsible for channeling the ore fluid and the final localization of the ore bodies in zones where the parent rocks were subjected to maximum fracturing and shearing. These faults divide the ore bodies into blocks, particularly in the western half of the deposit. Individual ore blocks thin and pinch eastwards and swell westwards, thus giving rise to boudinage structure (Figure 3).

Wrench faults which are common on the eastern flank of the deposit, post-date sulphide mineralization. They are east–west trending and nearly vertical. They displace the ore bodies eastwards through a distance of 20–80 m.

The contact between the ore bodies and the host rocks is sharp and well defined. Lenses of banded ironstone and sandstone are common in the ore bodies (Figure 3). Generally, sandstone forms the hanging wall of the ore bodies whereas metabasalt and banded iron formation occur on the footwall. However, along the western limbs of the anticlines, sandstones form the footwall while metabasalt and banded iron formation represent the hanging wall (Figure 4). A zone of chlorite forms at the contact of ore bodies and the host rocks. Calcite and ankerite form non-mineralized veins in host rocks. The sheet-like ore bodies have a strike length up to 500 m long and vary in width from a few centimetres to 17 m. Downwards the ore bodies pinch eastwards, changing into lenses, veins, and stringers. Ore lenses are common at the contact of metabasalt, banded ironstone, and sandstone. They often run parallel to the bedding planes of these rocks. Maximum strike length of ore lenses is 150 m with thickness ranging from 1·5 to 10 m and rarely up to 15 m. They are dipping to the northeast. On the eastern end of level 3 ore lenses of up to 12–15 m in length occur in banded iron formation (Figure 3). They are east–west trending and dip steeply to the north.

4b. Types of ores at Macalder

The Macalder ore deposit consists of five different types of ore: massive, banded, veined, disseminated and breccia. The massive ores are dominated by pyrite which may occur almost to the exclusion of other ores or may be accompanied by chalcopyrite and pyrrhotite. In the massive pyrite ore, the pyrite constitutes 85–95 per cent of the deposit. This may contain stringers of calcite and siderite. In the massive ore with pyrrhotite–chalcopyrite, pyrite constitutes 60–70 per cent of the deposit with 20–25 per cent pyrrhotite and 10–15 per cent chalcopyrite. Sphalerite, magnetite arsenopyrite, and carbonate are present as trace minerals. Polished ore sections of the pyrite–pyrrhotite–chalcopyrite ore reveal

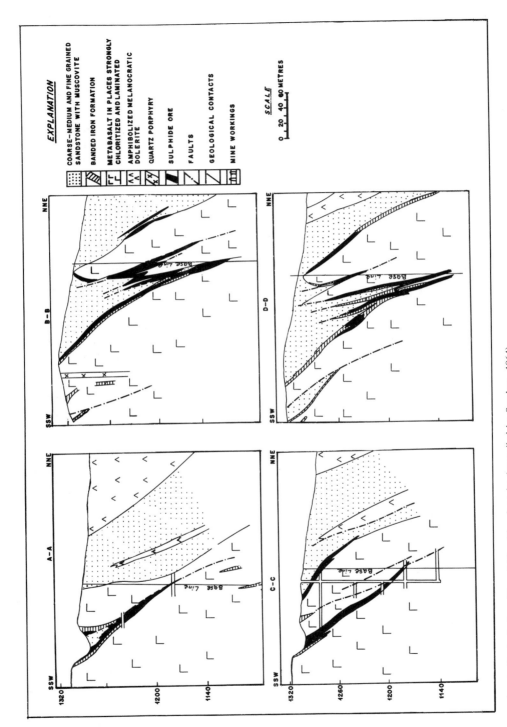

Figure 4. Geological profiles of Macalder deposits (compiled by Sanders, 1964)

that the pyrite is brecciated. The fractures in the pyrite breccias are infilled by sphalerite and calcite which also form the cement to the brecciated pyrite grains. This relationship indicates that pyrite formed first and was subjected to deformation and brecciation. As the ore fluid evolved it became rich in carbonates, which were deposited along fissures and around pyrite. At a later stage, sphalerite was deposited, and partly replaced earlier calcite or infilled fissures in pyrite or cemented pyrite grains. Pyrrhotite and chalcopyrite were deposited at a much later stage.

These massive ores are found in ore bodies which were localized at the contact of the host rocks and along fault planes where shearing and brecciation of the rocks were nominal.

The banded ore is characterized by alternating bands of pyrite, magnetite, sphalerite and minor chalcopyrite, pyrrhotite and galena. The bands, which are over 10 mm thick, are commonly cross-cut by carbonate stringers 1–5·0 mm thick. The banded ore contains coarse-grained pyrite which was formed in open spaces, crystallization of which started from the wall of the host rock inwards. Apparently, these fissures reopened periodically and at one stage when the ore solution was rich in zinc, sphalerite was deposited at their margins. This process was accompanied by the exsolution of chalcopyrite.

Quartz-rich ore veins commonly occur at the contact of host rocks and also within metabasalt, banded iron formation, and sandstones. In places, the ore occurs as pyrite stringers 0·2–5·0 mm thick. Magnetite, sphalerite, pyrrhotite, and chalcopyrite also occur with a magnetite content of up to 10–15 per cent.

Disseminated sulphide mineralization is common near the contact of the hanging wall and took place in metabasalts and sandstone. Characteristic minerals are pyrite, sphalerite, pyrrhotite, and chalcopyrite.

Breccia ore bodies occur in zones of maximum dislocation and shearing of the host rocks. Initially, fragments of brecciated and chloritized metabasalt, banded iron formation and sandstones were cemented by pyrite. Later sphalerite, chalcopyrite, and pyrrhotite enclose early pyrite and also cement rock breccias.

4c. Ore paragenesis

The process of ore formation occurred in two distinct stages: firstly the deposition of magnetite during the process of Banded Iron Formation and later sulphide mineralization:

— Magnetite occurs in massive, banded and veined ores and is believed to have originated from the Banded Ironstone Formation. The magnetite grains were corroded by the later hydrothermal solutions and relicts of magnetite with inclusions of pyrite are common in the ore. Magnetite is often replaced by pyrite, pyrrhotite, chalcopyrite, and calcite.

— The sulphide mineralization can be subdivided into four mineral associations on the basis of their mineralogical assemblages, textural characteristics, and structural relationships. The sequence of mineral formation is given in Table 2.

1. Quartz assemblage. Quartz-rich veins and stringers which conform to the regional strike of the host rocks are up to 300 m in length and 3 m in width. An early generation of quartz veins are rarely mineralized but later formed quartz stringers which are 0·3–5·0 mm thick often contain sphalerite, pyrrhotite, and chalcopyrite.

Table 2. Sequence of Macalder mineral formation

Minerals \ Assemblage	Quartz	Sphalerite–Pyrite	Pyrrhotite–Chalcopyrite	Carbonate
Quartz	DOMINANT			
Pyrite		DOMINANT – TRACE		
Cobaltine		TRACE		
Arsenopyrite		TRACE		
Sphalerite	TRACE	DOMINANT	TRACE – MAJOR	
Pyrrhotite	TRACE	TRACE	DOMINANT	
Chalcopyrite	TRACE	TRACE	DOMINANT	
Gudmundite			TRACE	
Bournonite			TRACE	
Galena			TRACE	
Native Bismuth			TRACE	
Boulangerite			TRACE	
Freibergite			TRACE	
Marcasite				MINOR
Calcite		MAJOR		MAJOR
Dolomite		TRACE		
Siderite				MINOR
Characteristic paragenetic associations		Pyrite-Cobaltine Aresenopyrite, Sphalerite–Chalcopyrite–Pyrrhotite	Pyrite-Pyrrhotite Chalcopyrite Bournonite Boulangerite Freibergite–Galena Gudmundite	Calcite Siderite
Ore structure	Massive Stringers	Massive Stringers–Brecciated	Massive Stringers Brecciated	Stringers
Rare elements in the primary minerals — %		FeS$_2$: Co, Ag, As (0·1)Au(0·001) ZnS:Co(0·02) Sn(0·1)Mn(0·3)	FeS: Co(0·01) As(0·5) FeCuS$_2$ Au(0·003)	
Temperature of ore formation	245–267°C	344–365°C	245–271°C	

EXPLANATION

▬ DOMINANT ╫╫ MAJOR ——— MINOR – – – TRACE

2. Sphalerite–pyrite assemblage. The sphalerite–pyrite assemblage is typical of massive and banded ores. Three paragenetic associations were established. (a) pyrite–cobaltine–arsenopyrite; (b) pyrite–carbonate; and (c) sphalerite–chalcopyrite –pyrrhotite.

(a) In the pyrite–cobaltine–arsenopyrite association pyrite is usually subhedral and partially or wholly replaced by marcasite, chalcopyrite, pyrrhotite, spalerite, and calcite, particularly in the massive ore where replacement of pyrite by marcasite is widespread. In places the pyrite was brecciated and fractured. The cobaltine which is pinkish-yellow in colour and weakly anisotropic, forms minute blebs 0·01–0·04 mm in size enclosed by pyrite. Where the pyrite has been replaced by chalcopyrite, the cobaltine remains unaffected. The arsenopyrite is rhombic and prismatic, rarely hexagonal, with crystals varying in size from 0·063 mm to 0·20 mm.

(b) The pyrite–carbonate association is also found in massive and banded ores. It consists of calcite, which is dominant and dolomite, with relicts of pyrite and magnetite. The dolomite forms stringers in calcite indicating that it formed later than the crystallization of calcite. The carbonates replace earlier formed quartz, pyrite and magnetite.

(c) The sphalerite–chalcopyrite–pyrrhotite association contains two generations of sphalerite. Polished ore sections show one generation of clear inclusion-free sphalerite and another with exsolved chalcopyrite. Treatment with aqua reaction on the sphalerite polished sections revealed the presence of polysynthetic twins (pressure lamellae) and zoning in sphalerite. This indicates, that local metamorphism of the ore occurred after the deposition of sphalerite. The sphalerite has a high content of Pb — 5·0 per cent, Sb — 0·1 per cent, As — 0·3 per cent and Sn — 0·1 per cent. In addition to forming exsolution blebs in sphalerite, the relatively coarse-grained anhedral chalcopyrite grains occupy the interstices between sphalerite grains. These represent some of the chalcopyrite emulsions in sphalerite which were expelled during metamorphism of the ore. Pyrrhotite forms plates in sphalerite and carbonate, ranging in size from 0·012 × 0·015 mm to 0·01 × 0·09 mm. In places, pyrrhotite forms lens-like grains of 0·4 mm in length with maximum width of 0·01 mm.

3. Pyrrhotite–Chalcopyrite assemblage. The dominant minerals in this association are chalcopyrite and pyrrhotite with minor galena and traces of gudmundite, bournonite, native bismuth, boulangerite and freibergite. Pyrrhotite was the first mineral to form in this association since it contains stringers of chalcopyrite. Both monoclinic and hexagonal pyrrhotite occur and these were studied

Table 3. Microprobe analysis of gudmundite

% Wt	1	2	3
Fe	25·67	24·95	24·60
Cu	0·04	0·02	0·03
Hg	0·04	0·10	0·07
Sb	57·37	57·38	57·06
S	15·83	15·77	16·43
TOTAL	98·95	98·22	98·19

using thermal and magnetic methods for domain structure identification. In the thermal method, polished pyrrhotite sections were heated for one hour at a constant temperature of 180°C. Domain structures were formed on the surfaces of hexagonal pyrrhotite grains, wheras, monoclinic pyrrhotite grains remained clear. Bands of monoclinic pyrrhotite were seen to be occurring within hexagonal surfaces, following the microfissures in the latter.

In the second experiment, a magnetic colloidal suspension of ferrous chloride, ferric oxide, and sodium iodide was applied to the polished pyrrhotite sections. Domain structures were formed on the surfaces of monoclinic pyrrhotite whereas hexagonal pyrrhotite grains remained clear. This is explained by the fact that monoclinic pyrrhotite is ferromagnetic, whereas hexagonal pyrrhotite is antiferromagnetic.

Some polished pyrrhotite sections were seen to exhibit domain structures when exposed to the atmosphere. This was, apparently, due to the oxidation of monoclinic pyrrhotite which oxidizes faster than hexagonal pyrrhotite as a result of an excess of sulphur.

X-ray powder diffractograms confirmed the coexistence of both hexagonal and monoclinic pyrrhotite. Hexagonal pyrrhotite is represented by a single peak wheras monoclinic pyrrhotite is represented by two peaks. Common peaks were those of hexagonal pyrrhotite.

The chalcopyrite content in this assemblage varies from 30–65 per cent. The mineral is anhedral, and coarse- to fine-grained. The coarse-grained aggregates are weakly anisotropic and platy with an irregular contact between such plates and magnetite, sphalerite, or pyrrhotite. Stringers and inclusions of chalcopyrite are common in both magnetite and pyrrhotite and more rarely chalcopyrite forms a rim around pyrrhotite. The fine-grained chalcopyrite commonly replaces pyrite, in places forming a net-like lattice.

Pyrite occurs as a secondary mineral formed within microfissures of monoclinic pyrrhotite. It forms rounded grains which vary in diameter from 0·003 to 0·012 mm.

The accessory minerals of gudmundite, bournonite, galena, native bismuth, boulangerite, and freibergite were formed more or less at the same time and apparently, from the same mineral solution as the pyrrhotite and chalcopyrite in the apparent paragenetic sequence as given.

In polished ore sections the gudmundite is seen as a pinkish-white mineral which is strongly anisotropic from yellowish-red to bluish-white. It is bireflectant, and has a higher reflective index and greater hardness than galena. The reflective index of the mineral at $\lambda = 593$ Hm was found to be 60 per cent. Individual gudmundite grains range in size from $0·02 \times 0·03$ to $0·06 \times 0·11$ mm. Occasionally, it forms polysynthetic twins which are from 0·06 to 0·09 mm long and from 0·003 to 0·022 mm wide. A microprobe analysis for the gudmundite is given in Table 3.

In polished ore sections, bournonite occurs as a white mineral with greenish tones. It exhibits bireflectance and is anisotropic (A2) with colours varying from light grey to bluish-grey. It has a higher relief than boulangerite and galena, but a lower relief than freibergite and gudmundite. It forms polysynthetic twins with individual laths ranging in length from 0·063 to 0·126 mm. Galena and boulangerite form rims around bournonite indicating that they were formed at a slightly later stage than the bournonite. A microprobe analysis for the bournonite is given in Table 4.

Galena forms less than 1 per cent of the assemblage and occurs infilling microfissures, or as inclusions in sphalerite. Where it occurs with other minerals it has a

Table 4. Analysis of bournonite and freibergite

	Bournonite % Wt.		Freibergite % Wt.
Cu	13.12	Cu	25.16
Sb	24.98	Ag	18.90
Pb	43.38	Fe	5.85
Bi	0.87	Zn	1.14
Se	0.54	Sb	27.27
Te	0.13	As	0.39
S	19.25	Bi	0.42
		Pb	0.26
		S	22.08
TOTAL	102.27		101.47

sharp contact with pyrrhotite and chalcopyrite but is intergrown with freibergite indicating that both minerals were formed more or less at the same time.

Inclusions of freibergite in galena up to 0.054×0.009 mm are common and galena also forms inclusions in freibergite.

In polished ore sections, native bismuth occurs as a pinkish yellow, anisotropic and soft mineral which exhibits bireflectance. It forms rounded inclusions of $0.003-0.017$ mm in diameter in galena. Its relationship with other minerals is not quite clear. However, considering its common occurrence in galena, it is concluded, that it was deposited more or less at the same time as galena.

Freibergite occurs as a white mineral with greyish brown tone in polished ore sections. It commonly occurs in association with galena, both of which are isotropic, but freibergite has a higher relief than galena. The chemical composition and structural formula for freibergite is shown in Table 4.

Boulangerite forming aggregates of anhedral grains, is white with bluish-green tinges in polished ores. The reflective index at $\lambda = 593$ Hm is 37.8 per cent. The chemical composition and structural formula of the mineral is given in Table 5.

Table 5. Microprobe analysis of boulangerite

	% Weight		
	1	2	3
Pb	54.29	54.16	54.56
Cu	0.14	0.12	0.20
Sb	20.87	20.65	21.15
Bi	7.94	8.18	7.45
As	0.13	0.27	0.31
Se	0.44	0.21	0.58
S	18.36	18.41	1832
TOTAL	102.17	102.0	102.47

4. Carbonate

The minerals of this assemblage are the last to be deposited in the paragenetic sequence. The carbonate is composed mainly of creamy white calcite together with a dull pinkish siderite. Carbonate minerals cross-cut and replace all earlier formed minerals. Carbonate veins vary in width from 0·007 mm to 0·10 mm with individual polysynthetic calcite plates ranging from 0·006 to 0·06 mm in size. There was no sulphide deposition during carbonate formation although pyrite was altered to marcasite at this stage. The marcasite is anhedral and is characterized by corrosive texture. The marcasite content in the ore is about 1 per cent; however, it reaches 10–15 per cent in the massive pyrite ore.

5. Fluid inclusion geothermometry

Fluid inclusion geothermometry of Macalder minerals involved decrepitation studies on ore and gange minerals coupled with homogenization of liquid/vapour inclusions in quartz.

5a. Decrepitation

The thermovacuum decrepitometer was used to analyse quartz, pyrite, sphalerite, and chalcopyrite. Decrepitation of quartz started at a temperature of 112°–195°C (Table 6). Temperatures marking the beginning and end of mass decrepitation were also recorded. From this data decrepitation temperatures were calculated. For quartz, this was found to vary from 258°C to 272°C.

Decrepitation temperatures of ore minerals showed a wide variation. Two types of fluid inclusions were recorded in pyrite and chalcopyrite with decrepitation temperature ranges between 83–145°C and 240–267°C respectively (Table 7). The first group of fluid inclusions was not detected in spharelite. At the same time,

Table 6. Decrepitation temperatures for quartz, from the Macalder ore deposit

No	Sample no	Decrepitation temperature °C				
		T_I	T_B	T_E	Δ_T	T_X
1	32	193	245	285	265	333
2	38	160	240	290	265	337
3	42	112	112	140	126	—
4	45	170	230	303	267	350
5	388	190	245	290	268	330
6	464	195	235	293	264	330
7	469	190	248	295	272	335
8	479	156	236	280	258	335

T_I — Initial temperature of decrepitation
T_B — Temperature marking the beginning of mass decrepitation
T_E — Temperature marking the end of mass decrepitation
Δ_T — Decrepitation temperature (mean)
T_X — Temperature marking the end of decrepitation

pyrite and sphalerite registered higher decrepitation temperatures of 344°C and 365°C respectively.

The variations in decrepitation temperatures indicated that sulphide mineralization occurred in different stages at variable temperatures. Initially pyrite and sphalerite were deposited at temperatures of 344°C and above. Later mineralization involved the deposition of different generations of pyrite, sphalerite and chalcopyrite. The decomposition of earlier chalcopyrite occurred at a temperature range of 245–267°C. This temperature range is in close agreement with the temperature of formation of the second generation of sphalerite (262°C) and may possibly refer to sphalerite with exsolution of chalcopyrite. Later formed chalcopyrite was deposited at temperatures between 98°C and 145°C.

The lowest decrepitation temperature of 83°C was registered for fluid inclusions in pyrite.

5b. Homogenization

Homogenization of fluid inclusions in quartz was accomplished by using a petrographic microscope and a heating stage. The analysis involved in situ observation of primary inclusions of gas and liquid phases. The size of fluid inclusions varied from 0·001 mm to 0·01 mm and the gas/liquid ratio ranged from 1:2 to 1:5. Four groups of fluid inclusions with homogenization temperature ranges of 100°C–132°C; 160–172°C; 200–292°C and 310–332°C respectively were detected (Table 8). Inclusions of the last two groups were dominant. Homogenization temperatures for quartz were found to be 276°C. This temperature is in close agreement with that of decrepitation ($T_x = 265°C$).

Table 7. Decrepitation of ore minerals, from the Macalder ore deposit

No	Sample No.	Mineral	Decrepitation Temperature °C					Disintegration Temperature °C
			T_I	T_B	T_E	Δ_T	T_X	
1.	396	Pyrite	30	65	100	83	—	
			195	215	290	253	290	442
			120	120	150	135	—	
2.	438	Pyrite	—	—	210	270	240	
			310	330	358	344	358	400
3	396	Sphalerite	266	266	298	262	—	
			—	335	395	365	395	442
4.	42	Chalcopyrite	60	130	160	145	—	
			—	230	260	245	305	429
5.	396	Chalcopyrite	30	80	100	90	—	
			—	230	253	142	270	380
6.	479	Chalcopyrite	65	85	110	98	—	
			—	130	140	135	180	
			—	245	288	267	332	420

T_I, T_B, T_E, Δ_T and T_X explained at the foot of Table 6.

Table 8. Homogenization temperatures for fluid inclusions in quartz from the Macalder ore deposit

No.	Sample no.	Gas/liquid ratio	Homogenization temperature (T^h)
		1:5	100
1	38	1:3	132
		1:4	302
2	42	1:2	204
		1:3	224
		1:2	301
		1:3	322
		1:2	200
3	45	1:3	203
		1:2	292
		1:4	302
		1:4	204
4	388	1:5	273
		1:3	324
5	469	1:3	252
		1:4	310
		1:4	100
6	479	1:5	160
		1:3	172

6. Discussion

Two major schools of thought exist about the genesis of the Macalder copper sulphide deposit.
   1. Volcanogenic–sedimentary (syngenetic) (Hutchinson, 1981).
   2. Hydrothermal–metasomatic (epigenetic) (Shackleton, 1946; Sanders, 1964). There are several factors which support the syngenetic hypothesis, such as sheet-like ore bodies, and concordant bedding of ore bodies with the host rocks. From this, Hutchinson (1981) concluded, that Macalder is a deformed volcanogenic massive sulphide deposit of a primitive type which was formed over a sea-floor hydrothermal vent. He concluded that the volcanogenic sulphide body was dynamically metamorphosed during tight isoclinal folding which resulted in the formation of a series of closely spaced anticlines and synclines.

   In contrast, the earlier work by Shackleton (1946) concluded that Macalder deposit was formed through hydrothermal–metasomatic processes. He considered the presence of apophyses and veined ore bodies within the host rocks to be indicative of such a process. On the flanks of the deposit the ore bodies split up into systems of subparallel veinlets. Ore solutions are considered to have originated from the granites which occur at a distance of less than 4 km from the deposit.

   While supporting this hypothesis, Sanders (1964) noted, that although metabasalt is dragged into the thrust zones, the sulphide ore at the sheared contacts shows no sign of equivalent drag or of brecciation. This suggests that the sulphide ore was emplaced in a pre-existing thrust structure and followed the contacts of the banded iron formation and the oblique shear cleavage.

Although stratigraphic or lithological control of the ore bodies is lacking, Hutchinson (1981) stated, that the sulphide ore represents an iron sulphide-rich facies and that the sandstones which form the hanging wall are barren and devoid of mineralization. But sulphide mineralization occurs in all the host rocks — metabasalts, banded iron formation, and sandstone. Some researchers (Smirnov 1982) have noted that the formation of the breccia ore within the pyroclastic rocks of the hanging wall is typical of volcanogenic sulphide deposits. In this case, breccias which are cemented by the wall rocks are composed of sulphide ore minerals. However, in the case of Macalder deposit, the opposite is observed. These observations indicate that the host rocks were brecciated and sheared prior to the sulphide mineralization. This is in support of the epigenetic model of formation.

According to Hutchinson (1981) the boudinage ore structure are manifestations of the tectonic deformational processes such as faulting and folding. However, in the author's opinion regional folding was responsible for the deformation of sheared layered zones along the contacts of the host rocks with different mechanical properties. During folding maximum tension and fracturing of these rocks occurred in the fold hinges. Thus the fold hinges were potentially more favourable for the localization of the sulphide ore bodies. The banded iron formation yielded by brittle failure during folding. This aided the penetration of the hydrothermal ore-bearing solutions and localization of the ore bodies within the fractured zones and at the contacts of banded iron formation with metabasalt and sandstone.

7. Conclusions

Macalder copper sulphide is a hydrothermal–metasomatic deposit which is epigenetic in relationship to the host rock and regional metamorphism. Structural studies, ore petrology, geochemical analysis, and fluid inclusion studies of the Macalder deposit indicate that:
1. Ore formation took place within a tectonic zone of intensively dislocated host rocks of the Nyanzian System.
2. Copper sulphide mineralization occurred at the contacts of metabasalts, banded iron formation and sandstones and also at the fold hinges and along faults and fissures.

Macalder copper sulphide is characterized by massive, banded, veined, disseminated and breccia ores. Massive and banded ores are predominant.

The process of ore formation occurred in two distinct stages: in the first place there was the deposition of magnetite during the diagenesis of iron-rich sediments, and accentuated during metamorphism of the banded iron formation: and secondly, localization of sulphide ore bodies from later circulating hydrothermal fluids from nearby granite intrusions.

Mineralization occurred at various stages at temperatures between 245°C and 271°C. However, pyrite and shalerite minerals indicated higher temperatures of formation (344°C–345°C). The sulphide ores can be divided into four main mineral assemblages; quartz, sphalerite–pyrite, pyrrhotite–chalcopyrite, and carbonate.

Each mineral assemblage is characterized by its own paragenetic association.

The most productive ores for copper are massive pyrrhotite–chalcopyrite and breccia ores which contain up to 10 per cent Cu and 3 per cent Cu respectively.

Zinc values in veined ore is up to 10 per cent. High quantities of Pb — 3 per cent, Sb — 0·1 per cent, Ag 0·05 per cent, Cd — 0·01 per cent and Sn — 0·03 per cent were established in banded ore. The cobalt and nickel values in massive pyrite and brecciated ore are 0·1 per cent and 0·01 per cent respectively. Gold was found to be occurring in pyrite — 0·001 per cent and chalcopyrite — 0·003 per cent. The presence of sulphosalts of Cu, Pb and Sb is unique to this deposit. These minerals were deposited towards the end of sulphide mineralization.

References

Allum, A. F. 1951. *Geological Mapping of the Mine Workings*. Mine records, Macalder mine, Kenya.
Baker, B. H. 1965. *An outline of the geology of the Kenya Rift Valley. East Africa Rift System*. UNESCO seminar, Nairobi, 1–19.
Cahen, L. and Snelling, N. J. 1966. *The Geochronology of Equatorial Africa*. North Holland Publishing Company, Amsterdam. 35–42.
Erringtone, K. 1963. *Geological Mapping of the Mine Workings*. Mine records, Macalder mine, Kenya.
Hitchen, C. S. 1936. *Geological Survey of No. 2 Mining Area, Kavirondo. Interim report and map of north-west quadrant, report No. 4*. Mining and Geological Department, Nairobi. 26 pp.
Homa Bay Report No. 32 1944. *Interim Report*. Mining and Geological Department, Nairobi. 24 pp.
Hutchinson, R. W. 1981. *Report on UNFNRE project, Migori Gold Belt, Kenya*. 22 pp.
Ogola, J. S. 1984. *Geological structure and mineral composition of massive sulphide ores of Macalder deposit, Kenya*.
Pulfrey, W. 1938. *Geology of No. 2 Mining Area, Kavirondo. Interim report and map of the south-west quadrant*. Mining and Geological Dept., Kenya. pp.
Sanders, L. D. 1964. *Copper in Kenya*. Geological Survey of Kenya, Memoir No. 4, Nairobi. 51 pp.
Shackleton, R. M. 1946. *Geology of Migori Gold Belt and Adjoining Areas*. Geological Survey of Kenya, Mining and Geological Department, Rept. No. 10, Nairobi. 60 pp.
—— **1969**. Upper Proterozoic orogenic belts in East Africa. *Ferrand. Lab., Geol. Miner. No. 41*.
Smirnov, V. I. 1982. *Geology of Mineral Deposits*. Mir Publishers, Moscow. 687 pp.

Stable isotopes in the carbonates and kerogen from the Archaean stromatolites in Zimbabwe

Paul I. Abell and Joseph McClory
Department of Chemistry, University of Rhode Island 02881, U.S.A.

A detailed isotopic analysis of a half metre core from the Cheshire Formation stromatolites of the Belingwe greenstone belt, Zimbabwe, is presented, including carbon isotopic ratio measurements on the kerogen fraction, and carbon and oxygen isotope ratio measurements on the limestone. An interpretation of these ratios is offered which views the limestone as primary, having been precipitated in fairly low energy environments, and entrapping varying amounts of organic debris in a warm marine embayment.

KEY WORDS Stromatolites Archaean Carbonates Kerogen Stable isotopes

1. Introduction

Early Earth history — the Hadaean and Archaean — is clouded with uncertainties as to the composition, quantity and temperature of the crustal material, oceans, and atmosphere. There are no remnants of the Hadaean, and only isolated fragments of the Archaean preserved on the face of the Earth, and of these, most have undergone substantial metamorphism (Schopf 1983). However, there are several pockets of Archaean sediments in the Belingwe greenstone belt of Zimbabwe which have been very well preserved, and these afford a chance to study Archaean environments with more likelihood of successful interpretations than for nearly any other known Archaean sediments. It was with these advantages and limitations in mind that we set out to examine the stromatolitic limestones of the Cheshire Formation of the Belingwe greenstone belt in Zimbabwe (Figure 1).

The field geology has been described by Martin (1978) and Martin et al. (1980), and a preliminary report on the textural preservation and stable isotope ratios of the stromatolites has been published (Abell et al. 1985). The stable isotope ratios measured produced a substantial range of values for both carbon and oxygen, and suggested that detailed studies of selected cores would be informative. The trends in these detailed analyses for a single such core (Core 23) and their interpretations are the subject of this report. A full report on the Cheshire Formation is in preparation.

0072–1050/87/TI0045–11$05.50
© 1987 by John Wiley & Sons, Ltd.

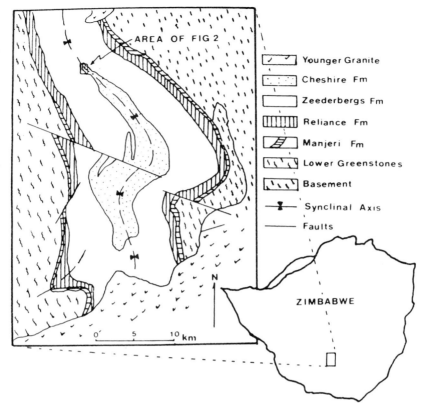

Figure 1. Geological map of the Shabani area of Zimbabwe, showing the synclinal character of the stromatolitic deposits.

As already noted, the climatic conditions prevailing in the Archaean eon are poorly defined. Available physical and chemical evidence upon which to argue for a particular set of conditions is scarce, and what geological residues we have may well have been drastically altered. No evidence can be cited that is absolutely unequivocal in defining environmental conditions in the Archaean, but gradually we hope to be able to eliminate some scenarios as inconsistent with the available evidence.

What do we indeed know? First, secular variations in the carbon and oxygen isotope ratios of carbonate sediments suggest that $\delta^{18}O$ values at about 2·7 billions years ago are likely to have been about −10 to −13 per mil, and $\delta^{13}C$ values near 0 per mil (both calculated vs. the PDB standard) (Veizer and Hoefs 1976). It has been suggested that these values may have been 'reset' by percolation of groundwater through the formations, and that the observed secular variations represent more or less complete exchange between groundwater and Archaean limestones and dolomites. It has also been suggested (Tucker 1982), from isotopic examination of primary carbonates vs. secondary cements in Precambrian dolomites and limestones, that the chemistry of the Archaean ocean might well have been different from that of the modern ocean, which could have yielded carbonate precipitates of somewhat different isotopic content than modern carbonates. Other authors (e.g. Perry 1967) would argue that the Archaean ocean was drastically different isotopically from today's oceans, and that the secular changes in oxygen isotope ratios represent additions of isotopically heavy juvenile water.

While we cannot definitely rule out the last two possibilities as explanations of the depleted oxygen isotope ratios in the Archaean, we believe that the concatenation of evidence suggests that water temperature was probably the principal factor controlling oxygen isotope ratios in those Archaean limestones which show little or no metamorphic change. We cite some evidence which argues against groundwater exchange later in this paper.

A second complication to be dealt with in interpreting isotope ratios is the problem of whether conditions were aerobic or anaerobic in the shallow waters where the cyanobacteria were producing sedimentary carbonates. Photosynthetic functionalities were firmly in place in many organisms by the middle to late Archaean (Hartman *et al.* 1985). These organisms may have produced locally oxidizing conditions, but the enormous quantity of iron in lower oxidation states in the ocean would have prevented any significant accumulation of oxygen in the atmosphere throughout the Archaean and well into the Proterozoic. In searching for clues to aerobic vs. anaerobic character during sediment depositions and early diagenesis, we have looked at the quantity and carbon isotopic ratios of the kerogen fraction of the stromatolites.

Thus, our examination of the Archaean sediments has focussed on the carbon and oxygen isotope ratios in the carbonates and their apparent preservation, and the preserved organic matter incorporated within these carbonates — the amounts, location relative to perceived sedimentary patterns, and carbon isotope ratios. From these data we draw a picture of stromatolite growth conditions at the edge of a warm sea, with varying intensities of organic productivity. We do not pretend that we have said the last word on Archaean environments, but only that we have drawn what we believe to be a self-consistent picture from the available evidence.

2. Geology of the Greenstone belt

The greenstone belts of Zimbabwe lie on either side of the Great Dyke, and include the Ngesi group as part of the Belingwe greenstone belt, near Shabani, Zimbabwe. The Ngesi group includes two stromatolitic limestone horizons, in a synclinal structure, which was probably formed initially as part of the subsidence history of accumulating sediments, but tightened by later deformation. The Cheshire Formation (Figure 2), which includes silts and shales as well as stromatolitic limestones, is underlain by the Zeederbergs basalts, which provide one of the constraining dates (2·7 billion years) for the stromatolites (Hamilton *et al.* 1977; Jahn and Condie 1976).

2a. Sampling procedures

The stromatolites of the Cheshire Formation are exposed on one side of a narrow hanging valley, the beds dipping at an angle of about 80°. The full 22 m thickness of the formation is well exposed for the 250 m length of the valley. Differential weathering of the strata makes it easy to observe the repetitive depositional cycles. Twenty-four cores were drilled with a diamond coring drill. Most of these were drilled nearly vertical to the bedding planes, and assignment of cycle and zonation within the cycles was possible in the field, but was verified

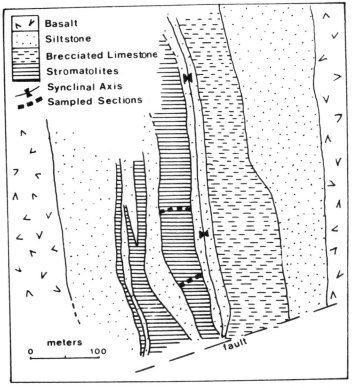

Figure 2. Cheshire Formation, showing locations of sampled sections.

by examination of the cores in the laboratory. Two sets of cores were drilled, duplicating samples cycles at a distance of about 100 m apart. The geologic section and sampling sites within the section are given in Figure 3.

2b. Laboratory procedures

We have examined seven cores (6, 9, 11, 15, 18, 19 and 23) at 5 mm sampling intervals, and three others (4, 17 and 22) at 20 mm intervals for oxygen and carbon isotope ratios in the carbonate fraction and total kerogen content and carbon isotope ratios in the kerogen (see Figure 3 for the locations of these cores in the section). For purpose of this paper we will discuss Core 23, which we consider to be typical of those analysed. It has portions of lower, middle and upper zones in its nearly half metre length. Core 23, taken from about 3·5 m above the bottom of the Cheshire Formation stromatolite sequence, was drilled perpendicular to the bedding planes, with a total length of 460 mm. Lower, middle and upper zones from Cycle 5 were clearly identified. The lower zone of Cycle 6 was either missing in this core, or too thin to recognize, and the sedimentation sequence seems to go directly from the upper zone of Cycle 5 to the middle zone morphology of Cycle 6. (Incomplete cycles have been observed in other cores as well.)

The samples for isotopic analysis of the carbonate were obtained by making parallel cuts about 1 mm apart into the edge of the half core set aside for this

STABLE ISOTOPES FROM ARCHAEAN STROMATOLITES 49

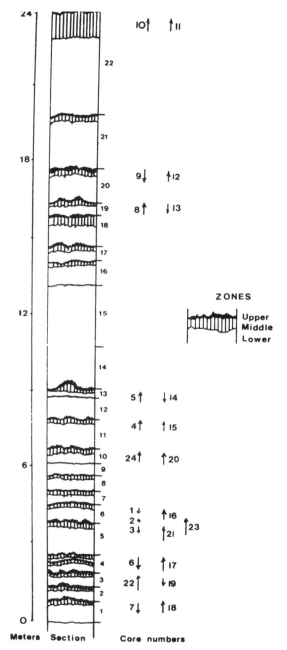

Figure 3. Geological section, zonation and core locations for the Cheshire Formation stromatolites.

use, and breaking out a triangular piece about 5 mm on each side. From this triangular piece a homogeneous sliver of appropriate size to give enough carbon dioxide for isotopic analysis was broken off using a scalpel. The interval between these parallel cuts for carbonate isotope samples was later cut out to give the larger samples necessary for the kerogen analyses. These were typically 0·5 to 1·0 g in total weight before removal of the carbonate. The samples for isotope analysis in the carbonate fraction were then transferred to a vacuum line, the carbon dioxide liberated by the action of 100 per cent phosphoric acid at 55°C, and the carbon and oxygen isotope ratios measured on a VG Micromass 602D mass spectrometer. The samples to be examined for kerogen content and carbon isotope ratio in the kerogen were weighed, dissolved in 10 per cent HCl and the insoluble residue (containing both kerogen and HCl-insoluble inorganic residues) was transferred quantitatively to a 6 mm glass tube, where it was washed extensively and dried. A few milligrams of copper oxide and manganese dioxide were added to each tube, the tube evacuated and filled with high purity oxygen to a pressure of 300 Torr, the tube sealed, and then heated to 800°C for about ten minutes, cooled to 300°C for another ten minutes, and transferred to a tube breaker on the mass spectrometer. The excess oxygen was bled off slowly through liquid nitrogen traps, and the residual carbon dioxide transferred to the mass spectrometer cavity, where the total ion current was read under standardized conditions, and compared with known volumes of carbon dioxide. The carbon isotope ratio was then measured. The quantitative reliability of this procedure was verified by combustion of powdered mixtures of charcoal and Ottawa sand of varying size and carbon content. Combustions were found to be complete and isotope ratios reproducible (to within 0·3 per cent) as long as an excess of oxygen was present. In the carbonate fraction the carbon and oxygen isotope ratios were reproducible on splits of the same sample to 0·1 per mil. The isotope ratios are reported with reference to the PDB standard, and all these data are plotted in Figure 4 together with our reading of the change in zonation along the core. The same detailed plots have been obtained for the other cores noted at the beginning of this section, as well as from selected sites on the other cores obtained for the Cheshire Formation stromatolites, and will be reported on in detail in a future publication.

3. Results and Interpretations, Core 23

3a. Oxygen isotope ratios in the carbonate fraction

The oxygen isotope ratios average near −10 per mil (PDB) in core 23, with two pronounced excursions to more positive values. One excursion is observed at 225 mm (Figure 4) (corresponding to the base of the middle zone of Cycle 5), the other (including all of the upper zone of Cycle 5) occurs at 300–350 mm. These may well reflect temporary changes in water conditions, either cooler water being supplied by the opening up of an off-shore bar, for example, or mild evaporative effects in an enclosed lagoon, concentrating the heavier isotope of oxygen. We incline toward an interpretation of the average value of the oxygen isotope ratio, −10 per mil, as a consequence of precipitation of limestone at a temperature near 70°C (Friedman and O'Neil 1977).

3b. Carbon isotope ratios in the carbonate fraction

There are no large scale variations in carbon isotope ratios along the core, the value of $\delta^{13}C$ being very close to 0 per mil (PDB). This means that no significant

STABLE ISOTOPES FROM ARCHAEAN STROMATOLITES

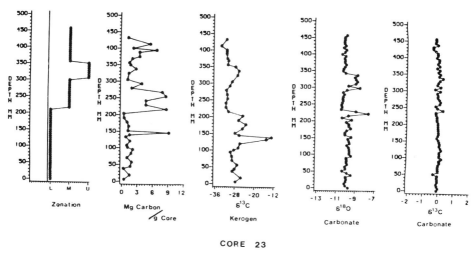

Figure 4. Analytical results for Core 23, including observed zonation, organic carbon content, $\delta^{13}C$ in the organic carbon and $\delta^{18}O$ and $\delta^{13}C$ in the carbonate fraction of the stromatolites.

contribution can have been made from oxidation of detrital organic matter, which would have pushed the values of $\delta^{13}C$ to negative values. Thus we construe the carbon isotope ratios as reflecting precipitation of calcium carbonate in good equilibrium with the dissolved carbon dioxide of the Archaean ocean. Having said this, however, we wish to point out that there are slight (0·2 to 0·4 per mil) excursions of the carbon isotope ratios at the same sites where we observed the oxygen isotope excursions noted above. These are of about the magnitude one would expect if temperature changes of a few degrees were affecting both isotopes. This by no means proves that temperature change was the causative factor, as both changes in temperature and evaporation of a body of water can produce changes in organic productivity, with changes in the carbon isotope ratios of precipitated calcite.

3c. Kerogen content

The kerogen content in this core 23, as with most others in this study, was highly variable. In general the lower zones have little kerogen, and we observed an average organic carbon content near 0·15 per cent. Occasionally, however, there are irregularities in a generally low-kerogen region of a core, but these are rare. Such an irregularity is observed here at about 150 mm. Such irregularities may well be the result of some singular environmental event, such as a storm that piles up large amounts of algal debris. The middle zones are typically higher in kerogen, often by factors of ten or more. The upper zones are rich in kerogen, too. This richness in kerogen is very apparent in thin sections, which can be nearly opaque because of high concentrations of the dark brownish-black organic matter. These variations in kerogen content can be primary, reflecting differing productivity, or secondary, reflecting differing preservation. We think the carbon isotope ratios discussed in the following section suggest that the former is the case.

3d. Carbon isotope ratios in the kerogen

The carbon isotope ratios in the kerogen of core 23 are widely variable. Just as the quantity of carbon varies from lower to middle zone, so do the carbon isotope ratios. Through most of the core the range is from -24 per mil to -32 per mil, with samples from the lower zone having an average value near -25.7 per mil, and those from the middle zone a mean value near -31.7 per mil. At 150 mm where there is a brief interval of very high kerogen content, the carbon isotope ratio in the kerogen reaches enrichments greater than -12 per mil.

The very negative (highly depleted in ^{13}C) values of carbon isotope ratios in Archaean kerogens are generally interpreted as reflecting the availability of large amounts of atmospheric carbon dioxide (Estep 1982). Inasmuch as the Archaean atmosphere was enriched in carbon dioxide by perhaps as much as 100 to 1000 times present levels (Walker 1983), these represent normal carbon isotope ratios derived from photosynthesizing organisms. Because the photosynthetic pathway discriminates against the heavier isotope of carbon, photosynthesizing species can be more selective as the supply of carbon dioxide increases. The supply of carbon dioxide will vary in a protected marine environment with depth of water and turbulence. In deep still water the rate of supply of carbon dioxide will be diffusion controlled, and highly variable. The carbon isotope ratios in the kerogen produced in such sediments should provide clues as to the depth and relative energy input, and thus we expect that there should be some correlation with the zonation. The isotopic ratios we observe indicate that the supply of carbon dioxide reaching the mat was greater during deposition of the middle zones as compared to the lower ones. The macroscopic texture of the stromatolites provide clues to validate this statement. The stromatolitic growth of the middle zones is characterized by irregular heads, similar in shape to modern algal growths in intertidal waters (Golubic 1976). The energy level of the water during the accumulation was quite high, and rapid mixing of the water would have provided a continuous fresh supply of carbon dioxide to the mat. In the lower zones the laminar nature of the stromatolitic growth suggests a low energy environment. The carbon dioxide supply would not be as great and might well be largely diffusion controlled. Therefore, the cyanobacteria would be forced to utilize a greater proportion of the carbon dioxide with less isotopic discrimination. The excursions to more positive values will, then, represent situations where discrimination in the selection of the carbon isotopes has become impaired.

4. General interpretations

The late Archaean at C.2.7 aeons ago was gradually acquiring diversity in its life forms, and was in the process of evolving from a largely anaerobic to an aerobic world. The atmosphere, weakly reducing, was composed largely of nitrogen and carbon dioxide, but photosynthesis was firmly esconced in the cellular processes of many early life forms (Hartman et al. 1985), and there were certainly local situations where oxidative conditions prevailed. It was still to be hundreds of millions of years, however, before the accumulated ferrous iron and sulphur and sulphide were largely oxidized, and important contributions of oxygen to the atmosphere were possible.

One can visualize a warm, humid climate, the consequence of the strong greenhouse effect of the carbon dioxide and water vapour in the atmosphere, which more than compensate for the lower luminosity of the early sun. The oceans, already nearing present volume, produced mainly by mantle degassing, were the home of the cyanobacterial population which was the dominant life form. The continents, highly fragmented, were scattered in an unknown configuration, but the Rhodesian craton appears from geomagnetic measurements, to have been in a subtropical latitude (McIlhenny 1973).

The cyanobacteria, requiring a protected environment for growth, were accumulating in shallow lagoons and bays. In the Cheshire Formation, the cyclical events and the absence of major weathering of erosional surfaces, indicates that there was a continual but not linear subsidence taking place as the sediments accumulated. The cyclical process can be interpreted as laminar growth in relatively deep, protected water, followed by growth of cabbage-like heads in a slightly higher energy environment. In this environment nutrient levels were higher and mat growth was thicker than the lower zones. There was then a final stage of the cycle involving detrital input, some oxidation and an evening off of the depositional surface before the laminar growth resumed. In some cases the boundaries between the zones are precise and unequivocal, but in others the transitions are not readily apparent.

In the lower zones of sediment accumulation, the build up of calcareous sediment was gradual and the strata were essentially horizontal, and there was little accumulation of organic matter. Photosynthesis was slow and controlled by carbon dioxide diffusion and nutrient supply.

With shallowing of the water, either through sediment accumulation or tectonics, we enter the middle zones. The hemispherical calcareous 'heads' of the middle zones were porous and provided a good growth surface, and a good surface for retention of organic debris. The limestone accumulation kept pace with algal growth, and provided equilibrium precipitation conditions with ocean water and dissolved carbon dioxide.

The upper zones represent transitions back to the lower zone conditions and reversion to laminar growth, low nutrient supply and low energy conditions. To a large degree they represent periods of infilling of the middle zone irregularities, and are seldom episodes of significant sediment accumulation. (Note in Figure 3 that the upper zones are typically very thin.)

Having a picture of the original growth conditions derived in part from the geology and in part from the isotope ratios, we must now see if this is compatible with what we know of the diagenetic processes which followed the initial sedimentation. As the calcite and algal debris accumulated there was compaction and lithification, but this was likely to have been largely an infilling of fenestrae with recrystallized calcite, as preservation of the stromatolite macrostructure seems to be nearly complete. The carbon isotopes in the carbonate remained unchanged, as there is no evidence that they went through any re-equilibration with dissolved carbon species in the interstitial water. Although it is possible that there may have been some minor modifications in the oxygen isotope ratios by water in the sediments, the water is still essentially isotopically the same as the shallow ocean water in which the original precipitation took place. 'Closing' of the beds is then perceived as being fairly rapid (Veizer and Hoefs 1976), and the stromatolites were preserved both texturally and isotopically. While pervasive exchange of isotopes with ground water remains a possibility as an explanation of the very depleted oxygen isotope ratios, it has to have been extraordinarily linear with time, and insensitive to the lithology (Veizer and Hoefs 1976; Perry 1967).

In the organic component, there could have been some minor alteration in the carbon isotope ratios during early diagenesis. Carbohydrates, which are isotopically heavy, are the first to decay, and there may be a consequent decrease in the carbon isotope ratios of residual organic matter to more negative values by up to 2 to 4 per mil (Spiker and Hatcher 1984; Peters et al. 1981; Rohrback et al. 1984). While this may alter the details of the changes in the carbon isotope ratios of the kerogen, it is unlikely to have altered the perceived pattern. At no time was there a major re-equilibration with organic-rich water, as we see no reflection of this in the carbon isotope ratios of the carbonate. It is to be noted that this model requires precipitation of the calcite in the early stages of stromatolite growth, rather than the accumulation of thick, decaying cyanobacterial mats with small amounts of calcite interspersed, such as is seen in some modern situations. The latter would require that the carbon isotope ratios in the calcite reflect a major component of isotopically light carbon dioxide from decay.

Once the beds are closed, which seems likely to have been fairly early in the diagenetic history of these stromatolitic limestones, the insoluble kerogen would not be expected to show any significant changes in ^{13}C over time (Schopf 1983) and the values we observe must be very close to those of the original organic matter at the time of closure. Schopf (1983) and Deines (1980) suggest that unmetamorphosed kerogens may be expected to show carbon isotopes ratios near -27.5 per mil, while recent organic sediments show values near -25.0 per mil. Our carbon isotope ratios in these Archaean kerogens are slightly more depleted.

Finally, we want to call attention to the fact that this sort of interpretation, which we believe to be reasonably self-consistent, is only possible when detailed millimetre scale sampling is performed. The tendency to make conclusions from random outcrop sampling is to be deplored as being incapable of leading to correct interpretations of past environments.

Acknowledgements. We acknowledge NASA grant NAG 3–206 for financial support, and thank Dr. Tony Martin of Cluff Mineral Exploration (Zimbabwe) Ldt., Harare, Zimbabwe, for his assistance in drilling the cores.

References

Abell, P.I., McClory, J., Martin, A. and Nisbet, E.G. 1985. Archaean stromatolites, from the Ngesi Group, Belingwe Greenstone Belt, Zimbabwe: preservation and stable isotopes — Preliminary results *Precambrian Research*, 27, 357–383.

Deines, P. 1980. The isotopic composition of reduced carbon. *In:* Fritz, P and Fontes, J. Ch. (Eds), *Handbook of Environmental Isotope Geochemistry*. Elsevier, Amsterdam, 329–406.

Estep, M.L.F. 1982. *Stable isotope composition of algae and bacteria that inhabit hydrothermal environments in Yellowstone National Park*, Carnegie Inst. Washington Yearbook.

Friedman, I. and O'Neil, J.R. 1977. Compilation of stable isotope fractionation factors. *Geological Survey Professional Paper*, 440–KK, U.S. Government Printing Office. Figure 27.

Golubic, S. 1976. Organisms that build stromatolites. *In:* Walter M. R. (Ed.), *Developments in Sedimentology, vol. 20, Stromatolites*, Elsevier Scie. Publ. Co., Amsterdam, 113–140.

Hamilton, P.J., O'Nions, R.K. and Evenson, N.M. 1977. Sm–Nd dating of Archaean basic and ultramafic volcanic rocks. *Earth and Planetary Science Letters*, 36, 263–268.

Hartman, H., Lawless, J.G. and Morrison, P. 1985. *Search for the Universal Ancestors*. NASA SP–477. Washington, D.C. 38–39.

Jahn, B.-M and Condie, K.C. 1976. On the age of the Rhodesian greenstone belts. *Contrib. Mineral. Petrol.*, 57, 317–330.

Martin, A. 1978. The geology of part of the Belingwe greenstone belt and adjacent country. *Bulletin of the Geological Survey of Rhodesia*, 83, 220 pp.

Martin, A., Nisbet, E.G. and Bickle, M.J. 1980. Archaean stromatolites of the Belingwe Greenstone Belt, Zimbabwe (Rhodesia). *Precambrian Research*, 13, 337–362.

McIlhinny, M.W. 1973. *Paleomagnetism and Plate Tectonics*. Cambridge University Press, Cambridge. 358 pp.

Perry, E.C. 1967. The oxygen isotope chemistry of ancient cherts. *Earth Plan. Sci. Lett.*, 3, 62–66.

Peters, K.E., Rohrback, B.G. and Kaplan, I.R. 1981. carbon and hydrogen-stable isotype variations in kerogen during laboratory-simulated thermal maturation. *American Association of Petroleum Geologists Bulletin*, **65**, 501–508.

Rohrback, B.G., Peters, K.E. and Kaplan, I.R. 1984. Geochemistry of artificially heated humic and spapropelic sediments — II: Oil and gas generation. *American Association of Petroleum Geologists Bulletin*, **68**, 961–970.

Schopf, J.W. 1983. *Earth's Earliest Biosphere: Its Origin and Evolution.* Princeton Univ. Press, Princeton, N.J. 543 pp.

Spiker, E.C. and Hatcher P.G. 1984. Carbon isotope fractionation of sapropelic organic matter during early diagenesis. *Organic Geochemistry*, **5**, 283–290.

Tucker, M.E. 1982. Precambrian dolomites: Petrographic and isotopic evidence that they differ from Phanerozoic dolomites. *Geology*, **10**, 7–12.

Veizer, J. and Hoefs, J. 1976. The nature of $^{18}O/^{16}O$ and $^{13}C/^{12}C$ secular trends in sedimentary carbonate rocks *Geochim. Cosmochim. Acta*, **40**, 1387–1395.

Walker, J.C.R. 1983. Possible limits on the composition of the Archaean ocean. *Nature*, **302**, 518–520.

Polymetamorphic Al, Mg-rich granulites with orthopyroxene–sillimanite and sapphirine parageneses in Archaean rocks from the Hoggar, Algeria

J. R. Kienast
Laboratoire de Pétrologie Métamorphique, Université Pierre et Marie Curie, 4 Place Jussieu, Paris, France

and

K. Ouzegane
U.S.T.H.B. – I.S.T., B.P.9, Dar el Beida, Alger, Algeria

An orthopyroxene–sillimanite association occurs in Al–Mg rich granulites within Archaean aluminous metapelites at In Hibaou (NW Hoggar). Sapphirine is characteristically present in those rocks as a reaction product of a high pressure mineral association. Other reaction products include cordierite and a second generation of orthopyroxene, confirming the presence of two superimposed metamorphic events (M1 and M2). The application of mineral geothermometers and geobarometers to the observed parageneses in the two metamorphic events gives the following P–T ranges M1, 10 ±1.5 Kb — 800–900°C; M2, 5 ± 1.5 Kb — 800–700°C. These values are in general agreement with those deduced from associated pyrigarnites.

KEY WORDS Algeria Hoggar Granulites Archaean Sapphirine Opx–sillimanite Association

1. Introduction

Alumino-magnesian mineral parageneses are very rare in granulite facies terranes. They are characterized by the presence of sillimanite–orthopyroxene associations and the relative abundance of corundum, spinel and particularly sapphirine with or without quartz. These associations are found exclusively in Archaean granulitic rocks of South Africa (Windley *et al.* 1984), India (Grew 1982) and the U.S.S.R. (Bondarenko 1972; Perchuk *et al.* 1985). Most of the occurrences show polymetamorphic assemblages indicating an evolution from a high pressure (10–12 kbar) to a low pressure (5–6 kbar) facies. In the Hoggar massif this evolution is shown by spectacular corona growth of low pressure phases (M2, Table 1) around primary high pressure phases (M1, Table 1).

0072–1050/87/TI0057–23$16.50
© 1987 by John Wiley & Sons, Ltd.

2. Geological setting

The Archaean in the In Hihaou granulite complex forms part of the northern In Ouzzal craton which covers a large area (Figure 1). To the west and east, the contact between this craton and the adjacent Pan-African belt corresponds to a N–S shear zone. The western and eastern branches of the Pan-African belt (Gravelle 1969; Caby 1970) are both deformed and metamorphosed under amphibolite and greenshist facies conditions, contrasting with the matamorphic grade of the In Ouzzal craton. Towards the tectonic boundary, granulites are retrogressed as pseudotachylites, while within the craton they are unaffected. With regard to the Iforas craton (Mali), an important sinistral displacement has been recorded across the western part of the In Ouzzal shear zone (Caby 1968). The sketch map (Figure 1) shows some E–W folds which can be traced over tens of kilometres, and are characteristic of the In Hihaou region. They are followed by two distinct generations of isoclinal folding. The higher massifs of the region are essentially composed of post Pan-African (530 My) alkaline granites and ignimbrites (In Hihaou, Nahalet, Tihimatine) which cross cut the Archaean granulites. The granulite complex of In Hihaou (West Hoggar) consists of a wide range of metapelites, banded magnetite quartzites, leptynites and marbles commonly associated with norite are lenses of pyrigarnite and lherzolite. The pyrigarnites present a clear case of polymetamorphism with corona transformation affecting a garnet + clinopyroxene + plagioclase + quartz assemblages which are transformed to an orthopyroxene + plagioclase association. This is a widespread transformation in Archaean granulitic terrains (de Waard 1965; Wells 1979) attributed to pressure at high temperatures. Associated Al–Mg granulites discussed in this paper are interpreted in the light of this evolution.

Previous studies (Lelubre 1952; Giraud 1961; Le Fur 1966; Caby 1970; Fourcade and Javoy 1985) have stressed the unusual characteristics of the Hoggar structures. Two age groups have been observed in this region. Whole rock Rb–Sr ages in the range 2·9–3·3 By (Ferrara and Gravelle 1966; Allegre and Caby 1972) are confirmed by U/Th/Pb on zircons (Lancelot et al. 1976) and whole rock Sm/Nd ages (Ben Othman et al. 1984). However ages obtained on mineral separates, such as Rb–Sr on by biotite (Allegre and Caby 1972), U/Th/Pb on zircon (Lancelot et al. 1976), and Nd/Sm on garnet (Ben Othman et al. 1984) are much younger (2·1 By). It seems likely that the two ages represent the high and low pressure metamorphic phases suggested by Fourcade and Javoy (1985).

The most typical feature of this area is the abundance and variety of orthopyroxene–sillimanite granulites. The orthopyroxene–sillimanite association is the primary high pressure assemblage in these rocks whereas sapphirine may occur as a secondary mineral. Such granulites form lenticular bodies with variable thicknesses (50 cm to 100 cm) interstratified in various metasediments. They were observed in direct contact with quartzites, magnetite quartzites, metapelites, leptynites and charnockites. In some areas, they are immediately in contact with lherzolite or marble.

3. Mineral assemblages and reactions

3a. Primary assemblages

Quartz-bearing Al–Mg granulites are volumetrically the major rock types (A, B, Table 1). Quartz-free granulites (C, D, Table 1) are only present as conspicuous

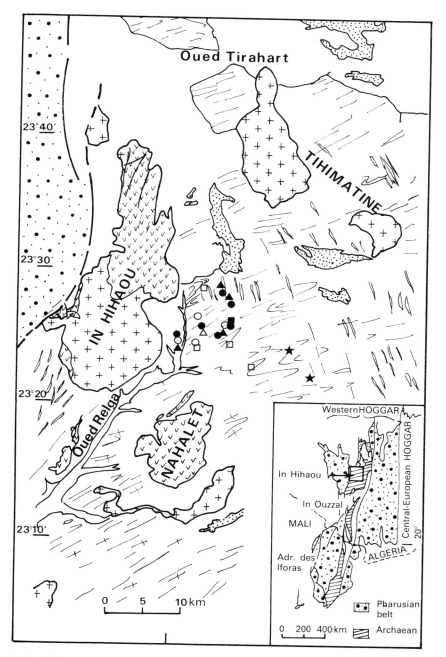

Figure 1. Simplified geological map of the In Hihaou granulite complex (western Hoggar, Algeria).

Table 1. Mineral assemblages in Al–Mg granulites from In Hihaou, Hoggar, Algeria

Rock type		Early assemblage M1	Second assemblage M2
Al–Mg granulites + quartz (Sapphirine free in M2 assemblages)			
A	1	Qz + Sill + Opx	Cd
	2	Qz + Sill + Gt	Cd
	3	Qz + Sill + Gt + Opx1	Cd + Opx2
	4	Qz + Sill + Gt + Opx1 + Bi	Cd + Opx2 + K.F + Pl
B	1	Qz + Gt	Cd + Opx2 (Figure 2)
	2	Qz + Gt + Opx1	Cd + Opx2
Al–Mg granulites–quartz (sapphirine-bearing in M2 assemblages)			
C	1	Sill + Opx	Cd + Sa (Figure 3)
	2	Sill + Gt	Cd + Sa
	3	Sill + Gt + Opx1	Cd + Sa + Opx2
	4	Gt + Opx	Cd + Sa + Opx 2
D	1	Sp + Opx1	Opx2 + Sa

Mineral abbreviations: Qz: quartz; Sill: sillimanite; Opx: Orthopyroxene; Cd: cordierite; Gt: garnet; Bi: biotite; K.F: K-feldspar; Pl: plagioclase; Sa: sapphirine; Sp: spinel.

nodules that may reach 1 m³ in volume.

Quartz-bearing granulites may be subdivided into sillimanite rich (A) and sillimanite free parageneses (B). The following primary phase assemblages have been observed: A1, Qz–Sill–Opx; A2, Qz–Sill–Gt; A3, Qz–Sill–Opx–Gt; B1, Qz–Gt; and B2, Qz–Opx–Gt. Opx-bearing parageneses are always richer in Mg than garnet-bearing ones which are richer in Fe.

Nodules of quartz-free Al–Mg granulites may show a variety of assemblages: C1, Sill–Opx; C2, Sill–Gt; C3, Sill–Opx–Gt and C4, Gt–Opx. The presence of orthopyroxene and/or garnet is also correlated with the relative abundance of Mg and Fe. Spinel or corundum may be present in some nodules where large blue sapphirine crystals occur as reaction products. Biotite may be present or absent in all the primary phase assemblages.

3b. Secondary assemblages

Corona reaction products of the high pressure primary assemblages are observed in all the Al–Mg granulites form In Hihaou. In the coronas, cordierite, orthopyroxene, and/or sapphirine have crystallized in symplectic intergrowths (Figures 2 and 3).

In quartz-rich rocks two types of mineral parageneses are recorded, depending on the presence or absence of sillimanite. In the first type, three primary minerals sillimanite, orthopyroxene, and quartz are systematically isolated from one another by an aureole of newly-formed cordierite, suggesting that the reaction orthopyroxene + sillimanite + quartz → cordierite (A1, Table 1) has taken place. This reaction, though rare in granulite facies metamorphism, has been described by several authors (Chinner and Sweatman 1968; Meng and Moore 1972; Grew 1982; Perchuk et al. 1985) and has also been experimentally investigated (Hensen and Green 1971; Annersten and Seifert 1981). In biotite-bearing granulite, complex symplectites of cordierite K-feldspar result from the breakdown of biotite in the presence of sillimanite and quartz. When garnet is present in the primary paragenesis, the symplectites result from the combination of several reactions: garnet + sillimanite + quartz → cordierite (A2, Table 1); garnet + quartz → orthopyroxene + cordierite (B1, Table 1) and orthopyroxene + sillimanite + quartz → cordierite (A1, Table 1). In the absence of sillimanite, garnet is isolated from quartz by an orthopyroxene and cordierite corona. Thin symplectites of orthopyroxene–cordierite are well developed along garnet fractures. Such types of associations and textures have been described as resulting from contact metamorphism (Berg 1977; Schreyer and Abraham 1978) and seldom from regional metamorphism (Vielzeuf 1980; Van Reenen 1983). The observed reaction garnet + quartz → orthopyroxene + cordierite (B1, Table 1, Figure 2) has also been experimentally investigated (Hensen and Green 1971; Holdaway 1976). When present, biotite is surrounded with coronas of orthopyroxene alone or in symplectites with cordierite which isolate it from quartz or garnet according to the reaction biotite + quartz → orthopyroxene + K-feldspar and biotite + quartz + garnet → orthopyroxene + cordierite + K-feldspar ± plagioclase, experimentally studied by Hoffer and Grant (1980).

In more aluminous quartz-free rocks, reactions where spinel is present, may be distinguished from spinel-free rocks. When spinel is absent, intergrowth of sapphirine and cordierite result from disequilibrium reactions of orthopyroxene–sillimanite or garnet–sillimanite. The reaction orthopyroxene + sillimanite → cordierite + sapphirine (C1, Table 1, Figure 3) is only known in the Limpopo belt (South Africa, Windley et al. 1984), whereas the reaction garnet + sillimanite → cordierite

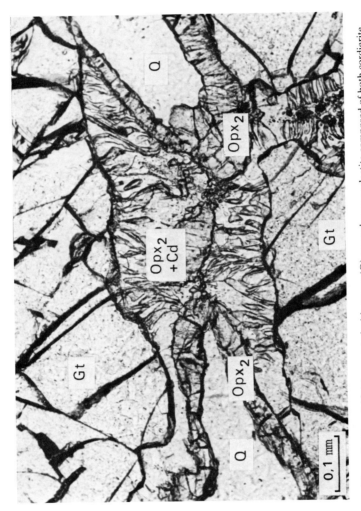

Figure 2. Garnet (Gt) has reacted with quartz (Q) to produce symplectite composed of both cordierite (Cd) and orthopyroxene (Opx). Opx_2 designates small grains of orthopyroxene belonging to the second generation (M2 in Table 1).

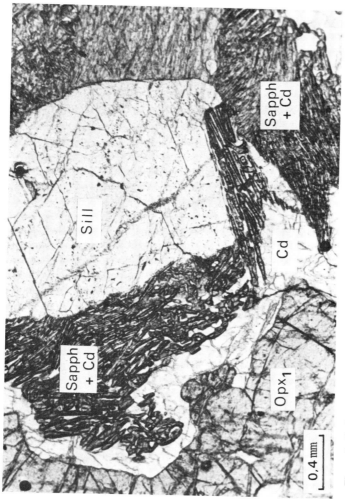

Figure 3. Orthopyroxene (Opx$_1$) and sillimanite (Sill) belonging to an earlier generation (M1 in Table 1), separated by intergrowth of cordierite (Cd) and sapphirine (Sapph). These unusual large Al-silicate grains were identified as sillimanite by XRD.

+ sapphirine (C2, Table 1) has apparently not been previously reported. When not in contact with sillimanite, garnet is surrounded by a symplectite of orthopyroxene + cordierite + sapphirine suggesting the reaction garnet → orthopyroxene + cordierite + sapphirine (C3, C4, Table 1).

Relict spinel is systematically surrounded by large crystals of sapphirine or by vermicular intergrowths of orthopyroxene and sapphirine. The observed reaction is spinel + Al-rich orthopyroxene → sapphirine + Al-poor orthopyroxene (D1, Table 1). Coronas of sapphirine around spinel are very common in many sapphirine bearing granulites worldwide, e.g. in Uganda (Nixon et al. 1973), in Brazil (Vicente et al. 1978), in India (Grew 1982), and in West Greenland (Herd et al. 1969).

Corundum–orthopyroxene associations, known only in a few other sapphirine localities (Windley et al. 1984), are rare in the In Hihaou region. The corundum–orthopyroxene parageneses have several common textural features with spinel–orthopyroxene parageneses. A first generation of sapphirine replaces corundum which may persist in the centre of large crystals. This first generation breaks down to symplectite of a second sapphirine + cordierite, while in spinel-bearing assemblages, the symplectites consist of sapphirine + orthopyroxene. Primary sillimanite, biotite and perthite also occurs in these rocks.

In summary, the Hoggar granulites are characterized by the following features:

(a) two successive mineral parageneses: M1 and M2;
(b) primary minerals are sillimanite, garnet, spinel, corundum, biotite and quartz;
(c) secondary symplectites of orthopyroxene–cordierite; sapphirine–cordierite; orthopyroxene–sapphirine, orthopyroxene–sapphirine–cordierite;
(d) sapphirine and quartz occur exclusive of each other.

4. Mineral chemistry

Representative microprobe analyses are given in Tables 2, 3, 4, 5, 6 and 7. They are used in thermodynamic calculations to determine the P–T conditions of granulite metamorphism. In samples with corona textures, compositional variations occur in some of the minerals as shown by differences between earlier and later generations and zoning in individual primary crystals.

4a. Orthopyroxene

Orthopyroxene occurs in Al–Mg granulites as well as in the pyrigarnite. The analysed orthopyroxenes (Table 2) in Al–Mg granulites have negligible amounts of CaO and MnO and are fairly Mg rich. Orthopyroxene Mg/Fe + Mg ratios are much lower in garnet-bearing granulites (0·65–0·75) than in garnet-free granulites (0·75–0·95). The higher Mg content of orthopyroxene in the garnet-free granulites is consistent with the high Mg content of associated sapphirine, cordierite, and biotite. In both cases, the Al_2O_3 content of primary orthopyroxene decreases from core (9 to 11 wt %) to rim (6 to 8 wt %) with increasing SiO_2 and MgO. This zoning is similar to the one observed by other authors (Chinner and Sweatman 1968; Ellis et al. 1980; Grew 1982; Droop and Nurminen 1984; Harley 1985) in orthopyroxene from Al–Mg granulites. Secondary orthopyroxene in symplectite with cordierite and/or sapphirine (A3, A4, C3, C4, D, Table 1) have lower Al_2O_3 content ranging from 5 to 7 wt %. However, the small compositional difference between the primary orthopyroxene edges and secondary orthopyroxenes is a good indicator of chemical equilibrium during the M2 stage. The orthopyroxene in reaction rims between garnet and quartz (B, Table 1) or biotite and quartz shows the lowest Al_2O_3 content (3 to 5wt%).

In pyrigarnites, orthopyroxene occurs in two habits, first as relict laths several centimetres long and second in symplectite with plagioclase resulting from the breakdown of garnet and clinopyroxene. Zoning is absent in orthopyroxene and there is no difference in composition between the two orthopyroxene generations. Their Al_2O_3 content (2 wt %) is similar to the Al_2O_3 content of orthopyroxene from meta–igneous rocks.

4b. Garnet

Garnets from Al–Mg granulite are mainly pyrope-almandine solid solutions characterized by very low CaO and MnO contents (Table 3). They are slightly zoned: the almandine content always increases with decreasing pyrope content from core to rim. The most Mg-rich garnet (50 to 60 per cent pyrope) coexists with sapphirine (C4, Table 1), while the garnet associated with sillimanite and quartz (A2, Table 1) are the least Mg-rich (30 per cent pyrope).

Garnets from pyrigarnite differ from Al–Mg granulite garnets not only in composition but also in type of zoning. Their pyrope and grossular contents decrease towards the contact with plagioclase–orthopyroxene intergrowths, while the almandine and spessartine contents increase (Table 3).

4c. Sapphirine

The ferric iron of the sapphirines was calculated from stoichiometry by the method of Higgins *et al.* (1979). The chemical compositions of sapphirine lie close to a tie-line representing the (Mg, Fe) Si→←Al Al tchermackite substitution. The large crystals of sapphirine containing spinel or corundum inclusions are systematically zoned in Al_2O_3, SiO_2 and MgO. The Al_2O_3 content is higher at the spinel contact than in the outer margin at the orthopyroxene contact (Table 4). When spinel (or corundum) is absent from the centre, the Al_2O_3 content increases from core to rim while the Mg/Mg + Fe ratio decreases. Orthopyroxene displays the opposite zoning pattern with respect to Al_2O_3 and Mg/Mg + Fe ratio. Compared to large grains, secondary sapphirines from sapphirine + cordierite + orthopyroxene (C, Table 1) or sapphirine + orthopyroxene (D, Table 1) symplectites are more aluminous and their composition (7MgO: $9Al_2O_3$: $3SiO_2$) is characteristic of low pressure growth conditions (Bishop and Newton 1975).

4d. Cordierite

Cordierite is homogeneous and no zoning has been detected. However, Mg/Mg + Fe ratios vary between 0·88 and 0·96 and are lower for garnet-bearing parageneses. As a rule Mg/Mg + Fe ratios are higher in cordierite than in coexisting sapphirine, biotite, orthopyroxene, or garnet. The low oxide totals (<100 per cent) are due to presence of H_2O or CO_2 in cordierite channels (Armbruster and Bloss 1980) (Table 5).

4e. Plagioclase

In sapphirine-bearing granulites, plagioclases contain from 30 per cent to 90 per cent anorthite, while in sapphirine-free granulites, their anorthite content is always less than 40 per cent (Table 6).

The composition of primary plagioclase laths enclosed in garnet from pyrigarnites is close in composition to labradorite (An_{60}). Very high anorthite contents (An_{80}–An_{90}) are restricted to the second generation. Calcium released by garnet

Table 2. Representative microprobe analyses of orthopyroxene

	Pyrigarnite		A1			A4				B2	
Ref. no.	508 1C	508 2	126 1C	126 1R	249 1R	249 2	391 1C	391 2	221 1C	221 2
SiO_2	52·17	51·82	50·73	51·87	47·87	50·11	49·67	51·05	49·10	51·00
TiO_2	0·00	0·02	0·18	0·12	0·20	0·10	0·17	0·03	0·09	0·00
Al_2O_3	2·03	2·46	9·55	6·28	10·15	6·76	8·79	6·92	9·93	5·90
Cr_2O_3	0·07	0·09	0·03	0·04	0·23	0·48	0·00	0·00	0·00	0·00
FeO	21·99	22·25	12·62	14·38	18·94	18·71	18·06	16·87	17·00	16·44
MnO	0·96	0·77	0·11	0·19	0·10	0·06	0·15	0·00	0·09	0·11
MgO	21·83	21·84	26·88	26·75	22·80	23·53	24·05	24·85	25·27	26·40
CaO	0·28	0·23	0·04	0·10	0·00	0·01	0·05	0·04	0·03	0·06
Na_2O	0·00	0·00	0·02	0·00	0·00	0·04	0·00	0·00	0·00	0·00
K_2O	0·00	0·00	0·01	0·00	0·00	0·15	0·00	0·00	0·00	0·00
Total	99·33	99·48	100·17	99·73	100·29	99·95	100·94	99·76	101·51	99·91

Structural formula calculated on the basis of 6 oxygens

	1C	1R					2			
Si	1·956	1·940	1·796	1·860	1·739	1·829	1·785	1·849	1·740	1·834
Al^{IV}	0·044	0·060	0·204	0·140	0·261	0·171	0·215	0·151	0·260	0·166
Σ	2·000	2·000	2·000	2·000	2·000	2·000	2·000	2·000	2·000	2·000
Al^{VI}	0·045	0·048	0·195	0·125	0·174	0·120	0·158	0·145	0·155	0·084
Ti	0·000	0·001	0·005	0·003	0·005	0·003	0·005	0·001	0·002	0·000
Cr	0·002	0·003	0·001	0·001	0·007	0·014	0·000	0·000	0·000	0·000
Fe^{3+}	0·000	0·009	0·000	0·007	0·069	0·042	0·048	0·004	0·100	0·000
Fe^{2+}	0·690	0·688	0·374	0·424	0·507	0·530	0·495	0·507	0·404	0·081
Mn	0·030	0·024	0·003	0·006	0·003	0·002	0·005	0·000	0·003	0·413
Mg	1·220	1·218	1·419	1·430	1·235	1·280	1·288	1·342	1·335	1·415
Ca	0·011	0·009	0·002	0·004	0·000	0·000	0·002	0·002	0·001	0·002
Na	0·000	0·000	0·001	0·000	0·000	0·003	0·000	0·000	0·000	0·000
K	0·000	0·000	0·000	0·000	0·000	0·007	0·000	0·000	0·000	0·000
Σ	1·998	2·000	2·000	2·000	2·000	2·000	2·000	2·000	2·000	2·000

1C: Core of large primary orthopyroxene
1R: Rim of large primary orthopyroxene
2: Secondary orthopyroxene
A1, A4, B2: See Table 1.

Table 3. Representative microprobe analyses of garnet

Ref. no.	Pyrigarnite			A4				B2	
	508 C	508 R	249 C	249 R	391 C	391 R	221 C	221 R
SiO_2	39·65	39·16	40·26	40·24	39·78	40·60	40·92	40·24
TiO_2	0·01	0·11	0·03	0·01	0·05	0·00	0·00	0·00
Al_2O_3	22·80	22·67	22·47	22·60	23·15	23·75	23·84	23·23
Cr_2O_3	0·00	0·10	0·21	0·13	0·00	0·03	0·02	0·00
Fe_2O_3	0·00	0·00	0·00	0·00	0·00	0·00	0·00	0·00
FeO	23·63	23·97	25·06	25·30	24·75	24·76	23·06	25·08
MnO	1·10	2·83	0·34	0·36	0·80	0·89	0·49	0·56
MgO	7·30	6·83	12·28	11·63	11·48	11·45	14·23	11·55
CaO	7·19	6·46	0·09	0·20	1·22	1·29	0·64	1·16
Na_2O	0·00	0·00	0·01	0·00	0·00	0·00	0·00	0·00
K_2O	0·00	0·00	0·00	0·00	0·00	0·00	0·00	0·00
Total	101·68	102·13	100·75	100·47	101·23	102·77	103·20	101·82

Structural formula calculated on the basis of 12 oxygens

Si		2.994	2.969	3.015	3.025	2.976	2.986	2.959	2.990
Al^{IV}		0.006	0.031	0.000	0.000	0.024	0.014	0.041	0.010
	Σ	3.000	3.000	3.015	3.025	3.000	3.000	3.000	3.000
Al^{VI}		2.024	1.996	1.984	2.003	2.019	2.046	1.991	2.026
Ti		0.001	0.006	0.002	0.001	0.003	0.000	0.000	0.000
Cr		0.000	0.006	0.012	0.008	0.000	0.002	0.001	0.000
Fe^{3+}		0.000	0.000	0.000	0.000	0.000	0.000	0.008	0.000
	Σ	2.024	2.009	1.999	2.011	2.022	2.048	2.000	2.026
Fe^{2+}		1.492	1.520	1.570	1.591	1.549	1.523	1.387	1.559
Mn		0.070	0.182	0.022	0.023	0.051	0.055	0.030	0.035
Mg		0.822	0.772	1.371	1.303	1.280	1.255	1.534	1.279
Ca		0.582	0.525	0.007	0.016	0.098	0.102	0.050	0.092
Na		0.000	0.000	0.001	0.000	0.000	0.000	0.000	0.000
K		0.000	0.000	0.000	0.000	0.000	0.000	0.000	0.000
	Σ	2.966	2.999	2.971	2.933	2.978	2.936	3.000	2.966
Pyrope		27.70	25.74	46.16	44.43	43.00	42.76	51.12	43.13
Alman		50.31	50.69	52.87	54.24	52.02	51.89	46.23	52.56
Spess		2.37	6.06	0.73	0.78	1.70	1.89	1.00	1.19
Gross		19.60	17.09	−0.23	0.27	3.19	3.40	1.35	3.11
Andr		0.02	0.21	0.06	0.02	0.09	0.00	0.26	0.00
Uvar		0.00	0.20	0.42	0.26	0.00	0.06	0.04	0.00

C : Core
R : Rim
A4, B2 : See Table 1

Table 4. Representative microprobe analyses of sapphirine

Ref. no.	D 423 1C	D 423 1R	D 423 1/Sp	D 423 2	C1 125 2	C4 392 2
SiO_2	14.61	14.01	13.56	13.42	12.39	12.79
TiO_2	0.07	0.04	0.00	0.08	0.00	0.00
Al_2O_3	60.99	61.94	62.25	63.60	63.30	64.65
Cr_2O_3	0.00	0.02	0.00	0.00	0.00	0.71
FeO	5.89	5.94	5.86	5.82	7.39	6.44
MnO	0.00	0.00	0.00	0.00	0.00	0.22
MgO	18.66	18.48	18.02	18.39	16.94	17.43
K_2O	0.06	0.03	0.00	0.00	0.00	0.00
Total	100.28	100.49	99.74	101.31	100.01	102.24

Structural formula calculated on the basis of 20 oxygens

Si	1.720	1.647	1.606	1.563	1.474	1.485
Al^{IV}	4.280	4.353	4.394	4.437	4.526	4.515
Al^{VI}	4.180	4.226	4.288	4.295	4.349	4.330
Ti	0.006	0.004	0.000	0.007	0.000	0.065
Cr	0.000	0.002	0.000	0.000	0.000	0.000
Fe^{3+}	0.101	0.125	0.106	0.142	0.177	0.120
Fe^{2+}	0.479	0.459	0.474	0.425	0.559	0.506
Mn	0.000	0.000	0.000	0.000	0.000	0.022
Mg	3.273	3.237	3.179	3.193	3.003	3.016
K	0.009	0.005	0.000	0.000	0.000	0.000
Total	14.05	14.06	14.05	14.06	14.09	14.06

1C: Core of large prismatic sapphirine
1R: Rim of large prismatic sapphirine
1/Sp: Large prismatic sapphirine at contact with Spinel
2: Secondary sapphirine
D, C1, C4: See Table 1.

and clinopyroxene in the pyrigarnite is incorporated in newly formed plagioclase intergrown with orthopyroxene.

4f. Other minerals

Clinopyroxene from the pyrigarnite is slightly zoned with respect to Al, Si, Na, Mg, and Ca contents (Table 7). This zoning pattern characterized by a decrease in Na and Al towards the edge is consistent with reequilibration during a fall in pressure.

The compositional variations in biotite from Al–Mg granulite is expressed by their colour change, directly related to the Ti content (1.0 to 5.6 wt %) increasing with decreasing Mg/Mg + Fe ratios (0.75 to 0.98). Substitution of Ti for Mg in the octahedral site requires a charge balance by tetrahedral substitution of Al for Si according to (Mg, Fe) + 2Si=Ti + 2Al (Hormann *et al.* 1980) (Table 8).

Spinels are, in most cases, hercynite–spinel solid solutions with negligible Cr_2O_3 content. The Fe^{3+} iron originally present in spinel has exsolved as magnetite.

Table 5. Representative microprobe analyses of cordierite

Ref. no.	A1 126	A4 249	A4 391	B2 221
SiO_2	50.47	49.81	50.17	50.42
TiO_2	0.02	0.00	0.00	0.00
Al_2O_3	34.32	34.21	34.37	34.68
Cr_2O_3	0.00	0.00	0.07	0.00
Fe_2O_3	0.00	0.00	0.00	0.00
FeO	1.94	3.02	2.86	2.20
MnO	0.01	0.02	0.05	0.00
MgO	12.30	11.73	11.79	12.38
CaO	0.01	0.01	0.04	0.00
Na_2O	0.02	0.04	0.00	0.00
K_2O	0.00	0.06	0.00	0.03
Total	99.09	98.90	99.35	99.71

Structural formula calculated on the basis of 18 oxygens

Si	5.001	4.974	4.982	4.973
Al^{IV}	0.000	0.026	0.018	0.027
Σ	5.001	5.000	5.000	5.000
Al^{VI}	4.010	4.003	4.006	4.007
Ti	0.001	0.000	0.000	0.000
Cr	0.000	0.000	0.005	0.000
Fe^{3+}	0.000	0.000	0.000	0.000
Fe^{2+}	0.161	0.252	0.238	0.182
Mn	0.001	0.002	0.004	0.000
Mg	1.817	1.746	1.745	1.820
Ca	0.001	0.001	0.004	0.000
Na	0.004	0.008	0.000	0.000
K	0.000	0.008	0.000	0.004
Σ	10.995	11.019	11.003	11.012

A1, A4, B2: See Table 1.

5. Temperature–pressure conditions of metamorphism

Paragenetic analysis indicates a polyphased granulite metamorphism, therefore the use of geothermometers and geobarometers requires some caution. We have assumed that the composition of mineral cores for primary zoned minerals represents the composition of equilibrium during phase M1 of metamorphism. Whereas primary mineral rims and compositions of newly formed minerals represent the conditions when the various reactions were arrested during phase M2.

Various lithological types were selected for the determination of the pressure and temperature conditions: pyrigarnites (Gt + Cpx + Opx + Pl + Qz) surrounding sapphirine-bearing granulites, and the quartz aluminomagnesian assemblages, Opx + Sill + Cd + Qz (A1, Table 1), Qz + Sill + Gt + Opx + Bi + Cd + Pl (A4, Table 1) and Qz + Gt + Opx_{1-2} + Bi + Cd (B2, Table 1).

Table 6. Representative microprobe analyses of plagioclase

	Pyrigarnite		A4
Ref. no.	508 1	508 2	391
SiO_2	53·03	47·36	59·62
Al_2O_3	30·09	33·75	25·18
Fe_2O_3	0·00	0·08	0·07
CaO	12·64	16·90	6·22
Na_2O	4·33	1·94	7·62
K_2O	0·12	0·02	1·01
Total	100·21	100·05	99·72

Structural formula calculated on the basis of 8 oxygens

Si	2·395	2·171	2·671
Al	1·603	1·824	1·330
Fe^{3+}	0·000	0·003	0·002
Σ	3·998	3·998	4·004
Ca	0·612	0·830	0·299
Na	0·379	0·172	0·662
K	0·007	0·001	0·058
Σ	0·998	1·004	1·019
An	61·30	82·70	29·32
Ab	38·00	17·18	65·01
Or	0·69	0·12	5·67

1: Primary plagioclase
2: Secondary plagioclase
A4: See Table 1.

5a. P–T path deduced from pyrigarnite

Pyrigarnite provide a basis for the determination of the P–T path in the In Hihaou complex. The primary crystallization conditions for pyrigarnite (Gt + Cpx + Pl (AN_{50}) + Opx1 + Qz) are evaluated at about 10 ±1·5 kbar (Perkins and Newton 1981) and temperature ranging from 800°C to 900°C from Opx–Cpx geothermometers (Wood and Banno 1973; Wells 1977) and Cpx–Gt geothermometers (Raheim and Green 1974; Ellis and Green 1979). In pyrigarnites the second metamorphic phase is characterized by the development of Opx–Pl symplectites at the expense of Cpx + Gt + Qz. P–T conditions for this transformation is evaluated from the orthopyroxene–plagioclase–garnet geobarometer (Perkins et al. 1981) and the geothermometers listed above. Phase M2 occurred at a pressure of 6 ±1·5 kbar and a temperature of 750 ±50°C. Temperature differences between the two metamorphic phases M1, M2, are apparently not very large and approximates to 100°C (Figure 4 and Table 9).

Table 7. Representative microprobe analyses of a clinopyroxene from pyrigarnite

Ref. no.	508 core	508 rim
SiO_2	47·99	51·57
TiO_2	0·89	0·38
Al_2O_3	6·29	3·25
Cr_2O_3	0·01	0·11
FeO	10·59	9·21
MnO	0·36	0·48
MgO	11·97	13·25
CaO	19·94	22·18
Na_2O	0·81	0·46
K_2O	0·00	0·00
Total	98·85	100·89

Structural formulae calculated to 6 oxygens

		508 core	508 rim
Si		1·810	1·905
Al^{IV}		0·190	0·095
	Σ	2·000	2·000
Al^{VI}		0·090	0·046
Ti		0·025	0·011
Cr		0·000	0·003
Fe^{3+}		0·108	0·058
Fe^{2+}		0·226	0·227
Mn		0·012	0·015
Mg		0·673	0·729
Ca		0·806	0·878
Na		0·059	0·033
K		0·000	0·000
	Σ	2·000	2·000
Ur		0·03	0·33
Aeg		6·08	3·04
Jd		0·00	0·00
	Σ	6·11	3·37
Ti–Ts		2·60	1·08
Al–Ts		9·30	4·73
	Σ	11·91	5·81
Wo		35·61	41·95
En		34·70	37·27
Fs		11·67	11·60
	Σ	81·98	90·82
Ca		46·95	47·47
Fe+Mn		13·85	13·08
Mg		39·20	39·44
Jd+Ur		0·03	0·33
Aeg		6·08	3·04
Rest		93·89	96·63

Table 8. Representative microprobe analyses of biotite

Ref. no.	A4 249 1	A4 249 2	A4 391 1	A4 391 2	B2 221 1
SiO_2	37.84	36.73	37.96	39.15	38.46
TiO_2	5.25	4.06	4.41	3.68	3.67
Al_2O_3	14.68	15.19	15.27	15.75	15.71
Cr_2O_3	0.33	0.47	0.00	0.02	0.00
Fe_2O_3	0.00	0.00	0.00	0.00	0.00
FeO	9.25	8.76	8.54	8.57	7.69
MnO	0.01	0.01	0.00	0.00	0.00
MgO	17.46	18.32	19.13	19.59	20.34
CaO	0.00	0.00	0.00	0.00	0.00
Na_2O	0.04	0.08	0.17	0.18	0.20
K_2O	10.11	10.33	11.13	10.95	10.60
Total	94.97	93.95	96.61	97.89	96.67

Structural formula calculated on the basis of 22 oxygens

Si	5.543	5.449	5.477	5.548	5.497
Al^{IV}	2.457	2.551	2.523	2.452	2.503
Σ	8.000	8.000	8.000	8.000	8.000
Al^{VI}	0.078	0.106	0.074	0.180	0.144
Ti	0.578	0.453	0.479	0.392	0.395
Cr	0.038	0.055	0.000	0.002	0.000
Fe^{3+}	0.000	0.000	0.000	0.000	0.000
Fe^{2+}	1.133	1.087	1.031	1.016	0.919
Mn	0.001	0.001	0.000	0.000	0.000
Mg	3.812	4.051	4.114	4.138	4.333
Ca	0.000	0.000	0.000	0.000	0.000
Na	0.011	0.023	0.048	0.049	0.055
K	1.890	1.955	2.049	1.980	1.933
Σ	15.543	15.731	15.794	15.758	15.779

1: Primary biotite
2: Secondary biotite
A4, B2: See Table 1.

5b. P–T path in aluminomagnesian granulites

Temperature conditions. The garnet–biotite geothermometer and garnet–cordierite geothermometer (Thompson 1976; Holdaway and Lee 1977) have been used to compute temperature for the presumed primary and secondary assemblages. However the temperatures obtained (560°C–680°C) are much lower than those for the associated pyrigarnites. Nevertheless, when Currie's (1971) garnet–cordierite geothermometer is used, temperatures of 750°C–850°C are obtained for the second metamorphism M2. These temperatures are in the same range found by other authors (Hollister 1977; Selverstone and Hollister 1980; Van Reenen 1983) for the Columbia and Limpopo Belts, using the same geothermometer.

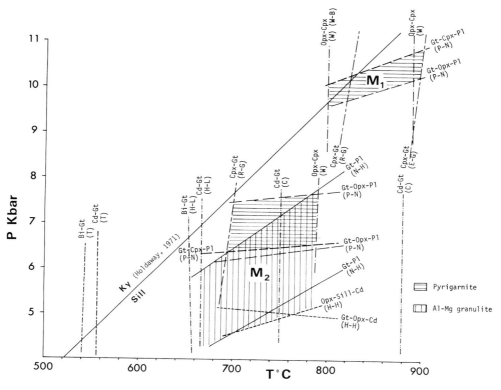

Figure 4. P-T path for Al-Mg granulites and pyrigarnite from the In Hihaou region. M1 and M2 areas result from the intersection of the constant KD lines for the different geobarometers and geothermometers (see summary of P-T calculations in Table 9).

A range of metamorphic temperature from 750°C to 850°C is also in agreement with the stability limits of sapphirine–enstatite associations (>765°C, >3 kbar; Seifert 1974) and the absence of the sapphirine–quartz pair stable above 900°C and 9 kbar (Newton 1972; Chatterjee and Schreyer 1972). It thus appears that the most reasonable range in temperature for the Al–Mg granulites is similar to that found for the pyrigarnites. The low temperatures obtained with the garnet–biotite and garnet–cordierite geothermometers may be in error due to the chemistry of the system. The Al_2O_3 content of orthopyroxene coexisting with sapphirine is a potential geothermometer.

However, experimental data (Danckwerth and Newton 1978) suggests that direct application of the experimental results of Anastasiou and Seifert (1972) gives temperature estimates that are too high (990°C–1000°C).

Pressure conditions. Concerning the first metamorphic event, the higher stability limit of the orthopyroxene + sillimanite + quartz (A1, Table 1) association at 1000°C is 17 ±0·5 kbar for a magnesian system (Annersten and Seifert 1981). The pressure boundary for the orthopyroxene + sillimanite → garnet + quartz reaction is however lowered with an increase in iron in the system (Newton 1978).

Table 9. Summary of the pressure and temperature estimates for the pyrigarnites (1) and Al–Mg granulites (2)

Coexisting minerals geothermometers		Method reference		M1 °C	M2 °C
(1)	Opx–Cpx	(W–B)	Wood and Banno 1973	797–836	769–786
	Opx–Cpx	(W)	Wells 1977	799–891	792–837
	Cpx–Gt	(R–G)	Raheim and Green 1974	821–893	689–707
	Cpx–Gt	(E–G)	Ellis and Green 1979	894–971	740–747
	Bi–Gt	(T)	Thompson 1976	614–689	629–541
	Bi–Gt	(H–L)	Holdaway and Lee 1977	643–712	600–657
(2)	Cd–Gt rim	(T)	Thompson 1976		557–682
	Cd–Gt rim	(H–L)	Holdaway and Lee 1977		577–686
	Cd–Gt rim	(C)	Currie 1971		750–860
Geobarometers				kbar	kbar
(1)	GT–Cpx–Pl	(P–N)	Perkins and Newton 1981	9·9–10·1	6·1–6·7
	Gt–Opx–Pl	(P–N)	Perkins and Newton 1981	9·6	6·5–8·2
	Opx–Sill–Cd	(H–H)	Harris and Holland 1984		5·0–5·7
(2)	Gt–Opx–Cd	(H–H)	Harris and Holland 1984		4·1–5·3
	Gt–Pl–Sill	(N–H)	Newton and Haselton 1981		5·8–8·0

The reaction orthopyroxene + sillimanite + quartz → cordierite (A1, Table 1) takes place at 11 kbar, 1000°C in a magnesian system but is also lowered to 9 ± 0·5 kbar with an increase of Fe/Mg + Fe ratio (Annersten and Seifert 1981). Water also plays a role in the equilibrium reaction orthopyroxene + sillimanite + quartz → cordierite. At 850°C the reaction occurs at 11 kbar stabilizing cordierite at higher pressures in the hydrous system but at only 8 kbar for the anhydrous system (Newton 1978). Moreover, the orthopyroxene + sillimanite → cordierite + sapphirine (C1, Table 1) reaction requires pressures similar to those of the sapphirine-free reaction (A2, Table 1). This is illustrated in the petrogenetic grid of Grew (1982). Concerning the second metamorphic event, in quartz alumino magnesian assemblages, Opx–Sill–Cd (A1, Table 1) and Gt–Opx–Cd (B1, B2, Table 1) geobarometers (Harris and Holland 1984) give pressures of 5 ±0·5 kbar for temperatures ranging from 700°C to 800°C. They are compatible with Gt–Pl–Sill (Newton and Haselton 1981) and Gt–Opx–Pl (Perkins and Newton 1981) geobarometers. Aluminium distribution between garnet and orthopyroxene was calibrated as a barometer (Harley and Green 1982). However this barometer is subject to considerable uncertainty as dP/dT slopes of the Al_2O_3 isopleths are approximately 4 kbar per 100°C. Estimated pressures for a temperature lower than 800°C are unrealistic (at 700°C, P = 1 kbar), whereas those obtained for a temperature ranging from 800°C to 900°C give 4–6 kbar consistent with other geobarometric methods. Pressures may also be estimated from garnet + sillimanite + quartz → cordierite (A2, Table 1) reaction provided temperature and the P_{H_2O}/P_{total} ratio is known (Lonker 1981; Martignole and Sisi 1981). However this is not the case here.

6. Conclusions

Geobarometric and geothermometric estimates on mineral assemblages from Al–Mg granulite and pyrigarnite indicate two stages of granulite facies metamorphism in the In Hihaou complex. The temperature (750°C–805°C) and pressure (10 ±1·5 kbar) of the primary stage (M1) crystallization constrains the thickness of the Archaean crust to a maximum of 35 km, assuming a geothermal gradient of approximately 25°C/km. The lower pressure of (6 ±1·5 kbar) for this second stage (M2) corresponds to a thickness of only 20 km for the Archaean crust. This implies a structural depth variation of no more than 15 km between the first and second stage metamorphic events over the time period between 3 and 2 billion years.

The P–T path, from high pressure and high temperature to low pressure without heat loss is similar to that recorded for other regions of granulite facies (Ouzegane 1981; Droop and Nurminen 1984; Windley et al. 1984; Perchuk et al. 1985).

On the basis of petrology and geochronology two hypotheses can explain this particular P–T path:

1. The maximum P–T conditions of the first stage (M1) occur red about 2·9–3·3 By ago. The second stage (M2) corresponds to the Eburnean orogenesis (2·1 By) known in the other Hoggar granulite regions (Bertrand and Caby 1978). The remarkable preservation of coronas textures is then explained by rapid isothermal uplift which affected the entire Archaean terrane during the Eburnean orogeny.
2. The development of reactions for all granulite rocks took place between 2·9 and 2·1 By ago after the climax of the M1 event. The P–T path is then interpreted as the result of a single metamorphic event. The thickness of the crust progressively decreased during a very long period of time due to a slow erosion rate since the Archaean.

The existing geochronological data which does not distinguish between primary and secondary minerals are not precise enough to choose between these alternatives.

Acknowledgements. The authors are grateful to J. Fabries for making available facilities for the microprobe analyses. Thanks go to P. Fourcade and G.J.H. Oliver who offered constructive suggestions. In addition we thank C. Mevel and M. Semet for the improvement of the English translation. Special thanks are due to E.RE.M., Algeria and to A. Slougui in particular for excellent working facilities in Hoggar.

References

Allegre, C. J. and Caby, R. 1972. Chronologie absolue due precambrien de l'Ahaggar occidental. *C. R. Acad., Paris, Serie D*, **275**, 2095–2098.
Anastasiou, P. and Seifert, F. 1972. Solid solubility of Al_2O_3 in enstatite at high temperatures and 1–5 kbar water pressure. *Contrib. Mineral. Petrol.* **34**, 272–287.
Annersten, H. and Seifert, F. 1981. Stability of the assemblage orthopyroxene–sillimanite–quartz in the system $MgO-FeO-Al_2O_3-SiO_2-H_2O$. *Contrib. Mineral. Petrol.* **77**, 158–165.
Armbruster, T. H. and Bloss, F. D. 1980. Channel CO_2 in cordierites. *Nature*, **286**, 140–141.
Ben Othman, D., Polve, M. and Allegre C. J. 1984. Neodymium–strontium composition of granulites, constraints on the evolution of the lower crust. *Nature*, **307**, 510–516.

Berg, J. H. 1977. Regional geobarometry in the contact aureoles of the anorthositic Nain complex, Labrador. *Journal of Petrology*, **18**, 399–430.

Bertrand, J. M. L. and Caby, R. 1978. Geodynamic evolution of the Panafrican orogenic belt: a new interpretation of the Hoggar shield (Algerian Sahara). *Geol. Rundschau*, **67**, 2, 357–388.

Bishop, F. C. and Newton, R. C. 1975. The composition of low pressure synthetic sapphirine. *Journal of Geology*, **83**, 511–517.

Bondarenko, L. P. 1972. Hypersthene–kyanite association in garnet–sapphirine granulites: thermodynamic conditions for their formation. *International Geological Review*, **14**, 466.

Caby, R. 1968. Une zone de décrochement a l'échelle de l'Afrique dans le précambrien de l'Ahaggar occidenal. *Bull. Soc. géol. Fr. Paris, sér 7*, **10**, 577–587.

—— 1970. La chaine pharusienne dans le nord-ouest de l'Ahaggar (Sahara central, Algerie); sa place dans l'orogenese du précambrien superieur en Afrique. *These Univ. Montpellier*.

Chatterjee, N. D. and Schreyer, W. 1972. The reaction enstatite+sillimanite=sapphirine+quartz in the system $MgO-Al_2O_3-SiO_2$. *Contrib. Mineral. Petrol.*, **36**, 49–62.

Chinner, G. A. and Sweatman, T. R. 1968. A former association of enstatite and kyanite. *Min. Mag.*, **36**, 1052–1060.

Currie, K. L. 1971. The reaction 3cordierite=2garnet+4sillimanite+5quartz as a geological thermometer in the Opinicon lake region, Ontario. *Contib. Mineral. Petrol.* **33**, 215–226.

Danckwerth, P. A. and Newton, R. C. 1978. Experimental determination of the spinel peridotite to garnet peridotite reaction in the system $MgO-Al_2O_3-SiO_2$ in the range 900–1100°C and Al_2O_3 isopleths of enstatite in the spinel field. *Contrib. Mineral. Petrol.*, **66**, 189–201.

De Waard, D. 1965. The occurrence of garnet in the granulite facies terrain of the Adirondack Highlands. *Journal of Petrology*, **6**, 165–191.

Droop, T. R. and Nurminen, B. 1984. Reaction textures and metamorphic evolution of sapphirine bearing granulite from the Gruf complex, Italiàn central Alps. *Journal of Petrology*, **25**, 3, 766–803.

Ellis, D. J. and Green, D. H. 1979. An experimental study of the effect of Ca upon garnet–clinopyroxene Fe–Mg exchange equilibria. *Contrib. Mineral. Petrol.*, **71**, 13–22.

——, Sheraton, J. W., England, R. N. and Dallwitz, W. B. 1980. Osumilite–sapphirine–quartz granulites from Enderby Land, Antarctica mineral assemblages and reaction. *Contrib. Mineral. Petrol.*, **72**, 123–143.

Ferrara, G. and Gravelle, M. 1966. Radiometric ages from western Ahaggar (Sahara) suggesting an eastern limit for the west African craton. *Earth and Planetary Science Letters*, **1**, 319–324.

Fourcade, S. and Javoy, M. 1985. Preliminary investigations of O^{18}/O^{16} and D/H compositions in rhyo-ignimbrites in the In Hihaou (In Zize) magmatic center, central Ahaggar Algeria. *Contrib. Mineral. Petrol.*, 2/3, **89**, 285–295.

Giraud, P. 1961. Les charnockites et les roches associées au Suggarien à facies In Ouzzal (Sahara algerien). *Bull. Soc. géol. Fr. Paris*, **3**, 165–170.

Gravelle, M. 1969. Recherches sur la géologie due socle précambrien de l'Ahaggar centro-occidental dans la region de Silet–Tibehaouine. *These doctorat d état, University of Paris*.

Grew, E. S. 1982. Sapphirine, kornerupine and sillimanite + orthopyroxene in the charnockitic region of south India. *Journal of the Geological Society of India*, **23**, 10, 469–505.

Harley, S. L. and Green, D. H. 1982. Garnet–orthopyroxene barometry for granulites and garnet peridotites. *Nature*, **300**, 697–700.

—— 1985. Garnet–orthopyroxene bearing granulites from Enderby land, Antarctica: metamorphic pressure–temperature–time evolution of the Archaean Napier complex. *Journal of Petrology*, **26**, 4, 819–856.

Harris, N. B. W. and Holland, T. J. B. 1984. The significance of cordierite–hypersthene assemblages from the Beitbridge region of the central Limpopo Belt; evidence for rapid decompression in the Archaean. *Am. Mineral.*, **69**, 1036–1049.

Hensen, B. J. and Green, D. H. 1971. Experimental study of the stability of cordierite and garnet in pelitic compositions at high pressures and temperatures. *Contrib. Mineral. Petrol.*, **33**, 309–330.

Herd, R. K., Windley, B. F. and Ghislair, M. 1969. The mode of occurrence and petrogenesis of the sapphirine bearing and associated rocks of west Greenland. *Gron. Geol. Undersogelse*, **24**, 1–44.

Higgins, J. B., Ribbe, P. H. and Herd, R. K. 1979. Sapphirine I: crystal chemical contributions. *Contrib. Mineral. Petrol.*, **68**, 349–356.

Hoffer, E. and Grant, J. A. 1980. Experimental investigation of the formation of cordierite–orthopyroxene parageneses in pelitic rocks. *Contrib. Mineral. Petrol.*, **73**, 15–22.

Holdaway, M. J. 1971. Stability of andalusite and the aluminium silicate phase diagram. *American Journal of Science*, **271**, 97–131.

—— 1976. Mutual compatibility relations of the Fe–Mg–Al silicates at 800°C and 3 kbar. *American Journal of Science*, **276**, 285–308.

—— and Lee, S. M. 1977. Fe–Mg stability in high grade pelitic rocks, based on experimental, theoretical and natural observations. *Contrib. Mineral. Petrol.*, **63**, 175–198.

Hollister, L.S. 1977. The reaction forming cordierite from garnet, in the Khtada lake metamorphic complex, British Columbia. *Canadian Mineralogy*, **15**, 271–299.

Hormann, P. K., Raith, M., Raase, P., Ackermand, D. and Seifert, F. 1980. The granulite complex of finnish Lappland: petrology and metamorphic conditions in the Ivalojoki–Inarijarvi area. *Geological Survey of Finland, Bulletin*, **308**, 1–95.

Lal, R. K., Ackermand, D., Seifert, F. and Haldar, S. K. 1978. Chemographic relationships in sapphirine bearing rocks from Sonahapar, Assam India. *Contrib. Mineral. Petrol.*, **67**, 2, 169–187.

Lancelot, J., Vitrac, A. and Allegre C. J. 1976. Uranium and lead isotopic dating with grain zircon analysis: a study of complex geological history with a single rock. *Earth and Planetary Science Letters*, **29**, 357–366.

Le Fur, Y. 1966. Nouvelles observations sur la structure de l'antécambrien du Hoggar nord accidental, région d'In Hihaou. *These 3éme cycle, Univ. Nancy.*

Lelubre, M. 1952. Recherches sur la géologie de l'Ahaggar central et occidental (Sahara central). *Bull. Serv. Carte geol. Algerie*, **22**, 2t.

Lonker, S. 1981. The P–T–X relations in the cordierite–garnet–sillimanite–quartz equilibrium. *American Journal of Science*, **281**, 1056–1090.

Martignole, J. and Sisi, J. C. 1981. Cordierite garnet H_2O equilibrium a geological thermometer, barometer and water fugacity indicator. *Contrib. Mineral. Petrol.*, **77**, 38–46.

Meng, L. K. and Moore, J. M. 1972. Sapphirine bearing rocks near Wilson lake, Labrador. *Canadian Mineralogy*, **11**, 777–790.

Newton, R.C. 1972. An experimental determination of the high pressure stability limits of magnesian cordierite under wet and dry conditions. *Journal of Geology*, **80**, 398–420.

—— 1978. Experimental and thermodynamic evidence for the operation of high pressures in Archean metamorphism. In: Windley, B.F. and Naqvi, S.M. (Eds), *Archean Geochemistry: Developments in Precambrian Geology*, **1**, 221–240, Elsevier, Amsterdam.

—— and Wood, B. J. 1979. Thermodynamics of water in cordierite and some petrological consequences of cordierite as a hydrous phase. *Contrib. Mineral. Petrol*, **68**, 391–406.

—— and Haselton, H. T. 1981. Thermodynamics of the garnet–plagioclase–Al_2SiO_5–quartz geobarometer. In: Newton, R.C., Navrotsky, and Wood, B.J. (Eds), *Thermodynamic of Minerals and Melts*, **1**, 131–147, Springer Verlag.

Nixon, P. H., Reedman, A. J. and Burns, L. K. 1973. Sapphirine-bearing granulites from Labwor, Uganda. *Mineralogical Magazine*, **39**, 420–428.

Ouzegane, K. 1981. Le métamorphisme polyphasé granulitique de la région de Tamanrasset. *These 3 éme cycle, Univ. Paris VII.*

Perchuk, L. L., Aranovich, L. Ya., Podlesskii, K. K., Lavrant eva, I. V., Gerasimov, V. Yu., Fed Kin, V. V., Kitsul, V. I., Karpasov, L. P. and Berdnikov, N. V. 1985. Precambrian granulites of the Aldan sheild eastern Siberia, U.S.S.R. *Journal of Metamorphic Geology*, **3**, 265–310.

Perkins, III D. and Newton, R. C. 1981. Charnockitic geobarometers based on coexisting garnet–pyroxene–plagioclase–quartz. *Nature*, **292**, 9, 144–146.

Raheim, A. and Green, D. H. 1974. Experimental determination of the temperature and pressure dependance of the Fe–Mg partition coefficient for coexisting garnet and clinopyroxene. *Contrib. Mineral. Petrol.*, **48**, 179–204.

Sandiford, M. 1985. The metamorphic evolution of granulites at Fyfe Hills; implication for Archean crustal thickness in Enderby land, Antarctica. *Journal of Metamorphic Geology*, **3**, 155–178.

Schreyer, W. and Abraham, K. 1978. Symplectitic cordierite–orthopyroxene–garnet assemblages as products of contact metamorphism of pre-existing basement granulites in the Vredefort structure, South Africa and their relations to pseudotachylite. *Contrib. Mineral. Petrol.*, **68**, 53–62.

Seifert, F. 1974. Stability of sapphirine: a study of the aluminous part of the system MgO–Al_2O_3–SiO_2–H_2O. *Journal of Geology*, **82**, 173–204.

Selverstone, J. and Hollister, L. S. 1980. Cordierite bearing granulites from the coast ranges, British Columbia: P–T conditions of metamorphism. *Canadian Mineralogy*, **18**, 119–129.

Thompson, A. B. 1976. Mineral reactions in pelitic rocks. II. Calculation of some P–T–X (Fe–Mg) phase relations. *American Journal of Science*, **276**, 425–454.

Van Reenen, D.D. 1983. Cordierite+garnet+hypersthene+biotite bearing assemblages as a function of changing metamorphic conditions in the southern marginal zone of the Limpopo metamorphic complex, south Africa. *Special Publication of the Geological Society of South Africa*, **8**, 143–167.

Vicente, A. V., Girardi, V. A. X. and Ulbrich, H. H. G. J. 1978. A sapphirine–orthopyroxene–spinel occurrence in the Pien area, Parana, southern Brazil. *Rev. Bras. Geoc.*, **8**, 284–293.

Vielzeuf, D. 1980. Orthopyroxene and cordierite secondary assemblages in the granulite paragneisses from Lherz and Saleix (French Pyrenees). *Bull. Mineral.*, **103**, 66–78.

Wells, P. R. A. 1977. Pyroxene thermometry in simple and complex systems. *Contrib. Mineral. Petrol.*, **62**, 129–139.

—— 1979. Chemical and thermal evolution of Archaean sialic crust, southern west Greenland. *Journal of Petrology*, **20**, 2, 187–226.

Windley, B. F., Ackermand, D. and Herd, R. K. 1984. Sapphirine–kornerupine bearing rocks and crustal uplift history of the Limpopo Belt, southern Africa. *Contrib. Mineral. Petrol.*, **86**, 342–358.

Wood, B. J. and Banno, S. 1973. Garnet–orthopyroxene and orthopyroxene–clinopyroxene relationships in simple and complex systems. *Contrib. Mineral. Petrol.*, **42**, 109–124.

Section 2
Proterozoic orogenies, magmatism, and related metallogeny

The only Proterozoic orogeny which was widespread throughout Africa was the Pan-African Orogeny. This orogeny is believed to represent the accretion of various terranes to form Gondwanaland.

Later Phanerozoic magmatism discussed in Section 3, and the controls on thedirections of continental rifting (section 4) are very much dependent on the structures created in late Precambrian times. It is not surprising therefore that earlier Proterozoic orogenies are often restricted to more localized regions of Africa confined to particular terranes. Nevertheless there is still considerable debate concerning the age relationships of old orogenic events.

The Kibaran cycle (approximately 1400 to 950 Ma ago) is yet another orogeny localized to central and south Africa, but which is responsible for the development of major structures, the evolution of characteristic magma types, and specific suites of ore minerals. There is also substantial evidence accumulating that just after the close of the Kibaran there was a period of alkaline magmatism developed along rift-like lineaments. These same lineaments were later exploited by Phanerozoic activity.

In this Section DeCarvahlo provides good evidence for Kibaran structures and magmatism in Angola. The evolution of the intracontinental Kibara belt is considered from the metallogenic point of view by Pohl. Its earlier history comprised of the deposition of a very thick psammo–pelitic sedimentary pile with some intercalated volcanics about 1200 Ma. This deformation created essentially upright structures, which appear to be strongly controlled by rigid basement blocks. Metamorphism and the intrusions of granites were more intense in anticlinal structures. From about 1000–950 Ma a renewed compressive phase was associated with the intrusion of LIL enriched 'tin' granites with pegmatites and quartz veins containing Sn,W,Nb/Ta,Be, and Li. This mineralization shows a pronounced structural control with a clear symmetry relation to N–S trending

© 1987 by John Wiley & Sons, Ltd.

folds and accommodation thrusts. Gold quartz veins occur in districts marginal to the Sn–W mineralization. There are also some mafic-ultramafic layered intrusions with Ni, Co, Cu, platinum group metals and Fe/Ti minerals which form a belt parallel to the eastern margin of the Kibaran towards the Tanzanian craton. Later subvolcanic alkaline and carbonatite complexes which are the source for some rare earth deposits are clearly post-Kibaran in age.

The paper by Baudet is concerned with three boreholes in the Bushimay Supergroup, (Upper Precambrian) from the Kasai and Shaba provinces of Zaire. The most fascinating aspect of this work is that there is evidence of a diachronism of the formations from northwest to the southeast in the Kasai–Shaba area because the top of the detrital formations and the stromatolites are younger towards the Kibaran continent. Consequently, the palaeogeographical model and the correlation scheme between Kasai and Shaba must be reconsidered. This diachronism links with the hypothesis of a transgressive Upper Riphean sea on an eroding Kibaran continent, with the transgressive cycle suggested by the vertical succession of the lithological facies of the Supergroup. This observation also helps to define the age of the lavas of the Supergroup.

The Tete Province of Mozambique is characterized by three kinds of gabbro–anorthosite intrusion. In contrast to most gabbro–anorthosite of the province, its western part is not associated with granulite facies rocks.

Barr and Brown discuss aspects of the petrography, chemistry, and age relationships of the complexes based on the origins and the evolution of this part of the crust during the Kibaran and Pan-African orogenies.

Vail presents evidence for an extensive terrane of continental gneisses and supracrustal metasediments which extends from the Nile valley westwards between southern Egypt and northern Kenya, and also occurs in southern Ethiopia, Somalia, and parts of the easternmost Arabian Nubian Shield. Mica–beryl pegmatites and associated kyanite are characteristically developed. The gneiss terranes were regionally affected by metamorphic overprinting and remobilization, thus masking their true age, which is considered to be Middle Proterozoic or older on the basis of isotopic and petrological evidence.

Between the continental terranes volcanosedimentary–ophiolite sequences and intrusive synorogenic calc-alkaline batholiths, post-tectonic and anorogenic calc-alkaline plutons occur in an area 700 km in width by over 2700 km long between Sinai and northern Kenya. Within this region of oceanic crust seven ophiolite- or shear-bounded terranes are recognized which developed around 900 to 600 Ma, through closure of island-arc and associated volcanism and sedimentation, the oldest being in the southeast. Hydrothermal gold mineralization is widespread but restricted to the oceanic terranes, and volcanogenic base metal sulphide deposits are lithologically–tectonically controlled and preferentially located parallel to certain ophiolite margined plate boundaries.

In the northern Red Sea Hills, of Sudan, deformation is largely concentrated into a network of shear zones. Several of these zones strike north–south, but the important Nakasib and Sol Hamid zones are northeast-trending, and there are also minor shear zones which strike east–west. Almond and Ahmed distinguish between massive shear zones, such as that of Nakasib, and the more diffuse braided shear zones, typified by the Oko zone. They suggest that the massive type owes its character to an origin as a reactivated oceanic suture, whereas the braided type was characterized by strike-slip shearing from its inception. Most of the shear zone rocks are mylonites formed under greenschist facies conditions, but early shearing along the Oko zone took place at higher temperature and resulted in gneissose mylonites with amphibolite facies mineralogy. The northeast-trending, Nakasib shearing appears to have preceded a phase of batholithic intrusion, whereas north–south shearing in the Oko and Abirkitib zones is younger than the batholithic intrusions and is in turn post-dated by emplacement of bimodal granite-gabbro complexes. These events cannot be dated precisely as yet, but took place between 99 ma and 700 ma, and probably towards the end of that interval. The pattern of shear and suture zones in northeast Sudan suggests that the northeast-trending Hijaz and Asir arcs, recognized in Saudi Arabia, terminate to the west against a north–south suture which was later rejuvenated to form the Abirkitib shear zone. West of this suture was an immature volcanic arc which may have lain along the margin of the African continent. Almond and Ahmed conclude that the rejuvenation which caused northeast and north–south strike-slip shear was probably a consequence of soft collisions in eastern Arabia.

Liégéois, Bertrand, and Black have provided a detailed study of a large composite calc-alkaline batholith, restricted to the Iforas region which occurs close to the Pan-African suture between the 2000 Ma old West African craton and the Trans-Saharan mobile belt. The Iforas batholith intrudes the western border of an old continental segment affected by early nappe tectonics (D1 event) and is flanked to the west by the Tilemsi palaeo-island arc. The batholith comprises several successive stages. The cordillera (>620 Ma), probably post-dating the D1 event, is essentially composed of volcanosedimentary sequences. The collision (620–580 Ma) is marked by the production of abundant granitoids mostly emplaced by the end of the D2 E–W compressional event. The post-collision tectonic stages (D3 and D4; 580–540 Ma) are characterized by strike-slip movements, reversals in the stress field, and a rapid switch from calc-alkaline to alkaline magmatism.

Similar dramatic changes in the chemistry of a Pan-African granite complex have also been recorded in the Hoggar by Azzouni-Sekkal and Boissonnas. In their short but pertinent paper they report the occurrence of a concentrically-zoned late Pan-African granite pluton intruded into the eastern branch of the Pharusian belt. The complex represents yet another example of the transition from calc-alkaline (monzonitic) to anorogenic alkaline magmatism during the terminal stages of the Pan-African orogeny.

In suitable terrain, Landsat and other satellite imagery can be used for the production of reconnaissance geological maps which contain a comprehensive range of lithological and structural information. Curtis and others describe an aerial photogeological interpretation that does not require extremely sophisticated image processing and is most effective when supported by a limited, controlled field survey. Using this procedure image-geological maps have been completed at 1:250 000 scale of an area of 92 000 km^2 of the Sudanese Nubian Desert, based on interpretation of Landsat MSS and RBV imagery and one swath of SIR-A radar imagery as well as on a rapid field check. Griffiths and others describe the main geological and mineral exploration results and the interprelation of lg satellite imagery.

The Kibaran cycle in Angola — a discussion

H. Carvalho, J. P. Crasto

Centro de Geologia, Instituto de Investigação Cientifica Tropical, Al. D. Afonso Henriques, 41-4º D. 1000 Lisboa, Portugal

Z. C. G. Silva

Depto de Geologia, Fac. de Ciências, Universidade de Lisboa, Rua da Escola Politécnica, 58, 1294 Lisboa CODEX, Portugal, Centro de Geologia do IICT, Al. D. Afonso Henriques, 41-4oº D. 1000 Lisboa, Portugal

Yves Vialette

Université de Clermont II et C.N.R.S.-U.A. 10, 5 Rue Kessler, 63038 Clermont Ferrand CEDEX, France

---

Three sedimentary formations of Kibaran age have been identified in SW Angola: the Cahama-Otchinjau Formation, the Chela Group and the Leba-Tchamalindi Formation.

The Cahama-Otchinjau Formation is intruded by red granites which are also found as boulders within the Chela Group basal conglomerate. These granites yielded the following Rb-Sr ages: Matala region t = 1350 ± 65 Ma, $^{87}Sr/^{86}Sr_i$ = 0.716 ± 0.003; Ompupa region t = 1407 ± 26 Ma, $^{87}Sr/^{86}Sr_i$ = 0.7058 ± 0.0020; Otchinjau region t = 1411 ± 24 Ma, $^{87}Sr/^{86}Sr_i$ = 0.7052 ± 0.0012; Chitado region t = 1302 ± 20 Ma, $^{87}Sr/^{86}Sr_i$ = 0.7075 ± 0.0016.

The red granites are cut by basic dykes of different compositions and ages. The noritic dolerites which are intrusive in the Chela Group yielded a good Rb-Sr isochron at t = 1119 ± 27 Ma, $^{87}Sr/^{86}Sr_i$ = 0.7044 ± 0.0010. This age postdates the deposition of the Chela Group and of the Leba Tchamalindi Formation.

These formations of Kibaran age extend to Namibia where they outcrop near the border. Their extension to the central and northeastern Angola is not yet well defined. Regarding the central-western region some metasediments outcrops are considered by several authors to belong to the Bale Group of Eburnean age or to the Oendolongo (s.s.) Group. It is assumed in this paper that some of the referred metasediments outcrops belong to the Bale Group while the others belong to the Oendolongo (s.s.) Group to which we suggest a Kibaran age. Concerning the eastern region we admit that the metasediments belonging to the Malombe and Luana Groups which are intruded by basic and granitoid rocks may be of Kibaran age. A more accurate establishment of the stratigraphy of this area will only be possible after isotope determinations, mainly on the basic rocks to define whether they belong to one or other of the two thermotectonic events already detected in Zaire at 2000–2220 Ma and 991 ± 30 Ma.

KEY WORDS   Kibaran   Angola   Lithology   Geochronology   Stratigraphy

1. General outline of the geology of Angola

The Central African craton extends from Zaire to Angola and is essentially composed of gneisses (not yet isotopically dated in Angola, probably older than 3000 Ma), gabbro-noritic and charnockitic complex, which underwent charnockitisation at about 2.820 Ma (Cahen *et al.* 1984) and granites and gneisses bearing an age between 2830 and 2600 Ma, volcano-sedimentary rocks outcropping in the northeastern and central-western regions of Angola, belonging to the Jamba Group and Chivanda – Negola Utende – Chela Supergroup, with probable ages lying between 2800 and 2200 Ma (Carvalho, 1983). The Eburnean granitoid and metasedimentary rocks are also abundant in NW, E and central-western Angola. In the northeastern and in the western margin of the craton the rocks have been reworked during various Precambrian cycles, mainly during the Eburnean, giving rise to vast granitoid areas. Large older gneisses and granite enclaves as well as discontinuous outcrops consisting of pre-Eburnean volcano-sedimentary rocks occur within those granitoid rocks.

In the southwestern margin pre-Eburnean metasedimentary rocks are spread throughout a considerable area. These rocks can be related to the greenstone belts of Southern Africa, older than 3000 Ma and are intruded by the 'complexo gabro-anortositíco do SW de Angola', also known as the Kunene Complex (Beetz 1933), with an estimated age of 2500 Ma or older (Carvalho 1983).

Kibaran rocks have been formally identified lithologically and geochronologically in the SW part of the craton in Angola (Carvalho 1983). As a whole they occur in outcrops as a strip about 600 km wide, having an approximate SW–NE trend (Carvalho 1982). In this region these rocks occupy a considerable area where cratonic rocks also crop out, wherever the differential erosion and the morphology permit. The extension of the strip to the east is masked by thick covers of the Kalahari group. Nevertheless, outcrops of probable Kibaran age are observed to the east of the 20°E and the 22°E meridians up to the Zaire border, where the cover is less prominent (Figure 2).

2. Kibaran rocks of southwestern region

1a. Metasediments

Kibaran metasedimentary rocks from southwestern Angola crop out in the Cahama-Otchinjau region and also to the west of that area, where they form the Chela Group and the Leba-Tchamalindi Formation (Figure 3). Although they present some similarities, the metasediments from the Cahama-Otchinjau and from the Chela Group have different ages and such will be discussed separately, according to their regional stratigraphic positions.

The metasediments cropping out in the Cahama-Otchinjau area are siliceous sandstones, showing a pink colour at the surface and grey at depth (Torquato and Salgueiro 1977), grey quartz schists, reddish and grey siliceous shales. They occur as subhorizontal beds, except in those areas where they are intruded by granitic and doleritic rocks, where they are tilted. These types of relationship are observed in the Otchinjau region and also in the Cahama region where they were also observed by Torquato and Salgueiro (1977). The geomorphologic expression of these metasediments is variable wherever they occur. In the Chela Group the subhorizontal beds form a gentle, flat relief; whereas the Cahama-Otchinjau

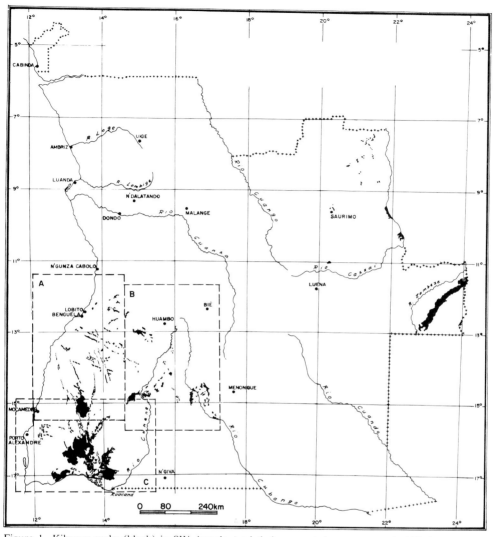

Figure 1. Kibaran rocks (black) in SW Angola, and their suggested counterparts in NE Angola.

metasediments have been eroded or exhibit sharp valleys. Whereas Rocks of the Cahama region have also been studied by Simões (in Carvalho 1969) who reports the existence of a basal conglomerate consisting of quartz and quartzite unsorted pebbles in a quartz-feldspathic matrix.

Intrusive into the Cahama-Otchinjau metasediments occur the SW red granites which at the contact become granite porphyries (1300–1400 Ma).

Taking into account the field evidence and the stratigraphic position proposed by Torquato and Salgueiro (1977), it can be assumed that the age of the Cahama metasediments lies between 1700 and 1300 Ma.

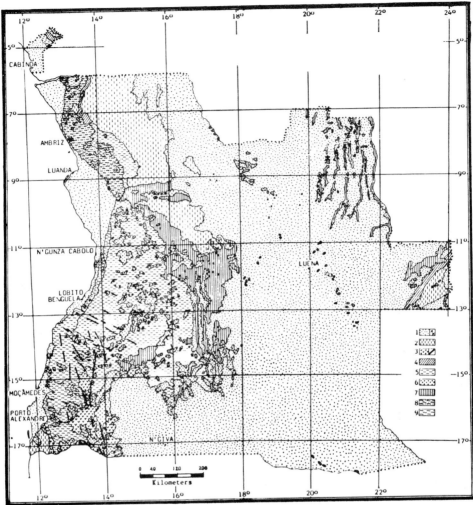

Figure 2. Sketch map of Angola showing major age groups of the Precambrian basement. 1–Phanerozoic: sand cover and other sedimentary rocks; Mesozoic magmatism—mainly cretaceaus (X); 2–Pan-african (c. 500–1000 Ma); 3–Kibaran, (c. 1000–1400 Ma); 4–Kibaran (?)–Eburnean (?); 5–Eburnean (c. 1800–2200 Ma); 6–Eburnean—Archaean interval (?) (c. 2200–2500 Ma), late Archaean (?) (c. 2500–2900 Ma); 7–late Archaeans; 8–reworked early Archaean (?) 9–early Archaean (>3000 Ma)

1b. Southwestern red granites and associated rocks

Granites, granodiorites, granite porphyries and syenites are abundant in the southwestern region of Angola and crop out generally according to NE–SW trends from Iona to Matala (Figure 3). They extend through the area as intrusive bands having tenths of kilometres of width and extending along an approximate NE–SW direction.

These rocks in general have a reddish colour due to the iron oxide content in the alkali feldspar which is also disseminated throughout the rock. The texture is usually porphyritic, and the matrix is coarse, hypidiomorphic-granular. The phenocrysts are microcline or orthoclase, in general kaolinized, well developed,

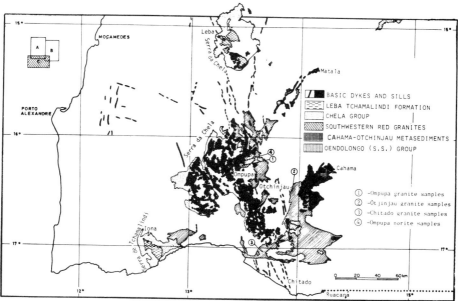

Figure 3. Geological sketch map of part of SW Angola showing locations of granite and norite samples.

reaching up to 3 and 4 cm in size. Occasionally they are perthitic. Quartz is always interstitial showing a bluish colour when observed in hand specimen. Accessory mafic minerals are green hornblende, more commonly, followed by biotite containing hematite impregnations. Apatite is observed as a minor accessory mineral. Locally these rocks are gneissic in texture near the contact with the Gabro-Anorthosite Complex, where they also exhibit a syenitic facies.

In other areas, variations in granularity and mineral composition are conspicuous. In the Ompupa region, these granites are finer-grained than the general type, perthites are more abundant and the rock colour is usually lighter. Alteration in the feldspars is more moderate and liberated iron ore is practically absent. Fluorite is the minor accessory. In the vicinity of Quedas de Montenegro (Epupa) in the Kunene river bed and its neighbourhood, these rocks occur, strongly tectonized, having a preferential E–W direction.

Granitic porphyries are confined to the contact zones with the host rocks. They vary in colour from brown to reddish and sometimes present textures with a vitreous matrix. In the microgranular matrix, quartz and alkali feldspar phenocrysts can be distinguished. Torquato and Salgueiro (1977) refer to intense impregnation of opaque minerals and to the typical habit of the phenocrysts bounded by rims of recrystallised material.

In the Otchinjau region intrusive in the metasediments, porphyries with breccia appearance (volcanic breccia) exhibit recrystallised siliceous matrix and phenocrysts consisting of agglomerates of quartz and commonly fractured plagioclase grains (Silva 1983).

Isotopic age determinations, using Rb-Sr whole-rock isochrons were carried out on various red granites. Rb and Sr contents were measured by mass-spectrometric isotope dilution. Repetitive analysis of the N.B.S. 987 $SrCO_3$ standard gives a value of 0.710263 ± 0.000030 (2σ). The Rb-Sr isochron ages and errors were computed by means of a least squares regression analysis according to York

(1966). The area of occurrence of these rocks are marked on the maps of Figure 1. They are identified from North to South as the massives of Matala, Ompupa, Otchinjau, and Chitado. Bassot *et al.* (1981) have reported an age of 1350 ± 65 Ma with initial $^{87}Sr/^{86}Sr$ = 0.7157 ± 0.0030 for a porphyritic facies of Matala granitoid. The authors have interpreted this age as that of the intrusion of the pluton in the metasediments. The high initial $^{87}Sr/^{86}Sr$ ratio shows an origin by anatectic melting of the crust.

The Ompupa red granite is in contact with the basement and the Gabbro-Anorthosite Complex; it is overlain by the Chela Formation. Analytical results

Table 1. Chemical analyses and normative compositions of Angolan red granites and noritic dolerites

	R 5667	R 5659	R 5665	R 5033	R 5030	R 6525	R 6526	R 7794
SiO_2	65·90	76·00	77·50	75·50	70·00	52·00	52·50	53·00
TiO_2	0·65	0·30	0·15	0·10	0·60	1·90	0·95	0·80
Al_2O_3	16·10	12·20	12·20	11·60	13·80	14·40	13·50	14·90
Fe_2O_3	2·20	1·20	1·70	2·35	2·85	3·46	3·34	3·17
FeO	2·25	1·15	—	—	1·30	9·50	7·15	6·43
MgO	0·50	0·20	0·05	—	0·40	5·90	7·70	6·85
CaO	1·70	0·85	0·25	0·60	1·70	10·10	9·80	11·10
Na_2O	3·20	2·65	2·85	3·45	3·60	2·30	1·90	1·80
K_2O	5·00	4·80	4·40	4·80	4·20	0·65	1·20	0·80
P_2O_5	0·15	0·25	—	—	0·15	0·20	—	0·10
MnO	0·06	0·04	—	—	0·08	—	—	—
H_2O+	0·88	0·09	0·04	0·64	0·57	0·21	0·19	0·15
H_2O-	0·10	—	0·08	0·22	0·10	0·47	1·97	1·83
							0·01	0·06
TOTAL	99·15	99·63	99·22	99·26	99·35	100·89	100·92	100·89
Q	23·39	40·63	43·49	36·02	29·15	5·50	5·33	7·23
Or	29·57	28·39	26·03	28·39	24·84	3·84	7·10	4·73
Ab	27·05	22·40	24·09	29·16	30·43	19·44	16·06	15·21
An	7·56	2·75	1·24	1·95	7·56	27·01	24·73	30·17
C	2·64	1·62	2·28	—	0·55	—	—	—
Di	—	—	—	—	—	18·01	16·31	29·66
En	—	—	0·13	—	1·00	—	—	—
Hy	3·15	1·20	—	—	—	17·72	29·05	15·76
Wo	—	—	—	0·43	—	—	—	—
Mt	3·19	1·74	—	—	2·17	5·02	4·84	4·60
Il	1·23	0·57	—	—	1·14	3·61	1·81	1·52
Hm	—	—	1·70	2·35	0·98	0·47	—	0·24
Ap	0·35	0·59	—	—	0·35	—	—	—

R 5667 – Red granite from Ompupa
R 5659 – Fine granite from Otchinjau
R 5665 – Fine red granite from Otchinjau
R 5033 – Porphyritic granite from Matala
R 5030 Porphyrite red granite from Matala
R 6525, R 6526, R 7794 – Noritic dolerites from Ompupa region

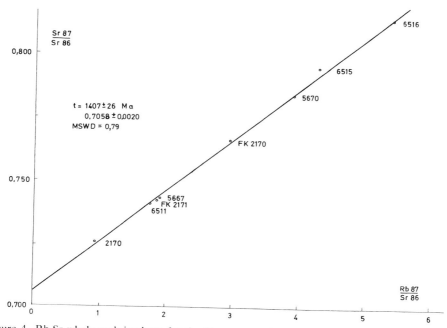

Figure 4. Rb-Sr whole rock isochron for the Ompupa granite.

are given in Table 1 and reported on a ^{87}Rb/^{86}Sr–^{87}Sr/^{86}Sr diagram (Figure 4), a regression analysis through all data points on isochron of 1407 ± 26 Ma with initial ^{87}Sr/^{86}Sr = 0.7058 ± 0.0020. Two feldspars (Mendes, unpublished data) fall on the regression line. This shows that after the intrusion of the granite there was no migration of Rb or Sr.

Two chemical analyses of the Otchinjau granite are reported in Table 1. The Otchinjau granite is richer in silica and poorer in alumina than the Ompupa granite. Seven rocks were investigated according to the Rb-Sr method (Table 2). The Otchinjau granite has an isochron correlation corresponding to an age of 1411 ± 24 Ma with initial ^{87}Sr/^{86}Sr = 0.7052 ± 0.0012 (Figure 5). The Rb contents of rocks are relatively low and consequently the K/Rb ratios, around 350, are higher than values of the 'main trend' of Shaw (1968) for crustal rocks. This fact associated with low initial ^{87}Sr/^{86}Sr ratio: (0.705) is in favour of a lower crustal melting origin.

The Chitado red granite is intrusive in the gabbro-anorthosite complex. This granitoid intrusion comprises syenite and alkaline granites. Analytical results and sample locations are given in Table 2. A regression analysis through all points produces an isochron of 1302 ± 20 Ma with initial ^{87}Sr/^{86}Sr = 0.7075 ± 0.0016 (Figure 6). These results demonstrate that the continental crust in the SW Angola has been affected by plutonic activity in Kibaran times. These numbers are quite similar to those corresponding to the age of an acid porphyite (1340 ± 27 Ma) and of a red porphyritic granite (1341 ± 202 Ma), both from Cahama region (Torquato and Salgueiro 1977), also intrusive in the metasediments.

Table 2. Rb-Sr whole rock isochron data for some SW Angolan granites

	Sample No	Longitude E	Latitude S	Rb (ppm)	Sr (ppm)	$^{87}Rb/^{86}Sr$	$^{87}Rb/^{86}Sr$
OMPUPA GRANITE	R 5667	13°28	16°12	114	174	1·907	0·7432
	R 6515	13°28	16°13	195	134	4·278	0·7951
	R 5670	13°28	16°11	198	147	3·927	0·7839
	R 6511	13°28	16°11	142	231	1·784	0·7413
	R 6516	13°28	16°11	248	135	5·328	0·8137
	R 2170	13°28	16°11	123	380	0·930	0·7255
	FK2170	13°28	16°11	325	321	2·940	0·7662
	FK2171	13°28	16°11	102	158	1·881	0·7440
OTCHINJAU GRANITE	R 5659	13°55	16°29	114	124	2·685	0·7601
	R 5660	13°55	16°29	89	130	1·996	0·7459
	R 5661	13°55	16°29	30	532	0·164	0·7086
	R 5662	13°56	16°30	88	132	1·931	0·7429
	R 6510	13°57	16°28	178	128	4·006	0·7857
	R 6513	13°57	16°26	42	260	0·464	0·7140
	R 6514	13°56	16°27	97	133	2·132	0·7478
CHITADO ALKALINE GRANITE	R 7946	13°38	17°03	92	135	1·975	0·7427
	R 7948	13°38	17°03	164	77	6·261	0·8222
	R 7949	13°39	17°04	13	783	0·049	0·7086
	R 7950	13°44	17°03	208	65	9·487	0·8841
	R 7951	13°51	17°06	110	58	5·504	0·8109
	R 6614	13°42	16°59	181	48	11·130	0·9107
OMPUPA NORITES	R 6521	13°22	16°24	33	126	0·752	0·7154
	R 6522	13°06	16°13	80	124	1·887	0·7344
	R 6523	13°06	16°12	48	119	1·159	0·7226
	R 6525	13°28	16°11	28	143	0·568	0·7129
	R 6526	13°28	16°12	68	90	2·190	0·7384

1c. Dolerites

Dolerite sills and dykes are intrusive in the SW red granites (Carvalho and Simões 1971) and in the Eburnean granitoid rocks (Figures 3 and 7). The main direction of these dykes is N 60°W. They present ophitic or sub-ophitic texture and contain as essential minerals labradorite, labradorite-andesine, titaniferous augite, pigeonite and olivine. Hornblende, sphene, epidote and scarce orthopyroxene are present in small amounts (Alfonso and Albuquerque 1972, Silva and Simões 1982).

Isotopic dating by K/Ar method carried out on rocks from Vila Paiva Couceiro by Silva et al. (1973) give the results shown in Table 3. Silva suggests that the probable age of the dolerite magmatism is around 1200 Ma. In fact, this age is reliable since field observations show that these dolerites are intrusive in the red granites (c. 1300 Ma) and are intruded by the noritic dolerites (c. 1100 Ma) discussed later.

1d. Leba-Tchamalindi Formation and Chela Group

The stratigraphy and lithology of the former 'Formação da Chela' has been reformulated in the Lubango (Sá da Bandeira) region (Correia 1976). Correia has

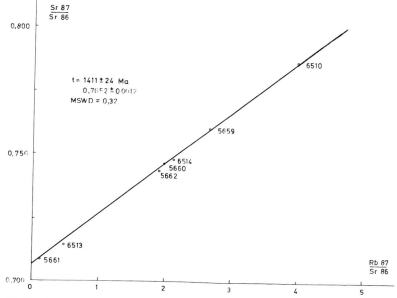

Figure 5. Rb-Sr whole rock isochron for the Otochinjau granite.

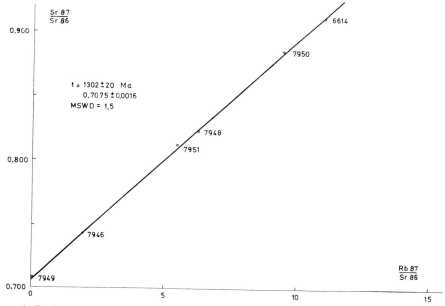

Figure 6. Rb-Sr whole rock isochron for the Chitado granite.

divided it into the 'Leba Formation' and the 'Chela Group', both occurring to the South, down to the Namibia border.

According to Carvalho et al. (1979), the Chela Group, the Leba Formation and the associated noritic dolerites form the Chela Supergroup. The limestones from

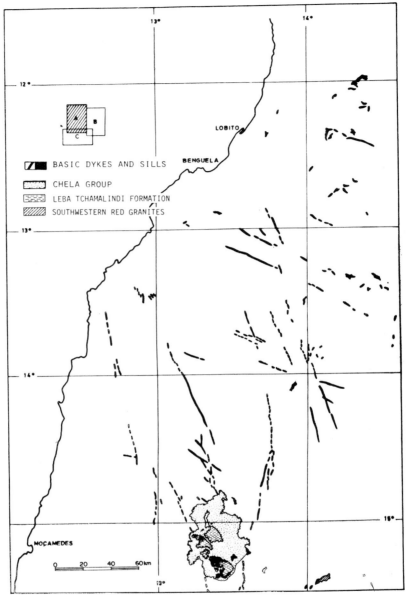

Figure 7. Geological sketch map of part of SW Angola (Area A, Fig. 1).

the Leba Formation occur also in the Tchamalindi region, and taking this into account those authors designated the whole litho-stratigraphic unit as 'Leba-Tchamalindi Formation'. The Chela Group and the Leba-Tchamalindi Formation occur from the north of parallel 15°S down to the Namibia border, where they still occur, cropping out in three distinct areas. The Leba-Tchamalindi Formation

Table 3. K-Ar whole rock data for Paiva Couceiro dolerites

Sample No	Longitude E	Latitude S	K %	rad ^{40}Ar c.c./g10^{-5}	atm ^{40}Ar	K/Ar age Ma
G 17/317	14°09'	14°13'	·6895	5·0370	2·76	1281 ± 22
G 54/317	14°15'	14°09'	·6605	4·2020	8·64	1157 ± 26
G 1/318	14°59'	14°08'	·5565	2·4840	34·35	880 ± 34
G 5/318	14°39'	14°06'	·3215	2·4040	13·59	1309 ± 99
G 16/317	14°09'	14°12'	·0920	0·5027	61·97	1301 ± 65
G 24/317	14°06'	14°04'	·1609	1·0440	36·13	1175 ± 69

* Data on plagioclase

has only been observed in the extreme areas (Figure 3), unlike the noritic sills which are more abundant in the central area.

The Chela Group is characterised by subhorizontal beds and by the development of fault scarps having NE–SW, NW–SE, NS and EW directions as main trends, as already noted by several authors. The lithology of the Chela Group is composed of conglomerates, quartzites, sandstones, siltstones, shales, acid-alkaline volcanics and volcano-sedimentary rocks (Correia 1976). At the base, both in the conglomerates and other rocks, clasts and other materials from the SW red granites are present. These occurrences can be observed in the southwestern region of Vila de Almoster. Near the Namibia border (Iona and Oncocua Explanatory Notice 1961), the basal conglomerate is overlain by quartzites, sandstones and shales. In that area the beds dip steeply (about 40°–50°) possibly due to their occurrence in a Mesozoic unstable tectonic zone which corresponds to the actual Kunene River bed (Carvalho 1983). The Chela basal conglomerate also contains clasts from the Gabbro-Anorthosite Complex, from the Schist-quartzitic-amphibolitic Complex and from pre-Eburnean granitoid rocks. Taking this information into account, it can be inferred that the age of the Chela Group lies between 1300–1400 Ma (age of the SW red granites whose clasts are enclosed in the basal conglomerate) and 1119 ± 27 Ma (age of the noritic dolerites, intrusive in the Chela Group).

The Leba-Tchamalindi Formation is underlain by the Chela Group, probably as a slight unconformity or as a disconformity (Correia 1976). According to Correia, in spite of the controversy about such unconformity, there is a lithologic discontinuity in the Leba region where no gradual transition between the two distinct types of sedimentation is observed. Lithologically the Leba Formation is composed of sedimentary rocks of chemical origin, such as stromatolitic dolomites, occurring as subhorizontal beds (Correia 1976). In the Tchamalindi region the limestone is dense and massive, being underlain by quartzites and conglomerates of the Chela Group (Iona and Oncócua Explanatory Notice 1961).

Several authors have correlated the Leba-Tchamalindi Formation with Pan-African rocks of NW Angola and Namibia. Nevertheless, the noritic dolerites are younger than the Chela Group and the Leba-Tchamalindi Formation (Correia 1976) and their age is 1119 ± 27 Ma (Figure 6). With this in mind, the assumption of a Pan-African age for the Leba-Techamalindi Formation does not seem to be plausible, since the deposition of those rocks took place before c. 1100 Ma.

1e. Noritic dolerites

These rocks crop out mainly from the Namibia border up to the north of the 13°S parallel and practically from the coast to the vicinity of the 16° 30'E meridian. They occur as sills within the Chela Group metasediments or as dykes in the granitoids, in the basic rocks or in the metasedimentary rocks of Kibaran, Eburnean or pre-Eburnean age (Carvalho 1981). The dykes are emplaced along faults or pre-existing fractures with a general trend between N–S and N 35°W. The noritic dolerites are also intrusive into older doleritic dykes emplaced in the SW red granites. Chemical analysis of 3 noritic dolerites from Ompupa are presented in Table 1.

Petrographic study of these norites (Carvalho and Simões 1971), show that they present a subophitic texture and contain plagioclase (mainly labradorite), orthopyroxene (enstatite-bronzite) and clinopyroxene (pigeonitic augite and pigeonite). The outcrops of dolerites whenever they cut the red granites and older rocks, confirms their relatively young age.

Isotope dating carried out on rock samples collected in the SW using the K/Ar method (Torquato and Amaral 1973, Torquato 1974, Silva 1980) yielded an approximate minimum age of 800 Ma. Five samples of norites from Ompupa region were investigated according to the Rb-Sr method (Table 2). The data points give an isochron with a slope corresponding to an age of 1119 ± 27 Ma with an initial $^{87}Sr/^{86}Sr = 0.7044 \pm 0.0010$ (Figure 8). Part of this sequence of Kibaran rocks from SW Angola, such as described in this paper, is similar, both in lithology, type of occurrence and age, to the stratigraphic sequence of the Sinclair Group of SW Africa, and as a whole, can be correlated to the Rehoboth Magmatic Arc, as defined by Watters (1977).

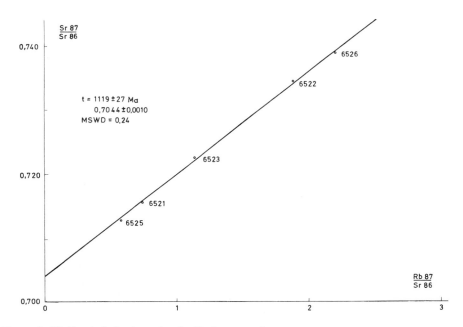

Figure 8. Rb-Sr whole isochron for the Norites near Ompupa.

Central-western region

The region comprises of the area between the parallels 12° and 16° S and the meridians 15° and 17° E. Though it has not been mapped in detail, in those areas where mining was being carried out by the late Companhia Mineira do Lobito, the documentation left by H. R. Korpershöek (unpublished) and by the geologists of Sofreminas provides valuable data. However, concerning the stratigraphic position of some units, especially those of sedimentary origin, some discrepancies still remain. Some outcrops which according to Korpershöek belong to the Bale Group of Eburnean age, in Bassot's opinion belong to the Oendolongo (s.s). It is maintained in this paper that not all outcrops of the Bale Group are Eburnean, as defended by Korpershöek, neither all Oendolongo (s.s) outcrops are post-Eburnean as considered by Bassot et al. (1981). Dolerites occur in this region according to the regional trend of the SW noritic dolerites (Figure 9). They have been correlated with the latest event by Korpershöek (unpublished).

3. North-eastern region

Between the north-eastern region and the west-central region there exists a thick Cenozoic cover which limits the observation of the continuity of the Kibaran rocks. However, in the eastern most part of Angola, in the valleys of the hydrographic basins of the Kasai and Zambeze Rivers, there are outcrops of older rocks, some of probable Kibaran age.

3a. Metasediments

The Luana and the Lower Malombe Groups consist of metasedimentary rocks, such as conglomerates, schists, graywackes, sandstones and quartzites which occur as beds with SW–NE strike and 25°SE maximum dip (Monforte 1960). Dykes or sills of basic and granitoid rocks are intrusive in these metasediments.

3b. Eruptive rocks

Eruptive rocks from northeastern Angola with probable Kibaran age comprise basic rocks and granitoids. According to Monforte (1960) and Rodrigues and Pereira (1973), based on field relations, the basic rocks are older than the granitoids.

Diabases and gabbros (Rodrigues and Pereira, op. cit.) are intrusive into the Luana Group in the Lunda region. Petrographically these rocks are similar: ophitic to sub-ophitic texture, in which the plagioclase usually shows saussuritization and the pyroxene, described as pigeonite, is often uralitized (Rodrigues and Pereira 1973; Rodrigues 1974). In spite of the similar petrographic characteristics, these rocks are ascribed as having different ages, related to three different phases of magmatism.

Granitoids are also intrusive into the Luana Group (Monforte 1960, Rodrigues and Pereira 1973). They are mainly alkaline and peralkaline granites which grade into porphyries and felsites. In general they are fine to medium grained, with porphyritic texture, occasionally gneissic, and in some areas, they show cataclastic features. Their colour varies from brownish pink to reddish. The alkali felspars are orthoclase, albite and microline, not uncommonly exhibiting alteration. Blue quartz has been identified in all types of granitoids described in the area. Green hornblende is the main mafic mineral, followed by biotite.

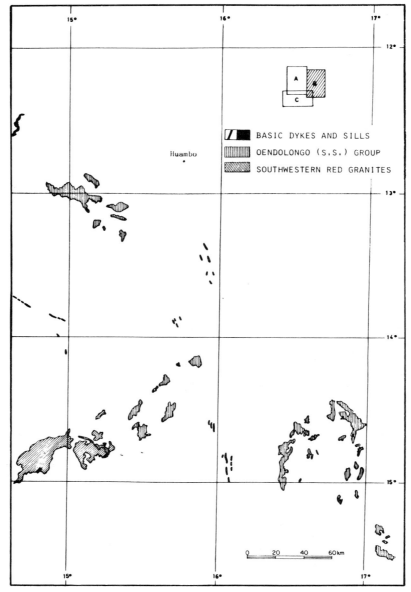

Figure 9. Geological sketch map of part of SW Angola (Area B, Fig. 1).

4. Discussion

The possible correlation between eruptive rocks from NE Angola and other rocks already described, allows an analysis of the data to be made.

Monforte (1960) and Rodrigues and Pereira (1974) consider the Luana Group as Pan African in age, and accordingly, basic and granitoid rocks intrusives in

Group would be also Pan-African in age or younger. On the other hand, Monforte (1960) refers to basic rocks (gabbros, diorites and dolerites) in North Lunda region which have different ages, though they have the same lithology and type of occurrence. Rodrigues (1974), based on field observations, also suggests that the different petrographic types result from intrusions which occurred in three distinct phases: pre-Cambrian, pre-Permian and Mesozoic.

Isotope ages determined for basic pre-Mesozoic rocks from Angola (Carvalho et al. 1983) and Zaire (Cahen et al. 1984) range from c. 1000 Ma to 2000 Ma, which has led Cahen et al. (1984) to admit the existence of an important thermotectonic event around 2200–2000 Ma and another less strong and later event, at 991 ± 30 Ma. Thus, the two events proposed by Cahen et al. (1984) could account for the occurrence of basic rocks lithologically similar, but different in age, in eastern Angola and southwest Zaire. The ages for the Zaire Lulua Group basic rocks are 1473 ± 33, 1461 ± 33, 1423 ± 20, 1356 ± 3, 1217 ± 29, 1130 ± 30, 1126 ± 27, 1092 ± 30 and 991 ± 30 Ma (Cahen et al. 1984). Besides, one sample from the Lulua Group analysed five different times yielded 2087 ± 36 Ma. The same authors are not certain about that age, not only due to the chemical composition of the sample but also because other ages between 1474 ± 23 Ma and 2078 ± 36 Ma were not found for the Lulua lavas. For basic rocks from the NE (South Lunda) in Angola (Carvalho et al. 1983), K-Ar ages of 1490 ± 40, 1338 ± 38, 1360 ± 40 and 1320 ± 36 Ma have been determined.

For the reasons given above it is admissible that the basic rocks older than c. 1300 Ma may correspond to the first event referred by Cahen et al. (1984) around 2000 Ma similar in age to the Eburnean orogeny. The basic Precambrian rocks of Monforte (1960) and Rodrigues (1974) may have undergone argon loss during the second event. The remaining basic rocks would correspond to this second event around 1000 Ma, corresponding to the Kibaran, and to the pre-Permian basic rocks of Monforte and Rodrigues. According to Delhal et al. (1966) the Lulua basic volcanism in Zaire is older than the cataclastic quartzites; on the other hand, the same quartzites of the Luana Group which occur in Angola (North Lunda) are older than the basic rocks (Rodrigues 1974). According to Monforte and Rodrigues and Pereira (1973) the Luana Group is related to the 'Schistogréseux of the Congo Occidental' sequence i.e., of Pan-African age. So, the basic and granitoid rocks intrusive into them might be Pan-African or even younger. Nevertheless, no Pan-African ages have been recorded in basic rocks from Zaire and Angola. On the other hand, correlation of the Luana and Lower Malombe Groups with Lulua Group of Zaire, seem to be only possible with the portion that outcrops in the Kasadi-Sadi region. In fact, Delhal (1958) considers the Kasadi-Sadi Formation as an entity younger than the Lulua Group. It is noteworthy that the Kasadi-Sadi pillow lava has yielded a K/Ar (whole rock) age of 926 ± 30 Ma (Cahen et al. 1984). While isotope dating for the basic rocks from Lunda is not available it seems to be reasonable to admit that they are pre-Permian in age and possibly Kibaran.

Isotope dating for granitoid rocks have not yet been undertaken. However these granitoid rocks are demonstrably younger than the Luana sedimentary and basic rocks (Monforte 1960, Rodrigues and Pereira 1973). Nevertheless, K-Ar isotopic dating carried out on a syenodiorite from Zaire, which is intrusive into the Lulua Group, and occurring as an extension to the NE of the Lunda (Angola) alkaline granitoids, yielded an age of 1155 ± 15 Ma (Cahen et al. 1984). On the other hand, in NE Angola the intrusion of granitoid rocks is related to epeirogenic movements with NW–SE trend (Rodrigues and Pereira 1973) which coincides with the directions of tectonic features of the SW red granitoids. So, the granitoid rocks of Lunda may also be Kibaran in age. Finally we admit that the Luana and Lower Malombe Groups may also have a Kibaran age, though only isotopic dating can clarify this point.

5. Conclusions and correlation with neighbouring countries

There are rocks of Kibaran age in Angola extending north to south from Lobito to Ruacana Falls with an approximate SW-NE trend. The outcrops are scarce from SW to NE, due in part to Cenozoic cover, thus rendering it difficult to establish the pre-Cenozoic stratigraphy. For the same reasons isotopic dating is incomplete from the west to the east. The Kibaran zone could extend through Zaire to Burundi and Rwanda. In the south this zone continues in Namibia near the border where it turns to the northeast, being again identified in Angola, in the Iona region.

Metasedimentary rocks occur in stable cratonic areas as subhorizontal beds. The Chela Group and the Leba-Tchamalindi Formation rocks which occur in the south near Namibia, are folded up, due to tectonism of probable Mesozoic age, along faults trending N 80°E. The actual path of the Kunene River follows this faulted area. Kibaran rocks are represented in the centre western region by the doleritic rocks and, at least, some outcrops of the Oendolongo Group (s.s.).

In the NE region the Luana Group and the Lower rocks occur as subhorizontal beds slightly inclined to the SE, in general.

In Table 4 an attempt is made in order to establish the correlation between the SW geological Kibaran units and those of possible Kibaran age occurring in centre-west and NE Angola.

Regarding the correlation with neighbouring countries, Kibaran rocks from NE Angola may continue to Zaire through the Shaba-Kibaran Supergroup and those

Table 4. Correlation of Kibaran rock types from different regions in Angola

SW Angola	West Central Angola	NE Angola
Noritic Dolerites 1120 ± 27 Ma	Noritic Dolerites (undated)	Later Precambrian basic rocks
Leba-Tchamalindi Formation		Luana and Lower Malombe (undifferentiated)
	Oendolongo (s.s.) Formation	
Chela Group		
Subophitic Gabbros c. 1200 Ma		Older Precambrian basic rocks
SW Red Granites 1300–1400 Ma		Alkaline and peralkaline granitoids
Cahama-Otchinjau 1300–1400 Ma c. 1700 Ma	Oendolongo (s.s.) Group of Matala region	Luana and Lower Malombe (undifferentiated)

from the SW Namibia in the Bairies Mountains region. The occurrence, lithology and chronology of the Kibaran Southern rocks are comparable to those of the Sinclair Group and the 'Rehoboth Magmatic Arc' from Namibia and Botswana (Carvalho 1983).

Acknowledgements. This research was developed at I.I.C.T., Lisbon (Portugal). Chemical analysis and isotopic determinations were performed at the Geochronology Laboratory of Clermont-Ferrand (France). The participation of one of the authors (Zenaide Silva) for presentation of the paper at the 13th Colloquim of African Geology was possible thanks to financial support provided by Fundação Calouste Gulbenkian, Portugal and the British Council.

References

Afonso, J.A. and Albuquerque, M.H. de 1972. Carta Geológica de Angola. Notícia Explicativa da folha 229 (Vila Sousa Lara). Serv. Geol. Minas, Luanda.
Bassot, J.P., Pascal, M. and Vialette, Y. 1981. Données nouvelles sur la stratigraphie, la géochimie et la géochronologie des formations précambriennes de la partie méridionale du Haut Plateau angolais. *Bull. Bur. Rech. géol. min. (2 série), sect. 4*, (1980–81), 285–309.
Beetz, P.F.W. 1933. Geology of South West Angola, between Cunene and Lunda Axis. Trans. *Geol. Soc. South Africa*, **36**, pp. 137–178.
Cahen, L., Snelling, N.J., Delhal, J. and Vaill, J.R. (Col. Bonhomme M; Ledent, D.) **1984**. The geochronology and evolution of Africa. Clarendon Press, Oxford, pp. 512.
Carvalho, H. de 1969. Cronologia das formações geológicas Precâmbricas da região central do Sudoeste de Angola e tentativa de correlação com as do Sudoeste Africano. *Bol. Serv. Geol. Minas Angola*, **20**, 61–71.
—— 1981. Breves considerações de natureza geológica e de cronologia absoluta sobre as rochas do soco antigo (Arcaico) de Angola. *Bol. Soc. geol. Portugal*, **22**, 307–14.
—— 1982. Geologia de Angola (folhas 1.2.3.4., escala 1/1.000.000). *Inst. Inv. Cient. Tropical (Centro Geol.), Lisbon.*
—— 1983. Notice Explicative préliminaire sur la géologie de l'Angola. Inst. Inv. Cient. Tropical. *Rev. Garcia de Orta, Lisboa, Sér. Geol.* **6**, 15–30.
—— **and Simões, M.C. 1971**. Carta Geológica de Angola. Notícia Explicativa da folha 376 (Macota). Serv. Geol. Minas, Luanda.
——, **Fernandez A. and Vialette, Y. 1979**. Chronologie absolue du pré-cambrien du Sud-Ouest de l'Angola. c.r. *Acad. Sci. Paris* **288**, 1647–50.
——, **Cantagrel, J.M. and Jamond, C. 1983**. Géologie de l'Angola oriental-Datation K-Ar de quelques roches basiques, leur place dans l'évolution de cette region. *Garcia de Orta, Sér. Geol. Lisboa* **6(1–2)**, 151–160.
Correia, H. 1976. O Grupo Chela e a Formação Leba como novas unidades litoestratigráficas resultantes da redefinição da 'Formação da Chela' na região do Planalto da Humpata (sudoeste de Angola). *Bol. Soc. geol. Portugal*, **20**, 65–130.
Delhal, J. 1958. Sur le volcanism ancien dans le sud Kasai (Congo Belge). *Bull. Soc. Belg. Géol. Paléontol. Hydrol.* **67**, 179–87.
——, **Lepersonne, J. and Raucq T. 1966**. Le complex sédimentaire et volcanique de la Lulua. Ann. Mus. R. Afr. centr., Tervuren (Belg.) Sci. Géol. **8**, no. 51.
Korpershöek (unpublished) The geology of Cassinga district, Angola, and its potential compared to that of the Serra dos Carajás, Brazil.
Monforte, A. 1960. Síntese Geral da Geologia do Nordeste da Lunda. Companhia de Diamantes de Angola. Unpublished.
Rodrigues, A. 1974. carta geológica de Angola. Notícia Explicativa da folha no 52 (Cassanguidi). *Serv. Geol. Minas, Luanda.*
—— **and Pereira, E. 1973**. Carta geológica de Angola. *Notícia Explicativa da folha* **69** (Cambulo). Serv. Geol. Minas, Luanda.
Shaw, D.M. 1968. A review of K/Rb fractionation trends by covariance analysis. *Geochimica Cosmochimica Acta* **32**, 573–601.
Silva, A.T.S.F. 1980. Idade radiométrica K-Ar do dique norítico de Vila Arriaga e sua relação com o Grupo Chela (Angola). *Mem. Acad. Ciên., Lisboa* **21**, 137–159.
—— **and Simões, M.C. 1982**. Geologia da região do Quipungo (Angola). Inst. Inv. Cient. Tropical. *Rev. Garcia de Orta, Lisboa, Sér. Geol.*, **5**, 33–58.
——, **Torquato, J.R. and Kawashita, K. 1973**. Alguns dados geocronológicos pelo método K/Ar da região de Vila Paiva Couceiro, Quilengues e Chicomba (Angola). *Bol. Serv. Geol. Minas Angola* **24**, 29–46.

Silva, Z. 1983. Relatório sobre os Pórfiros de Angola — estudo petrográfico e interpretação — regiões de Mupa, Ganda e Otchinjau. *Inst. Inv. Cient. Tropical (Centro Geol.).* Unpublished.

Torquato, J.R. 1974. Geologia do Sudoeste de Moçâmedes e suas relações com a evolução tectónica de Angola. *Tese Dout. Inst. Geoc. Univ. São Paulo.*

—— **and Amaral, G. 1973.** Idade K-Ar em rochas das regiões da Catamba e Vila de Almoster. *Inst. Inv. Cient. Angola,* **10,** Publ. IIC no 308.

—— **and Salgueiro, M.A.A. 1977.** Sobre a idade de algumas rochas da região da Cahama (Folha geológica no 399), Angola. *Bol. Inst. Geociências, Univ. São Paulo* **8,** 97–106.

Watters, B.R. 1977. The Sinclair Group: Definition and regional correlation. *Trans. geol. Soc. South Afr.,* **80,** 9–16.

York, D. 1966. Least square fitting of a straight line. *Can. J. Physics* **44,** 1079–1086.

Metallogeny of the northeastern Kibaran belt, Central Africa

W. Pohl

Institute of Geology, Technical University, D-3300 Braunschweig, West Germany

---

The evolution of the intracontinental Kibaran belt spans the period from about 1400 to 950 Ma. Its earlier history involved the deposition of a very thick psammitic–pelitic sedimentary pile with some intercalated volcanic rocks, which were folded around 1200 Ma ago. This deformation created essentially upright structures, which were controlled by rigid basement blocks. Metamorphism related to igneous granite intrusions were more intense in anticlinal structures.

From about 1000–950 Ma a renewed compressive phase equivalent to the Lomanian orogeny was associated with the emplacement of LIL-element enriched 'tin' granites. They formed sheets at basement/sediment and granite/sediment contacts, with cupolas at tectonic highs. Pegmatites and quartz veins with Sn, W, Nb/Ta, Be and Li are derived from the 'tin' granites. Generally, this mineralization shows a pronounced structural control with a clear symmetry relation to N–S trending folds and thrusts. Gold-rich quartz veins occurring in districts marginal to the Sn–W mineralization, are a distal and lower temperature phase.

Mafic–ultramafic layered intrusions with Ni, Co, Cu, platinum-group metals and Fe/Ti deposits form a belt parallel to the eastern margin of the Kibaran towards the Tanzanian craton. These rocks may be related to the Bukoban/Malagarasian basalts with an age of 1000–850 Ma.

Subvolcanic alkaline and carbonatite complexes which were the source for small Ree deposits are clearly post-Kibaran in age.

KEY WORDS   Mineralization   Kibaran   Proterozoic   Central Africa   Pegmatite

1. Introduction

The middle Proterozoic Kibaran belt extends from the Zambia/Angola/Zaire border in the south southwest, through Zaire, Burundi, Tanzania, and Rwanda to southwest Uganda in the north northeast. Its overall length is nearly 1500 km, and it attains widths of up to 300 km (Figure 1). Economic mineralization is known mainly from its northern sectors, within three geographically separated areas, comprising the Shaba and Kivu provinces west of the Central African Rift and a northeastern province covering Burundi, Rwanda, the west lake district of Tanzania, and southwest Uganda. The latter area is the main subject of this paper (Figure 2).

0072–1050/87/TI0103–17$08.50
© 1987 by John Wiley & Sons, Ltd.

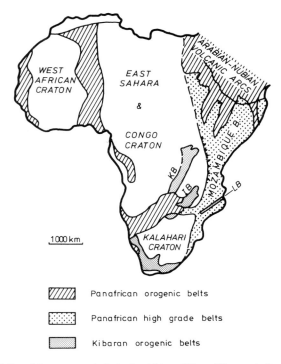

Figure 1. Kibaran and Panafrican orogenic belts in Africa. KB = Kibaran belt; IB = Irumide belt; LB = Lurio belt.

Since the discovery of the first cassiterite deposits after the First World War, the Kibaran belt has become a source of tin, tungsten, gold, beryllium, columbo–tantalite and lithium-ores. In addition, minor exploitation of bastnaesite (Ree), uranium ores, mica and semiprecious stones has taken place. On a worldwide scale, however, the belt has not been a major producer of any of these commodities. Cumulative cassiterite production in the northeastern province, for example, may be estimated at approximately 150,000 metric tons, which is roughly equivalent to maximum annual output in Malaysia. Nevertheless mineral production is an important economic factor for the countries concerned.

Since the metallogeny of any area can only be understood within the geological framework, a short introduction to the geological evolution of the Kibaran belt is given (Figure 3).

2. Geological evolution of the Kibaran belt

The Kibaran belt is the westernmost of three roughly synchronous and parallel orogenic belts in Central and Eastern Africa, the other two being the Irumides in Zambia (Daly 1984) and the recently recognized Lurio belt in Mozambique (Jourde and Vialette 1980; Sacchi et al. 1984). In the west the Kibaran belt is limited by the Congo craton (although the border area is not well exposed because of younger cover), and to the east its boundary with the Tanzanian craton is largely masked by a blanket of post-Kibaran platform sediments (Bukoban and

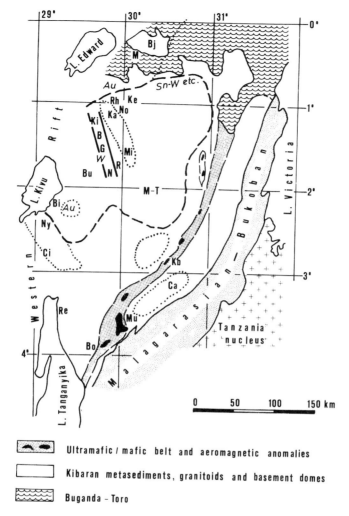

Figure 2. Sketch map of geology and metalliferous areas of the northeastern Kibaran Belt. Mines, mining districts and localities mentioned in the text: Uganda: Bj—Buhweju plateau; Ka—Kamena Fe; Ki—Kirwa W; Ke—Karenge G4 granite; M—Mashonga Au; No—Nyamulilo W; Rh—Ruhiza W. Rwanda: B—Bugarama W; Bu—Buranga pegmatite; Mi—Miyove Au; Bi—Bisesero Au; G—Gifurwe W; N—Nyamulilo W; Ny—Nyungwe Au; R—Rutongo Sn; M-T—Musha–Ntunga Sn. Burundi: Ci—Cibitoke Au; Ca—Cankuso Au; Bo—Buhoro gabbro; Mu—Musongati Ni; Re—Kayonze Ree. Tanzania: Kb—Kabanga Ni, Cu, Co |W| signifies the 'tungsten belt'; dotted lines enclose auriferous zones.

Malagarasian Supergroups). Further south, folded Katangan rocks (the Lufilian arc) occur to the east of the Kibaran belt. Within the belt, early Proterozoic schists and gneisses (very generally of Eburnean age) emerge in cores of antiforms and uplifted blocks. This relationship between the Kibaran rocks and their basement is exceptionally well exposed at its northern end in southwest Uganda, where the lower Proterozoic Buganda–Toro system occurs underneath a clear unconformity often marked by conglomerates. This is a unique area, since a virtually continuous

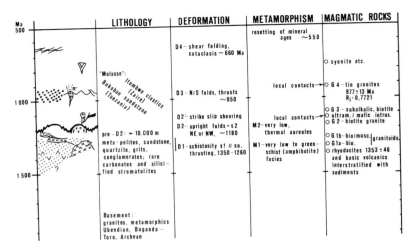

Figure 3. Kibaran evolution in Central Africa (data from Cahen et al. 1984; Klerkx et al. 1984; Lavreau 1984; Theunissen 1984).

cross-section of basement and folded Kibaran sediments is accessible here; geodynamic models of the Kibaran, therefore, must account for this situation.

2a. Lithology and lithostratigraphy

Kibaran rocks in the northern sector of the belt have been described as 'Burundian' in the francophone countries, and as 'Karagwe–Ankolean' or 'Muva–Ankole' in Uganda and Tanzania. The name 'Kibaran' was coined in Shaba/Zaire (Robert 1931). Everywhere in the belt, a threefold lithostratigraphic division is recognized (for a detailed discussion see Cahen et al. 1984).

The Lower group is characterized by dark, laminated pelitic sedimentary rocks, with intercalations of mature quartzites, sandstones and siltstones. Parts of the suite are probably of turbiditic origin. Fluvial conglomerates often occur near the base, with well rounded pebbles of quartz and of more resistant basement rocks (quartzite, pegmatite, gneiss etc.). Basic magmatic rocks ('dolerites') and acidic tuff bands are locally interbedded with the sediments. Silicified stromatolites have also been tentatively identified (van Straaten 1984).

Sedimentary structures abound in the clastic rocks (graded bedding, ripple marks, slumping etc.). Some of the quartzites persist over very wide areas and are useful regional marker horizons, while others are lenticular and rather local features. Carbonates are very rare. The lower group attains its maximum thickness in Burundi (about 8500 m).

The Middle group is clearly more arenaceous and of a lighter, often reddish, colour compared with the older rocks underneath. An important marker bed is taken as its base, which consists of 250 (Rwanda) to 1000 m (Burundi) of banded fine grained and occasionally conglomeratic white to pinkish quartzite. In the west, minor basaltic and dacitic volcanic rocks have been mapped.

The Upper group occurs in major synclinorial structures only. It consists of rather immature clastic sediments, in some areas with a slight erosional unconformity at its base marked by polymictic coarse conglomerates containing pebbles of quartz and underlying rocks. Quartzites are frequently ferruginous, and reddish or grey colours predominate.

2b. Deformation and structural evolution

The earliest deformation (D1) recognized to date is restricted to deeper levels of exposure near granite intrusions in anticlines or above basement cores. It comprises a schistosity parallel to bedding (s1), minor often isoclinal folds, and thrusts which appear to control the intrusion of syntectonic granite sheets (Theunissen 1984).

The ensuing major deformation (D2) which affected all Kibaran rocks is characterized by wide anticlinoria and rather narrow synclinoria. Folds are upright and of a cylindrical type, with cleavage (s2) at deeper levels. The axes trend approximately N–S, but NW–SE, NE–SW and even E–W directions are not rare. This as well as the variable directions of vergence and of local accommodation thrusts, reflects a strong control of deformation by basement morphology and structures. Broad shear-zones always appear, subparallel to the major folds, (D2' Theunissen 1984) which indicate movement between basement blocks after the main compressive phase. Mafic and ultramafic intrusions forming a belt parallel to the eastern margin of the Kibaran towards the Tanzanian nucleus (Radalescu 1982; van Straaten 1984), are probably controlled by one of these shear zones (Klerkx 1984).

Renewed upright folding (D3) with N–S axes cutting D2-folds at sharp angles, was deduced from tectonic control of vein mineralization in Rwanda (Pohl 1977; see also below). Generally, outcrop patterns created by fold interference are not rare and determine in many cases cupolas of G4-granites (Harpum 1970; Pohl 1977). This compressional phase may be identical with the Lomanian orogeny affecting the Lower Itombwe supergroup in Zaire, which is interpreted to be post-orogenic Kibaran molasse, and is intruded by the 'tin' (G4) granites (Cahen et al. 1984). Whether this deformation is still part of the Kibaran orogenic cycle (Pohl 1977) or rather belongs to the earliest tectonic activity of a Pan-African 'Western Rift Mobile Belt' (Cahen et al. 1984) appears to be a matter of interpretation.

Later shear folding and cataclastic deformation (D4) affecting alkaline and carbonatitic intrusions in the wider area, emphasizes the fact that structural instability persisted along the Western Rift. In fact the very existence of the Rift is further remarkable evidence for the longevity of ancient zones of crustal weakness.

2c. Metamorphism

Generally the metamorphic grade in the area is low. From rocks of very low grade in the synclinoria a steep gradient is obvious towards large batholithic granites and basement highs, where the amphibolite facies is attained and migmatites may occur. Two main phases of regional metamorphism (M1 and M2) are roughly time-equivalent with D1 and D2. Together with synorogenic granites (see below) they are due to high heat flow whose source is a matter of interpretation. Later intrusions produced variable aureoles of contact metamorphism. Many mineral ages of earlier rocks have been reset to about 550 Ma (Cahen et al. 1984) by a rather enigmatic tectonothermal event — obviously equivalent to the Pan-African *sensu stricto* (Kennedy 1964).

2d. Intrusions

The numerous granitoid intrusions which occurred over a wide time span are a remarkable feature of the Kibaran belt area. Four main age groups have emerged from published Rb/Sr data (Cahen et al. 1984; Klerkx et al. 1984; Lavreau 1984), the first two are approximately synorogenic, while the two latter are usually described as post-orogenic.

Porphyritic gneissose two-mica or biotite adamellites (G1) with low initial Sr-ratios (Klerkx et al. 1984) intruded the Lower Group subconcordantly around 1300 Ma. Usually non-porphyritic peraluminous adamellitic orthogneisses (G2) of S-type chemistry (Klerkx et al. 1984) are associated with the first, or formed separate domes dated at around 1200 Ma, which is considered to be time-equivalent to D2 (Lavreau 1984). As yet undated, are the mafic/ultramafic intrusions of the east, but field relations seem to support an age between G2 and G3. A genetic (and chronological) relationship with the basic volcanics of the Bukoban and Malagarasian supergroups dated at about 850–1000 Ma cannot be excluded (Tack 1984). Alkaline biotite granites (G3) are clearly post-tectonic to D2 and their ages vary around 1100 Ma.

Most important from the point of view of this paper are the leucocratic alkaline G4 'tin' granites with ages clustering at about 977 ± 13 Ma and with an initial Sr-ratio of $Ri = 0.7721$ (Cahen et al. 1984). They are equigranular or pegmatitic, often cataclastic and locally sheared, and clearly cross-cutting. Their country rocks include older granitoids and sediments. Typically, they consist of quartz, microcline, albite, and muscovite (biotite) with accessory apatite, zircon, and tourmaline. Their roof is often invaded by suites of pegmatites and quartz veins which host Sn, W, and Nb/Ta mineralization. Although these granites are clearly of crustal derivation, published data are insufficient to classify them in modern terminology (Hutchison 1982). In view of the pronounced structural control of the mineralization, and of many of the G4 intrusion sites themselves, it appears inappropriate to call them strictly anorogenic. In addition, their shape is mostly lobate and irregular (for example the Karenge granite, Uganda; Lowenstein 1966), which is not at all reminiscent of the well-defined cross-sections of typical subvolcanic A-type granites (Bowden and Karche 1984). The G4 granites are probably intruded as sheets along basement/sediment and granite/sediment contacts, forming cupolas in structurally controlled positions.

An important feature of the Kibaran granitoids in the area is their frequent occurrence in well-defined composite plutons, which may comprise cores of pre-Kibaran basement and then resemble mantled gneiss domes. Anticlinorial or tectonically elevated situations are the rule, and many of these have been sites of intrusive activity from G1 to G4. As earlier granites are roughly concordant with the country rocks, preferential erosion of the intrusion rocks led to the formation of wide circular depressions rimmed by more resistant metasediments; these are the spectacular 'arenas' so typical for the East of the area but absent in the morphologically immature Congo-Nile waterdivine to the west.

The syenites and carbonatites which are spatially associated with the Western Rift are clearly of post-Kibaran age; the Upper Ruvubu Complex in Burundi has been dated at 750–700 Ma (Tack 1984).

3. Mineralization

In the past, the mineral deposits of Sn, W, Nb/Ta and Au in the area have been discovered most efficiently by panning all alluvial material and by a physical search for mineralized outcrops (U, Be, Li, W). More recently, geochemical and geophysical methods have been used increasingly, and this has resulted in the recognition of the mafic/ultramafic belt in Burundi with its Ni, Co, Cu, Pt, Cr, Ti, Va and Fe mineralization (Radulescu 1982) and its equivalent in Tanzania.

The current mineral inventory (Ziserman et al. 1983; Barnes 1961; Tissot et al. 1982) allows the following generalized synthesis of geological evolution and mineralization to be made.

3a. Mineralization syngenetic with Kibaran sediments

Of this group, only gold has been exploited locally from Tertiary and Recent alluvial placers. The source of these placers may have been the basal conglomerates and quartzites of the Lower group (for example Cankuso/Burundi) or the intermediate volcanics interstratified with Lower group sediments (Cibitoke northwest Burundi).

Lenticular stratiform bodies of magnetite ($\pm$ pyrite) and specular haematite (for example Kamena/Uganda within the Middle group) which are uneconomic in terms of size occur in quartzites and metapelites. Such stratiform bodies are being increasingly recognized as magnetic surveys are carried out although their origin has not yet been investigated. It is tempting, however, to consider these ores as syngenic hydrothermal exhalites related to volcanism, which may indicate a potential for economically more interesting base metal mineralization.

3b. Mineralization associated with the mafic/ultramafic intrusions

Geochemical surveys prompted by earlier reports of Ni mineralization in southeastern Burundi led to the discovery of Ni-laterites of possible economic size and grade at Musongati (Radulescu 1982). Meanwhile, many other mafic/ultramafic bodies have been located within a belt 20 km wide and nearly 400 km in length which extends from the eastern shore of Lake Tanganyika to the Tanzanian/Ugandan border.

Anorthosites and leucogabbros of the complexes contain lenses of vanadium-bearing magnetite, Ti-magnetite and ilmenite (Buhoro/Burundi). Sulphide mineralization with Ni, Co, Cu, platinum group metals and some chromite have been found in ultramafic members (pyroxenites, peridotites, dunites) and their immediate country rocks in Burundi (Musongati, etc.; Niyondezo 1984) and in Tanzania (Kabanga; van Straaten 1984). A layered intrusion model seems to be applicable to these complexes, although no details have been published yet.

3c. Mineralization associated with the 'tin' granites (G4)

Until recently it was only this group that was of economic significance within the Kibaran belt. Accordingly, a large number of publications concerned with different aspects of the deposits has appeared. Unfortunately however, modern methods (fluid inclusions and isotope studies, Ree geochemistry, etc.) have not been applied yet, so that the present understanding is based on more traditional geological data and comparative interpretation with better known areas.

Three subgroups of granite-associated mineralization can be differentiated:

— Pegmatites with Sn and Nb/Ta (Li, Be, W, Bi, U/Th) and the non-metallics muscovite, kaolinite;
— Quartz veins with Sn and W (pyrite, siderite, Bi, Au, U), and
— Limonitic silicification zones and quartz veins with Au.

The pegmatites of the area have been studied by numerous authors from the mineralogical, geological and metallogenic points of view. Some of the pegmatites contain an exceptional number of rare minerals, especially those cutting basic country rocks (for example about 100 at Buranga/Rwanda; Bertossa 1965; Von Knorring 1969). With the exception of mineralogical observations the pegmatites have not been investigated in detail since Varlamoff (1975, 1972 and many earlier publications).

Varlamoff proposed the following zonation of different types of pegmatites and associated quartz veins in relation to granite cupolas in Rwanda, from the most distal to more proximal types (Figure 4).

— Type 8–9: quartz veins with scheelite (anthoinite), ferberite, with or without cassiterite grading with increasing depth to quartz veins with muscovite, cassiterite, and minor wolframite
— Type 7–8: quartz veins with crystals of microcline, frequently strongly albitized, with muscovite and cassiterite
— Type 7: strongly or completely albitized pegmatites with spodumene, muscovite, rare lepidolite, small prisms of white or greenish beryl, cassiterite and columbo–tantalite
— Type 6: zoned pegmatite, frequently with a quartz core, with gigantic crystals of amblygonite and spodumene, pockets of big crystals of beryl, low contents of cassiterite, columbo–tantalite, and microlite
— Type 5: muscovite pegmatites
— Type 4: muscovite and tourmaline pegmatites
— Type 3: muscovite, tourmaline and biotite pegmatites with graphic feldspars
— Type 2: biotite and tourmaline pegmatites with strong development of graphic structures (quartz/feldspar and quartz/tourmaline)
— Type 1: biotite pegmatite

In Rwanda the border between granite and country rocks would generally be at the level of type 3, but it is higher in Kivu and Shaba/Zaire. A pronounced albitization is ubiquitous in types 5–7, while greisenization is less frequent. Many of the pegmatites are strongly argillized, which may be an effect of supergene alteration in some cases. However the occurrence of unaltered pegmatite bodies close to completely kaolinized ones, makes hydrothermal alteration the more probable alternative. Argillization facilitates easier exploitation of low grade ore bodies where cassiterite and columbo–tantalite occur disseminated in the rock. High grade ore shoots are normally close to quartz cores or to contacts.

As it is presented above, Varlamoff's zonation model appears too schematic, and needs to be qualified. Firstly, the pegmatites may be of different ages and different granite generations have produced distinct pegmatites. Pegmatites related to younger 'tin' granites should be clearly differentiated from older ones. It can be assumed, that types 1 and 2 are among the latter, but even some members of types 3–5 may be unrelated to G4 granites. Secondly, pegmatites and major fields of quartz veins are probably independent, as they form different districts on metallogenic maps (for example Rwanda: Ziserman et al. 1983). This suggests either a variation of the timing of magmatic crystallization in relation to tectonic processes, or a higher ratio of internal versus external pressure during late crystallization of those granites producing pegmatites. This is in contrast to essentially post-crystallization liberation of fluids from less over pressurized granites producing quartz veins. Thirdly, a number of mineralized pegmatites are clearly a marginal facies of fine-grained leucocratic granite or 'aplites' bodies (as the pegmatite at Buranga, mentioned above) which most probably are G4 granite apophyses.

The Sn–W quartz veins are the major source of cassiterite in the area, and of practically all the tungsten ore. Important deposits are normally made up of a

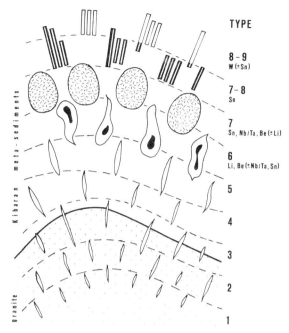

Figure 4. Distribution of pegmatite and quartz vein types associated with granite cupolas in Rwanda (redrawn from Varlamoff, 1972; for description see text).

multitude of single veins and veinlets forming stockworks and vein fields. As a rule these are hosted by deformed Kibaran metasediments; occurrences within granites are extremely rare. The major vein fields are clearly separated from pegmatite districts; some pegmatites, however, have been observed to be cross-cut by mineralized quartz veinlets, and others, as at Musha/Rwanda pass upwards into a zone of chaotic quartz–muscovite–kaolinite veins and pockets with cassiterite (Varlamoff 1969). Major tin fields are clearly separate from tungsten deposits, but there are numerous occurrences of quartz veins with both cassiterite and wolframite (Baudin *et al.* 1984), and traces of cassiterite have been found in ferberite deposits (Pohl 1975a). Both cassiterite and wolframite veins are known in host rocks ranging from quartzite (and sandstone) to metapelites, but in the largest tin district of the area, Rutongo/Rwanda, quartzites are preferentially mineralized while carbonaceous sandstones and schists are characteristic hosts for tungsten mineralization.

The spatial separation and differing host lithology of important tin and tungsten districts have been one argument for syngenetic models of the tungsten ores. To allow a more detailed discussion, typical features of both will be summarized below.

The tin-district of Rutongo (Pohl 1978) occupies the eastern limb of a large anticline, which plunges at a low angle to the NNW. The western flank is steeply dipping and locally overturned, while the eastern limb dips at about 30° to the east. The folded rocks are a suite of metapelites, sandstones and quartzites of the upper Lower group. The metamorphic grade increases to the south, where the core and the eastern limb of the anticline are intruded by the 'post-tectonic'

Kigali granite. Outcrops of the granite are few and strongly altered; fine-grained muscovite–tourmaline alaskite appears to be most typical.

The cassiterite–quartz veins at Rutongo occur almost exclusively in thicker bands of competent rocks. They form fields of parallel veins comprising up to several hundred single veins. Productive zones are those where the combined thickness of the veins is larger than 7–10 per cent measured in 50 m sections at a right angle to the strike of the veins. Five such zones each consisting of several vein fields can be differentiated; the most important one lies just to the east of the major north–south trending overthrusts near the crest of the anticline. More than 80 per cent of the district's cumulate cassiterite production of about 30,000 metric tonnes originated from this zone, which is up to 1·500 m wide and over 6 km long in a north–south direction. The veins strike north–south also and dip steeply to the west (Figure 5).

The veins contain quartz, muscovite and cassiterite; in addition, minor kaolinite (partly after feldspars), tourmaline, rutile/ilmeno-rutile, arsenopyrite, pyrite, chalcopyrite, galena and traces of gold have been found. Haematite/goethite coatings along late joints cutting the vein quartz are ubiquitous. Country rock alteration includes sericitization, tourmalinization, kaolinization, and silicification; disseminated cassiterite is rare. Most cassiterite occurs with coarsely crystalline muscovite at vein contacts, whilst the milky white or grey vein quartz is often nearly barren.

While the large Rutongo anticline is essentially a product of D2 folding along NNW–SSE directed axes, a later compressive event (D3) is responsible for a further tightening of the anticline producing the overthrusts of N–S strike and easterly dip, and the tensional fractures controlling mineralization (Pohl 1977). The syntectonic nature of the veins and their strict symmetry relationship with the anticline had been recognized earlier (Aderca 1957), without, however, differentiating the two deformation phases.

Three important tungsten ore mines in Rwanda (Nyakabingo, Gifurwe, and Bugarama) and Kirwa in southwest Uganda are aligned along a NNW-trending anticlinorium. Within this large structure, second order anticlines controlled the location of the mineralization. The roughly linear arrangement in conjunction with

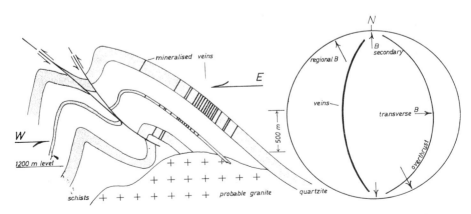

Figure 5. Rutongo tin district: idealized geological section and stereographic projection of structures controlling the quartz–cassiterite veins (after Pohl 1977). Arrows: Maxima of fold axes (B) measured in the field.

a similar stratigraphic position within the upper part of the Lower group prompted the designation 'tungsten belt' (De Magnee and Aderca 1960).

The ore bodies at Bugarama lie within the overturned limb of an anticline with a core of carbonaceous pyritiferous schists overlain by a thick quartzite member and grey sandstones with thin pelitic partings (Pohl 1975a, 1977; see Figure 6). Measured fold axes are nearly horizontal with a NNW–SSE direction. Their dispersal through about 30° may be interpreted as reflecting a later deformation of an originally more homogenous D2 fold pattern. The remarkably disharmonic fold set is obviously due to the variable lithology. An upward thrust and easterly 'overflowing' of the core is accentuated by a longitudinal crest fault. In view of this shape of the fold it is assumed to bottom out at about 500 m depth, where a pronounced structural unconformity is expected — this may be the roof of a larger granite body as yet uncovered by erosion in this area; a number of pegmatitic Be, Nb/Ta, and Sn occurrences at a distance of a few kilometres to the north support the assumption.

Mineralization consists of platy ferberite and porous, vuggy reinite (ferberite after scheelite) in thin quartz muscovite veins, which occupy bedding plane joints in the sandstones to some 40 m maximum distance from the hanging wall of the quartzite. The latter is not mineralized although it is strongly altered near the ore bodies (see below). Lower grade stockwork ore bodies consisting of numerous thin veinlets which may be both cross-cutting and bedding plane controlled, are found in the black schists and siltstones of the core, stratigraphically below the quartzite; these mineralized zones have a S–N elongation. The location of major ore bodies (concentrations of exploitable veins) appears to be controlled by structural highs along the hinge line of the D2 fold, which are interpreted as products of a later compressive (D3) phase producing S–N striking secondary folds.

The mine cumulative production is about 2000 metric tons of concentrate; 75 per cent of this originated from the conformable veins above the quartzite. Apart from ferberite, quartz, and muscovite, minor amounts of scheelite, tungstite, anthoinite, cassiterite, kaolinite, minute needles of tourmaline, and pyrite have been found. Late impregnations and coatings of goethite haematite are ubiquitous.

Country rock alteration includes widespread tourmalinization and silicification; kaolinization and sericitization occur more locally. Although not economically mineralized, the quartzite is brecciated, quartz veined, and variably silicified; irregular vitric masses with conchoidal fracture are the extreme endproduct. Geochemical tungsten haloes in country rocks near ore bodies are expressed by about 1000 ppm W in the quartzite (38 samples) and 600 ppm W in the grey sandstones (30 samples). This may be compared with data from Kifurwe, where 83 ppm W (maximum: 800 ppm, 59 samples) characterizes the mineralized dark schists, while only 4 ppm (10 samples) has been determined from similar rocks outside of the mine (Frisch 1975).

Ferberite nodules occur at Nyakabingo and Gifurwe (Frisch 1975) in Rwanda and at Nyamulilo and Ruhizha (Pargeter 1956) in Uganda. At Gifurwe, these flattened nodules consist of clay minerals, sericite, quartz, graphitic matter, and iron-hydroxides (at least in part after pyrite); they contain about 1100 ppm W (11 samples). Microveinlets of quartz and tourmaline were occasionally observed. The nodules form horizons in black schists and are clearly stratabound. Here, they are interpreted as diagenetic concretions, which were preferential sites of W-precipitation from pervading hydrothermal solutions (Frisch 1975). Apparently similar nodules at Ruhizha contain more ferberite and were actually exploited as

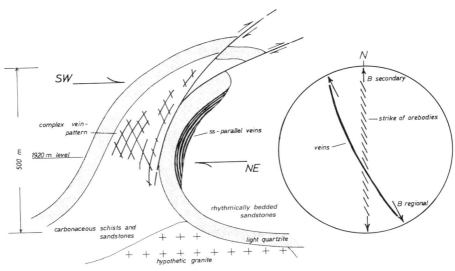

Figure 6. Bugarama tungsten mine: idealized geological section and stereographic projection of structures controlling the quartz–ferberite veins (after Pohl 1977).

low grade ore (Pargeter 1956). At Ruhizha and at Nyamulilo (De Magnee and Aderca 1960) the nodules were thought to have trapped tungsten diagenetically, which then would have been partially mobilized by metamorphism and/or granite derived fluids to produce the main mass of clearly epigenetic ores.

An interesting feature of the wolframite deposits of the area is the relationship between huebnerite and ferberite in different deposits. A rapid increase of ferberite content from granite hosted occurrences (about 50 mole per cent) to those in surrounding Kibaran sediments (about 97 mole per cent) had been detected at Kalima, Maniema (Varlamoff 1958). Similarly, earlier platy ferberite at Bugarama has more huebnerite (4·2–6·1 mole per cent, three samples) than later reinite from Bugarama and Gifurwe (2·4–5·4 mole per cent, x=3·0, 10 samples; data from Frisch 1975 and Pohl 1975a). This could be interpreted to reflect a temperature dependance (Oelsner 1944), but may equally well be simply due to prevailing Fe in the solutions which have metasomatically altered the primary scheelite to reinite. At Gifurwe, however, the huebnerite content increases with depth (from 2·4 to 5·4 mole per cent) indicating that temperature may be the essential control (Frisch 1975). Gold in wolframite placers was found adjacent to a number of ferberite deposits (Varlamoff 1958), notably at Gifurwe, where Au was first discovered and where the ferberite was recognized only later (Frisch 1975). This confirms the assumption of close metallogenic relationship between Sn, W, and Au mineralizations in the area (Frisch 1971; Pohl 1975b).

Gold-bearing quartz veins and silicified zones with secondary iron hydroxides in the superficial alteration zone and primary pyrite at depth are supposedly the source of Au placers, which have contributed practically all gold exploited in the area. Few of the primary deposits have been located (Ziserman et al. 1983; Niyondezo 1984; Barnes 1961), and most of these were found to be of uneconomic low grade and small size. Furthermore, detailed descriptions are not available.

Important gold districts are Nyungwe (Rwanda)–northwest Burundi, Bisesero and Miyove in Rwanda, and from there northwards to the outlier of Kibaran sediments (quartzites, sandstones and phyllites) in the Buhweju plateau (Ankole, Uganda). At Buhweju, the placers nearly always contain some cassiterite and monazite. Small auriferous quartz veins are not infrequent in both the underlying Buganda–Toro schists and in the Kibaran suite; pyrite, and more rarely Cu, Pb and Zn sulphides occur in the veins (Barnes 1961). At the southern Buhweju escarpment near the gold placers of Mashonga, leucocratic msucovite granite (G4 type?) intrudes Buganda–Toro rocks; its alteration halo with tourmalinization and quartz veining affects basement gneiss and schists, but also at least the basal Kibaran conglomerate (unpublished: Unesco geotraverse southwest Uganda 1984). Further south in the southwestern Kigezi district, some gold has been found in ferruginous (originally sideritic) veins with bismutite, pyrite, wolframite and cassiterite (Pargeter 1952) a type of mineralization which is unique in the area, and which should probably be classified as a skarn deposit.

The regional distribution of the gold districts (Figure 2) is generally peripheral to the Sn–W zones, although transitional overlap — mainly with ferberite — clearly occurs. This was interpreted as representing a more distal and lower temperature phase of mineralization in relation to the tin–tungsten ores (Frisch 1975).

3d. Mineralization related to alkaline complexes and carbonatites

The small bastnaesite deposit at Karonge and minor equivalents elsewhere in Burundi are probably associated with the Pan-African intrusive activity (Niyondezo 1984). At Karonge, bastnaesite occurs in veins and veinlets forming ore bodies of a stockwork type. It is accompanied by monazite and a gangue comprising quartz, barite and goethite. A hydrothermal derivation from carbonatite is assumed (Van Wambeke 1977).

4. Discussion and some tentative conclusions

Reviewing present knowledge concerning the evolution and mineralization of the northeastern Kibaran belt results in the conclusion, that available data allows few assertive statements to be made. Most of the present interpretations are little more than hypotheses and models, which need to be tested with more field and laboratory investigations.

An intracontinental setting for the Kibaran belt throughout its evolution (Klerkx et al. 1984) due to interaction between the Congo and Tanzania cratons appears quite well documented. It was preceded by Eburnean orogenic activity, and followed by repeated movements and mantle activity of rather anorogenic type during the Pan-African which is now thought to comprise the period from about 950 to 450 Ma (Kröner 1984). A major problem of understanding and definition is, however, the attribution of the G4 granites (1000–950 Ma) and the roughly time equivalent Lomanian folding to either a late Kibaran (Pohl 1977) or an early Pan-African phase (Cahen et al. 1984). Usually, the folding of molasse deposits is not questioned as representing a final stage of orogeny, but the large time lag between the main orogeny (about 1200 Ma) and this last compressive phase in the Kibaran certainly is unusual.

Similarly, the origin of the G4 'tin' granites needs clarification. Their mildly alkaline character, high LIL element contents, the extremely high initial Sr-ratios, the elevated gamma ray radiation levels (Unpublished: Unesco geotraverse southwest Uganda 1984), and some geological features may classify them as HHP-granites. A crustal derivation cannot be doubted, and the total absence of any contemporaneous basic magmatic activity appears to favour an origin by stacking of continental crust similar to the late-Hercynian 'tin' granites in Central and Western Europe (Windley 1984). As is obvious from the description above, the regional thrusts and/or nappes necessary for this model are unknown in the Kibaran belt. More field work is needed, probably along the western margin of the belt in Zaire, to resolve this uncertainty.

Mineralization associated with the tensional ('geosynclinal') stage of the Kibaran belt is restricted to minor occurrences of iron ores and traces of base metals. Prospecting until now has largely ignored this environment, but the increasing recognition of volcanic rocks in the sedimentary pile should encourage a search for stratabound ores.

It is very difficult to integrate the mafic–ultramafic belt of the Burundian and Tanzanian belt into the evolutionary history. The linear structural control, and its layered intrusion characteristics with the equivalent mineralization appear to favour a tensional setting. This may indicate a Pan-African age (possibly related to Bukoban and Malagarasian basalts) rather than a Kibaran age.

The origin of the tin–rare metal pegmatites and of the majority of metalliferous quartz veins is clearly associated with the G4 granites. A strong structural control of both the granites and the mineralization in related to a compressive deformation phase has generally been documented. In some respects, earlier metallogenic models (Varlamoff 1975; Frisch 1971 and 1975) appear too schematic, however.

To summarize the present understanding of the Kibaran 'tin' granite related mineralization, a cartoon is presented integrating important features of the deposits (Figure 7). This concentrates on the variable spatial association of pegmatites related to the granites, and on the comparable geology of tin and tungsten quartz veins, the latter not necessarily being more distal and lower temperature phases of the tin mineralization.

The fluids involved in Sn–W–Au mineralization have not yet been investigated. Accordingly, the mineralogical composition of the ores can only be used to approach this question. Most importantly, fluorine minerals are quite rare in all types of mineralization. The greisens described in earlier publications on the Kibaran consist of muscovite–quartz aggregates. Topaz and fluorite are so infrequent (Bertossa 1968), that the fluids would in most cases have contained little fluorine. Apatite and other phosphates are much more frequent (Bertossa 1968), which suggests that P was one of the more important volatiles, together with H_2O, boron (ubiquitous tourmaline!) and probably Cl and CO_2. High Na-contents of some fluids are indicated by frequent albitization of G4 granites and pegmatites. In view of the present understanding of fluids involved in this type of mineralization (Campbell *et al.* 1984; Manning 1984; Roedder 1984) a large variability in time and space has to be assumed. Fluid composition may be the prime agent in producing separate Sn and W deposits. For transport and deposition of gold, the availability of sulphur may have been a critical factor.

The teasing final question as to the original source of the metals remains a matter of interpretation in the Kibaran belt as elsewhere. Because of frequently elevated geochemical tungsten contents of Kibaran metasediments in southwest Uganda (Pargeter 1956; Jeffery 1959) a synsedimentary deposition of tungsten was proposed by some authors to have preceded concentration by metamorphism and/or granite intrusions (Pargeter 1956; Jeffery 1959; De Magnee and Aderca 1960; Reedman 1967). In view of the very low grade of metamorphism and the absence of quartz veins in unmineralized areas, however, a lateral secretion model

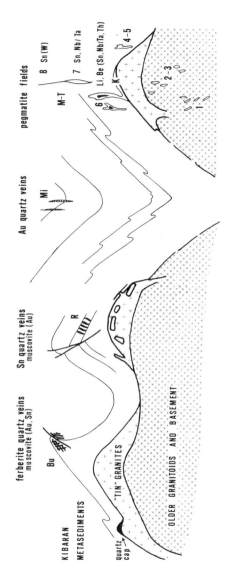

Figure 7. Cartoon of Kibaran mineralization related to G4 'tin' granites (figures denote pegmatite types after Varlamoff, 1972; Bu—Bugarama; K—Karenge granite; Mi—Miyove; M-T—Musha-Ntunga; R—Rutongo).

is unlikely. Furthermore, there is no indication of 'granitization' in the immediate surroundings of the ferberite deposits. Accordingly, the carbonaceous shales with high tungsten contents and ferberite nodules which are characteristic country rocks of ferberite quartz vein fields cannot be the source of the tungsten in the deposits. On the contrary, these anomalous geochemical contents are probably haloes of epigenetic mineralization as shown by Frisch (1975) at Gifurwe.

High geochemical tin levels in G4 granites (Lowenstein 1966) and the geological environment suggest that the granites (and not country rocks leached by hydrothermal convection systems) were the immediate source for the ores. Accordingly the crustal rocks affected by partial melting to produce the tin granites were most probably the original source. The previous geological history of this deep basement would then have determined the availability of the metals and at least of part of the volatiles. Whether this involves a previous geochemical enrichment, as demanded by many metallogenic models (Routhier 1980), or only more effective liberation of the metals from the normal crustal background should be one of the problems addressed by future research.

Acknowledgements. Unesco supported the author's participation in the Geotraverse through southwest Uganda in November 1984. Thanks are due to other participants, including J. P. Bicamumpaka, M. Daly, A. El-Etre, D. P. M. Hadoto, J. Lavreau and S. Sinabantu for lively discussions and mutual support. Mrs R. Beddies and Mr H. Stosnach kindly helped to prepare several typescripts and drawings of this paper.

References

Aderca, B. 1957. Un cas de 'boudinage' à grande échelle: La mine de Rutongo au Rwanda. *Ann. Soc. Géol. Belgique,* **80**B, 279–285.
Barnes, J.W. 1961. The mineral resources of Uganda. *Bull. Geol. Surv. Uganda,* **4**, 1–89.
Baudin, B., Zigirabili, S. and Ziserman, A. 1984. *Livre notice de la carte des gîtes minéraux du Rwanda (1:250,000),* 163 pp. Ed. Univ. Nat. Rwanda, Butare.
Bertossa, A. 1965. La pegmatite du Buranga. *Bull. Service Géol. Rwanda,* **2**, 1–6.
—— **1968.** Inventaire des minéreaux du Rwanda. *Bull. Service Géol. Rwanda,* **4**, 1967–68, 25–46.
Bowden, P. and Karche, J.-P. 1984. Mid plate A-type magmatism in the Niger-Nigeria anorogenic province: age variations and implications. In: Klerkx, J. and Michot, J. (Eds), *African Geology,* Tervuren, 167–177.
Cahen, L., Snelling, N.J., Delhal, J. and Vail, J.R. 1984. *The Geochronology and Evolution of Africa,* Oxford Science Publ., Clarendon Press, Oxford. 512pp.
Campbell, A. Rye, D. and Petersen, U. 1984. A hydrogen and oxygen isotope study of the San Cristobal mine, Peru: implications of the role of water to rock ratio for the genesis of wolframite deposits. *Economic Geology,* **79**, 1818–1832.
Daly, M.C. 1984. Preliminary report on the foreland structure of the Irumide belt of Northern Zambia. In: Klerkx, J. and Michot, J. (Eds), *African Geology,* Tervuren, 67–74.
De Magnee, I. and Aderca, B. 1960. Contribution à la connaissance du Tungsten-belt ruandais. *Acad. roy. Sci. Outre-Mer,* 8°, **11**/7, 1–56.
Frisch, W. 1971. Die Zinn–Wolfram-Provinz in Rwanda (Zentral-Afrika) aus montangeologischer Sicht. *Erzmetall,* **24**, 593–600.
—— **1975.** Die Wolfram-Lagerstätte Gifurwe (Rwanda) und die Genese der zentralafrikanischen Reinit-Lagerstätten. *Jahrb. Geol. B. –A.,* **118**, 119–191.
Harpum, J.R. 1970. Summary of the geology of Tanzania. *Mem. Mineral Resources Division,* **1**.
Hutchison, C.S. 1982. The various granitoid series and their relationship to W and Sn mineralization. In: Hepworth, J.V. and Yu Hong Zhang (Eds), *Tungsten Geology* Jiangxi, China, ESCAP/RMRDC Bandung, Indonesia, 87–114.
Jeffery, P.G. 1959. The geochemistry of tungsten, with special reference to the rocks of the Uganda Protectorate. *Geochim. Cosmochim. Acta,* **16**, 278–295.
Jourde, G. and Vialette, Y. 1980. *La chaine du Lurio (Nord Mozambique),* BRGM Int. Rep., 75 pp.
Kennedy, W.Q. 1964. The structural differentiation of Africa in the Pan-African ($\pm$ 500 m.y.) tectonic episode. *Ann. Rep. Res. Inst. Afr. Geol.,* Leeds University, **8**, 48–49.
Klerkx, J. 1984. Modèle d'évolution de la chaine Kibarienne. *UNESCO, Geol. for Development Newsletter,* **3**, 43–46.

——, Lavreau, J., Liegeois, J.-P. and Theunissen, K. 1984. Granitoides kibariens précoces et tectonique tangentielle au Burundi: magmatisme bimodal lié à une distension crustale. In: Klerkx, J. and Michot, J. (Eds), *African Geology*, Tervuren, 29–46.
Kröner, A. 1984. Late Precambrian plate tectonics and orogeny: A need to redefine the term Pan-African. In: Klerkx, J. and Michot, J. (Eds), *African Geology*, Tervuren, 23–28.
Lavreau, J. 1984. Apercu sur la géochronologie du Kibarien de l'Afrique Centrale. *UNESCO, Geology for Development, Newsletter*, 3, 31–35.
Lowenstein, P.L. 1966. Progress report on studies of pegmatite and tin mineralization in south-west Ankole, Uganda. *10th Ann. Rpt. Res. Inst. African Geol. Univ. Leeds*, 34–36.
Manning, D.A.C. 1984. Volatile control of tungsten partioning in granitic melt-vapour systems. *IMM Transactions, Section B*, 93, 185–189.
Niyondezo, S. 1984. Les ressources minérales du Kibarian au Burundi. *UNESCO, Geology for Development Newsletter*, 3, 37–41.
Oelsner, O. 1944. Über erzgebirgische Wolframite. *Ber. Freiberg. Geol. Ges.*, 20, 44–49.
Pargeter, R.C. 1952. The geology of Bismuth ores in Uganda, *Proc. 5th Intern. Geol. Congr. East Africa*.
—— 1956. The Ruhizha ferberite deposit, Kigezi. *Rec. Geol. Surv. Uganda*, 1954, 27–46.
Pohl, W. 1975a. Géologie de la mine de Bugarama et de ses environs (Rwanda, Afrique). *Bull. Serv. Géol. Rwanda*, 8, 13–42.
—— 1975b. Geologie und Lagerstätten Rwandas (Zentral-Afrika). *Berg. Huettenm. Monatsh.*, 120, 244–252.
—— 1977. Structural control of tin and tungsten mineralization in Rwanda, Africa. *Berg. Huettenm. Monatsh.*, 122, 59–63.
—— 1978. Die tektonische Kontrolle der Zinngänge von Rutongo, Rwanda (Afrika). *Mitt. Österr. Geol. Ges.*, 68, 89–107.
—— 1985. Geologie Zentral-und Ostafrikas, ein neuer Anfang?. *Mitt. TU Carolo-Wilhelmina Braunschweig*, 20, 1, 33–37.
Radulescu, J. 1982. Mineralization in the Karagwe–Ankolean System of East Africa/Burundi. In: *The Development Potential of Precambrian Mineral Deposits*. UN Dpt. Techn. Coop. Devel., 217–225.
Reedman, A.J. 1967. The geological environment and genesis of the tungsten deposits of Kigezi district, south-western Uganda (PhD thesis abstract, 1967), 11th Ann. Rpt. Res. Inst. African Geol. Univ. Leeds, p. 38.
Robert, M. 1931. Carte géologique du Katanga 1:1,000.00. *Nouv. Mém. Soc. Belge Géol. Paléont. Hydrog.*, 5, 1–14.
Roedder, E. 1984. Fluid inclusions. *Reviews in Mineralogy*, 12, 644 pp. Mineral. Soc. America, Washington DC.
Routhier, P. 1980. Où sont les metaux pour l'avenir? *Mém. BRGM*, 105, 1–408, 97 figs., 14 tables.
Sacchi, R., Marques, J., Costa, M. and Casati, C. 1984. Kibaran events in the southernmost Mozambique Belt. *Precambrian Research*, 25, 141–159.
Tack, L. 1984. Post-Kibaran intrusions in Burundi. *UNESCO, Geol. for Development Newsletter*, 3, 47–57.
Theunissen, K. 1984. Les principaux traits de la tectonique Kibarienne au Burundi. *UNESCO, Geology for Development, Newsletter*, 3, 25–30.
Tissot, F., Swager, C., Berg, R., Van Straaten, P. and Ingovatow, A. 1982. Mineralization in the Karagwe-Ankolean System of North-West Tanzania. In: *The Development Potential of Precambrian Mineral Deposits*, UN Dpt. Techn. Coop. Devel., 205–215.
Van Straaten, H.P. 1984. Contributions to the geology of the Kibaran belt in Northwest Tanzania. *UNESCO, Geology for Development, Newsletter*, 3, 59–68.
Van Wambeke, L. 1977. The Karonge rare earth deposits, Republic of Burundi. *Miner. Deposita*, 12, 373–380.
Varlamoff, N. 1958. Les gisements de tungstène au Congo et au Ruanda–Urundi. *Acad. roy. Sci. Coloniales*, 1958, 1–70.
—— 1969. Transitions entre les filons de quartz et les pegmatites stannifères de la région de Musha–N'tunga (Rwanda). *Ann. Soc. Gèol. Belgique*, 92, 193–213.
—— 1972. Central and West African rare metal granitic pegmatites, related aplites, quartz veins and mineral deposits. *Mineral. Deposita*, 7, 202–216.
—— 1975. Classification des gisements d'étain. *Acad. Roy. Sci. Outre-Mer*, 19, 5, 1–63.
Von Knorring, O. 1969. A note on the phosphate mineralization at the Buranga pegmatite, Rwanda. *Bull. Service Géol. Rwanda*, 5, 42–45.
Windley, B.F. 1984. *The Evolving Continents*, Second Ed., John Wiley, Chichester, 399 pp.
Ziserman, A., Zigirababili, J., Petricec, V. and Baudin, B. 1983. Données sur la métallogénie du Rwanda. Enseignements tires de la carte des gîtes minéraux. *Chron. rech. min.* 471, 31–40.

Implications of a palynological study in the Upper Precambrian from eastern Kasai and northwestern Shaba, Zaire

D. Baudet
Palaeobotanique et Palaeopalynologie, Université de Liege, Belgium

This study presents the results of three borehole correlations in the Upper Precambrian Bushimay Supergroup, from the Kasai and Shaba provinces in Zaire.

The most western borehole comes from the Lubi Valley (eastern Kasai) and cuts the lower part of the Bushimay Supergroup (BI). One palynological sample from the top of the BI Formation (BIe2) showed a microflora with Middle Riphean affinities that confirms the radiometric age previously found for the BI Formation in this area.

The borehole in the Kanshi valley (eastern Kasai) located at about 30 km to the southeast of the Lubi valley, cuts the lower part of the BII Formation which consists of stromatolitic dolomites with black shaly horizons. A study of the systematics and the biostratigraphy of the microflora from the shaly horizons enables four different assemblages to be defined which correspond with the base of the Upper Riphean from Siberia, the North Atlantic area and Australia.

The third borehole comes from Kafuku (Luembe valley, Shaba) located about 150 km to the southeast of the Kanshi core; it cuts detrital sedimentary rocks overlain by stromatolitic beds, respectively considered as BI and BIIa.

An assemblage of acritarchs found in a shaly horizon from the upper part of the BI (BIe2) is comparable with the upper part of the BIIc horizon from Kanshi and consequently with those from the Upper Riphean. Thus, there is evidence of diachronism in the BI and BIIa-c Formations from the North West to the South East in the Kasai-Shaba area because the top of the detrital Formations and the stromatolites are younger towards the Kibaran continent. Consequently, the palaeogeographical model and the correlation scheme between Kasai and Shaba must be reconsidered.

This diachronism combined with the classical scheme of a transgressive Upper Riphean sea on an eroding Kibaran continent, correlates with the transgressive cycle suggested by the vertical succession of the lithological facies of the supergroup. The age of the lavas in the supergroup needs to be reassessed.

KEY WORDS   Acritarchs   Upper Precambrian   Kasai   Shaba   Zaire Correlations

1. Introduction

The Bushimay Supergroup crops out in the type area of eastern Kasai and in northwestern Shaba, Zaire (Figure 1). In the eastern part of Shaba province, the

Bushimay Supergroup rests unconformably upon Kibaran rocks, while in southern Kasai it unconformably overlies the Kasai craton where the youngest rocks are 2000 Ma old. In the north, the Supergroup is overlain by Mesozoic sedimentary rocks belonging to the Zairian basin (Cahen 1963, 1973a; Cahen *et al.* 1984).

The Roan (Lower Katangan) is considered as the stratigraphical equivalent to the Bushimay Supergroup (Raucq 1957, 1970; Dumont 1971; Cahen 1982).

2. Tectonism

The tectonic influence on the supergroup is rather weak in Kasai with monoclinical beds showing long wavelength undulations orientated SW-NE and NW-SE. Folding is more intense in northern Shaba near the Lomani river, where the beds form part of the Lomamian orogeny (Lepersonne 1973). Moreover, near the Kibaran

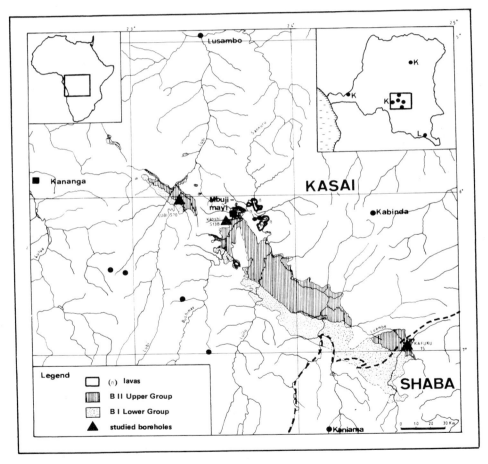

Figure 1. Map (after Raucq 1970) showing the outcrop locations of the Bushimay Supergroup at the boundary between Kasai and Shaba (Zaire). From the three studied boreholes, two (Lubi S70 and Kanshi S13B) come from eastern Kasai, the last (Kafuku 15) comes from northwestern Shaba.

belt, many faults affect the Supergroup (Dumont 1971). The Supergroup is unaffected by metamorphism.

3. Lithology

The Supergroup is divided into two groups, The lower group, BI, is essentially composed of conglomerates and quartzites at the base with siltstones, shales, dolomitic shales and some dolomitic beds at the top. Locally, interstratified breccias occur. Its thickness is estimated at about 550 m in Kasai and nearly 3000 m in Shaba. The coarse facies occurs only in the Shaba Province.

The upper group, BII is principally composed of carbonates which are generally dolomitic rocks containing stromatolites. In the lower part of BII there are breccia horizons, with some important shaly and pyritic horizons towards the middle. In Kasai, its thickness is about 1000 m.

The three boreholes that were studied were sunk during the 1950s to find copper and lead mineralization. They come from the Lubi Valley (S70), Kanshi valley (S13B), and Luembe valley (Kafuku no. 15). The three boreholes intersect various boundary horizons of the Bushimay supergroup known respectively as: (1) BIb,e – BIIa; (2) BIIb – BIIc; and (3) BIc, d-BIIa.

4. Volcanism

In Kasai, the Supergroup is overlain by amygdaloidal basaltic pillow lavas. Their relationship to the BII horizon has been argued but Raucq (1969, 1970, personal communication) considers that the lavas erupted immediately at the end of the deposition of BIIe or during the deposition of BII in some areas (Figure 2). For example, Raucq has found some volcanic fragments in the top calcareous bed near the Shankuru Bushimay confluence which he considers as volcanic bombs included in the sediment.

Similar lavas were recognized in the Mukulu Mountain, also overlying carbonate rocks. At this locality the thickness is less than half that of the Kasai area (Dumont 1971) Further to the east, at Bukama, lavas of a similar nature were described overlying detrital rocks considered as BI (Dumont 1971); Raucq *et al.* 1977). According to the authors this could be due to different volcanic events interstratified in the Supergroup or be related to only one event (Dumont 1971; Cahen *et al.* 1984). Many dolerite dykes cutting the Supergroup are considered as the feeder systems of the lavas.

5. Geochronology

Radiometric ages were measured by the K/Ar method on five samples of basaltic lavas from the Sankuru-Bushimay confluence, and an age of 948 ± 20 Ma was assigned (Cahen *et al.* 1974, 1984). Earlier, dating had been undertaken on syngenetic galena from sedimentary rocks (Cahen 1954; Holmes and Cahen 1955; Raucq 1957). Two samples from the BIe$_1$ horizon in the Lubi Valley and Songa-Songa gave respectively 1065 and 1040 Ma; the age was reassessed as 1125 Ma by Cahen and Snelling (1966). It is known that the beginning of the Bushimay

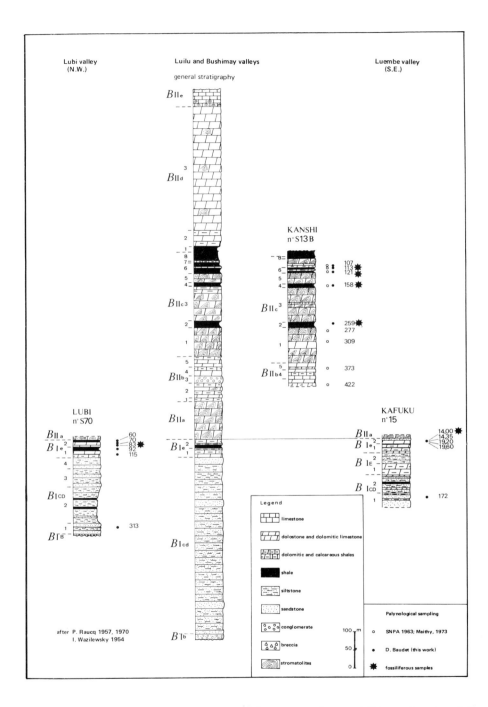

Figure 2. Lithostratigraphical column of the three boreholes and general lithostratigraphy of the Bushimay Supergroup. Location of the palynological samples.

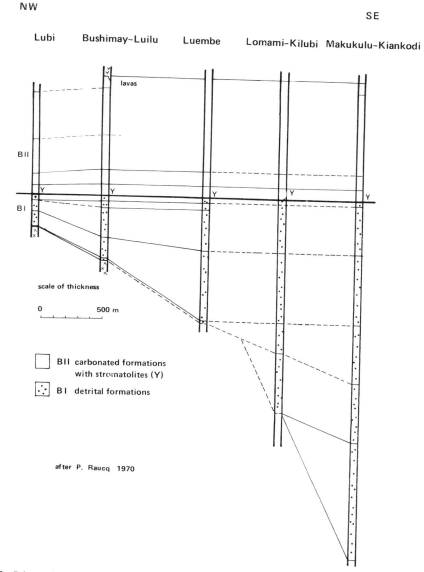

Figure 3. Schematic section given by Raucq (1970) showing the most plausible correlations between the formations (or the Supergroup) in Kasai and in Shaba; the appearance of stromatolites (boundary BI/BII) is taken as the main element of correlation through the basin.

Supergroup postdates the initial stages of the Kibaran orogeny dated at 1310 ± 25 Ma by Cahen et al. 1984.

6. Palaeontology

Following the work of Raucq (1957) and Dumont (1971), the stromatolites from the BII horizon were studied by Bertrand-Sarfati (1972b). She described different species of Conophyton, Baicalia, Tungussia, and Gymnosolen typical of the Upper Riphean base in Mauritania and U.S.S.R.. In fact, the stromatolitic assemblages from the BIIa–c and BIIe are equivalent to the ones of I4/I5 and I6 horizons from Mauritania. The early diagenesis of I5 and I6 units has been dated at 888 ± 16 Ma and 878 ± 15 Ma respectively (Clauer and Bonhomme 1971; Clauer 1973, 1976).

Cahen (1973b) discusses the divergence of the data and proposed that either the stromatolitic horizons were diachronous between Kasai and Mauritania or that there was an error in the value of the disintegration constant used to calculate the radiometric ages.

The palynological study of the BII levels by Boulouard and Calandra (1963) and Raucq (1970) described some 'Leiospheraeides'. This study was extended by Maithy (1975), who compiled an inventory of the microorganisms and compared them with the forms from the Upper Riphean in Australia and U.S.S.R. In this present work, the age of the Middle Riphean – Upper Riphean boundary is accepted as 1050 ± 50 Ma (Cahen 1973b).

7. Palaeogeography and geodynamic evolution

Based on the available data, the palaeogeographical evolution of the Kasai – Shaba area during the deposition of the Bushimay Supergroup can be summarized. Uplift of the Kibaran Mountains by successive paroxysmal tectonometamorphic phases of the Kibaran orogeny ($=1300$ Ma) was followed by erosion and deposition of molasse facies of the Lower BI into a trough bordering the continent ('avant-fosse'). Then during the deposition of the upper BI (1100 Ma), the trough largely filled and the sediments reached the outer margin of the trough. As the terrigenous sediments thinned outwards they were replaced progressively by thalassic sediments. When deposition of the continental-derived sediments ceased, the environment was suitable for colonization by stromatolites. They became established in the area around 1050 Ma ago and with chemical sedimentation built the BII Group. Some variations of the subsidence rate of the platform lead to changes in the bathymetry which influenced the stromatolite development until, locally, the stromatolites became completely silted up. Different breccia horizons interstratified in the supergroup suggest periods of tectonic activity.

In Kasai, the deposition of the Supergroup is terminated by a volcanic event ($\simeq 950$ Ma) linked to the Lomamian tectonism.

8. Stratigraphical palynology

As an introduction to this section on acritarchs it must be stressed that the main difficulty for Precambrian palynologists is the inconsistencies and inadequacy of descriptions and illustrations for many Precambrian acritarchs. This problem led Vidal and Siedlecka (1983) to suggest that we should redefine completely the systematics of Precambrian palynomorphs, with names of taxa totally different from the present ones, providing precise and detailed identifications created from a sufficient number of specimens in a perfect conservation state.

In response to this request, the palynological inventory of the deposits studied in this work, the precise description of the species, and the organization of these into a new coherent classification is a voluminous and specialized work to be published separately (Baudet thesis in preparation). Because a large part of the species are not yet published, numbers have been utilized to refer to species in this paper. In fact, only the unicellular organisms (acritarchs) are considered here because the pluricellular or filamentous micro-organisms are generally linked to stromatolites and are thus more benthonic than planktonic.

Maithy (1975) described 22 species of acritarchs in the four studied horizons (158 m, 121 m, 113 m and 107 m) from the Kanshi borehole, of which 21 species are new. Because of the difficulty in using definitions of species from earlier Russian workers, Maithy did not report any species recorded from Russia. He defined his own species but used Russian generic names, creating numerous synonyms. This prevents an easy use of his species. In the six fossiliferous levels there are acritarchs and filamentous cyanophytes of the *Taeniatum* sp. type (or *Eomycetopsis* sp. according to Schopf 1968 and Maithy 1973), but only the three levels 158 m, 121 m and 113 m from the Kanshi yield more complex cyanophytes. In these last three deposits, the two main types of associated microflora are abundant. We can note acritarchs (planktonic character) are often reported in the Russian and North European literature and cyanophytes (generally linked to the stromatolites and thus with a benthonic character) are often described in the American literature.

Whilst it is logical to find planktonic acritarchs in all the deposits, the other microflora should not be found in either the pre-stromatolitic BI levels or in the post-stromatolitic BII shales. The post-stromatolitic occurrence may be explained as residues of algal films in silts (Monty personal communication). The choice of planktonic organisms as guide assemblages for correlation severely limits the influence of sedimentary facies on correlation.

9 Biostratigraphy

The deposit at 83 m depth from the Lubi borehole is not very diversified and not very rich. The presence of species 2 (*Kildinella sinica* Timofeev 1966 or *Kildinosphaera chagrinata* Vidal and Siedlecka 1983) indicates a Riphean or Vendian age (Timofear, 1966, Vidal, 1981). The absence of typical Upper Riphean – Vendian forms suggests that this horizon should be assigned a Lower-Middle Riphean age. However, the presence of species 4 which, on account of its larger size, has an Upper Riphean affinity, enables this assemblage to be assigned to the top of the Middle Riphean.

Immediately succeeding this assemblage is a core (Kanshi 259 m), which is not rich or diversified in unicellular organisms, but contains acritarch species 7 (*Leiosphaeridia asperata* (Naumova) Lindgren 1982, *Kildinella hyperboreica* Timofeev 1966, *Kildinella timofeevii* Maithy 1975) which typically persisted in the Upper Riphean until Lower Cambrian times. Then a new species (6) of a larger size appeared.

The Kanshi 158 m horizon contains a very rich and diversified deposit. Many species developed among which is species 11 (*Chuaria circularis*) which is Upper Riphean to Vendian in age, species 15 (*Vavosphaeridium bharadwajii*), the only species not new in the work of Maithy (1975); species 17 (*Kildinella timanica* Timofeev 1969) of Upper Riphean age; species 19 (*Symplassospaeridiun bushimayensis* Maithy 1975) and species 21 (*Nucellosphaeridiwn magna* Maithy 1975 or *N. deminatum* Timofeev 1969).

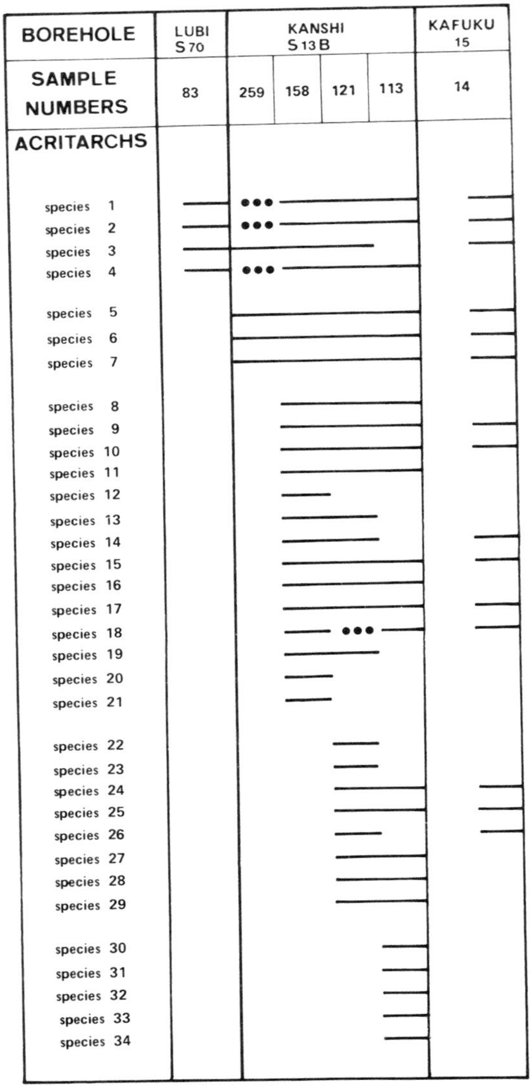

BOREHOLE	LUBI S70	KANSHI S13B				KAFUKU 15
SAMPLE NUMBERS	83	259	158	121	113	14
ACRITARCHS						
species 1	—	•••		—		—
species 2		•••		—		—
species 3	—			—		
species 4	—	•••		—		
species 5				—		—
species 6		—		—		—
species 7		—		—		—
species 8				—		
species 9				—		
species 10				—		
species 11				—		
species 12			—	—		
species 13				—		
species 14				—		—
species 15				—		—
species 16				—		—
species 17				—		—
species 18				•••	—	
species 19				—		
species 20				—		
species 21				—		
species 22				—		
species 23				—		
species 24				—		—
species 25				—		—
species 26				—		
species 27				—		
species 28				—		
species 29				—		
species 30						—
species 31						—
species 32						—
species 33						—
species 34						—

Figure 4. Stratigraphic distribution of acritarch species in the Supergroup in Kasai and acritarch contents of the fossiliferous levels from Kafuku (Shaba). From the comparison of these palynological assemblages, a distinct similarity appears between the levels 14 from Kafuku and 121 from the Kanshi.
Comments on the species:
Species 2: *Kildinella sinica* (Riphean – Vendian)
Species 7: *K. hyperboreica* (Upper Riphean – Lower Cambrian)
Species 10: *Trematosphaeridium* sp.
Species 11: *Chuaria circularis* Walcott 1899, *K. jakutica* Timofeev 1966 (Upper Riphean – Vendian)
Species 15: *Vavosphaeridium bharadwajii* (Vindyan)
Species 17: *K. timanica* Timofeev 1969 (Upper Riphean – Vendian)
Species 19: *Symplassosphaeridium bushimayensis* Maithy 1975
Species 21: *Nucellosphaeridium magna* Maithy 1975 *N. deminatum* Timofeev 1969
Species 22: *Kildinosphaera lophostriata* Vidal and Siedlecka 1983 (Upper Riphean)
Species 24: *Trachysphaeridium laminaritum* Timofeev 1966 (Upper Riphean – Vendian)
Species 26: *Archaeodiscina* sp
Species 31: *Protosphaeridium* cf. *flexuosum in* Vidal 1976 (Proterozoic – Middle Cambrian).

The Kanshi 121 m horizon, is also rich and gives new species, such as species 22 (*Kildinosphaera lophostriata* Vidal 1983) of Upper Riphean age and species 24 (*Trachysphaeridium laminaritum* Timofeev 1966) which appears typically in the Upper Riphean and continues until the end of the Vendian.

Finally, in the Kanshi 113 m horizon, new species occur among which species 31 (*Protosphaeridium* cf. *flexuosum* Timofeev 1966 in Vidal 1976) has also been recognized in the middle section and upper section of the Visingo beds in Sweden of Upper Riphean age, and Lower Vendian age. According to Timofeev (1966) *P. flexuosum* is considered as Middle Riphean to Upper Cambrian in age.

From the biostratigraphy, it is clear that we have similar assemblages to those from the top of the Middle Riphean (BI from the Lubi river) extending into the base of the Upper Riphean (BII from the Kanshi river) with the progressive development of species marking the transition to Upper Riphean times. Horizon 14 m from BI in the Luembe valley is rich and diversified although the organic matter is not in such a perfect state of preservation as in Kasai. The composition of its palynological assemblage is similar to the 121 m level in BII from the Kanshi river. On the basis of correlations between acritarch assemblages, the Bushimay Supergroup shows a clear diachronism between the Kanshi-Lubi area and the Luembe region, 150 km to the southeast.

10. Conclusions and discussion

Taking the palynological results into account, the Middle Riphean – Upper Riphean boundary occurs between the BIe_2 and the $BIIc_2$ horizons in the Lubi – Kanshi area. This is in accordance with a lead/lead age of 1125 Ma on galena from Lubi. The palynological composition of the overlying deposits, compared with the stratigraphy of the supergroup at the Kanshi river, shows by the progressively changing forms that we are at the base of the Upper Riphean stage. This places the Middle Riphean at about the base of BII (1050 Ma) and in agreement with the age of the lavas (950 + 20 Ma) at the top of BII in the Kanshi area.

There is a good correspondance of the palynological assemblages from the BIe_2 horizon in the Luembe valley (Kafuku 14 m) with the Kanshi valley $BIIc_6$ horizon (Kanshi 121 m). Together with the resultant diachronism it suggests that since the age of the BIe_2 in this area is younger than in the Lubi region, BIIa,b will be approximately 950 ± 20 Ma, corresponding to a lead/lead age of 910 Ma for galena.

Whereas in the Kasai, the boundary BI – BII could be more or less coincident with Middle to Upper Riphean (Cahen 1973b) this is no longer valid in the Shaba region. If this boundary is diachronous as represented by the incoming of stromatolites, between the Kanshi and the Luembe areas (150 km apart), similar diachronism would be expected over much greater distances (5,000 km) which separated Mauritania from Kasai. Furthermore, this BI – BII division which has been taken as a correlative marker between the Kasai and Shaba areas by Raucq (1970) for his geodynamic model which must therefore be revised (See Figure 3 and 5). The study of the lithological succession of the Supergroup reveals that during the whole period of deposition there was a general transgressive trend.

Since the continent lay to the southeast, the marine transgression was from the northwest to the southeast. This direction of transgression of the Riphean sea is confirmed by the isotopic dating and the palynological correlations. For example, while in the Kanshi area the deposition of black muds of BIIc was complete, the

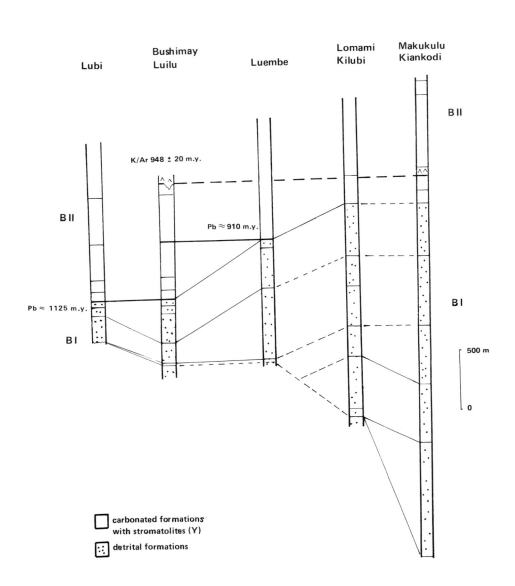

Figure 5: Proposition of a modified scheme of the correlation of the formations between Kasai and Shaba.
— Between the Lubi and Kanshi (30 km), the appearance of stromatolites (BI/BII) remains an approximative correlation element.
— Between Kanshi and Luembe, the common palynological assemblage is taken as a correlation basis.
— Between Kanshi and Mukukulu the hypothesis is made that the lavas are contemporary.

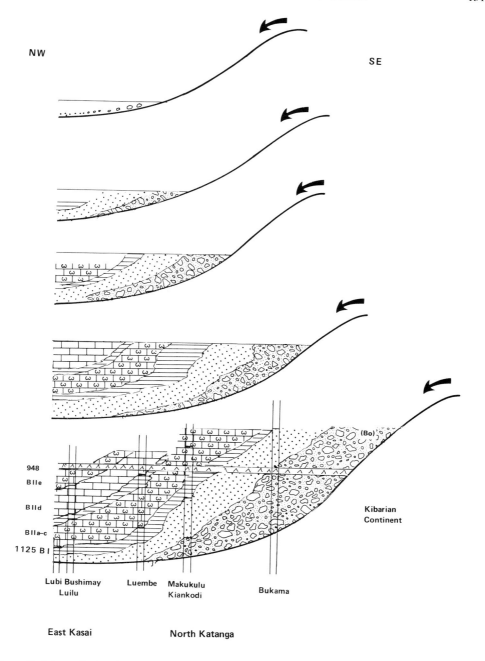

Figure 6: Schematic drawing of the proposed geodynamic model at the end of the Middle Riphean and beginning of the Upper Riphean: marine transgression from the northwest to the southeast on a Kibaran continent in erosive phase. The distribution of the lithological facies is mainly a function of the distance from the shore; perpendicular to the shore, the diachronism of formations is maximum but for the lavas which, considered as belonging to a same volcanic event, have a comparable age throughout the basin.

first stromatolites (BIIa) colonized the platform in the Luembe area. At the same time, thin terrigenous sediments (BId) were deposited in the southeast, and near the Kibaran continent the sediments became coarser (BIa). These observations support a classical scheme of infilling of a subsiding sedimentary basin during a marine transgression.

This type of model can be explained without referring to either a trough or complex epeirogenic movements which would be necessary to produce ridges on the basin floor. In this transgressive scheme, it is necessary to consider the volcanism in more detail. For example, the lavas overlie different sedimentary facies according to their distance from the continental margin. The Kanshi lavas overlie 1000 m of BII carbonate rocks above BI detrital sediments. At Makukulu (Shaba), nearer to the palaeocontinent, the thickness of the carbonate rocks under the lavas is reduced to less than half (Dumont 1971), and finally, at Bukama, further to the southeast, directly overlie detrital sediments belonging to the BI Formation (Dumont 1971; Raucq et al. 1977). In addition, an age of 926 ± 30 Ma (Cahen et al. 1984) given by amygdaloidal pillow–lavas from the Kasai – Sodi Formation of the Angolan border (Legrand and Raucq 1957; Delhal 1958) is surprisingly close to the 950 ± 20 Ma age for the Sankuru–Bushimay lavas (Cahen 1982). These observations are important. They suggest the occurrence of a synchronous volcanic event, throughout the Bushimay Supergroup basin. Outcrops of these lavas are distributed over a surface area not less than 150000 km.² One then poses the question: which important geological event could cause a volcanic episode of such magnitude.

Acknowledgements. I would like to thank Mrs. J. Vandandroye M.J. Delhal, P. Dumont, Cl. Monty, P. Raucq and J. Thorez for their advice and care in proof reading, and the editors and Dr Isles Strachan for improving the english presentation of the text.

References

Bertrand-Sarfati, J. 1972a. Les stromatolites columnaires du Précambrien supérieur du Sahara nord-occidental: inventaire, morphologie et microstructure des laminations; corrélations stratigraphiques. *CNRS, Paris, CRZA, sér. Geol.*, **14**, 245 pp.
—— **1972b**. Stromatolites columnaires de certaines formations carbonatées du Précambrien supérieur du Bassin congolais (Bushimay, Lindien, Ouest-Congolien). *Ann.Mus.roy.Afr.centr.*, Tervuren, Belg., sér. 8, Sc.géol., **74**.
Boulouard, Ch. and Calenda, F. 1963. Rapport de la S.N.P.A. sur la palynologie de quelques sondages du Congo belge. (unpublished). Report on the presence of *Leiosphaerids* with some microphotographs but without discussion or interpretation.
Cahen, L. 1954. Résultats géochronologiques sur des minéraux du Congo jusqu'en mai 54. *Ann.Soc.géol. Belg.*, **77**, 269–281
—— **1963**. Grands traits de l'agencement des éléments du soubassement de l'Afrique centrale. Esquisse tectonique au 1/500.000. *Ann.Soc.géol.Belg.*, **85**, 183 – 195.
—— **1973a**. L'uraninite de 620 m.a. post-date tout le Katangien (Mise au point). *Mus.roy.Afr.centr.*, Tervuren, Belg., Rapp.ann.Dépt.Géol.Min., **1972**, 35–38.
—— **1973b**. Corrélations de certaines séries du Précambrian supérieur du Zaïre à al lumière de l'étude des stromatolites et des données de géochronologie radiométrique. *Mus.roy.Afr.centr.*, Tervuren, Belg.Rapp.ann.Dépt.Géol.Min., **1972**, 38–51.
—— **1974**. Geological background to the copper-bearing strata of southern Shaba, Zaire. *Gisements stratiformes et provinces cuprifères. Centr.Soc.géol.Belg.*,Liège, 57–77
—— **1982**. Geochronological correlation of the Late Precambrian sequences on and around the stable zones of Equatorial Africa. *Precambrian Research.*, **18**, 73–86.
—— **and Lepersonne, J. 1967**. The Precambrian of the Congo, Rwanda and Burundi. *In:* Rankama, K. (Ed.), *The Precambrian, volume 3*, Interscience Publ., New York. 143 – 290.

——, Ledent, D. and Snelling, N.J. 1974. Données géochronologiques dans le Katangien inférieur du Kasai oriental et du Shaba nord-oriental (République du Zaïre). *Mus.royAfr.,centr., Tervuren, Belg., Rapp.ann. Dépt.Géol.Min.*, **1974**, 59–70.
—— and Mortelmans, G. 1948. Le système de la Bushimaie au Katanga. *Bull.Soc.belge géol.*, **56**, 217–253
—— and Snelling, N. J. 1966. *The Geochronology of Equatorial Africa.* North Holland Publ.C., Amsterdam. 196 pp.
Cahen, L., Snelling, N.J. Delhal, J. and Vail, J.R. 1984. *Geochronology and Evolution of Africa.* Clarendon Press, Oxford. 512 pp.
Clauer, N. 1973. Utilisation de la méthode rubidium–strontium pour la datation de niveaux sédimentaires du Précambrien supérieur de l'Adrar mauritanien (Sahara occidental) et la mise en évidence de transformations précoces de minéraux argileux. *Sci.géol., Strasbourg*, **45**, 256 pp.
—— 1976. Géochimie isotopique du strontium des milieux sédimentaires. Application à la géochronologie de la couverture du craton ouest-africain. *Mém.Sci.Géol. Univ. Louis Pasteur, Strasbourg*, **45**.
—— and Bonhomme, M. 1971. Preliminary Rb/Sr dating in the Upper Precambrian near Atar (Mauritania) (Abstract). *Ann.Soc.geol.Belg.*, **94**, 109.
Delhal, J. 1958. Sur le volcanisme ancien dans le sud-Kasaï (Congo belge). *Bull.Soc.belge. Géol.Paléont.Hydrol.*, **67**, 179–187.
Dumont, P. 1971. Révision générale du Katangien. Le plateau des Biano. Les phases précoces de l'orogenèse katangienne. *Unpublished Thèse doctorat, Université Libre de Bruxelles.*
—— and Cahen, L. 1978. Les complexes conglomératiques de la bordure sud-orientale de la chaîne kibarienne et leurs relations avec les couches katangiennes de l'arc Lufilien. *Mus.roy.Afr.centr., Tervuren Belg., Rapp.Ann. Dépt. Géol.Min.*, **1977**, 111 – 135.
Garris, M.A., Kazakoff, G.A., Keller, B.M., Polevaye, N.I. and Semikhatov, M.A. 1964. Geochronological scale of Upper Proterozoic, Riphean and Vendian. *XXII Int.geol.Congress, New Delhi*, **X**, 568–584
Holmes, A. and Cahen, L. 1955. African geochronology. Results available to 1st September 1954. *Colon.Geol.Miner.Resour., London*, **5**, 3–39.
Keller, M. 1964. The Riphean Group. *XXII Int.Geol.Congress, New Delhi*, X, 323–338.
Legrand, R. and Raucq, P. 1957. La faille de la Malafudi et son cadre géologique (Kasai). *Bull.Soc.belge.Géol.*, **66**, 109 – 133.
Lepersonne, J. 1973. Structure plissée du Bushimay dans le degré carré Mani (S7/25) (Zaïre) et les régions voisines. *Mus.roy.Afr.centr., Tervuren,Belg., Rapp.ann. Dépt.Géol.Min.*, **1972**, 20–29.
Lindgren, S. 1982. Algal Coenobia and Leiospheres from the Upper Riphean of the Turukhansk region, eastern Siberia. *Stock.Contr.Geol.*, **38**, 35–45.
Maithy, P.K. 1975. Micro-organisms from the Bushimay system (Late Precambrian) of Kanshi, Zaire. *The paleobotanist*, **22**, 133–149.
Polinard, E. 1924 – 1925. Constititution géologique des régions de la Bushimaie et de la Lubi. *Ann.Soc.géol.Belg., Publ.rel.au Congo belge, XXXIV*, 42–132.
—— 1949. Constitution géologique du bassin de la Bushimaie entre le Mvi et la Movo (Congo belge). *Inst.roy.col.belge.* **7**, 50 pp.
Raucq, P. 1957. Contribution à la connaissance du système de la Bushimay (Congo belge). *Ann.Mus.Congo belge,Tervuren, Belg., 8°, Sci. géol.*, **18**
—— 1969. Etat des connaissances sur la Bushimay dans le bassin du Sankuru (Rép.Dém. du Congo). *Ann.Soc.géol.Belg.*, **92**, 293 – 306
—— 1970. Nouvelles acquisitions sur le système de la Bushimay (Rép. Dém. du Congo). *Ann.Mus.roy.Afr.centr., Tervuren, Belg., 8°, Sci. géol.*, **69**.
——, Ladmirant, H. and Delhal, J. 1977. République du Zaïre. Carte géologique au 1.200.000. Feuille Mbufi-May. Degré carré S7/27. *Serv.Géol.Zaïre* (1977). Notice (1979) by L. Cahen *et al.* 1984, 50 pp.
Salubja *et al.* 1971.
Schopf, J.W. 1968. Microflora of the Bitter Springs Formation, Late Precambrian of Australia. *Journal of Paleontology*, **42**, 651–689.
—— and Blacic, J.M. 1971 New micro-organisms from the Bitter Springs Formation (Late Precambrian) of the north-central Amadeus Basin, Australia. *Journal of Paleontology.*, **45**, 925–960.
Timofeev, B.B. 1959. The ancient flora of the Baltic regions and its stratigraphic significance (in Russian). *V.N.I.G.R.I., Leningrad, Mem.*, **129**, 350 pp. (French translation, 1960).
—— 1966. Micropaleontological research into ancient strata (in Russian). *Inst.Nauka Moskou* 147 pp., Leningrad 240 pp. (English translation 1974).
—— 1969. Proterozoic Sphaeromorphida (in Russian). *Akad. Nauk. SSSR, Inst.geol., Leningrad*, 146 pp.
Vidal, G. 1976. Late Precambrian microfossils from the Visingsö Beds in southern Sweden. *Fossils and Strata, Oslo*, **9**, 57 pp.
—— 1981. Micropalaeontology and biostratigraphy of the Upper Proterozoic and Lower Cambrian sequence in Eastern Finmarks, Northern Norway. *Bull.Norges Geol.Undersøk.*, **362**, 1–53
—— and Siedlecka, A. 1983. Planktonic, acid-resistant microfossils from the Upper Proterozoic strata of the Barents Sea region of Varanger Peninsula, East Finmark, Northern Norway. *Bull, Norges Geol.Undersøk.*, **382**, 45–79
Wasilewski, I. 1954. Exploration en profondeur des formations du Système de la Bushimaie (Bakwanga, Kasai). *Mém.Inst.Geol.Univ.Louv.*, **19**, 145–172.

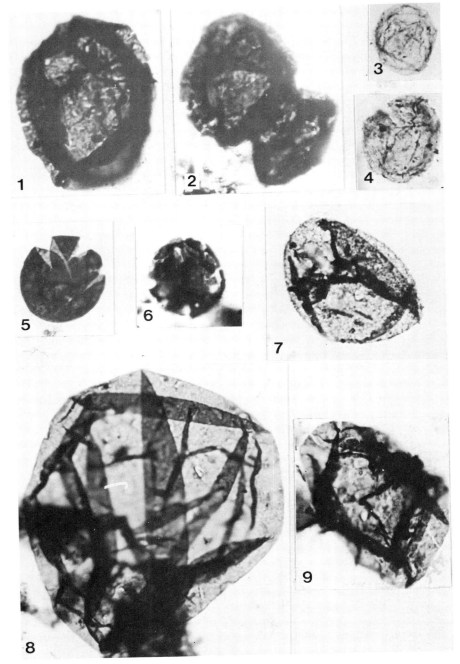

PLATE I
Magnification: 1000 ×

1,2 species 2 from Kanshi 113 m and kafuku (*Kildinosphaera chagrinata* Vidal and Siedlecka 1983, formerly *Kildinella sinica* Timofeev 1966).

3,4 species 5 from Kanshi 121 m and Kafuku.

5,6 species 1 from Kanshi 121 m and Kafuku.

7   species 3 from Kanshi 158 m.

8,9 species 7 from Kanshi 158m and Kafuku. (*Leiosphaeridia asperata* (Naumova) nov. comb. Lindgren 1982, *Kildinella hyperboreica* after Timofeev 1966).

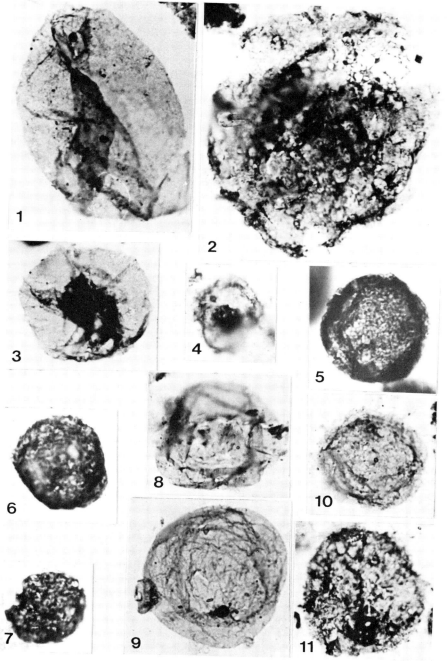

Plate II
Magnification: 1000 ×
1,2 species 6 from Kanshi 121 m and Kafuku.

3,4 species 26 from Kanshi 121 m and Kafuku.

5   species 14 from Kanshi 158 m.
6,7 species 15 from Kanshi 121 m and Kafuku. (*Vavosphaeridium bharadwajii.*)

8,9 species 17 from Kafuku and Kanshi 113 m. (*Kildinella timanica* Timofeev 1969).

10,11 species 18 from Kanshi 113 m and Kafuku.

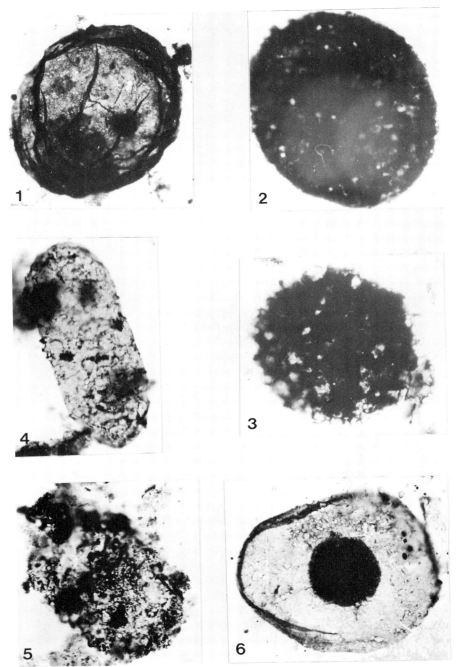

Plate III
1 species 11 from Kanshi 113 m — Magnification 250 ×. (*Chuaria circularis* Walcott 1899, *Kildinella jakutica* after Timofeev 1966).

2.3 species 10 from Kanshi 121 m and Kafuku Magnification 500 ×.

4.5 species 25 from Kanshi 121 m and Kafuku Magnification 500 ×.

6 species 21 from Kanshi 121 m — Magnification 250 ×. (*Nucellosphaeridium magna* Maithy 1975 or *N. deminatum* Timofeev 1969).

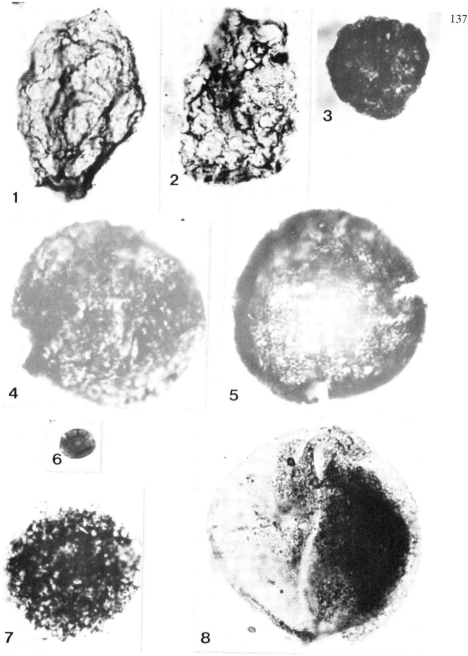

Plate IV
Magnification 1000 ×

1,2 species 9 from Kanshi 121 m and Kafuku.

3   species 19 from Kanshi 158 m.

4,5 species 24 from Kafuku and Kanshi 121 m. (*Trachysphaeridium laminaritum* Timofeev 1966).

6   species 30 from Kanshi 113 m.

7   species 31 from Kanshi 113 m. (*Protosphaeridium* cf. *flexuosum* in Vidal 1976).

8   species 34 from Kanshi 113 m.

Precambrian gabbro–anorthosite complexes, Tete Province, Mozambique

M. W. C. Barr and M. A. Brown

Hunting Geology and Geophysics Ltd., Elstree Way, Borehamwood, WD6 1SB, U.K.

The Tete Province of Mozambique is characterized by three kinds of gabbro–anorthosite intrusion. In the northeast anorthosite forms bands or layers infolded with the country rock gneisses and granulites with which it has been intensely deformed and recrystallized. In the northwest gabbro–anorthosite complexes form distinct though irregular intrusions in the country rock granulites but here have largely resisted later deformation. In the south, the Tete gabbro–anorthosite complex is distinguished from the other complexes by its large size and lopolith-like form and by being everywhere floored by carbonate country rocks. In contrast to most gabbro–anorthosite of the province, its western part is not associated with granulite facies rocks.

Aspects of the petrography, chemistry and age relationships of the complexes are discussed based on the origins and the evolution of this part of the crust during the Kibaran and Pan-African orogenies.

KEY WORDS Anorthosite Leucocratic gabbro Norite Tete Province Mozambique

1. Introduction

The Tete Province of Mozambique occupies an important area for an understanding of the Precambrian geology of east–central Africa. It covers the zone where the Zambezi Belt of strongly deformed gneisses and metasediments bounding the Zimbabwe Craton on the north impinges on the Mozambique Belt (Figure 1). Within it the predominantly east–west structural trends on the Zambezi Belt, the northeast trends of the Irumide Belt in eastern Zambia and the meridional trends of the Mozambique Belt in southern Malawi come together in a region of complex structure.

The area is little known to English-speaking geologists. The first systematic work was carried out by the Longyear Company (1956) concentrating mostly on the western half of the province but the area northwest of Tete town was moderately well known before that time as a result of the discovery of uranium mineralization in the Mavudzi valley (Davidson and Bennet 1950; Luna and Freitas 1953). Since the fifties much useful work has been carried out in parts of the province but with a few notable exceptions (Real 1966; Vail and Pinto 1966; Coelho 1969) remains unpublished. Besides the Tete Complex, where previous knowledge was

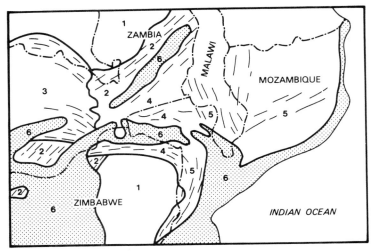

Figure 1. Regional setting of Tete Province. 1, Archaean Cratons; 2, Irumide Belt; 3, Lufilian arc; 4 Zambezi Belt; 5, Mozambique Belt; 6, Phanerozoic cover.

brought together in an excellent review by Svirine (1980), the gabbro–anorthosite complexes are virtually undescribed.

New knowledge of the area has resulted from a comprehensive mineral exploration and reconnaissance geological mapping project carried out by the Mozambiquan Government through the agency of Hunting Geology and Geophysics Limited between 1980 and 1984.

2. Outline of the geology

Relevant aspects of the geology of Tete Province are illustrated in Figure 2. The Precambrian rocks are divided into two contrasting segments by a major zone of cataclasis and shearing the Sanangoe Zone (Hunting 1984; Barr *et al.* in preparation). The northern zone is further subdivided into western and eastern parts by the Desaranhama Granite.

The country north of the Sanangoe Zone and west of the Desaranhama Granite is underlain by belts of gneisses and granulites separated by large tracts of granite. These rocks are overlain unconformably by the Fingoe Group and intruded by later suites of granite and ultramafic rocks. Gabbro–anorthosite complexes are concentrated in the eastern area of gneisses and granulites which have complex structures lacking a regional grain (Luia Group, Figure 2). They also occur as resistors in later granites including the Desaranhama Granite.

That part of the province east of Desaranhama Granite is also composed of gneisses and granulites (Angonia Group, Figure 2) but contrasts with areas further west in having regular north-northwest trends, apparent not only on the geological map but also in the airborne magnetic and spectrometric data. These rocks are continuous with those in the Kirk Plateau and Dedza areas of Malawi (Bloomfield and Garson 1965; Thatcher 1965).

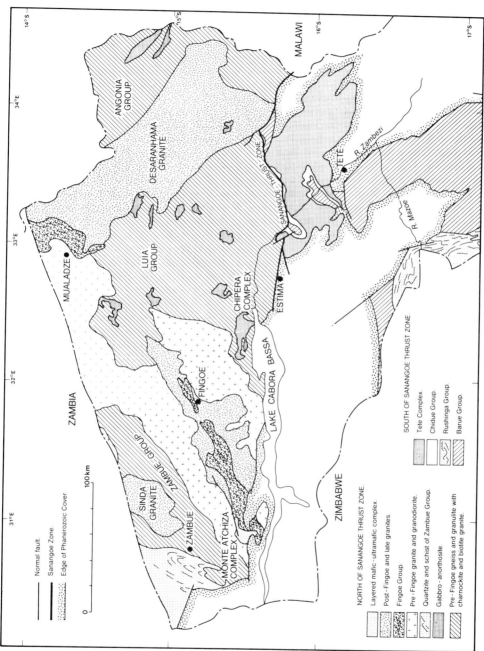

Figure 2. Precambrian geology of Tete Province.

The country rocks south of the Sanangoe Zone are strongly contrasted with those to the north. Granite which is very widespread and plentiful to the north is subordinate to biotite and hornblende gneiss. The gneisses and associated metasediments are involved in periclinal folding which is the main structural feature of southern Tete Province and indeed of the whole vast tract of country extending down and eastern margin of the Zimbabwe Craton (Vail 1964; Hunting 1984). The Tete Complex appears to form part of this gneissic sequence as described below.

3. Gabbro–anorthosite intrusions in the Luia Group

Bodies composed predominantly of leucocratic gabbro are widespread in the Luia Group (Figure 2). They form irregular masses or are interlayered with the enclosing granulites and migmatites. Similar bodies occur as resistors or inclusions within the biotite granites and charnockites which enclose the Luia Group and less frequently in the Zambue Group to the West.

The dominant rock-type in most of the complexes is a medium to coarse-grained leucocratic gabbro or norite composed of anhedral or lath-shaped crystals of grey plagioclase and subordinate equant grains of pyroxene and opaque minerals. Dark minerals locally make up a larger proportion of the rock and olivine gabbro has been reported at a few localities. Elsewhere, especially in the Chipera Complex, dark minerals are present in accessory proportions only and the rock passes into an anorthosite. The plagioclase is typically sodic labradorite or andesine. Both augite and hypersthene are present in most gabbros. Layering is typically absent on the scale of the outcrop, although some of the lithological differentiation within the complexes may be due to segregation of early formed crystals. Segregations rich in magnetite and ilmenite occur but these are on a small scale compared to the Tete Complex. A marginal facies of monzonite composition (mangerite) is typical of the Chipera Complex and may be present in some other bodies but is generally extremely difficult to distinguish from the darker rock-types of the envelope rocks.

The basic rocks in the cores of the complexes are typically unfoliated and unbanded and lack clear field evidence of metamorphic recrystallization. In thin section, however, replacement of original pyroxenes by hornblende and biotite is widespread. Towards the margins of the complexes, the gabbro–anorthosite grades into mafic granulite of the Luia Group by a reduction in grain size and a change towards a more clearly granoblastic texture, or through coronites into amphibolite which is hard to distinguish from other parts of the enclosing gneisses.

In the Chipera Complex and in the River Capoche in the northwestern part of the Luia Group outcrop, large parts of the bodies exhibit a parallel orientation of euhedral plagioclase laths taken to be an igneous texture reflecting flow in the partly consolidated state.

3a. Field relationships

The contacts of the complexes are seldom well exposed and even less often provide information about their age relationships. The contact relationships of the Chipera Complex, one of the more intensively studied intrusions is believed to be typical in many ways. The contact is conformable with the banding of the country rock for much of its length and in places leucocratic gabbro is interlayered with the country rock gneiss or granulite. The large-scale disposition of the banding in the country rocks strongly suggest that it has been deformed after the intrusion of the basic rocks yet the gabbro shows very little textural or mineralogical modification at its margin. The rocks immediately in contact with the gabbro on its northern margin are in some places biotite gneiss, in others biotite granite or charnockite. The last named contains inclusions of basic igneous rocks which can be matched with lithologies within the complex. Towards the west of the complex hypersthene granite cuts across the trace of the layering. These data are interpreted in terms of emplacement of the basic intrusion early in the history of the Luia Group country rocks followed by deformation and high grade (granulite facies) metamorphism with charnockite intrusion. The retention of igneous mineral assemblages and texture within the complex is attributed in part to the strength of the basic igneous rocks which resulted in the deflection of the strain during deformation around the margins of the intrusion and in part to the lack of volatiles and high grade of metamorphism during recrystallization which has blurred the distinction between igneous and metamorphic assemblages. Leucocratic gabbro in the Luatize river southwest of Mualadze grades outwards into two-pyroxene granulite with the granoblastic texture more typical of granulite facies rocks and corroborates this interpretation.

The relationships described above are believed to be typical of those complexes intruded into high-grade rocks. However, parts of the Luia Group are of lower grade (amphibolite facies) and here the basic intrusives shows much more evidence of marginal deformation and downgrading to amphibolite although in many cases a core of gabbro, norite, or anorthosite is retained.

4. Plagioclase gneiss of Angonia District

Concordant bodies of foliated leucocratic gneiss called 'granite gneiss' or 'Angonia Granite' in the older Mozambiquan literature occur in the Angonia Group. They are characterized by low natural radioactivity and few magnetic anomalies on the airborne geophysical maps (Hunting 1984), and were first shown as anorthosites on a geological map by geologists of Brodoimpeks-Geozavod (1983). There is now a consensus based on mineralogy amongst geologists who have worked in the area that these rocks do indeed represent anorthosites (de Decker, Lacheldt personal communication 1984). The main bodies, selected mostly on the basis of low natural radioactivity on the airborne spectrometric maps, are shown in Figure 2.

Adequate descriptions of these rocks from Mozambique are extremely sparse; the area has not been visited by the authors. They are medium-grained, even-grained foliated rocks composed of andesine and very subordinate dark minerals. The northern body is extensively kaolinized and contains important occurrences of graphite. A narrow zone extending almost the entire length of the body contains anastomosing layers of ilmenite (de Decker 1983) which invites comparison with the iron–titanium oxide segregation of other anorthosite intrusions described in the literature. From available descriptions, the bodies have conformable contacts with the country rock gneisses and granulites and are deformed and recrystallized throughout.

Reference to descriptions of the geology of adjacent parts of Malawi show that these anorthosites form a link between the Linthipe anorthosite complex of the Dedza area to the north (Thatcher 1968) and the layers of anorthosite gneiss occurring in the Kirk Range (Bloomfield and Garson 1965) and Nsanje area (Andreoli 1984) to the south (Figure 3). Plagioclase gneiss mapped by Walshaw

(1965) immediately to the east of the Mozambique occurrences is also now believed to be an anorthosite intrusion (O'Connor personal communication 1984). Like the Mozambiquan occurrences, these bodies are extensively kaolinized at the surface and in places contain stringers and veinlets of graphite which have been prospected in the past.

Reference to Figure 3 shows that the anorthosite occurrences in Angonia District share with the Malawi occurrences a spatial association with intrusive syenites (the perthite syenite complexes of Bloomfield 1968), as well as granulite facies country rocks. All the bodies appear to have been folded and thoroughly recrystallized with them during later deformation and metamorphism.

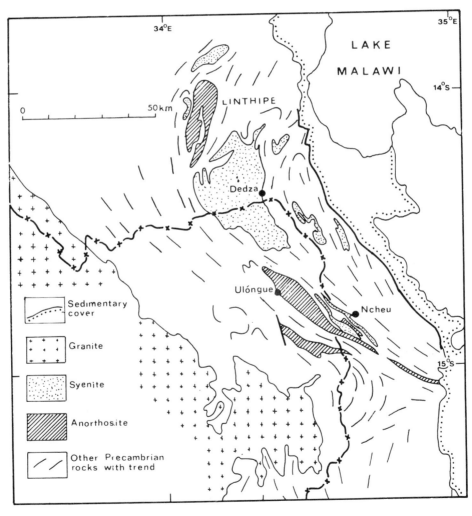

Figure 3. Anorthosite complexes of Angonia District and southern Malawi.

5. Tete Complex

The Tete Complex extends from near Estima in the west, across the Zambezi river north of Tete town to the Malawi border in the east (Figure 4). Its margins are largely overlain unconformably by Karroo rocks. However aeromagnetic data indicate that it does not extend far beyond its present outcrop beneath these rocks.

The complex is composed predominantly of light-coloured gabbro and norite with subordinate, anorthosite as layers or later intrusions. Ultramafic rocks, mostly pyroxenite, are a widespread but usually minor rock type as are rocks composed predominantly of iron–titanium oxides. The complex is cut by very numerous basic dykes which are believed to be related to it genetically. Gabbro bodies forming mountains at the north margin of the Complex, and at some other localities, appear to lack later dykes and are interpreted as late stage intrusions. Towards the east of the Complex two-pyroxene granulites with layers of biotite gneiss occur.

Textures are medium to very coarse grained or pegmatitic, except in the dykes which are predominantly fine grained. Widespread replacement of the original minerals and the imposition of orientated fabrics occur in various places throughout the Complex but are most common near its margins. Zones of cataclasis and shearing are typical of the central parts of the Complex.

5a. Gabbro and norite

Fresh gabbro and norite of the Tete Complex is a light, homogeneous unfoliated granular rock varying in grain size from medium to very coarse grained. The main minerals are plagioclase, pyroxene, and iron–titanium oxides. The plagioclase is sodic labradorite. The pyroxene is either augite or hypersthene. In many gabbros and norites both pyroxenes occur, often intergrown. In a few cases they are accompanied by olivine (Svirine 1980). There appears to be no firm evidence to support the contention of Davidson and Bennet (1950) that these basic rocks are dominantly noritic. Augite so far as is known is about as plentiful and widespread as hypersthene (Coelho 1969; Svirine 1980). The textures of the rock are allotriomorphic; both plagioclase and pyroxene form anhedral interlocking grains.

In many places the original minerals have been partly or completely replaced. The pyroxenes are replaced marginally and along cleavage traces by aggregates of hornblende in places with minor biotite. The plagioclase is replaced by scapolite or less commonly by epidote. The abundance of hornblende-bearing rocks led early workers to refer to the Complex as the Tete diorite–gabbro Complex. At nearly all outcrops of otherwise entirely fresh-looking gabbro, the rock is crossed by a network of hairline cracks in the walls of which pyroxene and plagioclase are replaced. These cracks stand out sharply because of their darker colour.

There is a general but by no means exclusive tendency for alteration to be more noticeable in foliated gabbro and norite, suggesting that the replacement of the original minerals is related to later deformation and retrogressive metamorphism. This is certainly the case in strongly foliated parts of the Complex where it is converted to hornblende gneiss and amphibolite. Elsewhere replacement has occurred without the formation of an orientated fabric and could represent pneumatolytic alteration as easily as later retrogressive metamorphic. Indeed in some cases the main dark mineral is hornblende as coarse single crystals and there is no textural evidence that the rocks ever contained pyroxene.

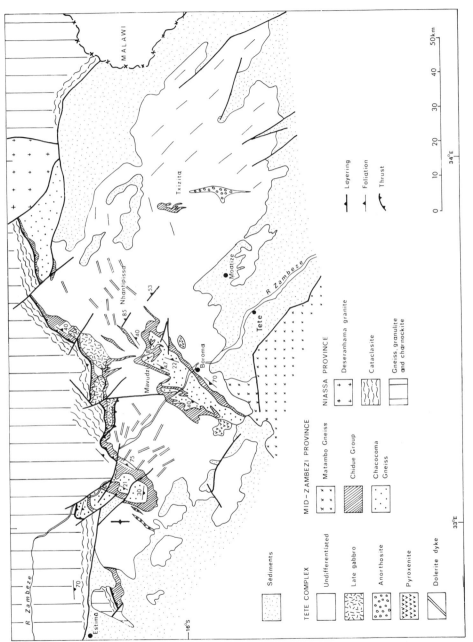

Figure 4. Geological map of the Tete Complex.

5b. Late gabbro

Gabbro believed to be late in the sequence of intrusion forms a range of hills along the north margin of the Complex (Figure 4). It is much less strongly sheared than the main part of the Complex and appears to lack basic dyke intrusions. It cannot be differentiated petrographically from other parts of the Complex.

5c. Anorthosite

Grey or white anorthosite forms a subsidiary but substantial proportion of the Tete Complex. It is composed of andesine or sodic labradorite with minor pyroxene and opaque minerals. Alkali feldspar, biotite, sphene, sulphides, and garnet occur as accessory minerals. The textures are granular with a tendency towards a bimodal size distribution, larger grains of plagioclase being enclosed in a mosaic of smaller grains. As with the gabbros, the plagioclase of the anorthosites is in places replaced by scapolite and the pyroxene by hornblende.

Anorthosite forms either layers within the undifferentiated gabbro–anorthosite sequence or discrete bodies with sharp but unchilled margins against adjacent rock-types, interpreted as later segregations or intrusions. The latter tend to be coarse grained and form piles of blocks and boulders on the Tete Complex plateau particularly noticeable in north–central parts of the Complex. The former rocks grade through leucogabbro and norite into the more normal basic rocks described above, by an increase in the proportion of dark silicates.

5d. Iron–titanium oxide enriched rocks

The Tete Complex is rich in opaque minerals and these locally become the main constituent of the rock. In some places these oxide rocks can be clearly seen to form layers parallel to the general layering of the Complex. Elsewhere they form irregular bodies and have been variously interpreted as dykes or segregations. They are black, very dense rocks, locally very coarse grained, and are composed predominantly of magnetite and ilmenite. They contain up to 0·66 per cent of vanadium oxide.

5e. Pyroxenite

Ultramafic rocks are poorly represented in the Tete Complex but do occur at widely scattered localities. They form bands within the gabbro–anorthosite sequence and one or two more substantial bodies (Figure 4).

They are green granular medium grained rocks when fresh with a soft orange-weathered outer skin. Pyroxenites composed of granular augite and hypersthene with minor opaque minerals predominate. In some areas the pyroxenes are enclosed by plagioclase, partly replaced by scapolite, in a cumulate texture, the only unequivocal identification of such textures in any rock type of the Complex made in the present survey. Olivine is rather common in the pyroxenites in which case they grade into peridotites. Replacement of plagioclase by scapolite and the dark silicates by mixtures of hornblende, serpentine, and talc is commonplace.

5f. Basic dykes

The Tete Complex is cut by numerous basic dykes which do not extend beyond its outcrop and are therefore believed to be genetically related to it. At some localities 20 per cent of the outcrop is of dykes and four separate intersecting sets have been recognized although the last of these may be of Karoo age.

The dykes are predominantly of grey, brown-weathered dolerite composed of plagioclase, partly altered to epidote, and augite, usually extensively altered to hornblende and chlorite. Phenocrysts of plagioclase and less commonly pyroxene can be recognized in outcrop at many localities, and are clearly distinguishable in thin section. A very characteristic and widespread dyke lithology consists of blue metadolerite or diorite with a glomeroporphyritic texture of aggregates of white altered euhedral plagioclase phenocrysts set in a fine-grained groundmass. The descriptions of Svirine (1980) show that this has in the past often been mistaken for a country rock-type.

The dykes are up to 10 m thick and form swarms parallel, or in many case oblique or at right angles to the banding. In the latter cases their margins are intensely sheared and the dykes are in places disrupted. Where foliation and recrystallization become intense it is difficult to distinguish dyke rock from enclosing gabbro.

5g. Pyroxene granulite

The southeast part of the Complex is composed of partly foliated medium-grained granoblastic pyroxene granulite consisting of plagioclase, alkali feldspar, ortho- and clinopyroxene with secondary or accessory hornblende, scapolite, biotite, apitite, sphene, garnet, and iron oxides (Afonso and Araujo 1970). The rocks are dark in colour and have a broadly dioritic composition (Afonso and Araujo 1970). They contain conformable bands or belts of granitic granulites composed of alkali feldspar, oliogoclase, and biotite and are intruded by anorthosite and norite.

5h. Syenites and Granites

Syenites and microgranites form dykes, lenses, and small stocks distributed in certain zones of the Tete Complex, usually near its margins or where it contains inclusions of the country rocks. They are composed of alkali feldspar and plagioclase, with or without quartz and minor dark silicates, typically either hornblende, augite, or aegirine–augite (Freitas 1953). The granitic rocks are fine grained (aplite) but the syenites range from fine grained to pegmatitic. Pegmatites carrying corundrum occur at a number of localities, and have been worked east of Moatize.

5i. Chemistry

No new major element were carried out on samples of the Tete Complex during this survey. Earlier analyses quoted by Coelho (1969) and discussed by Svirine (1980) are reproduced in Table 1.

They are compared with the trends typical of other gabbro–anorthosite complexes and with Archaean layered basic irruptives in an alkali–silica diagram (Figure 5). The number of Tete Complex analyses is regrettably inadequate to be representative; in particular, pyroxenites are greatly overrepresented and the granites, which have quite as much right as the syenites to be considered along with the basic rocks, are unrepresented. Nevertheless Figure 5 shows that in the Tete Complex, the proportion of alkalis increases rapidly with silica content so that the analyses fall within or close to the field typical of other gabbro–anorthosite complexes and well outside the field of layered basic intrusions such as the Great Dyke or the Bushveld Complex.

Figure 6 shows the same analyses plotted on an AFM diagram. Here the lack of analyses of intermediate composition prevents a clear picture of the chemical affinities of the complex to be displayed. The pyroxenites fall outside the typical calc-alkaline field and compare with early phases of the Bushveld Complex. This is almost certainly because they contain cumulate dark silicates enriched in the

Table 1. Chemical analyses and CIPW norms of Tete Complex samples (from Coelho 1969)

	1	2	3	4	5	6	7	8	9	10	11	12	13	14	15	16	17	18
SiO_2	54·62	52·43	52·58	47·33	48·80	50·50	51·39	52·01	46·52	47·07	45·52	44·08	45·56	58·56	63·69	61·46	48·79	45·27
TiO_2	0·23	0·29	0·26	2·01	1·02	1·05	0·36	0·43	1·22	2·07	0·56	0·72	1·11	0·10	—	0·11	0·44	2·33
Al_2O_3	26·44	26·65	27·40	14·97	17·74	18·37	17·83	22·72	15·04	12·32	10·09	11·15	4·26	20·77	20·90	20·66	17·47	16·58
Cr_2O_3	—	—	—	—	—	—	—	—	—	—	—	—	0·29	—	—	—	—	—
Fe_2O_3	0·92	1·91	0·72	4·34	2·43	1·61	1·39	1·03	2·19	2·15	2·79	3·80	2·95	3·32	0·49	0·60	3·43	3·53
FeO	0·39	0·63	0·64	7·72	5·04	5·59	5·99	3·64	9·97	11·88	10·29	8·26	7·73	1·34	0·55	0·41	4·72	9·22
MnO	—	0·01	0·04	0·14	0·21	0·10	0·13	—	0·24	0·23	0·19	0·23	0·24	0·03	—	—	0·09	0·21
MgO	0·20	0·15	—	7·31	6·56	7·24	8·31	5·40	7·89	10·51	18·56	22·33	23·32	—	0·39	—	3·22	7·11
CaO	9·23	9·64	9·77	9·38	15·42	10·84	9·98	10·80	13·79	12·00	8·47	5·57	10·71	2·71	1·72	1·51	13·04	10·40
Na_2O	4·57	4·22	5·56	4·23	1·97	3·27	2·92	2·42	2·16	1·52	1·30	1·48	0·96	5·67	9·47	5·19	3·45	3·38
K_2O	2·47	1·97	1·81	1·31	0·45	0·58	0·34	0·81	0·27	—	0·50	—	0·37	6·49	1·56	8·85	3·42	0·71
P_2O_5	0·07	—	0·05	0·19	—	0·03	—	0·02	0·24	—	—	0·04	0·16	0·12	0·10	0·07	0·39	0·17
Cl	0·22	0·10	0·29	0·14	0·32	0·30	0·10	0·20	—	—	—	—	—	0·10	0·12	0·32	0·87	0·70
$H_2O\pm$	0·59	2·49	0·92	0·80	0·61	0·64	0·74	0·58	0·61	0·50	1·69	2·04	1·71	0·72	0·60	0·84	0·66	0·52
CO_2	—	—	—	—	—	—	—	—	—	—	—	—	0·63	—	—	—	—	—
	99·95	99·49	100·04	99·87	100·57	100·12	99·48	100·06	99·90	100·27	99·96	99·75	99·90	99·84	99·86	100·02	99·99	100·13
qz	0·97	1·57	—	—	0·99	—	—	4·07	—	—	—	—	—	—	0·65	0·88	—	—
c	0·23	1·10	—	—	—	—	—	—	—	—	—	—	—	—	—	—	—	—
or	14·62	11·68	10·67	7·73	2·67	3·45	2·00	4·78	1·55	—	2·94	—	2·17	38·26	9·23	52·26	20·24	4·17
ab	35·37	34·16	33·43	23·37	11·95	23·21	23·16	17·50	16·40	12·84	10·95	12·47	8·12	38·20	76·45	34·06	11·05	18·23
an	45·35	42·92	46·62	19·04	40·70	36·00	35·30	50·18	30·32	26·77	20·24	23·77	6·20	12·81	7·89	7·00	28·82	33·41
ne	—	—	5·08	5·59	—	—	—	—	0·90	—	—	—	—	4·46	2·18	2·75	2·84	—
di	0·50	2·45	0·86	21·06	28·72	14·03	10·67	2·73	28·45	26·70	17·45	2·85	33·17	0·11	1·58	0·08	27·25	14·01
hy	—	—	—	—	8·57	14·46	22·61	17·29	—	15·98	9·72	16·60	8·98	—	—	—	—	8·47
ol	—	—	—	11·31	—	3·04	1·80	—	15·03	10·39	31·85	34·92	30·92	—	—	—	—	9·46
mt	0·48	1·18	1·04	6·26	3·53	2·32	2·02	1·48	3·18	3·11	4·03	5·50	4·27	4·13	0·69	0·85	4·96	5·10
ilm	0·44	0·55	0·49	3·80	1·93	2·05	0·68	0·82	2·31	3·95	1·06	1·37	2·09	0·18	—	0·21	0·83	4·41
hm	0·51	1·11	—	—	—	—	—	—	—	—	—	—	—	0·46	—	—	—	—
ap	0·17	—	0·13	0·43	—	0·07	—	—	—	—	—	—	0·37	0·14	0·24	0·16	0·91	0·40
sn	0·59	0·27	0·78	0·39	0·88	0·82	0·27	0·53	1·11	0·50	—	0·10	—	0·27	0·33	0·88	2·39	1·92
cm	—	—	—	—	—	—	—	—	—	—	—	—	0·42	—	—	—	—	—
cc	—	—	—	—	—	—	—	—	—	—	—	—	1·43	—	—	—	—	—

1–3, anorthosite; 4–5, gabbro; 6–8, norite; 9–13, pyroxenite; 14–16, syenite; 17–18, metadolerite.

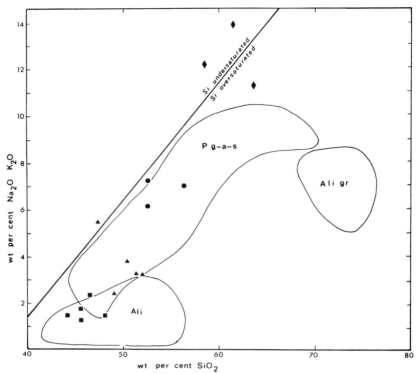

Figure 5. Alkali–silica diagram for Tete Complex samples. Diamonds syenite; circles, anorthosite; triangles, gabbro and norite; squares, pyroxenite. Pg–a–s, field of gabbro–anorthosite-syenite suites from the Elsonian of Labrador (Emslie, 1978) and the Grenvillian of the Adirondacks (Buddington 1939; de Waard 1970), 56 analyses. Ali, field of Archaean layered intrusions from the Bushveld (Daly 1929, du Toit 1926), Fiskenaesset (Windley et al. 1973), Muskox (Irvine 1970), Stillwater (Hess 1960), Messina (Barton et al. 1979) and Sittampundi (Janardhanan and Leake 1975) Complexes; 47 analyses. Ali gr; field of granitic rocks associated with the Bushveld Complex (Daly 1929, du Toit 1954); 22 analyses.

magnesian end-members. No analyses show the exceptional iron-enrichment characteristic of the dry magmas of layered complexes and the more aluminous rocks, syenites, and anorthosites, fall at the aluminous end of the calc-alkaline field.

Figure 7 illustrates the rare earth element (REE) content of a suite of samples from the Tete Complex collected for radiometric dating using the samarium–neodymium method. The gabbroic samples (5157B and C) show REE enrichment compared to chondritic material which is similar to that of gabbroic rocks of mantle origin. The anorthosite (Sample 5157A) has in general lower REE contents and a positive europium anomaly. Because europium is strongly fractionated in favour of plagioclase where that mineral is growing in a silicate melt, this observation is taken as evidence of the occurrence of cumulate plagioclase in the sample — indeed it has a bimodal texture which suggests the presence of larger cumulate plagioclase grains. By using the distribution coefficients for REE of Arth (1976), it is possible to calculate the REE content of a melt from which Sample 5157A could have formed assuming that it is entirely of cumulate origin ('melt' of Figure 7). These contents are closely similar to those of the other two

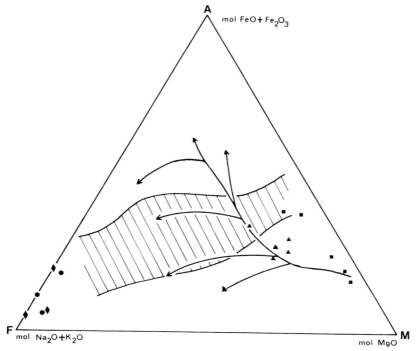

Figure 6. AFM tertiary diagram for Tete Complex samples; hatched, field of Grenville–Elsonian gabbro–anorthosite–syenite suites from Buddington (1939), de Waard (1970) and Emslie (1978), 56 analyses; branching lines, Bushveld Complex compositional trends from Willemse (1969). Symbols for Tete Complex analyses as Figure 5.

samples. The presence of a minor negative europium anomaly suggests that the assumption of a 100 per cent cumulate origin is incorrect and that the rock is in fact composed of a large proportion of cumulate grains enclosed in minor intercumulus material with a 'whole melt' REE concentration pattern.

5j. Internal structure

The Tete Complex is distinguished from many large basic complexes described in the literature by the general absence of original layering at outcrop scale. Nearly all the outcrops are of a single homogeneous rock-type except where refoliation and segregation has led to banding of metamorphic origin.

Layering, believed to be original, has however been observed at a few outcrops. Metre-scale interbanding of hornblende gabbro and anorthosite has been observed north of Tete. Interlayering of anorthosite, pegmatitic gabbro, pyroxenite, and iron–titanium oxide rock occurs at Nhantipissa. The layering is not regular and individual rock types appear to be homogeneous or at least not regularly banded at outcrop. The layering is steeply dipping and trends north, almost at right angles to the main lineament direction on the aerial photographs which is due to swarms of dolerite dykes. In the River Chirodze and the headwaters of the River N'Comadzi west of Boroma, layering is defined by bands 2 to 10 cm thick with greater proportions of oxide minerals, and numerous elliposoidal aggregates 10 cm long by 3 cm wide with a greater proportion of dark silicates than the enclosing gabbro.

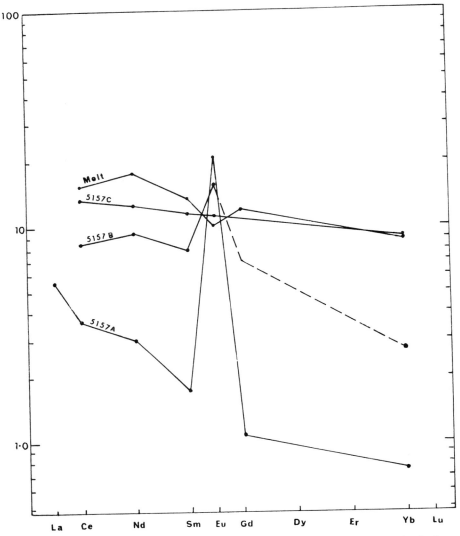

Figure 7. REE content normalized to average chondrite composition for Tete Complex basic rocks.

These are evenly distributed through the rock and give to it a crude foliation. Parallel to this fabric are dark-coloured hairline cracks and thicker zones along which preferential alteration of pyroxene to hornblende has taken place. Adjacent to the contact with Chidue Group rocks at Boroma, centimetre scale interbanding of normal gabbro and pyroxenite was observed. These few examples emphasize:
  (a) the general lack of igneous layering in the Complex.
  (b) that trends visible on the aerial photographs do not in general define the orientation of such layering as exists.

Parts of the Complex, for example in the Mavudzi valley, contain a wide variety of rock-types other than those of basic igneous origin. These include syenites, granites, pegmatites, quartz veins and carbonate and calc-silicate rocks of various types. Where these rock-types occur, the structures and textures of the rocks are extraordinarily complex. They indicate at least two distinct episodes of post-consolidation deformation and retrogressive metamorphism besides any earlier changes associated with the late magmatic stage. The carbonate and calc-silicate rocks have in the past been interpreted as metasomatized carbonate veins (Davidson and Bennet 1950). Some of them undoubtedly have this origin. However some retain a banding which appears to be of sedimentary origin, form extensive zones which are large enough to map at 1:250,000 scale and are associated with psammites, schists and biotite gneiss. It is clear that these zones represent inclusions of the underlying Chidue Group.

Turning now to gross structural features of the Tete Complex, it has been noted by several workers that the Complex has a generally arcuate form, convex northwards. The airborne geophysical survey shows that the Complex does not extend far beneath the Phanerozoic cover so the general form of the Complex is confirmed. Because of strong short wavelength anomalies, the thickness of the body is hard to estimate from the aeromagnetics, but it is of substantial thickness, perhaps between 10 and 20 km. The north margin of the Complex dips steely south at the surface and then shallows to a more gently south-dipping attitude.

The aerial photolineaments in the western part of the Complex are convex northwards broadly parallel to the form of the body — a feature noted by previous workers (Svirine 1980). However, these lineaments do not represent any fundamental structure and are in general not parallel to the layering; moreover there is a discontinuity in their distribution and attitude at the longitude of Txizita. West of this location the aerial photolineaments mainly represent dykes swarms and associated fractures. Access to areas to the east was too restricted to determine the nature of the lineaments there. However, the descriptions of Afonso and Araujo (1970) strongly suggest that they represent the layering and foliation of pyroxene granulites with enclosed belts of granitic gneiss. Variations in the attitude of igneous layering from place to place within the Complex, the feature most fundamental to an understanding of its overall structure, are therefore virtually unknown. The most that can be said is that, where it has been observed during the present survey, the layering is steeply dipping.

In contrast to igneous layering, deformation fabrics are widespread within the Complex: river bed exposures show that irregular intersecting shear zones, along some of which dolerite dykes have been intruded, cut the Complex in great abundance. At some exposures their average spacing is less than one metre. Sheared gabbro is easily weathered and examination of eluvial material gives the erroneous impression that the rocks of the Complex are generally massive and undeformed.

5k. Contact relationship

The Tete Complex is in contact with the Chidue Group along its northern edge and around the margins of gneiss domes which occur within it. It appears to be structurally overlain by the Matambo Gneiss (Figure 4).

The contact with the Chidue Group is marked by very complicated relationships, with numerous faults and thrusts and zones of refoliation and retrogressive metamorphism of the basic rocks extending in places for tens of kilometres into the Complex. The Chidue Group rocks contain coarse-grained calc-hornfelses, magnetite segregations, and small bodies of sulphides which have in the past been called skarns. However, it is very difficult to decide whether the development of these unusual rock types is the result of contact metamorphism related to the intrusion of the Tete Complex basic rocks or arises during later regional metamorphism and intrusion of dykes and veins of syenite, granite, pegmatite and quartz or carbonate minerals. Contacts of the Tete Complex with the Chidue Group have been observed north of Boroma where there is no marginal modification of the basic rocks and no evidence of contact metamorphism or metasomatism in the adjacent marble and striped amphibolite of the Chidue Group. However, several clear examples of an intrusive relationship between the Complex and the Chidue Group have been seen elsewhere. Our interpretation of some of the carbonate rocks within the Complex as inclusions leaves no doubt that the Tete Complex is intrusive into the Chidue Group. The identification of a foliation in some marginal parts of the Complex with the first foliation in the Chidue Group and the development of the banding in some of the adjacent gneisses shows that this intrusion took place early in the structural and metamorphic evolution of the region.

Against this body of evidence of an intrusive relationship with the surrounding rocks must be set the distribution of dolerite dykes. These are very abundant within the Complex and in some places strike into its margin but are almost completely absent in the surrounding metasediments and gneisses. Although they show no chilled margins they are clearly intruded into brittle fractures in the Complex after its consolidation. If this took place *in situ* then the dykes should reasonably be expected to extend into the country rock. The fact that they do not suggests that the Complex was emplaced tectonically on top of the Chidue Group after the intrusion of the basic dykes.

In spite of the attractions of this hypothesis, the evidence that the Tete Complex is intruded into the Chidue Group appears to be unassailable and at two exposures, dolerite dykes extending a short way into the Chidue Group were observed. The general failure of the dolerite dykes to extend into the country rocks is perhaps explained by a difference in ductility during their intrusion. At that time, the temperatures and pressures were such that brittle fracture took place in the Tete Complex allowing the intrusion of dykes but more plastic deformation took place in the surrounding ductile marbles and access by dolerite magma was prevented.

6. Discussion

One of the outstanding problems of the geology of east–central Africa is the relationship between the Irumide fold belt of northeast Zambia (Figure 1) and the Zambezi and Mozambique belts which have traditionally been regarded as Pan-African in age and therefore younger than the Irumide. There is a body of evidence which suggests that the main fabric in eastern Zambia is Irumide in age and this dominant fabric can be traced southwards into the gneisses and granulites of Tete Province north of the Sanangoe Zone (Figure 2). Similarly, the main deformation within the Zambezi Belt of northern Zimbabwe can be traced continuously into Tete Province south of the Sanangoe Zone where it is represented by the dominant gneissic fabric.

Hence the Sanangoe Zone appears to separate regions where the deformation is Kibaran (to the north) from those where it is Pan-African. The attraction of this proposition is that it explains the marked contrasts in structural style, lithology, and metamorphism which exist across the Sanangoe Zone as well as the difference in age. Clearly, if the two regions have entirely different early histories and have

Table 2. Features common to Tete Province gabbro–anorthosite complexes

Crustal setting:	Granulite–upper amphibolite facies metamorphism; spatially associated with syenites
Contact relations:	Conformable
Internal structure:	Layering poorly developed
Mineralogy:	Rich in plagioclase and iron-titanium oxides; poor in pyroxenes and olivine
Mineral chemistry:	Plagioclase is andesine–labradorite; opaque oxides rich in Ti, V, poor in Cr, Co

Table 3. Significant differences between Tete Province gabbro–anorthosite complexes

Tete Complex	In Luia Group	In Angonia Group
Large 6000 km²	Small Typically 100 km²	
Associated with amphibolite facies gneisses and metasediments	Associated with granulites and charnockites	Associated with granulites and gneisses
Igneous mineralogy and texture retained in core		Deformed and recrystallized throughout
Leucogabbro predominant		Anorthosite predominant
More plentiful FeTi oxide segregations and pyroxenite	Insignificant FeTi oxide segregations and pyroxenite	
Abundant dykes	Few dykes	

been brought together only late in their geological development, then these contrasts should be expected.

This argument is considerably weakened if it can be shown that the two segments of crust have unusual features in common, particularly if these features reflect their early development. It is therefore pertinent to review the similarities and differences between the three groups of gabbro–anorthosites described in the preceding sections. Tables 2 and 3 list the main similarities and differences between the two groups. While there are of course differences, they do not seem to us to be sufficient to outweigh the similarities. Gabbro–anorthosite complexes are not very common in the Proterozoic crust. Much stronger evidence than we have been able to find would be required to make a convincing case that the Tete Complex was formed independently of the rather similar gabbro–anorthosite complexes north of the Sanangoe Zone. It seems to us more likely that all are related and were formed approximately contemporaneously during an early stage of development common to all of the rocks now exposed in Tete Province.

Knowledge of the crustal conditions during this stage of development requires discussion of the origin of the Tete Province gabbro–anorthosite through a comparison with similar, better studied bodies elsewhere in the world.

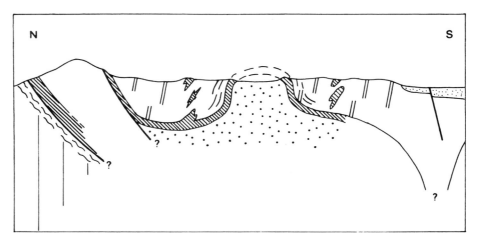

Figure 8. Schematic cross-section of the Tete Complex. Symbols and patterns as for Figure 4.

Some of the features of the Tete Complex relevant to a discussion of its origin are illustrated schematically in Figure 8 and summarized below:

(a) It is a large sheet or lopolith between 10 and 20 km thick composed of basic igneous rocks.

(b) Its western part is everywhere floored by the Chidue Group, parts of which have become detached during intrusion and have floated up through the consolidating magma to form zones of inclusions. Skarns have formed at the contacts of the complex with some of the carbonate rocks.

(c) It contains numerous dolerite dykes which are believed to be cogenetic.

(d) It is composed of gabbro, norite, and anorthosite with minor ultramafic rock-types.

(e) It contains a plagioclase rich in the albite end member (andesine–sodic labradorite) compared to most layered intrusions — for example the Bushveld Complex. It is poor in olivine and pyroxene compared to these intrusions and these minerals are richer in the iron end member. It contains very plentiful opaque minerals and these are enriched in titanium and vanadium and impoverished in chromium and cobalt compared to the opaque minerals of such layered intrusions.

(f) These mineralogical characteristics are reflected in the overall chemistry of the Complex. It has high calcium and low magnesium compared to, for example, the Bushveld Complex. The pattern of alkali enrichment with silica content compares closely with other suites of gabbro–anorthosites and contrasts with layered intrusions. The rare earth element pattern indicates that the anorthosite units of the Complex are plagioclase cumulates derived from magmas similar to the gabbros of the complex in composition.

(g) So far as can be judged, in the west it retains its igneous texture, partly modified by pneumatolytic alteration but unaffected by regional metamorphism except in refoliated zones near its margin and along brittle shear zones. However, in the east, complete foliation and regional metamorphism to the granulite facies has taken place. In the west, there is no evidence whatever, either from within the Complex itself or from the surrounding Chidue Group and Chacocoma Gneiss, that regional metamorphism at the granulite facies ever took place; all the evidence points to amphibolite facies metamorphism.

(h) It is poorly banded and lacks the rhythmic layering characteristic of many large basic intrusions like the Bushveld and Skaergaard Complexes and the Great Dyke. Cumulate textures are restricted to ultramafic rock types, although the REE data indicate that some of the recrystallized units do contain cumulate phases especially plagioclase.

The Tete Complex shares many of these characteristics with other gabbro–anorthosite complexes and is chemically, mineralogically, and structurally distinct from the Bushveld Complex and Great Dyke with which it was sometimes grouped in the older literature. Its inferred form, of a lopolith intruded at a particular horizon in the country rock sequence (Figure 8), the occurrence of skarns, zones of inclusions of country rock and of cataclasis are similar to features of the Adirondack and Morin anorthosites of the Greenville Province of the United States and Canada (Stockwell et al. 1970). Its eastern part in contrast more closely resembles the Archaean and Proterozoic gabbro–anorthosite reviewed by Windley et al. (1981) in being recrystallized throughout at the granulite facies and in being isoclinally infolded with the enclosing country rocks. These characteristics it shares with the complexes north of the Sanangoe Zone.

There appears to be a general concensus based on their coarse grain size, structural content and geobarometry that many north American anorthosites were formed in an anorogenic environment in the deep crust (Emslie 1978). This is in part implied in the origin suggested by Andreoli (1984) for the southern Malawi anorthosites mentioned in section 4. A deep crustal origin is consistent with the association with high grade country rocks and charnockites characteristic of the Luia and Angonia Group occurrences and to a certain extent with the eastern part of the Tete Complex. It is however more difficult to envisage the western part of the Tete Complex as having been emplaced in the deep crust. Because it is everywhere floored by the Chidue Group, these metasediments must have been flat lying and undeformed during intrusion, otherwise the basal contact of the Complex would surely cut across other rock groups. It is hard to see how the Chidue Group could have reached the deep crust in this undeformed state. Similar inconsistencies, for example the occurrence of well-marked thermal aureoles, have been noted around some Grenvillian gabbro–anorthosites. In these cases, it has been suggested that intrusion in the partly consolidated state took place much higher in the crust than the depth of the origin of the bodies and there is in one case strong supporting evidence from the hydrogen and oxygen isotopes of skarns formed during emplacement. A similar origin is suggested for the western part of the Tete Complex. It is possible therefore that the Complex as now exposed represents a section through a large and complicated intrusion with the western part representing the upper portion and the eastern, the root zone.

This conclusion, if substantiated, implies differential uplift of the eastern part of the Tete Complex when compared with the western part. Attempts to explain the evolution of the Mozambique Belt will have to take this difference into account.

Acknowledgement. This article is published by kind permission of the Director, Instituto Nacional de Geologia, Maputo, Mozambique.

References

Afonso, R. S. and Araujo, J. R. 1970. Geologia dos regions de Doa e Moatize. *Instituto Nacional de Geologia, Maputo; unpublished report.*

Andreoli, M. A. G. 1984. Petrochemistry, tectonic evolution and metasomatic mineralisation of Mozambique Belt granulites from S. Malawi and Tete (Mozambique). *Precambrian Research,* 25, 161–186.

Arth, J. G. 1976. Behavior of trace elements during magmatic processes — a summary of theoretical models and their applications. *United States Geological Survey Journal of Research,* 4, 41–47.

Barr, M. W. C., Downing, K. N., Harding, A. R. and Loughlin, W. P. in preparation. Regional correlation near the junction of the Zambezi and Mozambique Belts, east–central Africa.

Barton, J. M. Jr., Fripp, R. E. P., Horrocks, P. and McLean, N. 1979. The geology, age and tectonic setting of the Messina Layered Intrusion, Limpopo mobile belt, southern Africa. *American Journal of Science,* 279, 1108–1134.

Bloomfield, K. 1968. The pre-Karroo geology of Malawi. *Memoir of the Geological Survey of Malawi,* 5.

—— and Garson, M. S. 1965. The geology of the Kirk Range–Lisungwe Valley area. *Bulletin of the Geological Survey of Malawi,* 17.

Brodoimpeks-Geozavod, 1983. The final report on the geological prospecting and exploration works of the Jugoslav team in 1981–1982. Project: geological investigations of graphite, Angonia. *Instituto Nacional de Geologia, Maputo, unpublished report.*

Buddington, A. F. 1939. Adirondack igneous rocks and their metamorphism. *Memoir of the Geological Society of America,* 7, 354 pp.

Coelho, A. V. P. 1969. O complexo gabro–anortositico de Tete (Mocambique). *Bol. Serv. geol. min. Mocambique,* 35, 63–78.

Daly, R. A. 1929. Bushveld Igneous Complex of the Transvaal. *Bulletin of the Geological Society of America,* 39, 703–768.

Davidson, C. F. and Bennet, J. A. E. 1950. The uranium deposits of the Tete district, Mozambique. *Mineralogical Magazine,* 32, 291–303.

De Decker, M. 1983. Mineral occurrences of Macanga–Angonia region, Tete Province. *Instituto Nacional de Geologia, Maputo.*

De Waard, D. 1970. The anorthosite–charnockite suite of rocks of Roaring Brook Valley in the eastern Adirondacks (Marcy Massif). *American Mineralogist,* 55, 2063–2075.

Du Toit, A. L. 1926. *The Geology of South Africa.* Oliver and Boyd, Edinburgh.

Emslie, R. F. 1978. Anorthosite massifs, rapakivi granites, and late Proterozoic rifting of North America. *Precambrian Research,* 7, 61–98.

Freitas, A. J. 1953. Geologia e metalogenia dos depositos de Uranio de area das concessoes Celestino–Bingre distrito de Tete. *Instituto National de Geologia, Maputo, unpublished report.*

Hess, H. H. 1960. Stillwater Igneous Complex, Montana, a quantitative mineralogical study. *Memoir of the Geological Society of America,* 80, 230 pp.

Hunting Geology and Geophysics Limited 1984. Mineral Inventory Project, Final Report. *Instituto Nacional de Geologia, Maputo, unpublished report.*

Irvine, T. N. 1970. Crystallisation sequences in the Muskox intrusion and other layered intrusions. *Geological Society of South Africa Special Publication,* 1, 441–476.

Janardhanan, A. S. and Leake, B. E. 1975. The origin of the meta-anorthositic gabbros and garnetiferous granulites of the Sittampundi Complex, Madras, India. *Journal of the Geological Society of India,* 16, 391–408.

Longyear Company 1956. Report to the Government of Portugal on the Tete area, Mozambique, Volume V. *Instituto Nacional de Geologia, Maputo, unpublished report.*

Luna, I. R. and Freitas, F. 1953. Geologia e Metalogenia des depositos de ijranio de Vale de Mavudzi *Bol. Sen. geol. min.* 11.

Real, F. 1966. *Geologia da Bacia do Rio Zambeze, (Mocambique).* Junta de Investigacoes do Ultramar, Lisboa.

Stockwell, C. H., McGlynne, J. C., Emslie, R. F., Sanford, B. V., Norris, A. W., Donaldson, J. A., Fahreg, W. F. and Carrie, K. L. 1970. Geology of the Canadian Shield. In: Douglas, R.J.W. (Ed.), *Geology and Economic Minerals of Canada. Geological Survey of Canada Economic Geology Report,* 1, 43–150.

Svirine, G. 1980. Geologia e jazigos minerais do Complexo gabro–anortositico de Tete. *Instituto Nacional de Geologia, Maputo, unpublished report.*

Thatcher, E. C. 1968. The geology of the Dedza area. *Bulletin of the Geological Survey of Malawi,* 29.

Vail, J. R. 1964. Esboco geral da geologia da regiao entre os rios Lucite e Revue — distrito de Manica e Sofala, Mocambique. *Bol. Surv. geol. min. Mocambique,* 32.

—— and Pinto, M. S. 1966. Contribuicao para o estudo das rochas da area do Fingoe (Tete, Mocambique). *Bol. Surv. geol. min. Mocambique,* 33, 33–40.

Walshaw, R.D. 1965. The geology of the Ncheu–Balaka area. *Bulletin of the Geological Survey of Malawi,* 19.

Willemse, J. 1969. The geology of the Bushveld Igneous Complex, the largest suppository of magmatic ore deposits in the world. In: Wilson, H.D.B. (Ed.), *Magmatic Ore Deposits. Econ. Geol. Mon.,* 4, 1–22.

Windley, B. F., Herd, R. K. and Bowden, A. A. 1973. The Fiskenaesset complex, West Greenland. Part 1. A preliminary study of the stratigraphy, petrology and whole rock chemistry from Qeqertarssuatslaq. *Grønlands Geol. Unders. Bull.*, **106**, 1–80.

——, **Bishop, F. C. and Smith, J. V. 1981.** Metamorphosed layered igneous complexes in Archaean granulite–gneiss belts. *Ann. Rev. Earth. Planet. Sci.* **9**, 175–198.

Late Proterozoic tectonic terranes in the Arabian–Nubian Shield and their characteristic mineralization

J.R. Vail
Department of Geology, Portsmouth Polytechnic, Portsmouth, U.K.

An extensive terrane of continental gneisses and supracrustal metasediments extends from the Nile valley westwards between southern Egypt and northern Kenya, and also occurs in southern Ethiopia, Somalia and parts of the easternmost Arabian Shield. Mica–beryl pegmatites and associated kyanite are characteristically developed. The gneiss terranes were regionally affected by metamorphic overprinting and remobilization thus masking their true age which is considered to be Middle Proterozoic or older on the basis of isotopic and petrological evidence.

Between the continental terranes volcanosedimentary–ophiolite sequences and intrusive synorogenic calc-alkaline batholiths, post-tectonic and anorogenic calc-alkaline to alkaline plutons occur in an area 700 km in width by over 2700 km long between Sinai and northern Kenya. Within this region of oceanic crust seven ophiolite- or shear-bounded terranes are recognized which developed around 900 to 600 Ma, through closure of island-arc and associated volcanism and sedimentation, the oldest being in the southeast. Hydrothermal gold mineralization is widespread but restricted to the oceanic terranes, and volcanogenic base metal sulphide deposits are lithologically–tectonically controlled and preferentially located parallel to certain ophiolite margined plate boundaries.

KEY WORDS Proterozoic tectonic terranes   Nile valley   Egypt   Sudan   Arabian–Nubian Shield
Mineral deposits   Ophiolite assemblages.

1. Regional geology

The Arabian–Nubian Shield consists of those pre-Phanerozoic rocks, previously referred to as Precambrian Basement Complex, that crop out in the Yemen, western Saudi Arabia, Sinai, the Eastern Desert of Egypt, the Sudan, Ethiopia, and Somalia. In the last three areas the limits of the Shield are less clear since basement rocks are exposed as inliers far to the west and south. The Shield is therefore arbitrarily taken to extend westwards to about 28°E longitude and southwards to the equator, and hence to include parts of Uganda and Kenya and the north coast of Somalia (Figure 1).

0072–1050/87/TI0161–14$07.00
© 1987 by John Wiley & Sons, Ltd.

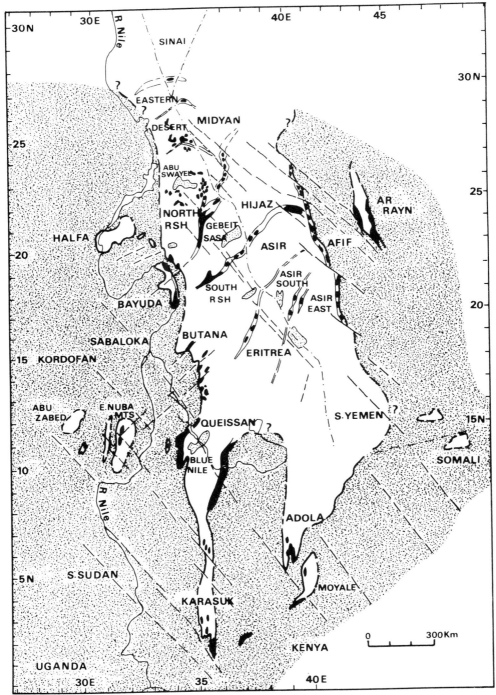

Figure 1. Interpretive map of the sub-Phanerozoic Arabian–Nubian Shield, showing continental type gneiss–metasedimentary terrane (stippled) and oceanic volcanosedimentary terranes (unornamented bordered by heavy lines); ophiolites (black) and tectonic belts (double lines) mark terrane boundaries. Faults (dashed lines), Red Sea closure (dash–dot). The terranes are named following Stoeser and Camp (1985) and Vail (1985b).

Geologically there is a coherence to this area, yet its true tectonic significance has only recently been recognized; far from being an undifferentiated extent of similar basement rocks, in recent years modern plate tectonic concepts and interpretations have led to the recognition of important differences. Much of the basement is covered by Phanerozoic deposits and consequently the nature of the underlying rocks remains speculative over large areas. The extent of this cover is shown elsewhere (Vail 1978, 1985b).

Many lithostratigraphical units are present within the Shield and largely unsuccessful attempts had been made to correlate the rocks on the basis of similar lithology, metamorphic grade, environment, and locality, using many local names. However, modern concepts have revealed significant inconsistencies, and have indicated the need to reconsider the stratigraphic nomenclature used in the region.

Despite the Phanerozoic cover, the patchy geological information, and insufficient isotopic, geochemical and structural data for many areas it is now possible to propose a regional tectonic synthesis based on the considerable detailed mapping and reconnaissance work done throughout the region during the past decade.

For present purposes, four major lithological groups will be considered.

1a. Gneiss Group

Quartzofeldspathic para- and orthogneisses, variously banded with contrasting mineralogy, occur between Egypt and Uganda mainly along the Nile river valley and westwards (Figure 1). In the Sudan, they crop out south of the Halfa district, in the Bayuda desert, Sabaloka inlier, Kordofan, Nuba Mountains inlier, Blue Nile Valley and in Southern Sudan. They are also to be found in western and southern Ethiopia and in northern Uganda, Kenya, and Somalia. In the last named three areas they constitute the northern extent of the Mozambique belt of eastern Africa.

Metamorphic grade is generally amphibolite, rising to granulite facies in some places (Vail 1978); marble bands, rare quartzites, and amphibolites are present and migmatization and granitic emplacement are evident, but because of intense foliation original rock types and internal contacts are not everywhere recognizable.

Age determinations of these rocks are sparse (Figure 2). In central Sudan Sm–Nd model ages suggest an event around 1550 Ma (Harris et al. 1984). In southern Sudan K–Ar ages up to 1845 Ma have been obtained (Civetta et al. 1980). In northern Uganda, away from the Tanzanian craton, isotopic investigations (Leggo 1974) have revealed rocks of this group to be early Proterozoic to Archaean in age, whereas in southern Egypt detrital zircons have yielded an age of 1800 Ma (Abdel Monem and Hurley 1979) and granite pebbles have yielded ages of up to 2060 Ma (Dixon 1981), indicating that an early Proterozoic source is not far distant.

A second major occurrence of quartzofeldspathic gneisses is in the eastern part of the exposed Shield in the Afif region of Saudi Arabia and in northern Somalia. Stacey and Hedge (1984) and Stoeser et al. (1984) have recognized that at least some of the rocks of the eastern Shield are mobilized and reactivated gneisses of Early to Middle Proterozoic age; they constitute the Afif and Ar Ryan terranes (Figure 1).

1b. Metasedimentary Group

In places, mainly near the Nile valley, the quartzofeldspathic gneisses are overlain by extensive supracrustal metasediments of quartz psammites, garnet–mica pelites, white or pink marbles, and amphibolites. They are highly

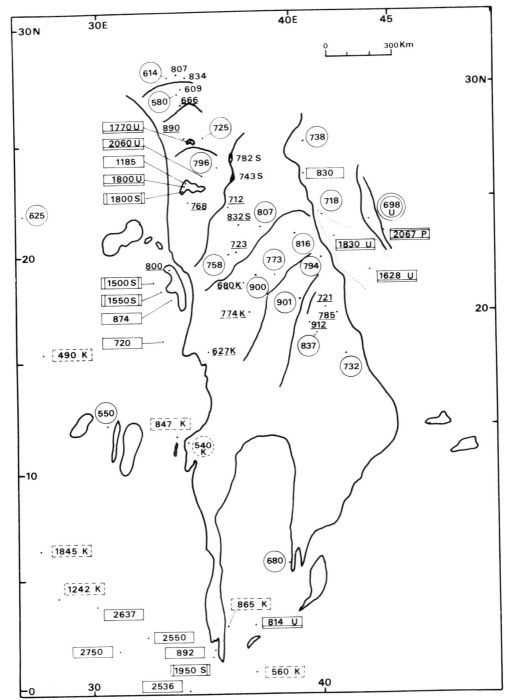

Figure 2. Selected isotopic age determinations for the Arabian–Nubian Shield. Maximum ages are shown for each terrane in Ma, and by method. (U = U–Pb, S = Sm–Nd, K = K–Ar, P = Pb–Pb, unlettered = Rb–Sr isochron). Rectangles indicate results from continental type areas, circles denote plutonic events, others are volcanic or ophiolitic suite ages. Dotted outline is Afif block area of ancient rocks (Stoeser and Camp 1985).

folded and metamorphosed to amphibolite facies, and have unclear relationships with the rock units on either side due to later tectonism.

They are well developed between the Halfa and Red Sea Hills terranes (RSH on Figure 1), in the eastern Bayauda desert, west of the Blue Nile, and in west Karasuk regions where they form miogeosynclinal-type, continental shelf deposits.

Small inliers of high grade metasedimentary rocks within the lower grade volcanosedimentary sequences (see below) also may belong to this group. Such inliers in Egypt are near Abu Swayel and possibly core the Hafafit and Mitiq domes. In the Sudan patches of metasediments form the Sasa plains, an area of paragneisses in the Southern Red Sea Hills, and the Queissan block on the Blue Nile; isolated patches of metasedimentary rocks occur about the Red Sea closure line in Ethiopia and Arabia. They might be detached microcontinental blocks, basement floor to the younger units, or locally higher grade regions of metamorphism and tectonism.

There are a few direct age determinations on these rocks, but intrusive and metamorphic events affecting them may be in the order of 1185–874 Ma (Cahen *et al.* 1984; Meinhold 1979; Ries *et al.* 1985). It seems appropriate to interpret the main outcrop as representing a metasedimentary prism along the margin of the gneissose continent of northeast Africa (Vail 1983).

1c. Volcanosedimentary–ophiolite assemblages

Most of the Eastern Desert of Egypt, the Red Sea Hills of northeast Sudan, much of Ethiopia, and the greater part of the western Arabian Shield in Sinai, Saudi Arabia and the Yemen are underlain by a heterogeneous accumulation of island-arc and ocean floor calc-alkaline to tholeiitic basalts, andesitic and rhyolitic lavas, interbedded pyroclastics volcanogenic sediments, thin limestones and lesser shales and impure sandstones. Numerous names have been used for these units (e.g. El-Baz 1984; Fitch 1978; Jackson 1980; Kabesh and Lofti 1962; Whiteman 1971) but the very nature of their genesis, similarity of lithology and common depositional environment is such that they do not usually extend over great distances, and local unconformities and dislocations are not regionally significant, hence stratigraphic nomenclature other than purely local, and regional correlations are not appropriate.

Within the volcanosedimentary rocks a number of linear, fragmented, and tectonically altered ultramafic masses are present, many of which have been recognized as remnant ophiolite suites (reviewed for example in Vail 1983, 1984b; Bentor 1985; Kröner 1985). Serpentinized pyroxenites, layered gabbros, basic dykes, pillow lavas and rare cherts are preserved along narrow bands, which occur in several belts within the volcanic assemblages and also along the contacts with the gneisses or metasedimentary groups. Tectonic disruption, thrusting, mylonization, and melange development affect these ultramafics, particularly in the Eastern Desert of Egypt where they form dismembered fragments in a regional melange (Ries *et al.* 1983). Most of the Northern Red Sea Hills plate consists of autochthonous material (Shackleton *et al.* 1980) but over the rest of the Shield the ophiolite suites belong to well-defined linear belts.

Within the gneiss terranes in parts of E. Sudan, N. Kenya and Saudi Arabia outliers of the volcanic–sedimentary assemblages, some containing ophiolite fragments, occur as exotic blocks. The most notable are near Halfa, in the eastern Bayuda along the Nile valley, the Abu Zabed and East Nuba Mountain block of central Sudan, the Moyale and other fragments in northern Kenya, the Al Amar–Idsas region between the Afif and Ar Rayn terranes in Saudi Arabia and possible blocks in south Yemen and north Somalia (Figure 1).

The volcanosedimentary ophiolite assemblages are usually in greenschist to lower amphibolite grade of regional metamorphism, in contrast to the gneisses and metasedimentary groups. Blueschist facies are not present, the only possible remnant being in the Nuba Mountains (Hirdes and Brinkmann 1985), but widespread synorogenic batholithic granitoid emplacements and horizontal structural displacements may have obliterated evidence of such metamorphism.

It has been proposed (Kröner 1985; Stoeser and Camp 1985; Vail 1985a, 1985b) that the volcanosedimentary rocks comprise a mosaic of ophiolite or tectonically bounded terranes of oceanic, partly island-arc and subduction derived material. Despite uncertainties of interpretation and incomplete data in some places, these terranes can be recognized on both sides of the present Red Sea and it is apparent that prior to its opening the separate blocks continued across the closure line (Figure 1).

It is not yet clear whether these tectonic terranes or plates have characteristic and distinguishable ages of formation. Bentor (1985) has discussed the problem of the apparently large spacing between ophiolite belts, and the long history of volcanism and plutonic activity in the Shield, whilst subsequent geological processes have continued within and across the various tectonic terranes thus confusing the evolutionary pattern of events. Maximum isotopic ages have indicated that volcanism in the terranes was active at least between c. 912 Ma to 609 Ma (Figure 2).

If the terranes were formed as discrete units, then any regional consideration of geochronological results (e.g. Bentor 1985) must take this into account, for except for later Shield-wide activity, the geochronology of each plate should be different.

Stoeser and Camp (1985) considered it likely that the oldest rocks in Saudi Arabia were in the southeast, and that there was a progression from about 900–800 Ma in the Asir terrane, through 800–700 Ma in the Hijaz, to 700–600 Ma in the Midyan terrane. Similarly Stern and Hedge (1985) considered the mean age of the Egyptian basement to decrease from south to north; igneous activity in the North Red Sea Hills (RSH) began prior to 765 Ma and ended by 540 Ma in the Eastern Desert terrane. The maximum isotopic age determinations of intrusive events across the whole Shield shows a similar northwards decrease (Figure 2).

1d. Syn- orogenic and post-orogenic and anorogenic plutonic igneous activity

Cutting across all volcanic and sedimentary rock units and their bounding ophiolites are large heterogeneous batholithic and plutonic complexes of diorites–gabbros, granodiorites, tonalites, and adamellites–granites, previously referred to as Batholithic Granites. They form syn- to late orogenic emplacements spread over a wide time span and an extensive area. Such plutonic rocks in the Arabian Shield have been comprehensively reviewed by Stoeser (1986) who has shown how the rocks progressively evolved in both ensimatic island-arc and continental-margin environments from around 900 Ma to 600 Ma (Figure 2).

This granitization is regarded as a cratonization process following the subduction related evolution of the oceanic crust (Greenwood et al. 1976). Consequently the distribution of particular units and their subsequent history is complex, and detailed investigations are required to understand the sequence of events. Such information is largely lacking in Ethiopia and the Sudan, and is incomplete in Egypt, Yemen, and parts of Saudi Arabia.

Distinguishable from the batholithic granitoid emplacements are smaller plutons, many in the form of bimodal calc-alkaline ring-complexes, of gabbros and granites which pierce all preceding rock types (Figure 4). In the Sudan, Vail (1984a) recognized bands of these plutons parallel to the ophiolite belted terrane margins in certain areas, suggesting their location to be tectonically related.

The plutons are distinct from anorogenic alkaline plugs and ring-complexes of mainly alkali granite and quartz syenite, very rarely foid syenites, which postdate orogenic events in the Shield (Figure 4). These latter intrusions are comparable to the Younger Granites of West Africa. In Saudi Arabia, syenitic plutonic rocks were emplaced from about 620–550 Ma and alkali granite between 680–517 Ma (Stoeser 1986). The anorogenic rocks in Egypt and the Sudan can be grouped in a number of provinces distinguished by age and location. Alkali granite–syenite ring complexes formed almost entirely in the gneisses–metasedimentary terrane west of, and along the Nile Valley (Figure 4) between 250 Ma and 222 Ma. Another group in the Northern Red Sea Hills, Gebeit, and Blue Nile terranes were intruded between 150 and 89 Ma (Vail 1985a, in press).

2. Regional structure

Stoeser and Camp (1985) and Vail (1984a, 1985b) have defined a number of late Proterozoic tectonic terranes across the Arabian–Nubian Shield. These comprise blocks of similar rock types separated by linear belts along which the ultramafic–mafic fragments are aligned. Thrust faulting, shearing, mylonitization, and tectonic melange are characteristic of many of the contacts, along which ophiolite remnants are not always present (Kröner 1985; Kröner et al. in press).

Two categories of terrane are distinguishable:
1. Terranes of reworked ancient continental type gneiss and supracrustal metasediments in the east, south, and west.
2. Oceanic affinity volcanosedimentary greenschist assemblages with associated granitoid emplacements in a central zone of younger terranes.

Within the latter, seven or eight plates can be distinguished separated by oblique ophiolite belts or fracture zones, which are considered (Stoeser 1986; Kröner 1985; Vail 1985a, 1985b) to represent probable Proterozoic suture zones. Blocks of gneisses and metasediments within the oceanic terrane are suspected as being microcontinental fragments while the exotic blocks of volcanic terrane within the gneissic area, some with contained ophiolite fragments, could be possible thrust slices.

It is believed that closure of island-arcs and continent-arc collision severely disrupted the sequences. Horizontal movements, major thrusting, transform and strike-slip transcurrent faulting with significant displacements and igneous intrusions will all have complicated the pattern (Figure 1).

3. Tectonic controls of economic mineralization

There are five important groups of economic mineral deposits occurring in the Proterozoic rocks of the Arabian–Nubian Shield (Al Shanti et al. 1978; Al Shanti and Roobol 1979; Pohl 1979, 1982; Vail, 1979, 1985a). Several deposits have been exploited and numerous small showings are known. Detailed studies have been made of many of the mineral occurrences, and also of their regional distribution but it is only with the establishment of a tectonic framework of adjacent terranes that it is possible to reassess the controls on the economic mineral deposits distribution in the Shield.

3a. Pegmatite and metamorphics refractory minerals

Quartz and K-feldspar pegmatites with muscovite mica, beryl, and tourmaline occur as small, usually zoned bodies. They are particularly numerous in the Bayuda desert of Sudan, (Meinhold 1983). Other, minor occurrences are known in northern Kenya, southern Ethiopia (Assefa 1985), northern Somalia, and in the gneiss–metasedimentary areas along the Red Sea closure line in Ethiopia, and in the Afif terrane of Arabia. It is noteworthy that none occur in the volcanosedimentary terranes (Figure 3).

Kyanite is concentrated in some of the mica schists in the Bayuda Desert in the vicinity of the pegmatites, and is also present as a regional metamorphic mineral in the basement rocks of northern Kenya and southern Ethiopia. Minor occurrences of graphitic schists and extensive bands of crystalline marbles also occur in these areas.

Although the pegmatites and high grade metamorphic minerals occur exclusively in the ancient gneiss terranes their age of formation, at least in the Bayuda desert, is probably around 640 Ma (Vail 1978) which is the peak of Pan-African tectonic–thermal activity (Kennedy 1965).

3b. Tin, tungsten, rare earth elements

A few localities are known where anomalous concentrations of Sn–W–REE have been found, although so far economic mineral deposits of these minerals are lacking. Some of the post-orogenic to anorogenic granitic plutons have minor amounts of cassiterite and scheelite but none to compare with the Nigerian Younger granites. In Saudi Arabia several of the granite complexes have anomalously high U background counts (Stoesser *et al.* 1984) and the Afif terrane is characteristically tungsten rich. Carbonatites and nepheline syenites are quite rare in the area although traces of REE minerals are to be expected. The distribution of the igneous complexes hosting these minerals is shown in Figure 4. Further details are in Stoeser (1986) and Vail (in press).

3c. Chromite, asbestos, magnesite, nickel

Magmatic minerals occur associated with the mafic–ultramafic ophiolite suite rocks (Figure 1), but usually in disseminated or trace amounts. Chromite has been mined in the Ingessana Hills of the Blue Nile Province in eastern Sudan, where large reserves of short fibre asbestos are also available. Minor mineralization is known from most of the other occurrences in the Arabian–Nubian Sheild, which are confined to the ultramafic bands or fragments occurring within or adjacent to the volcanosedimentary terranes.

3d. Gold

Occurrences of gold are widespread throughout the Arabian–Nubian Shield. They are usually in the form of auriferous quartz veins, or as disseminations in the wall rocks, less commonly associated with volcanogenic sulphide mineralization in the basic volcanic rocks, or in shear zones in the granitoid batholiths, and as related placer deposits.

Many hundreds of small gold deposits are known (Egyptian Geological Survey, 1979; Van Daalhoff 1982; Ahmed 1983; Fletcher 1985), and most were exploited by the ancient miners, particularly during Pharaonic times. A few mines have been worked to the present day.

The primary gold occurrences are almost entirely confined to the volcanosedimentary plutonic terranes and appear to be more numerous in some blocks

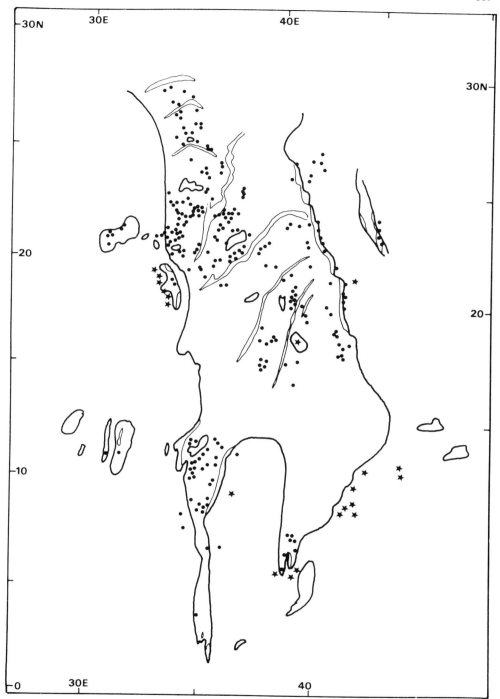

Figure 3. Distribution of gold mineralization (dots) and pegmatite occurrences (stars) within the various tectonic terranes in the Arabian–Nubian Shield.

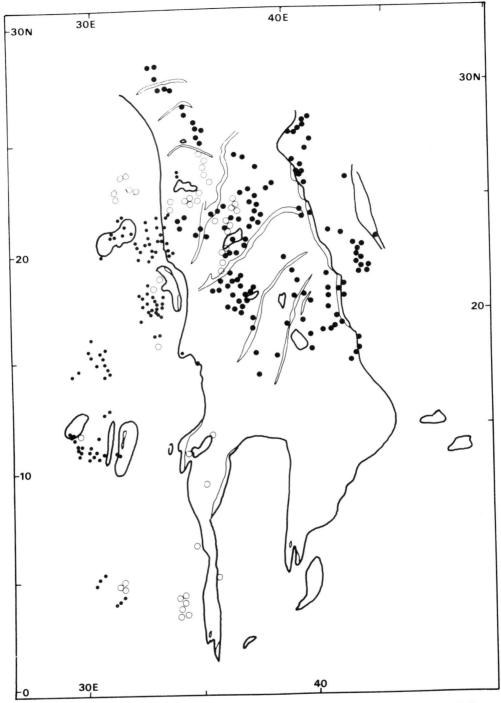

Figure 4. Distribution of post- and anorogenic igneous complexes in the Arabian–Nubian Shield. Post-orogenic complexes 720–517 Ma (large dots), alkali ring complexes 250–220 Ma (small dots), alkali complexes 150–89 Ma (open circles). Tectonic terrane boundaries heavy lines and narrow double lines.

MINERALIZATION IN THE ARABIAN–NUBIAN SHIELD

compared with others (Figure 3). However, lack of ground data, and cover by younger rocks might be significant limiting factors. The Northern Red Sea Hills–Midyan, Gebeit–Hijaz, Eritrea-Asir, and Blue Nile terranes appear to have the most numerous deposits.

3e. Volcanogenic massive and stratiform base metal sulphide deposits

Massive and disseminated pyrite, chalcopyrite, sphalerite, some galena, and associated gold and silver mineralization occurs within restricted volcanic and sedimentary formations of the oceanic terrane in several areas.

Until recent years the presence of such mineralization was not fully realized but recent studies have indicated the extent of this type of mineralization (Al Shanti *et al.* 1978; Johnson and Vranas 1984; Aye *et al.* 1985; Tawfig *et al.* 1985). Because of the environment of formation within oceanic terranes it is to be expected that further deposits will be found.

The most favourable terrane for base metal sulphide volcanogenic deposits is clearly a band 600 km in length along the north margin of the Asir–Southern Red Sea Hills plate. The South and East Asir and Eritrean plates also carry this type of mineralization, and the volcanosedimentary–ophiolite strip between the Ar Rayn and Afif terranes is also mineralized in base metals (Figure 5).

4. Conclusions

Tectonic terranes comprising continental type gneisses and continental margin metasediments have been identified westwards from the Nile valley from central Sudan to northern Uganda, in Kenya and Somalia, and in the easternmost Arabian Shield. Within these terranes characteristic mineralization is restricted to mica–beryl pegmatites.

The age of formation of the continental terranes has not been established with certainty, due mainly to the widespread effects of metamorphism and tectonic remobilization. The gneisses in northern Kenya have generally been regarded as an extension of the Mozambique mobile belt of east Africa, but whether those in the Sudan and Saudi Arabia can be similarly regarded remains questionable.

Between the continental terranes there is an extensive area of oceanic crust approximately 700 km in width, extending from beneath a sedimentary cover southwards for over 2700 km. Within this area of approximately one and a half million square kilometres seven ophiolite- or shear-bounded volcanosedimentary terranes have been recognized. Geological processes during the evolution of the rocks were repetitive and similar across the area, so correlation of units is difficult. Nevertheless, future investigations into the detailed petrography, geochemistry, and geochronology should enable individual terranes to be better characterized and consequently the tectonic pattern to be modified.

The distribution of certain mineral deposits, in particular the volcanogenic base metals, appears to be lithologically–tectonically controlled. However, it is not certain whether the terranes are internally consistent or whether systematic changes take place along strike, or across the width of each plate. It would seem likely that if the bordering ophiolite belts mark suture zones then subduction related magmatism would progressively change away from the plate margins and parallel bands of magmatic and hydrothermal events should be present. Stoeser (1986) has already identified belts of intermediate plutonism close to and parallel to the proposed suture zones in the Arabian Shield; the distribution of base metal

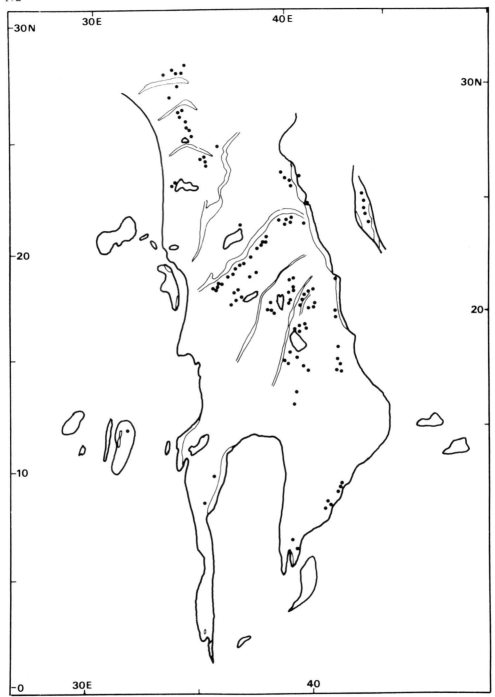

Figure 5. Distribution of volcanogenic base metal deposits in the Arabian–Nubian Shield. Tectonic boundaries between continental and oceanic terranes (heavy lines) and within oceanic terranes (light double lines).

deposits is another indication of such controls, and the recognition of tectonic terranes and internal structure can therefore be utilized in the search for further deposits.

Acknowledgements. John Davidson and Jackie Duggua at Portsmouth are both thanked for assisting with the draughting of the text figures and the typing of the manuscript respectively.

References

Abdel Monem A.A. and Hurley, P.M. 1979. U–Pb dating of zircons from psammitic gneisses, Wadi Abu Rasheid–Wadi Sikait area, Egypt. In: Tahoun, S.A. (Ed.) *Evolution and Mineralization of the Arabian–Nubian Shield, Vol. 2.* Pergamon Press, Oxford. 165–170.

Ahmed, F.I. 1983 Sudan's mineral deposits. *Mining Magazine*, **148**, 31–35.

Al Shanti, A., Frisch, W., Pohl, W. and Abdel Tawab, M.M. 1978. Precambrian ore deposits in the Nubian and Arabian Shields and their correlation across the Red Sea. *Oster. Akad. Wissen.*, **3**, 37–44.

Al Shanti, A.M.S. and Roobol, M.J. 1979. Some thoughts on metallogenesis and evolution of the Arabian–Nubian Shield. In: Tahoun, S.A. (Ed.), *Evolution and Mineralization of the Arabian–Nubian Shield, Vol. 1.* Pergamon Press, Oxford. 87–96.

Assefa, G. 1985. The mineral industry of Ethiopia; present conditions and future prospects. *Journal of African Earth Science*, **3**, 331–336.

Aye, F., Cheze, Y. and El Hindi, A. 1985. Discovery of a major massive sulphide province in northeastern Sudan. In: *Prospecting in Areas of Desert Terrain.* Inst. Min. Metal., London. 43–48.

Bentor, Y.K. 1985. The crustal evolution of the Arabo-Nubian massif with special reference to the Sinai Peninsular. *Precambrian Research*, **28**, 1–74.

Cahen, L., Snelling, N.J., Delhal, J., and Vail, J.R. 1984. *The Geochronology and Evolution of Africa.* Oxford Univ. Press, Oxford. 512 pp.

Civetta, L., De Vivo, B., Giunta, G., Ippolito, F., Lima, A., Orsi, G., Perrone, V. and Zuppetta. A. 1980. Geological and structural outlines of the southern Sudan. *Atti Conv. Lincei No. 47, Geodynamic evolution of the Afro-Arabian Rift System, Rome 1979.* 175–183.

Dixon, T.H. 1981. Age and chemical characteristic of some pre-Pan African rocks in the Egyptian Shield. *Precambrian Research*, **14**, 119–133.

Egyptian Geological Survey and Mining Authority 1979. Mineral Map of Egypt. 2 sheets, Scale 1:2 million, explanation. *Geol. Surv. Egypt, Cairo.* 44 pp.

El-Baz, F. 1984. *The Geology of Egypt, an Annotated Bibliography.* E.J. Brill, Leiden, 778 pp.

Fitch, F.H. 1978. Informal lithostratigraphic lexicon for the Arabian Shield, Saudi Arabia. *Dir. Gen. Miner. Resour. Tech. Rep. TR–1970–1*, 163 pp.

Fletcher, R.J. 1985. Geochemical exploration for gold in the Red Sea Hills, Sudan. In: *Prospecting in areas of desert terrain.* Inst. Min. Metal., London. 79–94.

Greenwood, W.R., Hadley, D.G., Anderson, R.E., Fleck, R.J. and Schmidt, D.L. 1976. Late Proterozoic cratonization in southwestern Saudi Arabia. *Philosophical Transactions of the Royal Society of London*, **A280**, 517–527.

Harris, N.B.W., Hawkesworth, C.J. and Ries, A.C. 1984. Crustal evolution in north-east and east Africa from model Nd ages. *Nature, London*, **309**, 773–776.

Hirdes, W. and Brinkmann, K. 1985. The Kabus and Balula serpentinite and metagabbro complexes — A dismembered Proterozoic ophiolite in the northeastern Nuba Mountains, Sudan. *Geol. Jahrbuch*, **B58**, 3–43.

Jackson, N.J. 1980. Correlation of Late Proterozoic stratigraphies, NE Africa and Arabia: summary of a IGCP project 164 report. *Journal of the Geological Society of London*, **137**, 629–634.

Johnson P.R. and Vranas, G.J. 1984. The geotectonic environments of late Proterozoic mineralization in the southern Arabian Shield. *Precambrian Research*, **25**, 329–348.

Kabesh, L. and Lotfi, M. 1962. On the Basement Complex of the Red Sea Hills, Sudan. *Bull. Inst. Désert d'Egypte*, **12** 1–19.

Kennedy, W.Q. 1965. The influence of basement structure on the evolution of the coastal (Mesozoic and Tertiary) basins of Africa. In: *Salt basins around Africa.* Institute of Petroleum, London. 7–16.

Kröner, A. 1985. Ophiolites and the evolution of tectonic boundaries in the late Proterozoic Arabian–Nubian Shield of northeast Africa and Arabia. *Precambrian Research*, **27**, 277–300.

——, Greiling, R., Reischmann, T., Hussein, I.M., Stern, R.J., Krüger. J., Dürr, S. and Zimmer, M. 1987. Pan-African crustal evolution in the Nubian segment of northeast Africa. *American Geophysical Union, Geodynamic Series*, **17**, 235–237

Leggo, P.J. 1974. A geochronological study of the basement complex of Uganda. *Journal of the Geological Society of London*, **130**, 263–277.

Meinhold, K.D. 1979. The Precambrian basement complex of the Bayuda Desert, northern Sudan. *Geogr. phys. Geol. dyn.*, **21**, 395–401.

—— 1983. Summary of the regional and economic geology of the Bayuda Desert, Sudan. *Bull. geol. Miner. Resour. Sudan*, **33**.

Pohl, W. 1979. Metallogenic–mineralogenic analysis. Contribution to the differentiation between Mozambique basement and Pan-African superstructure in the Red Sea region. *Annals of the Geological Survey of Egypt.*, **9**, 32–44.

—— 1982. Large scale metallogenetic features of the Precambrian in northeast Africa and Arabia. *Precambrian Research*, **16**, A32–33.

Ries, A.C., Shackleton, R.M., Graham, R.H. and Fitches, W.R. 1983. Pan-African structures, ophiolites and mélange in the Eastern Desert of Egypt: a traverse at 26°N *Journal of the Geological Society of London*, **140**, 75–95.

——, Shackleton, R.M. and Dawoud, A.S. 1985. Geochronology, geochemistry and tectonics of the NE Bayuda Desert, N. Sudan: implications for the western margin of the late Proterozoic fold belt of NE Africa. *Precambrian Research*, **30**, 43–62.

Shackleton, R.M., Ries, A.C., Graham, R.H. and Fitches, W.R. 1980. Late Precambrian ophiolitic mélange in the eastern desert of Egypt. *Nature, London*, **285**, 472–474.

Stacey, J.S. and Hedge C.E. 1984. Geochronologic and isotopic evidence for early Proterozoic continental crust in the eastern Arabian Shield. *Geology*, **12**, 310–313.

Stern, R.J. and Hedge, C.E. 1985. Geochronologic and isotopic constraints on late Precambrian crustal evolution in the Eastern Desert of Egypt. *American Journal of Science*, **285**, 97–127.

Stoeser, D.B. 1986. Distribution and tectonic setting of Arabian Shield plutonic rocks. *Journal of African Earth Science*, **4**, 21–46.

—— and Camp, V.E. 1985. Pan-African microplate accretion of the Arabian Shield. *Bulletin of the Geological Society of America,*. **96**, 817–826.

——, Stacey, J.S., Greenwood, W.R. and Fischer, L.B. 1984. 'U/Pb zircon geochronology of the southern portion of the Nabitah mobile belt and Pan-African continental collision in the Saudi Arabian Shield. *Saudi Arabia Deputy Min. Miner. Resour. Tech. Rec.*, **USGS–TR–04–05**. 88 pp.

Tawfig, M.A., Legg, C.A. and Last, B.J. 1985. A multidisciplinary exploration programme in the Samran area, Kingdom of Saudi Arabia. In: *Prospecting in Areas of Desert Terrain*. Institute of Min. Metal. London. 111–120.

Vail, J.R. 1978. Outline of the geology and mineral deposits of the Democratic Republic of the Sudan and adjacent areas. *Overseas Geol. Miner. Resour. London*, **49**, 1–67.

—— 1979. Outline of geology and mineralization of the Nubian Shield east of the Nile valley, Sudan. In: Tahoun, S.A. *Evolution and Mineralization of the Arabian–Nubian Shield*, Vol. 1. Pergamon Press, Oxford 97–108.

—— 1983. Pan-African crustal accretion in north-east Africa. *Journal of African Earth Science*, **1**, 285–294.

—— 1984a. Distribution and tectonic setting of post-kinematic igneous complexes in the Red Sea Hills of Sudan and the Arabian–Nubian Shield, *Bulletin of the Faculty of Earth Science, King Abdulaziz University, Jeddah*, **6** (1983), 259–269.

—— 1984b. The nature of the Basement Complex in the Nubian Shield in north-east Africa: addendum, *Journal of African Earth Science*, **2**, 389–390.

—— 1985a Relationship between tectonic terrains and favourable metallogenic domains in the central Arabian–Nubian Shield, *Trans, Inst. Min. Metal., Appl. Earth Sci.*, **B94**, 1–5.

—— 1985b. Pan-African (late Precambrian) tectonic terrains and the reconstruction of the Arabian–Nubian Shield. *Geology*, **13**, 839–842.

—— in press. Ring complexes and related rocks in Africa. *6th Conference of African Geology, Gaberone 1985, Geological Society of Africa.*

Van Daalhoff, H. 1982. Mineral locality maps of Saudi Arabia. *Saudi Arabian Deputy Min. Miner. Resources*. Geologic Map GM–66. Scale 1:2 500 000.

Whiteman, A.J. 1971. *The Geology of the Sudan Republic*. Clarendon Press, Oxford. 290 pp.

Ductile shear zones in the northern Red Sea Hills, Sudan and their implication for crustal collision

David C. Almond
Department of Geology, Kuwait University, Kuwait
Farouk Ahmed
Department of Geology, University of Khartoum, Khartoum

In the northern Red Sea Hills, of Sudan, deformation is largely concentrated into a network of shear zones. Several of these zones strike north–south, but the important Nakasib and Sol Hamid zones are northeast-trending, and there are also minor shear zones which strike east–west. We distinguish between massive shear zones, such as that of Nakasib, and more diffuse braided shear zones, typified by the Oko zone. The massive type owes its character to an origin as a reactivated oceanic suture, whereas the braided type was characterized by strike-slip shearing from its inception. Most of the shear zone rocks are mylonites formed under greenschist facies conditions, but early shearing along the Oko zone took place at higher temperature and resulted in gneissose mylonites with amphibolite facies mineralogy. The northeast-trending, Nakasib shearing appears to have preceded a phase of batholithic intrusion, whereas north–south shearing in the Oko and Abirkitib zones is younger than the batholithic intrusions and is in turn post-dated by emplacement of bimodal granite–gabbro complexes. These events cannot be dated precisely as yet, but took place between 99 Ma and 700 Ma, and probably towards the end of that interval. The pattern of shear and suture zones in northeast Sudan suggests that the northeast-trending Hijaz and Asir arcs, recognized in Saudi Arabia, terminate to the west against a north–south suture which was later rejuvenated to form the Abirkitib shear zone. West of this suture was an immature volcanic arc which may have lain along the margin of the African continent. The rejuvenation which caused northeast and north–south strike-slip shear was probably a consequence of soft collisions in East Arabia.

KEY WORDS   Ancient sutures   Collision lineaments   Sudan   Shear zones

1. Introduction

The existence of a widespread system of shear zones within the Proterozoic rocks of the Red Sea Hills, has been discussed briefly in other papers (Almond 1982; Almond et al. 1984a, 1984b). It was there pointed out that these structures are responsible for characteristic lineament patterns (Ahmed 1983) and represent one of the chief deformational episodes in this part of the Arabian–Nubian Shield. Furthermore, the shear zones guided late Proterozoic gold mineralization and several periods of magma intrusion between late Proterozoic and Tertiary times.

0072–1050/87/TI0175–10$05.00
© 1987 by John Wiley & Sons, Ltd.

This paper describes the nature of the shear zones in more detail, and explores their general relationship to the tectonic evolution of the Shield.

2. Regional geology

The region between the Nile and the Red Sea coast can be described in terms of three major units, or blocks (Figure 1). The western unit is a belt of strongly deformed and metamorphosed rocks about 100 km wide, extending from the Nile at Abu Hamed to about 34° 30'E. Within it, deformation is expressed by north-striking foliations which dip steeply to the east and west which are related to tightly compressed folding around axes which plunge 10–20°N. Towards the Nile, the rocks are mainly shelf sediments intruded by granites and metamorphosed in the amphibolite facies. In this section, the folds verge eastwards.

The eastern half of the metamorphic belt is composed of volcanic rocks, greywackes and phyllites with intercalations of marble and siliclastic metasediments, the latter only near the base. Metamorphism is in the greenschist facies and the folds verge westwards, in the opposite direction to the folds near the Nile. The contact between high and low-grade rocks is strongly tectonized and in many places coincides with a narrow, vertical zone of shearing (the Keraf zone of Figure 1). Locally, the low-grade rocks structurally overlie the high-grade rocks. Opinions differ about the general relationship of these two rock groups in northern Sudan. Some workers (including ourselves) have supposed them to represent basement and cover, with a tectonized plane of unconformity between, but recent geochron-

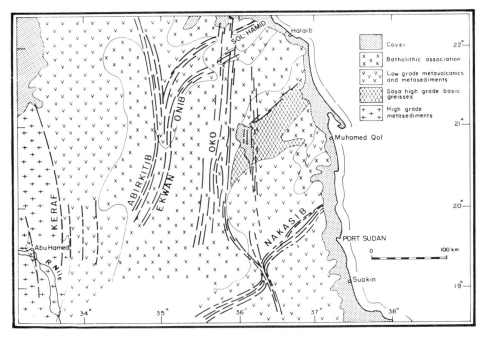

Figure 1. Geological sketch-map of the Northern Red Sea Hills of Sudan, showing the major shear zone.

ological work suggests that the two groups are similar in age (e.g. Ries et al. 1985). Consequently, it has also been supposed by others that the metamorphic, structural, and lithological differences between the two groups merely reflect different depths within the late Proterozoic crustal section. Apart from shearing along the contact between high- and low-grade rocks, there are also at least two north–south shear zones within the zone of low-grade rocks. These are narrow, subvertical and contain small lenses of retrograde ultramafic rocks. Similar features occur around an isolated mass of low-grade rocks near Abu Hamed and have previously been reported along zones of dislocation, in the northern Bayuda Desert, to the west (Dawoud 1980). We believe that the ultramafic lenses have been tectonically derived from ophiolites initially emplaced at other structural levels. One final characteristic of the metamorphic belt is the almost complete absence of syn- to post-tectonic igneous intrusions, in strong contrast to the central and eastern blocks.

The central block extends from 34° 30′ to about 36° 00′E and its main element is a large composite pluton of calc-alkaline rocks here called the Serakoit batholith. Rock types range from gabbro to diorite and adamellite, but true granites are rare. The batholith has a complex history which has not been worked out in detail. Some of the early intrusions are strongly foliated and gneissose, whereas the youngest are unfoliated. Screens of metavolcanic rock locally separate adjacent intrusions and have in places been rotated during emplacement. In the north, and west of Onib (Figure 1), the batholith opens out to enclose a large tongue of metavolcanic and metasedimentary rock containing lenses of what appear to be dismembered ophiolites (Hussein et al. 1984). Shear zones and ophiolites define a Y-shaped pattern, one branch trending north northwest and the other northeast, towards Halaib. Although it is clear that these volcanic rocks and ophiolites antedate the Serakoit batholith, it is also evident from our work that some of the characteristic tectonism of the region around Onib took place after intrusion of the batholith. Thus, further south, the major shear zones of Abirkitib and Ekwan (Figure 1) transect the batholith, reducing plutonic rocks to mylonite (Almond et al. 1984a). Moreover, lenses of carbonated ophiolite still occur as far south as Abirkitib, within the sheared portions of the batholith. It is probable, therefore, that dismemberment of the Onib ophiolites in part took place after intrusion of the Serakoit batholith. At this time, several masses of ophiolite were transported up, or along, the shear zones. The ophiolites of Onib may well have originated within the oceanic suture zones as claimed by Hussein et al. (1984), but subsequently these sutures were reactivated and became part of the shear zone network. A similar multistage history is also likely for the Nakasib shear zone (Figures 1 and 2), which has been cited as a likely suture zone by Vail (1983) and by Embleton et al. (1984).

Extensive north–south shearing occurs within a wide zone towards the eastern side of the Serakoit batholith, mainly affecting the igneous plutons and their country rocks to the east. This 'Oko shear zone' has had a great influence on post-batholithic magmatism, and both post-tectonic calc-alkaline complexes and anorogenic alkaline complexes are aligned along it. In addition, numerous north–south dykes of calc-alkaline and alkaline composition intrude along the shear plane, while the less common east–west dykes cut across the shears.

In the Sasa area, east of the Serakoit batholith (Figure 1), extensional faults of Tertiary age are structurally controlled by a branch of the Oko shear zone and preserve a sequence of clastic cover sediments within a complex graben. The sediments are tentatively correlated with the Cretaceous Nubian Formation and

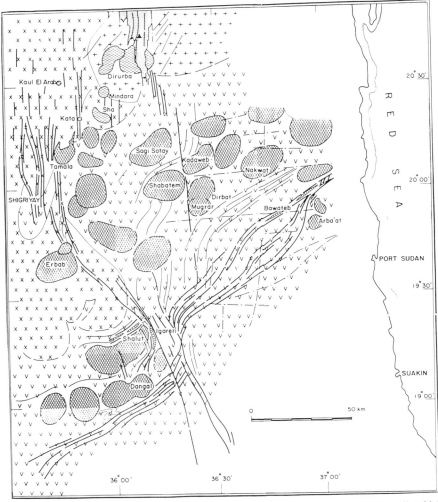

Figure 2. Relationship of the N to NNW-trending shears of the Oko zone to the NE-trending Nakasib shear zone. The V-ornament marks low-grade rocks; +-ornament the high-grade rocks of Sasa; X-ornament the Serakoit batholith. Intrusions shown cross-hatched are calc-alkaline, broken lines show peralkaline intrusions. Ophiolites within the Nakasib zone are lined. Stippling marks downfaulted cover rocks.

the graben has acted as a locus for Tertiary alkali basalt–trachyte volcanicity.

Apart from the Sasa graben, the eastern block of the Red Sea Hills is largely composed of weakly metamorphosed, calc-alkaline volcanic rocks. There is a local basement of high-grade gneisses in the Sasa area. Abundant calc-alkaline plutonic rocks have intruded the volcanic rocks and range from gabbro and diorite to granodiorite, adamellite, and granite. Bimodal gabbro–granite complexes are common. Some of the complexes are syntectonic and may be offshoots of the Serakoit batholith. Others, including the bimodal complexes, are post-tectonic. In contrast to the western metamorphic belt, the main structural and lithological features follow northeasterly trends and, except in the shear zones and Sasa basement, the level of deformation is low. Late east-west dykes are numerous and probably related to those within the Serakoit batholith. In many cases these dykes follow small shear zones, which form the youngest element of the Red Sea Hills shear zone network.

The most prominent shear zone within the eastern block is that of Nakasib, west of Port Sudan. Within it, deformation is intense and has reduced the volcanic rocks to subvertical belts of green mylonite. Large lenses of ophiolite rocks occur scattered along the length of the zone but appear to have been emplaced into their present position late in the deformation sequence. Tight major and minor folds also occur within the Nakasib zone and relate to an early stage of its development. All this is very different from the situation in the Nakasib zone, where cleavage is rarely seen and deformation is limited to open folding.

3. Shear zone characteristics

Two types of shear zones are distinguished here. In the massive type, exemplified by Nakasib, shear deformation is penetrative, severe, and continuous over bands that are relatively well defined, broadly rectilinear, and in the order of 10 km wide. Younger bands obliquely cut off the foliation in older bands. The Oko is representative of the second, braided type of shear zone. In this type, shearing is concentrated into bands only a few metres wide. These follow sinuous courses and define an anastomosing pattern in which local strike may deviate as much as 20° from the overall trend. Lenses of little-deformed country rocks are preserved between the shear bands. Where structurally competent rocks have been affected, some of these lenses are very large, as is the case west of Tamala, where a lens of gabbro is 30 km long and up to 7 km wide, but the same pattern persists on a much smaller, outcrop scale. It is almost impossible to place lateral boundaries to these braided zones. Most of the shearing within the Oko zone is confined within a width of about 30 km, but some shears occur outside these limits and die out along strike.

Both types of shear zone contain internal evidence of polyphase development. In the case of Nakasib, this is most obvious from the cross-cutting relation between shear bands of different generations, although without perceptible change in metamorphic grade. The youngest shearing in this zone is locally expressed in small-scale shear structures akin to widely spaced crenulation cleavage. On the other hand, within the braided Oko shear zone some early bands have amphibolite facies mineralogy whereas the younger bands are entirely in greenschist facies. In both Oko and Abirkitib zones, an early generation of quartz veins has been disrupted by late shearing whereas the youngest veins are unaffected. Gold mineralization associated with the shear zones is generally found within the early generation of veins (Almond et al. 1984b).

Lenses of ophiolitic rocks are more numerous within the massive shear zones than within the braided zones. Indeed the common occurrence of these rocks is the main reason for the common supposition that they mark oceanic sutures. The full ophiolitic sequence is nearly intact at Halaib (Fitches et al. 1983) and considerably dismembered at Onib (Hussein et al. 1984), while in the Nakasib all the common elements of the association except sheeted dykes occur but again are highly dismembered. The ultramafic rocks at Nakasib are mostly either completely serpentinized or carbonated. Where ophiolite lenses occur within braided shear zones, as they sometimes do, it is likely that they derive from suture zones but have been transported tectonically into their present position. At the intersection of the Oko and Nakasib zones (Figure 2),the ophiolite mass forming Jebel Igareri belongs to the Nakasib zone but has been dragged into conformity with the later shears of the Oko zone. A continuation of this process can account for the presence of local ultrabasic lenses within the braided zones elsewhere.

The Oko shear zone has offset the subvertical Nakasib zone by some 30 km in a sinistral sense at their point of intersection, 100 km west southwest of Port Sudan (Figures 1 and 2). Since subhorizontal lineations are dominant in the Oko zone it is likely that this offset represents a true strike-slip displacement. No equally clear examples of offset have been found elsewhere in this shear zone network, but minor structures (late shear bands, drag folds, tension gashes) indicate a dextral sense of strike-slip movement on the Nakasib zone. However, lineations in this zone locally plunge as steeply as 65° northeast, so vertical displacements were probably large along parts of this zone. In the Birkitib zone, lineations vary in plunge from 25°N to 25°S, suggesting predominantly strike-slip displacements, but there is only equivocal evidence of the sense of movement. Further north, the continuation of the Abirkitib shearing appears to have offset the Halaib–Onib belt of volcanic rocks and ophiolites in a sinistral sense. Our provisional conclusion is that movement on the shear zone network has been mainly strike-slip, with sinistral displacement on northerly trending shears and dextral displacement on northeast-trending shears. If the Oko and Nakasib shearing was penecontemporaneous, this pattern suggests a west northwest–east southeast regional compressional stress at the time of strike-slip displacement. Earlier movements on former suture zones, such as Nakasib and Onib, no doubt had a larger vertical component.

The essential characteristics of the massive type of shear zone follow from their origin by strike-slip reactivation of pre-existing zones of shearing and faulting, some of them probably oceanic sutures. The braided type, on the other hand, were strike-slip shear zones from their inception, and followed less well defined crustal weaknesses.

The rocks of the shear zones are mylonites rather than cataclasites, using 'mylonite' in the sense of White (1976) for rocks in which the mechanism of shear deformation has involved more or less thorough recrystallization. The mineralogy in most rocks indicates greenschist facies conditions during shearing. Chlorite, epidote, and sodic plagioclase, with or without actinolite, are major constituents of sheared basic rocks, while quartz, alkali feldspars, chlorite and minor epidote predominate in more acidic rocks. Where these low-grade shears affect the volcanosedimentary sequences, metamorphism within the mylonites is more or less isofacial with that in their country rocks, but country rock metamorphism preceded shearing, as evidenced by porphyroclasts of epidote within some of the mylonites. Where low-grade shears transect plutonic rocks there has been mineralogical retrogression, with chlorite displacing biotite, saussuritization of plagioclase, and replacement of hornblende and pyroxene by actinolite. However, there are also a few localities where shearing occurred at higher temperatures, resulting in mylonites composed of, for example, quartz, potassium feldspar, calcic oligoclase and hornblende. Such rocks have been found only within the Serakoit batholith, and it is therefore suspected that they reflect high ambient temperatures retained within igneous intrusions during early phases of shearing.

4. Age relations

Relative age limits on shearing, such as that along the Oko zone, are provided by sheared plutons of the Serakoit batholith on the one hand and by igneous complexes which cut short the shears on the other. Post shear intrusions into the Oko zone include some which are bimodal gabbro–granite complexes, such as those of Tamala and Erbab (Figure 2). Others are alkaline/peralkaline complexes, including Sha and Mindara. There are no radiometric ages on the bimodal complexes, and early Cretaceous ages on the alkaline complexes (Vail et al. 1984) obviously refer to anorogenic events much later than the shearing. However, two bimodal complexes some 50 km east of the Oko zone have been given ages of c. 720 Ma by Rb–Sr isochron methods (Klemenic 1985), and north–south dykes intruded into an offshoot of the Serakoit batholith have been assigned an age of c. 724 Ma (Vail et al. 1984). Our conclusion is that late shearing along the Oko zone probably occurred before 720 Ma. The age range of the intrusions comprising the Serakoit batholith is even more doubtful. In the eastern Desert of Egypt, comparable rocks were emplaced within the interval 715 – 665 Ma (Stern and Hedge 1985), whereas the batholiths of western Arabia range in age from 900 to 740 (Stoeser and Camp 1985). The Arabian batholiths may provide the better guide, because the mean age of the Arabian Shield rocks decreases from south to north (Stern and Hedge 1985), and the Arabian and Serakoit batholiths have similar latitudes.

The northeast-trending Nakasib zone developed in two main stages, both of them older than Oko shearing. During the first stage, ophiolites were emplaced along an oceanic suture, accompanied by shearing. Reworking of this suture by the second stage of shearing caused dismemberment of the ophiolites and replacement of the fragments within broad mylonite bands. We suggest that the first of these stages took place along a subduction zone, whereas the second accompanied collision between arcs. Current models of Arabian–Nubian Shield evolution, based on evidence from Arabia, suggest that subduction was well under way by 900 Ma, with interarc collisions during the interval 700 to 640 Ma. Field evidence from near the intersection of the Oko and Nakasib zone suggest that some, at least, of the granitoid members of the batholith association postdate the full development of the Nakasib zone but are older than Oko shearing. Pending aquisition of more radiometric dates we can conclude provisionally that the general sequence was:

1. Nakasib shearing
2. Batholith intrusion
3. Oko shearing took place within the interval 900–700 Ma, and probably towards the latter end of that time.

5. Tectonic Interpretation

Geological interpretation in the Red Sea Hills of Sudan is heavily dependent on comparisons with the better known and better dated geology in adjacent parts of the Arabian Shield. The recent model of Stoeser and Camp (Camp 1984; Stoeser and Camp 1985) for the western part of the Shield is particularly relevant here. These authors propose an evolutionary sequence in which volcanic arcs began to

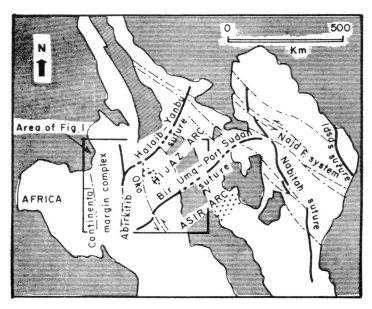

Figure 3. Arcs and related structures in the central Arabian–Nubian Shield (Red Sea closed), modified after Camp (1984) in the area outlined. Cover rocks horizontally ruled; stipple shows distribution of high-grade rocks in the Hijaz and Asir arcs.

grow at c. 950 Ma and, after reaching maturity, accreted together by a series of collisions during the interval c. 715–630 Ma. In what is now western Arabia and northeast Sudan, an older 'Asir arc' (900–700 Ma) collided with a younger' Hijaz Arc' (800–700 Ma) at c. 700 Ma, along a former subduction zone, to form the 'Bir Umq – Port Sudan suture (Figure 3). The Sudanese section of this suture lies along the line of the Nakasib shear zone, but the initial suture was modified by shearing with a large strike-slip component. The Hijaz Arc is bounded to the northwest by the 'Halaib–Yanbu subduction zone', corresponding with the line of the ophiolite-bearing shear zone extending from Onib to Halaib in Figure 1. Most of the area of Figure 1 thus lies within the 'Hijaz Arc' of Stoeser and Camp (1985; see also Vail this volume). The northeast-trending arcs are truncated in east Arabia by the north–south sutures of Nabitah and Idsas (Figure 3), which record the accretion of two other microplates at c. 650 Ma.

We accept the Stoeser and Camp model (Camp 1984; Stoeser and Camp 1985) as a working hypothesis, and our main contribution to its development is to point out that the northeast-trending Hijaz arc is truncated by northeast structures to the west, in Sudan as well as to the east in Arabia (Figure 3). The line of truncation is the north northwest shear zone west of Onib which cuts short the Onib–Halaib suture (Figure 1). This line is followed by dismembered ophiolites (Hussein *et al.* 1984) and by shearing which continues southwards as the Abirkitib shear zone. We regard the Abirkitib section of this line as a part of the north–south suture rejuvenated after intrusion of the Serakoit batholith. West of the Abirkitib line is the north–south metamorphic belt which forms the western block of the Red Sea Hills. This belt comprises a zone of low-grade metavolcanic and metasedimentary rocks backed by an area of high-grade metasedimentary rocks. Vail (1979, 1985) and Almond (1982, 1984) have argued that the high-grade rocks (around Abu Hamed, Figure 1) have the field characteristics of an older basement in respect to the low-grade rocks. This view is disputed by Ries *et al.* (1985), since they can find as significant age differences between the two groups. In any case, the metamorphic rocks of the western block can be interpreted as an immature island arc lying west of the Abirkitib suture. If the high-grade rocks do represent the basement of this arc, they could also mark the margin of continental Africa in the Proterozoic. Collision of this western arc with the Hijaz arc must have occurred later than 700 Ma, and might have been coeval with accretion of the eastern Arabia microplates at 650 Ma.

The Stoeser and Camp model offers an explanation for the general distribution of volcanosedimentary units and ophiolite-bearing sutures in Sudan, but does not specifically account for the episode of late shearing which rejuvenated the former sutures and created new shear zones in the Red Sea Hills. Nor does it adequately explain the major system of northwest sinistral strike-slip faults (Najd) which transects all the Arabian microplates (Figure 1). Stoeser and Camp (1985) regard Najd faulting as one facet of an intracratonic tectonism which continued for at least 80 Ma after collisional orogeny had ceased.

Like many previous writers, they draw an anology with the intracratonic phase which followed collision of the Indian and Asian plates. Almond (1984) has argued that this analogy is inappropriate, since there is no evidence for the deep erosion which would inevitably follow a Himalayan-type orogeny. This point is also made by Stern (1985), in a recent discussion on the origin of the Najd fault system. On the other hand, there is no reason to suppose that late strike-slip fault systems only arise in cases of 'hard' collision, such as between India and Asia. 'Soft' (i.e. less energetic) collisions will include those between island arcs, and between island arcs and continental margins. Strike-slip faulting will be favoured by collisions which are oblique, rather than head on. The north–south sinistral/northeast dextral phase of ductile shearing in northeast Sudan may have been a consequence of soft collisions in eastern Arabia, but in order to define this possibility we need more precise data on the age of the shearing. At the present time it seems likely that the north Sudan shearing is somewhat older than formation of the Najd faults (dated at around 540–620 Ma; Stern 1985), and took place at a deeper structural level. The northwest-trending Aswa shear zone of Uganda and southern Sudan has been placed within the age range 788 ± 15 to 558–590 Ma by Cahen *et al.* (1984). It could be of similar age to either Najd or the northern Sudan shear zones, but was probably of deeper level origin than either.

Acknowledgements. This work was supported by a grant from the University of Kuwait (Project SG 017), for which the authors are grateful. Field work in Sudan also benefited by help from the Department of Geology in the University of Khartoum and from the Sudanese Ministry of Energy and Mining. The figures were redrawn by Mustafa Kambal of the University of Kuwait.

References

Ahmed, F. 1983. Relationships of mineral deposits and lineament analysis of the Red Sea region, north-eastern Sudan. *Advances in Space Research*, **3**, 71–79.

Almond, D.C. 1982. New ideas on the geological history of the Basement Complex of north-east Sudan. *Sudan Notes and Records*, **59**, 106–136.

—— 1984. The concepts of 'Pan-African Episode' and 'Mozambique Belt' in relation the geology of east and north-east Africa. *King Abdulaziz University, Jeddah, Faculty Earth Science Bulletin*, **6**, 71–78.

——, **Ahmed, F. and Dawoud, A.S. 1984a.** Tectonic metamorphic and magmatic styles in the northern Red Sea Hills of Sudan. *King Abdulaziz University, Jeddah, Faculty Earth Science Bulletin*, **6**, 450–458.

——, **Ahmed, F. and Shaddad, M.Z. 1984b.** Setting of gold mineralization in the northern Red Sea Hills of Sudan. *Economic Geology*, **79**, 389–392.

Camp, V.E. 1984. Island arcs and their role in the evolution of the western Arabian Shield. *Geological Society America Bulletin*, **95**, 913–921.

Cahen, L., Snelling, N.J., Delhal, J. and Vail, J.R. 1984. *The Geochronology and Evolution of Africa.* Clarendon Press, Oxford. 512pp.

Dawoud, A.S. 1980. Structural and metamorphic evolution of the area SW of Abu Hamed, Nile Province, Sudan. Unpublished Ph.D. Thesis, University of Khartoum.

Embleton, J.C.B., Hughes, D.J., Klemenic, P.M., Poole, S. and Vail, J.R. 1984. A new approach to the stratigraphy and tectonic evolution of the Red Sea Hills, Sudan. *King Abdulaziz University, Jeddah, Faculty Earth Science Bulletin*, **6**, 101–112.

Fitches, W.R., Graham, R.H., Hussein, I.M., Ries, A.C., Shackleton, R.M. and Price, R.C. 1983. The late Proterozoic ophiolite of Sol Hamed, NE Sudan. *Precambrian Research*, **19**, 385–411.

Hussein, I.M., Kröner, A. and Durr, St. 1984. Wadi Onib — A dismembered Pan-African ophiolite in the Red Sea Hills of Sudan. *King Abdulaziz University, Jeddah, Faculty Earth Science Bulletin*, **6**, 320–327.

Klemenic, P.M. 1985. New geochronological data on volcanic rocks from the northeast Sudan and their implication for crustal evolution. *Precambrian Research*, **30**, 263–276.

Ries, A.C., Shackleton, R.M. and Dawoud, A.S. 1985. Geochronology, geochemistry and tectonics of the NE Bayuda Desert, N. Sudan: implications for the western margin of the Late Proterozoic fold belt of NE Africa. *Precambrian Research*, **30**, 43–62.

Stern, R.J. 1985. The Najd fault system, Saudi Arabia and Egypt: A late Precambrian rift-related transform system. *Tectonics*, **4**, 497–511.

—— **and Hedge, C.E. 1985.** Geochronologic and isotopic constraints on Late Precambrian crustal evolution in the Eastern Desert of Egypt, *American Journal of Science*, **285**, 97–127.

Stoeser, D.B. and Camp, V.E. 1985. Pan-African microplate accretion of the Arabian Shield. *Geological Society of America Bulletin*, **96**, 817–826.

Vail, J.R. 1979. Outline of geology and mineralization of the Nubian Shield east of the Nile Valley, Sudan. In: Al-Shanit, A.M.S. (Convenor), *Evolution and Mineralization of the Arabian–Nubian Shield*. Pergamon Press, Oxford, 97–107.

—— **1983.** Pan-African crustal accretion in north-east Africa. *Journal of African Earth Science*, **1**, 285–294.

—— **1985.** Pan-African (Late Precambrian) tectonic terrains and the reconstruction of the Arabian–Nubian Shield. *Geology*, **13**, 839–842.

——, **Almond, D.C., Hughes, D.J., Klemenic, P.M., Poole, S., Nour, S.E.M. and Embleton, J.C.B. 1984.** Geology of the Wadi Oko–Khor Hayat area, Red Sea Hills, Sudan. *Ministry of Energy & Mineralogy Resources Department Bulletin*, **34.**

White, S. 1976. The effects of strain on the microstructures, fabrics and deformation mechanisms in quartzites. *Philosophical Transactions of the Royal Society, London*, A.238, 69–86.

The subduction- and collision-related Pan-African composite batholith of the Adrar des Iforas (Mali): a review

J.P. Liégeois
Service de Géochronologie, Musée royal de l'Afrique centrale, 1980 Tervuren, Belgium

J.M. Bertrand
Centre de Recherches pétrographiques et géochimiques, B.P.20, 54501 Vandoeuvre–lès—Nancy, France

and

R.Black
Laboratoire de Pétrologie et U.A. 728, Université Pierre et Marie Curie, 75230 Paris Cedex 05, France

---

A large composite calc-alkaline batholith, in the Iforas region, Mali, occurs close to the Pan-African suture between the 2000 Ma old West African craton and the Trans-Saharan mobile belt. Its location in an embayment of the West African craton is probably responsible for the important production of magma. The Iforas batholith intrudes the western border of an old continental segment affected by early nappe tectonics (D1 event) and is flanked to the west by the Tilemsi palaeo–island arc. The batholith comprises several successive stages. The cordillera (>620 Ma), probably post-dating the D1 event, is essentially composed of volcanosedimentary sequences. The collision (620–580 Ma) is marked by the production of abundant granitoids mostly emplaced by the end of the D2 EW compressional event. The post-collision tectonic stages (D3 and D4; 580–540 Ma) are characterized by strike-slip movements, reversals in the stress field, and a rapid switch from calc-alkaline to alkaline magmatism. Magmas corresponding to each step show distinctive geochemical trends but all share low $^{87}Sr/^{86}Sr$ initial ratios (0·7035–0·7061). The possible successive sources have been evaluated from different entities in the Inforas region: Eburnean granulites for lower crust, Tilemsi palaeo–island arc for depleted subduction source and the Tadhak undersaturated province for asthenospheric more primitive mantle. A geodynamic model is proposed where all the calc-alkaline groups originated from a classical subduction source (depleted upper mantle modified by hydrous fluids from the subducted oceanic plate) which, some fifty million years after the beginning of the collision, was taken over by an asthenospheric source producing the alkaline province.

KEY WORDS   Wilson cycle   Calc-alkaline to alkaline magmatism   Contrasting stress regimes   End Pan-African   Iforas batholith, Mali

0072–1050/87/TI0185–27$13.50
© 1987 by John Wiley & Sons, Ltd.

1. Introduction

The Pan-African orogeny (± 600 Ma) is the oldest period where the complete Wilson cycle of large-scale plate tectonics is well established. The complete cycle occurred in the Trans-Saharan belt along the eastern margin of the West craton (see review by Cahen *et al.* 1984; Black *et al.* 1979; Caby *et al.* 1981), and accreted island arc material in the Arabian-Nubian shield (Greenwood *et al.* 1976; Duyverman *et al.* 1982). A striking feature of the Adrar des Iforas is the presence, 100 km to the east of the suture zone, of a large batholith (Figure 1) which displays a complete record of calc-alkaline magmatism exposed at different structural levels, linked respectively to subduction, collision, and finally post-collision stages, with a rapid switch to alkaline within-plate magmatism 50 Ma after oceanic closure.

Since the excellent pioneering work by Karpoff (1961) and Radier (1957), research over the last ten years has led to the publication of a new geological map (Fabre *et al.* 1982) and in this paper reference is made to some key areas where problems have been studied in some detail. The joint gravity survey carried out in the region provided some good constraints especially on the shape and extent of buried basic bodies which outline the suture zone and on the structure of the batholith (Bayer and Lesquer 1978; Ly *et al.* 1984).

The tentative synthesis of the batholith presented here is the result of two convergent approaches. JPL and RB, after detailed studies of the alkaline ring-complexes and associated dyke swarms, have worked backwards in time focusing their attention on the calc-alkaline–alkaline transition; JMB, starting from the early nappe structure, has moved forward in time, paying particular attention to the structural control of the early stages of emplacement of the batholith.

2. Geological Setting

2a. The setting of the batholith in the Iforas segment

The Iforas segment may be subdivided from west to east into three domains (Figure 1):

- The Tilemsi 'accretion domain' which is the result of the tectonic evolution of an island arc (Caby 1981); metabasalts and serpentinites of oceanic origin also occur to the west of the suture in nappes translated onto the West African craton (Taounnant, Timetrine; Caby 1978; Caby *et al.* 1981).
- The central Iforas domain, which is characterized at least in its eastern part, by the occurrence of early intracontinental deformation involving basement nappes and associated metamorphic reactivation (Boullier *et al.* 1978; Boullier 1982). The Iforas composite batholith, whose relationships with the surrounding gneisses show complex tectonic and plutonic interactions (Bertrand and Wright 1986), occupies the western half of the Central Iforas. It incorporates pendants of gneiss and belts of low-grade volcanosedimentary rocks (e.g. Tafeliant Group, Oumassene Group; Fabre 1982).
- The Eastern Iforas, delimited by the Adrar fault, forms the southern prolongation of the eastern branch of the Pharusian belt defined in the Hoggar (Caby *et al.* 1981; Bertrand and Caby 1978; Davison 1980; Caby *et al.* 1985).

2b. Lithological environment of the batholith

High-grade and low-grade formations are juxtaposed in the Central Iforas (Figures 1 and 2) and are separated by stratigraphical unconformities (Tafeliant,

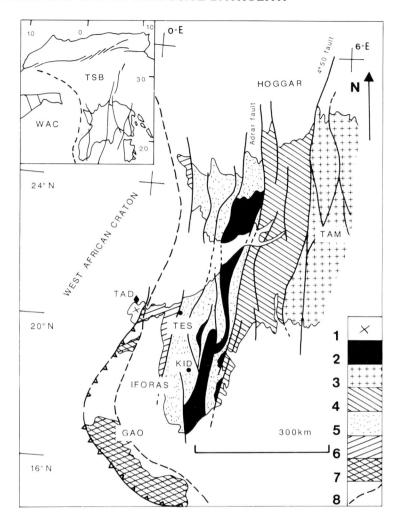

Figure 1. Sketch map of the northern part of the Trans-Saharan belt. 1) Eburnean gneisses of the West African craton. 2) Eburnean granulites within the mobile belt. 3) Reactivated pre-Pan-African basement of the Central Hoggar. 4) Eastern branch of the Pharusian belt. 5) Western branch of the Pharusian belt. 6) Tilemsi island arc. 7) Gourma and Timetrine nappes. 8) Suture zone. WAC= West African craton; TSB=Trans-Saharan belt; TAM=Tamanrasset; TES=Tessalit; KID=Kidal; TAD=Tadhak.

Ourdjan), tectonic contacts, or large areas of plutonic intrusives.

Amongst the high-grade rocks, the Iforas Granulitic Unit (IGU) is of Eburnean age (Lancelot et al. 1983) and is generally non-penetratively deformed and retrogressed during the Pan-African orogeny (Davison 1980; Boullier 1982). According to these authors the IGU forms a large-scale early nappe emplaced onto another high-grade complex, the so-called Kidal Assemblage. This latter unit is a 'sack term' for an association of different lithologies whose common features are polyphased deformation and high-grade metamorphism. It comprises, together with ubiquitous gneisses (reactivated granulites), abundant pretectonic intrusives and some highly deformed lenses of metasediments (quartzites, marbles and meta-

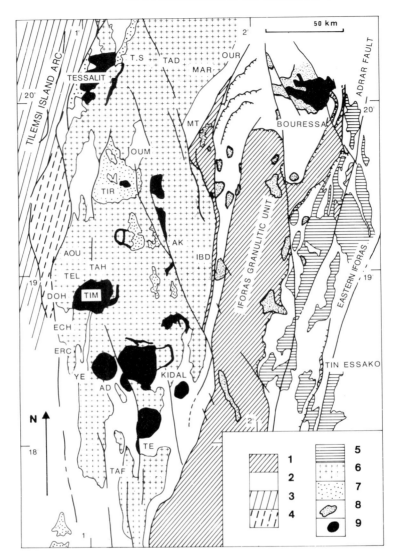

Figure 2. Sketch map of the Central Iforas. 1) Eburnean granulites (Iforas granulitic unit (IGU), Bezzeg unit, Tin Essako unit). 2) Undifferentiated schists and gneisses. 3) Volcanic and associated plutonic bodies of the Tilemsi island arc. 4) Aguelhoc gneisses. 5) Pan-African plutonic bodies of the eastern part of Central and Eastern Iforas, undifferentiated. 6) Iforas composite batholith. 7) Rhyolite flows. 8) Post-tectonic plutons in the Kidal Assemblage and in the IGU. 9) Alkaline ring-complexes. Localities quoted in the text: AD=Adma; AOU=Aoukenek; AK=Akoumas; DOH=Dohendal; ECH=Echaragalen; ERC=Erecher; IBD=Ibdeken; MAR=Mareris; MT=Mareris-Tichedait volcano-sedimentary unit; OUM=Oumassene; OUR=Ourdjan; TAD=Tadjoudjemet; TAF=Tafeliant; TAH=Tahrmert; TE=Teggart; TEL=Telabit; TIM=Timedjelalen; TIR=Tiralrar; T.S.=Tin Seyed; YE=Yenchichi.

pelites) assoicated with orthoamphibolites. Old supracrustals rocks which may have been affected by amphibolite facies metamorphism, include a sequence of aluminous quartzites, pelites and metarhyolites or alkaline leptynites which have been dated at 1837 ±13 Ma (Caby and Andreopoulos-Renaud 1983) and a unit characterized by the association quartzite–marble, with some pelites, meta–arkoses and minor volcanics, similar to the 'Série à Stromatolites' of northwest Hoggar (Caby 1970) traditionally attributed to the early Upper Proterozoic. This sequence is well developed on the eastern edge of the batholith and to the east of the IGU where it occurs as high level nappes overlying the granulitic unit; its deformation history is similar to that of the Kidal Assemblage. A similar unit displaying greenschist facies metamorphism may also be observed as a narrow monocline resting stratigraphically on top of the IGU 50 km ENE of Kidal.

The low-grade rocks comprise several volcanosedimentary units: the Tafeliant Group, a shallow sea clastic sequence containing a tillite, felsic and basic flows and dykes, which occurs as a N–S belt measuring 150 × 10 km between the Timedjelalen and Takellout complexes (Fabre 1982; Ball and Caby 1984); The Oumassene Group, a continental sequence composed essentially of andesites and subordinate pyroclastics; a volcanic sequence consisting mainly of dacites and rhyolites in the Mareris–Tichedait region and greywackes in the Ourdjan area. Along the western margin of the batholith, long and narrow belts of volcanics and greywackes have not been described in detail. They are unconformably overlain by molassic type formations (Echaragalen, Fabre et al. 1982). Large unconformable plateaux of rhyolite flows and ignimbrites (Tiralrar, Figure 2) are part of the late alkaline province and were previously called Nigritian by Karpoff (1960). Small undeformed basins of molassic red arkoses and conglomerates attributed to the Cambrian by comparison with the similar 'Série Pourprée' of northwest Hoggar (Caby 1970) are preserved along the major N–S shear zones.

2c. The structural evolution of Central Iforas

A succession of four main tectonic events has been recognized in the Central Iforas (Boullier et al. 1978; Davison 1980; Boullier 1979, 1982; Boullier et al. 1986; Champenois et al. 1986).

— The first event, D1, is characterized by the emplacement of large northward verging nappes, involving the 2000 Ma old Eburnean granulites (IGU). The lower and higher grade part of the nappe pile corresponds to the Kidal assemblage, while the upper nappes (Ibedouyen area) are metamorphosed under greenschist facies conditions. A possible equivalent of these upper nappes is known in the Ourdjan area (see Figure 2). The dating of the D1 event is a major problem as Rb–Sr and U–Pb isotopic systems have apparently been reset during a D2 event (Bertrand and Dautel unpublished data) and as a consequence are poorly constrained.

— The D2 event is compressional and varies in time from a SE–NW to a E–W shortening direction. It is clearly related to the collision and took place c. 600 Ma ago (Bertrand and Davison 1981; Bertrand et al. 1984b; Liégeois abnd Black 1984). In the Tafeliant area, the compressional deformation is accompanied by a sinistral N–S movement which can be confined to an age range between 620 + 8/−5 Ma (U/Pb on zircon age) and 595 ± 24 Ma (Rb–Sr whole rock isochron). These ages have been obtained from the late-tectonic Adma pluton (Figure 2) emplaced in the volcanosedimentary formation (Andreopoulos-Renaud in preparation; Liégeois and Black 1984). In the northern part of the batholith, Bertrand and Wright (1986) describe a large-scale dome-like structure, which is attributed to this phase of deformation.

— The D3 event corresponds to a reversal in the stress field expressed by dextral strike-slip movements along N–S to N20° shear zones and faults whose age has been estimated in the range 566–535 Ma (Lancelot *et al*, 1983). The shortening direction is NE–SW but rotates with time to E–W (e.g. in the Abeibara– Rharous shear zone; Boullier 1985).

— The last D4 tectonic event occurred in brittle conditions producing a conjugate set of strike-slip faults, the shortening direction having swung back to NW–SE. The spectacular intrabatholitic E–W and N–S acid dyke swarms were emplaced between the D3 and D4 event and are related to the rotation of the stress field (Boullier *et al.* 1986). The age of 545 ±16 Ma obtained in the Yenchichi area (Liégeois and Black 1984) probably dates the D4 event. The D4 event can be attributed to the last stage of the collision between the West African craton and the mobile belt (Ball 1980) even if these structures have been reactivated during the Phanerozoic.

An alternative interpretation has been proposed by Ball and Caby (1984), who dismiss the distinction between D1 and D2. They consider that the N–S trending open folds in the Tafeliant volcanosedimentary formation are the shallow expression of recumbent folds at depth, the entire edifice having resulted from continuous deformation. These features are interpreted as the result of underthrusting of the Kidal Assemblage below the Tafeliant unit. More precise geochronological constraints for the early stages are necessary to resolve this problem.

3. Petrology, geochronology and geochemistry

A summary of the magmatic succession is given in Table 1.

3a. Pre-D1 intrusives

The Kidal Assemblage, which outcrops along the eastern margin of the batholith (Figure 2), includes metatonalites associated with metatrondhjemites, gneisses, metasediments, metavolcanites, metagabbros, and ultramafites. All these rocks show polyphase structures only compatible with a pretectonic character. Between the northeastern margin of the batholith and the northern tip of the IGU, metatonalites occur as thick sheets displaying an easterly gently dipping foliation (30° to 50°) and have undergone complete recrystallization in amphibole facies conditions. No syntectonic intrusives have been recognized so far; metatonalites display only local partial melting during the D1 event. D1 features, which have been intensely overprinted by the D2 deformation and annealed by high-grade low-pressure metamorphism, can be found in numerous gneissic pendants within the batholith, east of a N–S line joining Tadjoudjemet and Kidal (Figure 2).

Even if the D1 nappe emplacement tectonics is related to the main collision (*ca.* 600 Ma), it is not clear whether or not the metatonalies of the Kidal Assemblage are linked to the subduction process. Indeed, all attempts to date these intrusions have failed, either by Rb–Sr whole rock dating or by U–Pb on zircon (Bertrand and Dautel unpublished results). Only the $^{87}Sr/^{86}Sr$ ratios 600 Ma ago can be estimated to be between ·0·705 and 0·707. The geochemistry of the D1 pretectonic intrusions is characterized by low-K_2O content (1·3 to 2·0 per cent) and very little correlation with silica (Bertrand *et al.* 1984b). The trend shown on the SiO_2 versus K_2O diagram (Figure 3) and in the Q–F diagram (Figure 4) may be compared with those of the tonalite–trondhemite suite (Barker 1979).

Furthermore, geochemical data (major, trace, Ree elements) on the surrounding amphibolites which are regarded as metavolcanics show clear affinities suggesting

Table 1: Magmatic succession in the Iforas batholith

(R)Yenchichi 1 granite	S	544±16Ma	(0·7063±5)	(D4 tectonic)
(I)Timedjelalen ring-complex	C	549±6Ma	(0·7051±5)	
(I)Kidal ring-complex Lavas	S	561±7Ma	(0·7061±7)	WEAK EXTENSION
(I)N–S dykes	C	543±9Ma	(0·7050±3)	
(I)Tahrmert alkaline granite	C	541±7Ma	(0·7061±4	
(I)Yenchichi 2 syeno-granite	S	577±14Ma	(0·7038±10	
(I)E–W dykes:				
Telabit	C	544±12Ma	(0·70505±10)	UPLIFT
Dohendal	C	556±10Ma	(0·70511±12)	
Yenchichi	S	565±14Ma	(0·7048±5	
(EC)Aoukenek monzo-granite	C	591±18Ma	(0·7035±5)	(D3 tectonic)
(EC)Iforas monzogranite	C	"error chron"	(0·7042 to 0.7053)	
(EC)Adma granodiorite	S	595±24Ma	(0·70482±26)	COLLISION
(EC)Tin Seyed Q-monzodiorite	N	581±15Ma	(0·70530±14)	(D2 tectonic)
(E) Erecher tonalite	S	602±13Ma	(0·70590±8)	
(R)Ibdeken granodiorite	S	613±29Ma	(0·7057±2)	
(BC)Adma granodiorite	S	620±6Ma	(Andreopoulos-Renaud in preparation)	
(I)Yenchichi 1 granite	S		(<0·704)	
(E)Erecher tonalite	S		(<0·705))	SUBDUCTION
(I)Ibdeken granodiorite	S		(<0·705)	
(I)Tafeliant dykes	S	634+15Ma	(0·70505+6)	
(I)Teggart Q-diorite	S	696±5Ma	(Caby and Andreopoulos-Renaud, 1985)	
(I)Pre-D1 intrusives	K	?		(D1, tectonic)

(Q) = model quartz
(R) = rehomogenization
(EC) = end of crystallization
(BC) = beginning of crystallization
(I) = intrusion

The chronological order of this table is based on field relationships.
N=northern part of the batholith; C=central part; S=southern part.
K=Kidal Assemblage (just east of the batholith).

an extensional environment for the Kidal Assemblage pre-D1 metavolcanics and intrusives (Leterrier and Bertrand 1986). It should be noted, however, that there are examples of deformed plutons along the eastern margin of the batholith such as in the Mareris area which have slightly higher K_2O content and display trends intermediate between the Kidal metatonalites and the batholith in the K_2O versus SiO2 and Q – F diagrams. Although texturally indistinguishable from the Kidal Assemblage metatonalites, a cordilleran origin cannot be entirely dismissed.

3b The cordillera episode

This stage is characterized by volcanoclastic or volcanic deposits (Fabre 1982) representing marine, aerial or subaerial environments (Tafeliant Group, Oumassene Group, Ourdjan area, respectively). The Oumassene group is composed mainly of continental andeites with intercalated pyroclastics (Chikhaoui 1981).

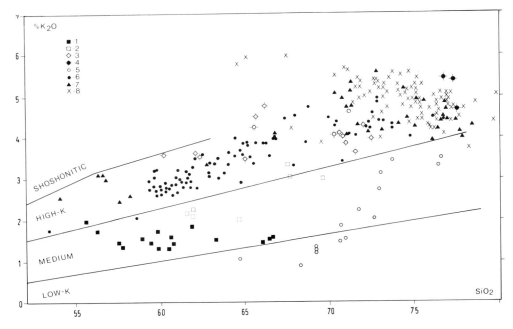

Figure 3. K_2O versus SiO_2 diagram. The field boundaries are from Ewart (1979) 1) Pre-tectonic metatonalites from the Kidal Assemblage. 2) Pre-D1 Mareris quartz-metadiorites. 3) Pre-D2 Ibdeken quartz-monzodiorites and granodiorites. 4) Pre-D2 RAPU. 5) Pre-D2 Erecher tonalite. 6) Late-D2 granitoids (from quartz-monzodiorites to monzo-granites). 7) Post-D2 high-level plutons and E–W dykes. 8) Post-D2 alkaline N–S dykes and ring-complexes

Dykes and sills, intruded in the Tafeliant formation, have been dated at 634 ± 15 Ma (Liégeois, Fabre and Caby work in progress).

The degree of the diachronism among these basins is unknown; however, all the volcanosedimentary sequences found within the batholith are monocyclic and, as in the case of the Tafeliant Group, are believed to lie unconformably upon deformed gneisses and on Upper Proterozoic sediments.

By contrast, large cordilleran-related plutons are scarce, the Iforas batholith being essentially late- to post-tectonic in character, with emplacement occurring after oceanic closure. Essentially three plutons have been recorded and studied, which share the pre-D2 character. The Teggart quartz-diorite (Figure 2) is the older, situated beneath the unconformity of the Tafeliant Group and has been dated at 696 ± 6 Ma (Caby and Andreopoulos-Renaud 1985). The detailed chemistry of this pluton has not yet been studied. This intrusion has been previously considered as having a syn-D1 character (Boullier et al. 1978; Caby et al. 1981) but recent observations show it to be earlier than the nappe structures (Caby and Andreopoulos-Renaud 1985).

The other two plutons are widely separate, one near the island arc boundary to the west, the other on the eastern edge of the batholith, not far from the Iforas Granulitic Unit. If the former (Erecher tonalite, Liégeois and Black 1984) is clearly intrusive into a volcanosedimentary sequence, the latter (Ibdeken quartz-monzodiorites and granodiorites) has no field relationships with such sequences.

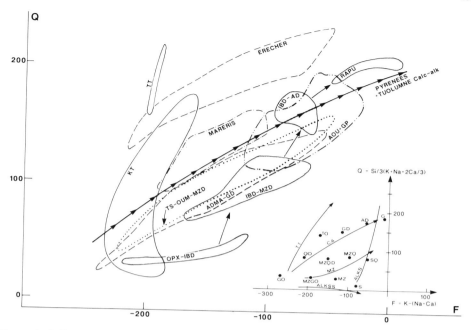

Figure 4. Q–F diagram (expressed in millications = gram-atoms × 10^3 in 100 g of rock or mineral, La Roche (de) 1962). Standard locus and terminology are from Debon and Le Fort (1983). Some typical suites are given in the insert where: Tr=tholeiitic (tonalite) suite from Afghanistan (Debon and Le Fort, 1983); CA=calc-alkaline suite from French Pyrénées (Debon, 1980); MZ=high-K suite from Vosges (Pagel and Leterrier 1980); ALKS=alkaline oversaturated suite from Afghanistan; ALKSS=alkaline saturated suite from Afghanistan (Debon and Le Fort 1983). In the main diagram: KT=metatonalites from the Kidal Assemblage; TT=particular metatonalites from the Kidal Assemblage (Bertrand et al. 1984a); OPX–IBD=OPX-bearing Ibdeken quartz-monzodiorites; IBD–MZD=Ibdeken quartz-monzodiorites; IBD–AD=Ibdeken granodiorites; TS–OUM–MZD=Tin Seyed quartz-monzodiorites; ADMA–GD=Adma granodiorite; AOU–GP=porphyritic monzogranite and Aoukenek fine-grained monzogranite; RAPU=red aplo-pegmatitic unit.

The two plutons give similar Rb–Sr whole rock ages, attributed to a resetting of the isotopic system during the D2 deformation related to the collision with the West African craton, the initial ratios being then considered as maximum values: Erecher tonalite: 602 ±13 Ma (Ri = 0·7059 ±8, 9WR, MSWD=1.0, Figure 7; Liégeois and Black 1984); Ibdeken quartz-monzodiorite: 613 ±29 Ma (Ri= 0·7057 ±2, 7WR, MSWD=1.6, Figure 5, Table 2).

The two units display different geochemical trends which may be explained by the variation of the K_2O associated with the spatial distribution of granitic plutons (Dickinson and Hatherton 1967). The main discrimination is the content in K_2O vs h where h is a measure of the vertical distance from the inferred subduction trench zone. More precisely, the Erecher tonalite (Liégeois and Black 1984) is characterized by a relatively high SiO_2 mean content and by the evolution of the more basic terms in the low-K calc-alkaline field and by a rapid enrichment in K_2O in the more felsic samples (Figure 3). In a general way, the Erecher tonalite displays a particular trend intermediate between a classical cordilleran magmatism and various island arc patterns. The location of this pluton along the western border of the batholith, near the island arc, could explain this feature. All

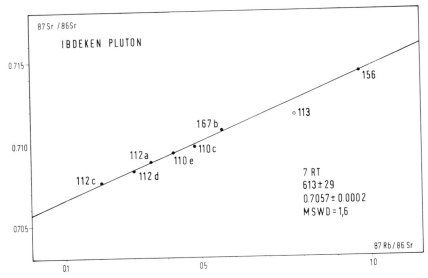

Figure 5. Rb–Sr isochron of the D2 pretectonic Ibdeken pluton (rehomogenization age). See Table 1 and appendix.

Table 2. Rb–Sr isotopic results for the Ibdeken and the Tin Seyed plutons

Sample	Rb ppm	Sr ppm	$^{87}Sr/^{86}Sr$	$\pm 2\sqrt{m}$	$^{87}Rb/^{86}Sr$
Ibdeken pluton					
Q110c	113	678	0·70979	0·00004	0·4824
Q110e	94	648	0·70941	0·00006	0·4198
Q112a	76	622	0·70887	0·00004	0·3536
Q112c	70	960	0·70761	0·00007	0·2110
Q112d	85	813	0·70829	0·00004	0·3025
Q113	138	513	0·71167	0·00004	0·7787
Q156	158	463	0·71423	0·00006	0·9946
Q167b	106	545	0·71080	0·00006	0·5630
Tin Seyed pluton					
JPL 499	96	629	0·70884	0·00006	0·4417
JPL 501	86	620	0·70854	0·00004	0·3995
JPL 502	97	525	0·70976	0·00004	0·5320
JPL 503	115	442	0·71140	0·00004	0·7531
JPL 505	124	460	0·71181	0·00004	0·7803
JPL 506	90	656	0·70859	0·00004	0·3983
JPL 507	152	343	0·71597	0·00004	1·283
JPL 508	174	311	0·71866	0·00005	1·621
JPL 509	136	374	0·71500	0·00005	1·053
JPL 510	78	685	0·70807	0·00004	0·3287
JPL 511	103	535	0·71005	0·00004	0·5572
JPL 512	157	330	0·71636	0·00004	1·378

the geochemical and geological characteristics indicate that the Erecher tonalite originated during the building of the cordillera; its geochemical and isotopic signature can then be taken to infer the characteristics of the subduction source near the trench. More data are necessary to define the problem, but the Ibdeken pluton is a good candidate to constrain the subduction source far from the trench.

In the same area as the Ibdeken pluton, there is an intricate altered and often highly deformed complex composed of aplitic and pegmatitic dykes and of leucogranitic pegmatitic irregular stocks. There is a possible genetic link between the RAPU (Red Aplo-Pegmatitic Unit) and the Ibdeken pluton following Bertrand and Wright (1986) and in this case, the RAPU may be considered to be similar to the Jaglot complex in the Kohistan arc of the Himalayas which is interpreted as the root of a cordillera (Petterson and Windley 1985). In view of the lack of constraints, however, a crustal origin for the leucocratic material cannot be excluded.

3c. The collision period: late-tectonic magmas

Granitoids of this group make up the bulk of the batholith. Although an absolute diachronism has possibly occurred in different areas of the batholith (250 × 100 km in size), the same relative chronology of intrusive phases has always been observed with an evolution from more basic to acid compositions. The whole group is intrusive into volcanosedimentary sequences with the frequent development of contact aureoles except when, as in the northern part of the batholith (Tin Seyed area), pendants of volcanites have been previously migmatized during D2.

Unlike the country rocks most of these intrusions have not undergone the climax of the D2 collision tectonics and are considered as 'late tectonic'. Liégeois and Black (1984) have described a two-stage crystallization, the older one being broadly contemporaneous with deformation while the younger was post-tectonic (Adma granodiorite, porphyritic granite).

New geochemical and isotopic data have been obtained on the Tin Seyed quartz-monzodiorite N–S plutonic alignment which stretches from the Oumassene andesites in the south to Tin Seyed. This band, comprising of three units generally showed undeformed or slightly deformed textures, except in narrow zones, and is composed of plagioclase, microcline, green hornblende and biotite, and rare quartz. Sphene, apatite and Fe–Ti oxides are common accessory minerals. An 11 point Rb–Sr isochron on the southern part of the Tin Seyed alignment, gave an age of 581 ± 15 Ma (Ri = 0.70533 ± 0.00014, MSWD=0.9, Figure 6, Table 2), which is within the range of other late-tectonic pluton ages shown in Figure 7 (Liégeois and Black 1984). The Sr initial ratios are all similar ($0.7053-0.7048$).

The main Iforas porphyritic granite which occurs all over the batholith and is particularly well developed in its western part, also shows, although less clearly, the two steps of crystallization, but is intrusive in the Tin Seyed quartz-monzodiorites. It yields only an errorchron for various reasons (discussed in Liégeois and Black 1984). However before the central part of the batholith was completely consolidated, it was intruded by the fine-grained Aoukenek granite with an estimated Sr initial ratio in the range $0.7042 - 0.7053$. Bertrand and Davison (1981) have recognized subhorizontal layering of granodiorites and porphyritic monzogranites in the Tadjoudjemet area of the northern part of the batholith. This feature suggests sheet-like intrusion, parallel to the main foliation in the high-grade metavolcanics (interpreted as S2 by Bertrand and Wright 1986), with the porphyritic granite emplaced after the granodiorites. The area is cut by small bodies of fine-grained biotite granite. The heterogeneity of this area (from

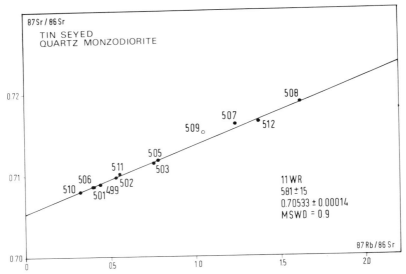

Figure 6. Rb–Sr isochron of the Tin Seyed late-tectonic quartz-monzodiorite. See Table 1 and appendix.

hornblende-bearing granodiorite to leucocratic varieties) probably explains the errorchron defined by the Rb–Sr method on seven whole rock samples (613 ±45 Ma, Bertrand and Davison 1981) and the scatter of the samples on most of the geochemical diagrams. Two samples of the porphyritic granite seem to suggest higher initial ratios, but more field and geochemical data are needed to evaluate the extent of the crustal contamination proposed by Bertrand and Davison (1981).

All the late-tectonic plutons show a well-defined high-K calc-alkaline trend (Figure 8.). Moreover, the Adma granodiorite, the porphyritic granite and the Aoukenek plutons determine a very good calc-alkaline array on the AFM diagram (Figure 9). A distinct and unique trend is yielded on various diagrams involving trace elements e.g. Sr versus Rb (Figure 10).

In conclusion, the different studied complexes, representative of the large collision-related magmatic group, have similar ages (Rb/Sr isochrons: 595 ±25 Ma, 591 ±18 Ma, 581 ±15 Ma), similar Sr initial ratios (0·7048, 0·7035, 0·7053), and similar geochemical evolution. Although they are not strictly cogenetic, these plutons can be considered as sharing the same source and evolution that characterize the collision environment. This geochemical evolution is very similar to that described for modern plate boundary series and closely resembles the cordilleran magmatism of western U.S.A. (Ewart 1979, 1982; Figure 8). This similarity to subduction-related magmatism rather than collision-related magmatism will be discussed in the last section.

3d. The calc-alkaline – alkaline transition

This transition occurred in post-tectonic conditions relative to D2 after considerable uplift of the belt. The relationships with D3 (N–S to N20° dextral strike-slip shear zones and faults) and D4 tectonics (brittle sinistral NNW–SSE and dextral ENE–WSW conjugate faults) are discussed in details by Boullier et al. (1986) while the petrographic and petrogenetic point of view is discussed in Liégeois and Black (1984, 1986).

In this post-tectonic environment the calc-alkaline magmatic production (post-tectonic I group) is rapidly taken over by alkaline to peralkaline magmas (post-tectonic II group). The calc-alkaline post-tectonic I group is represented by extensive E–W dyke (e.g. Telabit, Dohendal swarms, Figure 2) which are cross-cut by circular high-level plutons (e.g. Yenchichi 2). Quartz–two–feldspar porphyry is the most common rock type in the E–W dykes, the majority of which are acid and high-K calc-alkaline. However, the largest swarm (Telabit) consists of dykes with a more basic composition (down to 54 per cent SiO_2) and the Dohendal swarm, west of the Timedjelalen alkaline ring-complex, shows calc-alkaline dykes with alkaline affinities such as perthitic alkali feldspar, the presence of fluorite,

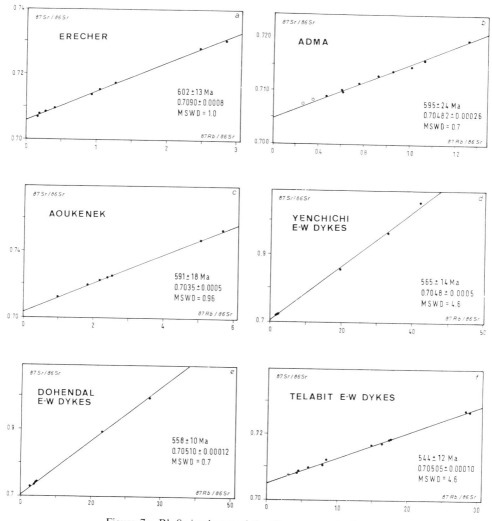

Figure 7a. Rb-Sr isochrons of the Iforas composite batholith.

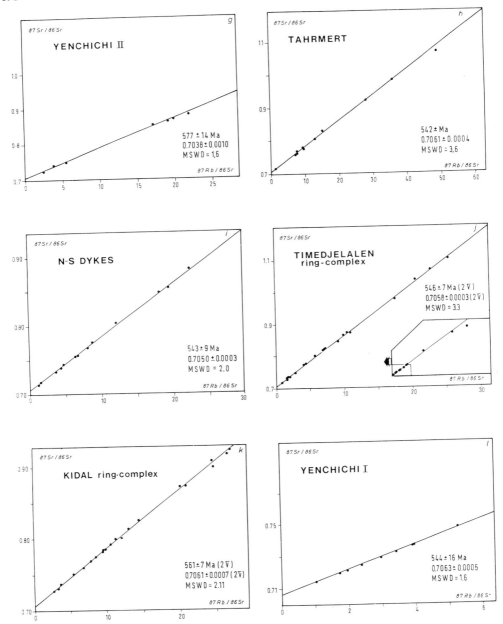

Figure 7b. Rb–Sr isochrons of the Iforas composite batholith (Liégeois and Black 1984). See appendix.

and the occurrence of red aphanitic felsites. The Yenchichi-type pluton is an homogeneous coarse-grained leucocratic biotite syenogranite (zoned oligoclase, perthitic microcline, quartz, and biotite).

The geochemical evolution of the post-tectonic group I is based essentially on the Telabit swarm. The trends defined by the major elements are similar to the

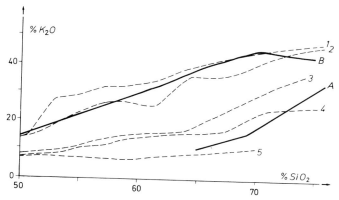

Figure 8. Schematic K$_2$O versus SiO$_2$ diagram comparing reference trends to the pre- (A) and late-tectonic (B) trends of the Iforas batholith. Ref. 1 to 4 from Ewart (1979) and 5 from Miyashiro (1974). 1. Western U.S.A., eastern zone; 2. Western U.S.A., western zone; 3. Northeastern Pacific (Cascades–Aleutians); 4. Northwestern Pacific (Kamchatka–Mariannes); 5. Central Kurilles.

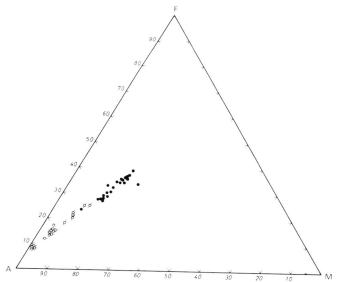

Figure 9. AFM diagram (A=Al$_2$O$_3$, F=Fe$_2$O$_3$+FeO, M=MgO) showing the calc-alkaline trend of the late-tectonic group of the southwestern part of the batholith. Filled circles: Adma granodiorite; open circles with inclined line: porphyritic monzogranite; open circles with horizontal line: Aoukenek fine-grained monzogranite.

calc-alkaline late-tectonic trend, with alkaline affinities for some samples as shown on the K$_2$O vs SiO$_2$ diagram (Figure 3). Trace element signatures clearly discriminate the group, placing it in an intermediate position between the late-tectonic and the alkaline groups, but always nearer to the former (see for example Figure 10).

Three dyke swarms (Yenchichi, Dohendal, Telabit) and one circular pluton (Yenchichi 2) have been studied isotopically and have yielded Rb–Sr isochrons

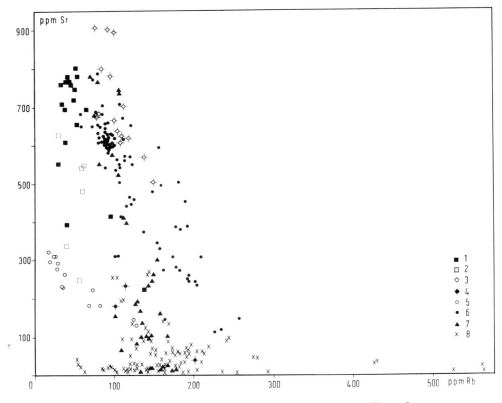

Figure 10. Sr versus Rb diagram. Same symbols as in Figure 3.

(Liégeois and Black 1984, see Figure 7). These data indicate an age trend from south (Yenchichi swarm: 565 ±14 Ma, Yenchichi 2 pluton: 577 ±14 Ma) to north (Dohendal swarm: 556 ±10 Ma, Telabit swarm: 544 ±12 Ma). This tendency is also apparent in the subsequent alkaline group (see below). The Sr initial ratios are in the range 0·7038–0·7051, which are very similar to the late-tectonic range (0·7035–0·7053).

This group, of relatively restricted occurrence, is considered as representing the end of the collision-source magma production with an incipient and erratic participation from a deeper alkaline source emplaced at high structural level during the uplift of the whole belt (from 575 Ma to 545 Ma).

In contrast, the alkaline post-tectonic group II is well represented in the western and central parts of the batholith. Field relationships have shown that the different terms of this group are always later than the post-tectonic group I, but the relative chronology of the different alkaline rocks is intricate. Indeed, the general succession of the Iforas alkaline province consists of early hybrid plutons, followed by giant N–S dykes swarms, rhyolitic and ignimbritic lavas and ring-complexes, but some N–S dykes cut the lavas and even some of the early phases of the ring-complexes. Likewise, the D4 brittle deformation does not affect all the ring-complexes. The hybrid plutons such as Tahrmert type (Ba *et al.* 1985) however are early as they mark the beginning of the unroofing accompanied by brittle behaviour, and the formation of an erosion surface on which the lavas have poured out. In detail, the Tahrmert pluton is a coarse-grained granite with alkaline-type felsic and accessory minerals (rounded quartz, perthites, zircon, sphene, Fe–Ti oxides and abundant fluorite) but with magnesian-rich mafic minerals.

The N–S dykes are typically alkaline, represented essentially by quartz- and quartz-feldspar porphyries and by some microsyenites with basic inclusions. Devitrified rhyolitic dykes are also abundant. Both metaluminous and peralkaline varieties are present. Most of these N–S dykes occur along the axial zone of the batholith centred on the major rhyolitic and ignimbritic plateau of the Iforas (Tiralrar) and may have been feeders to the lavas (Liégeois and Black 1984).

The alkaline complexes, about fifteen in number, include typical large size ring-complexes (20 to 30 km in diameter, e.g. Kidal and Timedjelalen ring-complexes). Several complexes are described in detail in Ba et al. 1985. All the rocks are rich in silica. Two distinct trends have been recorded, (in Figure 11) as a sodic peralkaline trend and a potassic aluminous trend (see Ba et al. 1985 and Liégeois and Black 1986). The peralkaline trend shows an evolution from quartz-syenite with an increase of Na+K/Al with constant K/Na represented by fractionation of plagioclase and early fayalite, Ca-pyroxene, and amphibole. There then followed a decrease in K/Na, controlled by substantial fractionation of K-feldspar, as the liquid line of descent crosses the peralkaline field and descends down the thermal valley (Bailey and MacDonald 1969). Late deuteric recrystallization has occurred in the presence of hydrothermal F-bearing fluids giving albite–microcline arfvedsonite granites.

The aluminous trend appears between the granite porphyries and the subsolvus biotite–chlorite granite, the trend being marked by a decrease of Na+K/Al and Na/K ratios reflecting the appearance of biotite and perthite with K>Na. A possible explanation can be provided by fractionation of basic oligoclase, Fe-hornblende, and clinopyroxene, if the dark inclusions, described by Ba et al. 1985, are cumulates. The two trends separated before the syenite stage. Nevertheless, a common deep source is likely for both granitic trends as their initial ratios are identical.

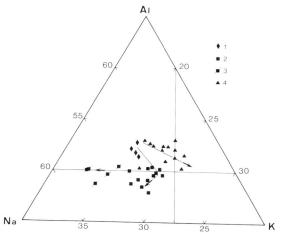

Figure 11. Atomic Na–K–Al diagram for the Kidal ring-complex (Ba et al., 1985). 1. Syenite. 2. Peralkaline hypersolvus granite. 3. Albitic peralkaline granite. 4. Metaluminous granite.

The Iforas ring-complexes display then all the main petrological and geochemical characteristics of a typical anorogenic oversaturated province (e.g. Niger–Nigeria province; Black 1963; Black et al. 1967; Jacobson et al. 1958; Bowden and Turner 1974). The principal difference is the absence of volcanism centred on the ring-complexes. In the Iforas the Tiralrar volcanic plateaux have a fissural origin, related to the N–S dyke swarms. There is also a distinctive lack of basic rocks and of any economic mineralization in the Iforas. Although the Rb–Sr geochronology shows a general south to north migration in age, it is unable to separate the various overlapping alkaline intrusions (Figure 12).

A Pb–Sr–O study of the Timedjelalen ring-complex and of the Tiralrar N–S dyke swarm (Weis et al. 1986b), shows that meteoric alteration has been intense but only locally, namely at the ring dyke contacts, and that the ^{18}O values and Pb initial ratios are, as with the Sr initial ratios, only compatible with a deep subcrustal origin.

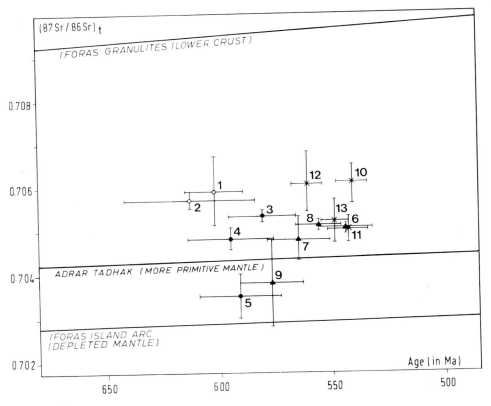

Figure 12. $^{87}Sr/^{86}Sr$ initial ratios versus time diagram for the Iforas composite batholith. Empty circles: pre-D2 plutons (rehomogeneization values); filled circles: late-tectonic plutons; triangles: post-tectonic I calc-alkaline dykes and plutons; crosses: post-tectonic II alkaline dykes and plutons. The different plutons and dyke swarms are: 1: Erecher tonalite; 2. Ibdeken quartz-monzodiorite; 3. Tin Seyed quartz-monzodiorite; 4. Adma granodiorite; 5. Aoukenek fine-grained monzogranite; 6. Telabit E–W dyke swarm; 7. Yenchichi E–W dyke swarm; 8. Dohendal E–W dyke swarm; 9. Yenchichi 2 syenogranite; 10. Tahrmert alkaline granite; 11. Tiralrar N–S dyke swarm; 12. Kidal ring-complex; 13. Timedjelalen ring-complex.

THE ADRAR DES IFORAS COMPOSITE BATHOLITH

4. Discussion and conclusions

4a. The geodynamic evolution from structural data

The external shape of the composite batholith is controlled by the position of the suture zone. The batholith forms a large elongated body parallel to the suture, and at a distance of *ca.* 100 km from the suture (see Figure 2.). This feature implies an overall control by subduction and collision processes. In the batholith, the cordilleran-type volcanism and plutonism postdates a large-scale tectonometamorphic event whose age and significance are still in question (D1 event), and the bulk of the volume of plutonic rocks in post-oceanic closure. They have been affected by collisional processes operating either at a deep ductile level (D2 and D3 events) or in shallow and brittle conditions (D4 event).

The existence of the D1 event implies an early collision. Its intensity and kinematic characteristics are quite different from the second D2 collision which may be considered in the Iforas as a docking rather than a Himalayan-type confrontation (Boullier 1982: Liégeois and Black 1984). The poorly-constrained age of the D1 deformation leads one to envisage two geodynamic possibilities. If the D1–D2 interval is short, implying a genetic link between the two collisions, the shoulder shape of the suture near Niamey can play the role of a rigid promontory in an early phase, producing the nappe structure further north (Figure 13A). In the other case, i.e. if the D1 deformation occurred before e.g. 700 Ma, one must seek, an early collision elsewhere, as already proposed in the northern part of the Tuareg shield (Caby *et al.* 1981) or to the south with the Congo craton.

In this connection it should be recalled that stacking of crystalline nappes indicative of N–S shortening occurs in the Tassendjanet area in northwest Hoggar (Caby 1970) and is a characteristic feature of the Central Hoggar–Air domain of the Tuareg shield (Bertrand *et al.* 1985; Latouche 1986) where tangential tectonics occurred in the 630–580 Ma range (Bertrand *et al.* 1984d, 1986).

The D2 deformation marks the collision with the West African craton, which was oblique (Ball and Caby 1984). An important feature is the gentle dip of the main foliation (both in high-grade volcanosedimentary facies and in plutonic rocks) occurring in the batholithic domain and contrasting sharply with the N–S upright folds and cleavage affecting the margins and the low-grade volcanosedimentary basins. This contrast could be related either to a flattening induced by the interface between large-scale plutonic domes and surrounding schists (Brun 1981; Bertrand and Wright 1986) or to a difference in exposed structural level. A striking feature is the N–S subhorizontal stretching related to sinistral shear observed along the western margin of the Tafeliant belt.

The D3 deformation, still ductile in the mylonites near the granulites (Boullier 1985), marked a reversal in the stress field and now corresponds to a NE–SW shortening direction (Figure 13c). After uplift leading to the removal of at least 10 km of material (low-pressure sillimanite outcropping at the pre-rhyolitic flow erosion surface), the D3–D4 transition reflected a complete reversal in direction of the maximum compressive stress (from NE–SW back to ESW–NNW; Figure 13d). Boullier *et al.* (1986) suggest that such reversals can have a profound disruptive effect without necessarily large strike-slip movement and in this case correspond to the emplacement of alkaline magmas (Black *et al.* 1985).

4b. Origin and evolution of the magmas

From the field, petrological, geochemical, and geochronological data a model can be proposed for the petrogenetic evolution of the Iforas batholith. The

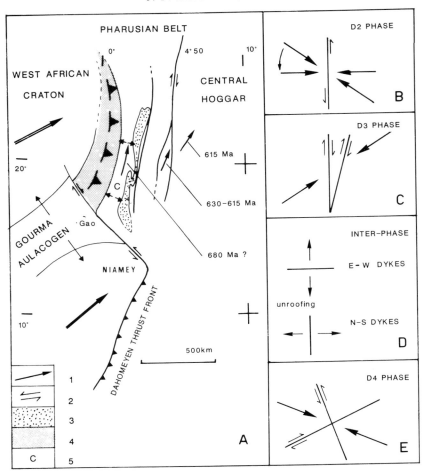

Figure 13. (A): Sketch map showing the Trans-Saharan belt just after an early oblique collision (680 Ma?) producing the nappe structures. The Iforas island arc and cordillera would have been protected by the embayment of the West African craton around the Gourma aulacogen. In this area, the collision (in fact a docking) has occurred later, around 600 Ma. The two eastern thrust ages are from Bertrand et al. (1984c, 1986), the 680 Ma (?) is hypothetical (see text). Arrows in the West African craton indicate the postulated movement of the craton relative to the mobile belt and the extension resulting in the Gourma aulacogen. 1. Early D1 thrusts and wrench faults; 2. Post-D1 relative movements. 3. Eburnean granulitic units, 4. Oceanic domain with subduction; 5. Cordillera. (B) to (E) (based on Boullier et al. 1986): shortening (B, C, E) and extension (D) directions during the successive collision (ca. 600 Ma) and post-collision (580–540 Ma) events.

studied plutonic bodies and complexes have relatively low $^{87}Sr/^{86}Sr$ initial ratios (0·.7035–0·7061, Figure 12) which preclude an origin from an old continental crust. The remaining possible sources are late-Proterozoic upper crustal rocks, lower continental crust or depleted or more primitive mantle. Isotopic characteristics of these reservoirs can be inferred in the Iforas region.

The lower crust. The Iforas granulitic unit (IGU) which is interpreted as a basement nappe (Boullier et al. 1978) can be taken as representing the lower continental crust. Granulite facies metamorphism has been dated ca. 2100 Ma (Lancelot et al. 1983) but in a northern extension of the IGU in Algeria (In Ouzzal unit), the material is Archaean in age (Ferrara and Gravelle 1966; Allègre and Caby 1972). Nine representative whole rocks have been measured for their Sr isotopic composition (Table 3).

Even if the data for sample 5513 is ignored, the mean ($^{87}Sr/^{86}Sr$) 600 Ma ratio is 0·7095. These results preclude an origin from the lower crust as represented by the IGU since the Iforas batholith is less radiogenic. To infer a hypothetical more depleted lower crust is difficult because the Aoukenek granite, for example, has a $^{87}Sr/^{86}Sr$ initial ratio as low as 0·7035 and such a depleted old lower crust is not propitious for producing a huge quantity of magma. Thus, we think that the Iforas lower crust has not played a role as a magma source, but may have acted as a contaminant.

The young upper crust. The Pre-cordilleran young crust consists of Proterozoic shelf sediments and the Kidal Assemblage. The former are very poorly-represented, particularly in the western batholith and can be ignored; the latter is not known with certainty west of the Tadjoudjemet–Kidal N–S line and in any case, its ($^{87}Sr/^{86}Sr$) 600 Ma mean ratio is also too high as shown by the representative samples of the Table 3 (mean: .07062). Its role would then be similar to that of a lower crustal contaminant.

The depleted mantle. The composition of the depleted mantle can be inferred from the magmatism of the Iforas palaeoisland arc (Caby 1981), where no basement remnants have been found. It is composed of trench-type volcanosedimentary sequences intruded by basic and ultrabasic rocks. The $^{87}Sr/^{86}Sr$ initial ratios 700–600 Ma ago are around .07030 (Dupuy, Liégeois and Caby in preparation), and are interpreted as the lithospheric depleted mantle with alkali-rich hydrous fluids coming from the subduction oceanic slab (Thorpe et al. 1976).

Table 3. Rb–Sr isotopic results of representatives of the Iforas Granulites and of the Kidal Assemblage

Ech.	Rb. ppm	Sr ppm	$^{87}Sr/^{86}Sr \pm \sqrt{m}$	$^{87}Rb/^{86}Sr$	($^{87}Sr/^{86}Sr$) 600 Ma
Iforas Granulites					
Q178b	34.4	253	0·71112±00005	0.3936	0·7078
Q178c	35.4	597	0·70718±00004	0.1716	0·7057
Q180d	124	1048	0·71541±00005	0.3426	0·7125
IC5510	72.9	3027	0·70704±00004	0·0697	0·7064
IC5513	133	179	0·78796±00006	2·167	0·7694
JPL484	2.9	194	0·70468±00011	0·04818	0·7043
JPL485	28.4	179	0·72092±00005	0·4597	0·7170
JPL486	37.2	359	0·71722±00004	0·3001	0·7147
JPL487	9.1	386	0·70800±00004	0·06822	0·7074
Kidal Assemblage.					
JPL478	157	635	0·70708±00006	0·2504	0·7049
JPL491	33·5	409	0·70768±00005	0·2370	0·7053
JPL492	22·4	260	0·71018±00004	0·2494	0·7080

The enriched mantle. One hundred kilometres west of the batholith, the Permian undersaturated alkaline province (Figure 1; Liégeois *et al.* 1983) was intruded along the eastern margin of the 2 Ga old West African craton, just west of the palaeosuture. It has been demonstrated from Pb and Sr isotope studies (Weis *et al.* 1986a) that the Adrar Tadhak ring-complex is a pure mantle product and thus represents the deep mantle under the Iforas region 270 Ma ago. This deep mantle is assumed to be the asthenosphere (Liégeois and Black 1986). Indeed, along the Cameroon line, Fitton and Dunlop (1985) have shown that the lithospheric mantle cannot be the source of the Tertiary to Recent alkaline magmatism as their geochemical and isotopic characteristics are identical in the continental segment as well as in the oceanic domain. On the other hand, it is likely that the location of the alkaline anorogenic complexes was controlled by the structure of the lithosphere (Black *et al.* 1985; Boullier *et al.* 1986). It is clearly the case in the Iforas for the Pan-African province (along the plate margin, just after a collision) as well as for the Permian province which developed along the palaeosuture zone. These two complimentary arguments are most easily reconciled with an asthenospheric source or, at least, with an asthenospheric intermediate reservoir.

The source regions for different magmas which crystallized to form the Iforas batholith combined to form a dynamic model divided into three periods.

The subduction period. The pretectonic subduction-related magmatism in the cordillera seems to be essentially represented by the volcanosedimentary or volcanic sequences. The only well-established plutons related to subduction are the Erecher tonalite situated close to the island arc and the Ibdeken pluton located along the eastern edge of the batholith, compatible with a classical subduction origin (e.g. Thorpe *et al.* 1976). The characteristics of the Erecher tonalite, which is intermediate between continental and oceanic arc magmatism, are clearly distinguishable both in the field and on the geochemical diagrams (Figure 8). Such characteristics favour a relatively shallow mantle source for this tonalite. The Ibdeken pluton would represent the other extreme in the cordillera (Figure 14). The possibility of a marginal sea at this stage between the island arc and the cordillera, represented in part by the Tafeliant Group, is currently being studied (Liégeois, Fabre and Caby in preparation).

The collisional period. The collision in the Iforas is characterized by the apparent absence of S-type granites and by the preservation of the pretectonic subduction-related volcanites and plutonites. These two features are probably linked to the lack of nappe structures in the island arc and batholith domain during the D2 event. Here the collision is best explained by a 'docking' between the Tuareg shield and the West African craton. On the other hand, the geochemistry and the isotopic ratios are very similar to that of subduction-related magmatism, particularly, the Pb initial ratios (measured on feldspars, Liégeois and Lancelot, unpublished data). In addition the Ree patterns are identical (Liégeois and Hertogen, unpublished data). The main geochemical difference between the two groups seems to be, at a given distance from the palaeotrench, a greater potash content (and related elements such as Rb) for the late-tectonic group, which can be related to a K-h relationship (Dickinson and Hatherton 1967; Dupuy *et al.* 1978; Arculus and Johnson 1978). Thus we propose the same lithospheric mantle-source as for the pretectonic magmas with alkali-rich aqueous fluids coming from the dehydration of the subducted slab, but originating from a greater depth, due

THE ADRAR DES IFORAS COMPOSITE BATHOLITH

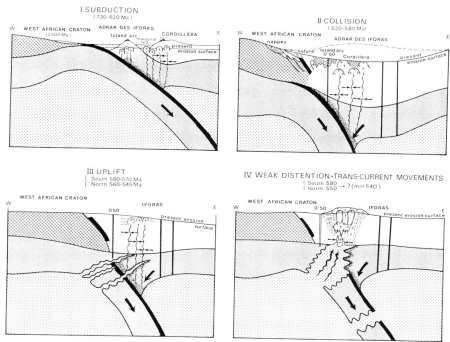

Figure 14. Schematic evolution model for the Iforas batholith during the Pan-African orogeny.

to the crustal thickening during the collision. These mantle products had to be slightly contaminated by the lower crust during ascent (Figure 14II).

The 'within-plate' period. As the post-tectonic group is poorly represented and very similar to the late-tectonic units, except for some erratic alkaline affinities, we consider it as the late manifestation of the source mobilized during the collision, linked to the initiation of an alkaline source. This group is contemporaneous with the rapid uplift of the belt and are only known as high level plutons and dykes (Figure 14III).

The alkaline group followed very rapidly, particularly in the Timedjelalen area where no break in time has been found. Its emplacement was contemporaneous with the end of the uplift, with the unroofing of the batholith when isostatic equilibrium was reached. A thick tabular sequence of rhyolites and ignimbrites fed by the N-S dykes, covered the erosion surface, into which the ring-complexes were emplaced. This is supported by the rhyolitic compositions of Tiralrar-type boulders in the late molasses of the belt (Fabre 1982). The emplacement of alkaline rocks is also roughly contemporaneous with the reactivation of the near N-S shear-zones and with the rotation of the stress field between the D3 and D4 tectonic events (Figures 13C, 13D, 13E; Boullier *et al.* 1986). The apparent diachronism between the emplacement of the two post-tectonic groups in the Kidal and the Timedjelalen area could be related to a different rate of uplift in the two regions, controlled by the oblique character of the collision (Boullier 1982; Ball and Caby 1984; Boullier *et al.* 1986).

During the uplift and the post-tectonic period, the subduction source continued to generate calc-alkaline magma (post-tectonic group I). The abrupt transition to alkaline magmatism suggests a new source region. The general tendency with time for the late calc-alkaline and post-tectonic group I is an evolution towards more acid types accompanied by a diminution of the $^{87}Sr/^{86}Sr$ initial ratios in the plutons (Oumassene quartz-monzodiorite: 0·7053; Adma granodiorite: 0·7048; Aoukenek fine-grained granite: 0·7035, Yenchichi 2 granite: 0·7038). As the alkaline group is younger, granitic, and emplaced as circular high-level plutons, the expected Sr initial ratios would be 0·7035–0·7040 rather than 0·7050–0·7061.

Crustal contamination cannot explain these higher initial ratios. Indeed, the upper crust is essentially represented by the batholith which did not have high enough initial Sr ratios 550 Ma ago. Moreover Weis et al. (1986b) have shown, using oxygen isotope data, that hydrothermal alteration could be intense but only locally. On the other hand, the Pb initial ratios (measured on K-feldspars; Liégeois and Lancelot, unpublished data) are similar for all the calc-alkaline groups but distinctly more radiogenic for the alkaline magmas. This precludes a greater involvement of the lower crust at the alkaline stage.

An asthenospheric mantle source seems to be then the best candidate for the Iforas alkaline family, with the enrichment in radiogenic Sr and Pb contributed from the mantle and not in the crust. We suggest that the asthenosphere originally underlying the subducted plate could be remobilized during its arise to shallow depth in the lithosphere after the rupture of the cold oceanic plate (Figure 14d.) Such fracturing could have occurred during reactivation of D3 shearing with a complete reversal in the stress field compared with D2 (Boullier et al. 1986; Black et al. 1985). The Iforas alkaline province is petrographically and geochemically very similar to the anorogenic Jurassic Nigerian province (see Ba et al 1985; Liégeois and Black 1986) which is totally unrelated to subduction processes and which probably shared the same asthenospheric source. But the two alkaline provinces are different because of the absence of Sn–W mineralization in the Iforas. This difference may be due to the nature of the basement which in Central Nigeria is partly crustal in origin (Van Breeman et al. 1977). In contrast the calc-alkaline Iforas batholith is mantle in origin, and its granulitic basement (IGU) in the Iforas is depleted in ore-bearing elements.

References

Allègre, C. J. and Caby, R. 1972 Chronologie absolue due Précambrien de l'Ahaggar occidental. *C.R. Acad. Sci. Paris*, **D 275**, 2095–2098
Andreopoulos-Renaud, U. Age U–Pb sur zircon d'un massif syncinématique pan-africain dans l'Adrar des Iforas, Mali, *C.R. Acad. Sci. Paris*, submitted.
Arculus, R. J. and Johnson, R. W. 1978. Criticism of generalised models for the magmatic evolution of arc-trench systems. *Earth and Planetary Science Letters*, **39**, 118–126
Ba, H., Black, R., Benziane, B., Diombana, D., Hascoet-Fender, J., Bonin, B., Fabre, J., and Liégeois, J. P. 1985. La province des complexes annulaires sursaturés de l'Adrar des Iforas, Mali, *Journal of African Earth Science*, 3, 123–142.
Bailey, D.K. and MacDonald R. 1969. Alkali-feldspar fractionation trends and the derivation of peralkaline liquids, *American Journal of Science*, **267**, 242–248.
Ball, E. 1980 An example of very consistent brittle deformation of a wide intracontinental zone: late Pan-African fracture system of the Tuareg and Nigerian shield; structural implications. *Tectonophysics*, **61**, 363–379.
—— **and Caby, R. 1984.** Open folding and constriction synchronous with nappe tectonics along a megashear zone of Pan-African age, *In*: Klerkx, J. and Michot, J. (Eds), *African Geology*. Tervuren 75–90.

Barker F. 1979. Trondhjemite: definition, environment and hypothesis of origin. *In*: Barker, F. (Ed.), *Trondhjemites, Dacites, and Related Rocks*, Elsevier. 1–12.

Bayer, R. and Lesquer, A. 1978. Les anomalies gravimétriques de la bordure orientale du craton ouest-african: géométrie d'une suture pan-africaine. *Bull. Soc. Geol. Fr,* **XX**, 863–876.

Bertrand, J. M., Dupuy C., Dostal, J. and Davison, I. 1984b. Geochemistry and geotectonic interpretation of granitoids from central Iforas (Mali, W. Africa). *Precambrian Research,* **26**, 265–283.

——, Merium D., Lapique, F., Michard, A., Dautel, D. and Gravelle, M. 1986a. Nouvelles données radiométriques sur l'âge de la tectonique pan-africaine dans le rameau oriental de la chaîne pharusienne (région de Timgaouine, Hoggar, Algérie). *C. R. Acad. Sci.,* **302**, 437–440.

——, Michard, A., Carpéna, J., Boullier, A.M., Dautel, D., and Ploquin, A. 1984a. Pan-African granitic and related rocks in the Iforas granulites (Mali). Structure, geochemistry and geochronology. *African Geology. In:* Klerkx, J. and Michot, J. (Eds), Tervuren 147–166.

——, Michard, A., Dautel, D. and Pillot, M. 1984. Ages U/Pb éburnéens et pan-africains au Hoggar central (Algérie). Conséquences géodynamiques. *C. R. Acad. Sci. Paris,* **D298**, 643–646.

—— and Wright, L. I. 1986. Structural evolution of the Pan-African Iforas batholith (Mali). Early units, *Journal of the Geological Society of London* submitted.

—— and Caby, R. 1978. Geodynamic evolution of the Pan-African orogenic belt: a new interpretation of the Hoggar shield (Algerian Sahara). *Geol. Runds.,* **67**, 357–388.

—— and Davison, I. 1981. Pan-African granitoids emplacement in the Adrar des Iforas mobile belt (Mali) — a Rb/Sr isotope study. *Precambrian Research,* **14**, 333–362.

Black R. 1963. Note sur les complexes annulaires de Tchouni–Zarniski et de Gouré (Niger). *Bull. Bur. Rech. Géol. Minière,* **1**, 31–45.

——, Caby, R., Moussine-Pouchkine, A., Bayer, R., Bertrand, J. M. L., Boullier, A.M., Fabre, J. and Lesquer, A. 1979. Evidence for late Precambrian plate tectonics in West Africa. *Nature,* **278**, 223–227.

——, Jaujou, M. and Pellaton, C. 1967. Notice explicative de la carte géologique de l'Aïr à l'échelle 1/500 000. *Dir. Mines Géol. Niger.*

——, Lameyre, J. and Bonin, B. 1985. The structural setting of alkaline complexes. *Journal of African Earth Science,* **3**, 5–16.

Boullier, A.M. 1979. Charriage et déformation de l'unité granulitique des Iforas au cours de l'orogenèse pan-africaine. *Rev. Geol. Dyn. Geogr. Phys.,* **21**, 377–382.

——, 1982. Etude structurale du centre de l'Adrar des Iforas (Mali). Mylonites et tectogenèse. *Thèse d'Etat, INPL Nancy,* 327 pp.

—— 1986. Sense of shear and displacement estimates in the Abeibara–Rarhous late Pan-African shear zone (Adrar des Iforas), Mali. *Journal of Structural Geology,* **8**, 47–58.

——, Davison, I., Bertrand, J. M. and Coward, M. 1978. L'unité granulitique des Iforas: une nappe de socle d'âge pan-african précoce. *Bull. Soc. Géol. Fr.,* **20**, 877–882.

——, Liégeois, J. P., Black, R., Fabre, J., Sauvage, M., and Bertrand, J. M. 1986. Late Pan-African tectonics marking the transition from subduction-related calc-alkaline magmatism to within-plate alkaline granitoids (Adrar des Iforas, Mali). *Tectonophysics,* **132**, 233–246.

Bowden, P. and Turner, D. C. 1974 Peralkaline and associated ring-complexes in the Niger–Nigeria Province, West Africa. *In:* Sorensen, J. (Ed.), *The Alkaline Rocks.* J. Wiley, London. 330–351.

Brun, J. P. 1981. Instabilité gravitaire et déformation de la croûte continentale. Application au développement des dômes et des plutons. *Thèse d'Etat, Rennes.* 197 pp.

Caby R. 1970. La chaîne pharusienne dans le nord-ouest de l'Ahaggar (Sahara central, Algérie); sa place dans l'orogenèse du Précambrien supérieur en Afrique. *Thèse d'Etat, Montpellier.*

—— 1978. Paléodynamique d'une marge passive et d'une marge active au Précambrien Supérieur: leur collision dans la chaîne pan-africaine du Mali. *Bull. Soc. Geol. Fr.,* **7**, 857–862.

—— 1981. Associations volcaniques et plutoniques prétectoniques de la bordure de la chaîne pan-africaine en Adrar des Iforas (Mali): un site d'arc-cordillère au Protérozoique supérieur. *11th Coll. Afr. Geol. Milton Keynes,* **30**.

—— and Andreopoulos-Renaud, U. 1983. Age à 1800 Ma du magmatisme sub-alcalin associé aux métasédiments monocycliques dans la chaîne pan-africaine du Sahara central. *Journal of African Earth Science,* **1**, 193–197.

—— and Andreopoulos-Renaud, U. 1985. Etude pétrostructurale et géochronologique d'une métadioritе quartzique de l'Adrar des Iforas pan-africaine de l'Adrar des Iforas (Mali). *Bull. Soc. Geol. Fr.*

——, Andreopoulos-Renaud, U., and Lancelot, J. R. 1985. Les phases tardives de l'orogenèse pan-africaine dans l'Adrar des Iforas oriental (Mali): lithostratigraphie des formations molassiques et géochronologie U/Pb sur zircon de deux massifs intrusifs. *Precambrian Research,* **28**, 187–199.

——, Bertrand, J. M. and Black, R. 1981. Pan-African closure and continental collision in the Hoggar–Iforas segment, Central Sahara. In: Kröner, A. (Ed.), *Precambrian Plate Tectonics.* Elsevier, Amsterdam. 407–434.

Cahen, L., Snelling, H. J., Delhal, J. and Vail, J. R. 1984. *The Geochronology and Evolution of Africa.* Clarendon Press, Oxford. 512 pp.

Champenois, M., Boullier, A. M., Sautter, V., Wright, L.I. and Barbey, P. 1986. Tectonometamorphic evolution of the gneissec Kidal Assemblage related to the Pan-African thrust tectonics (Adrar des Iforas, Mali). *Journal of African Earth Science,* **6**, 19–27.

Chikhaoui, M. 1981. Les roches volcaniques du Protérozoïque supérieur de la chaîne pan-africaine du

NW de l'Afrique (Hoggar, Anti-Atlas, Adrar des Iforas). Caractérisation géochemique et minéralogique — implications géodynamiques. *Thèse d'Etat USTL Montpellier,* 195 pp.

Davison, I. 1980. A tectonic, petrographical and geochronological study of a Pan-African belt in the Adrar des Iforas and Gourma (Mali). *Ph.D. thesis Leeds and Montpellier,* 337 pp.

Debon, F. 1980. Genesis of the three concentrically zoned granitoids plutons of Cauterets–Panticosa (French and Spanish western Pyrénnées). *Geol. Runds.,* **69,** 107–130.

—— **and Le Fort, P. 1983.** A chemical–mineralogical classification of common plutonic rocks and associations. *Transactions of the Royal Society of Edinburgh: Earth Science,* **73,** 135–150.

Dickinson, W. R. and Hatherton, T. 1967. Andesitic volcanism and seismicity around the Pacific. *Science,* **157,** 801–803.

Dupuy, C., Dostal, J. and Vernières, J. 1978. Genesis of volcanic rocks related to subduction zones, geochemical point of view. *Bull. Soc. Geol. Fr.,* **19,** 1233–1244.

——, **Liégeois, J. P. and Caby, R. 1987.** A late Proterozoic island arc in northern Mali. In preparation.

Duyverman, H. J., Harris, N. B. W. and Hawkesworth, C.J. 1982. Crustal accretion in the Pan-African: Nd and Sr isotope evidence from the Arabian shield. *Earth and Planetary Science Letters,* **54,** 313–326.

Ewart, A. 1979. A review of the mineralogy and chemistry of tertiary-recent dacitic, latitic, rhyolitic, and related salic volcanic rocks. In: F. Barker, F. (Ed.), *Trondhjemites, Dacites, and Related Rocks,* Elsevier. 13–122.

—— **1982.** The mineralogy and petrology of Tertiary–Recent orogenic volcanic rocks: with special reference to the andesite–basalt compositional range, In: Thorpe, R.S. (Ed.), *Andesites,* J. Wiley, London. 25–98.

Fabre, J. 1982. Pan-African volcano-sedimentary formations in the Adrar des Iforas, *Precambrian Research,* **19,** 201–214.

——., **Ba, H., Black, R., Caby, R., Leblanc, M. and Lesquer, A. 1982.** Notice explicative de la carte géologique de l'Adrar des Iforas. *Dir. Nat. Géol, Mines, Bamako.*

Ferrara, G. and Gravelle, M. 1966. Radiometric ages from western Ahaggar (Sahara) suggesting an eastern limit for the West African craton. *Earth and Planetary Science Letters,* **1,** 319–324.

Fitton, J. G. and Dunlop, H. M. 1985. The Cameroon line, West Africa, and its bearing on the origin of oceanic and continental alkali basalt. *Earth and Planetary Science Letters,* **72,** 23–38.

Greenwood, W. R., Anderson, R. E., Fleck, R. J. and Schmidt, D. L. 1976. Late Proterozoic cratonization in southwestern Saudi Arabia. *Philosophical Transactions of the Royal Society of London,* **A280,** 517.

Jacobson, R. R. E., MacLeod, W. N. and Black, R. 1958. Ring-complexes in the Younger Granite province of northern Nigeria. *Memoir of the Geological Society of London,* **1,** 72 pp.

Karpoff, R. 1960. La géologie de l'Adrar des Iforas. *Thèse Paris 1958, Publ. Bur. Rech. Géol. Min. Dakar,* **30.**

—— **1961.** Le granite nigritien d'Ileathène dans l'adrar des Iforas (Sahara central). *C.R. somm. Soc. Géol. Fr.,* **6,** 166.

La Roche, H. de 1962. Sur l'expression graphique des relations entre la composition chimique et al composition mineralogique quantitative des roches cristallines. *Sci. de la Terre, Nancy,* **IX,** 293–337.

Lancelot, J. R., Boullier, A. M., Maluski, H., and Ducrot, J. 1983. Deformation and related radiochronology in a late Pan-African mylonite bearing shear zone, Adrar des Iforas, Mali. *Contr. Min. Petrol.,* **82,** 312–326.

Latouche, L. 1986. Les collisions intracratoniques et la tectonique intracontinentale dans le Pan-Africain du Hoggar central. In: Black (Ed.) Évolution Geólogique de l'Afrique Occaisional Publication CIFEG, 143–158.

Leterrier, J. and Bertrand, J. M. 1986. Pretectonic tholeiitic volcanism and related transitional plutonism in the Kidal Assemblage (Iforas Pan-African belt, Mali). *Journal of African Earth Science,* submitted.

Liégeois, J. P., Bertrand, H., Black, R., Caby, R. and Fabre, J. 1983. Permian alkaline undersaturated and carbonatite province, and rifting along the West African craton. *Nature,* **305,** 42–43.

—— **and Black, R. 1984.** Pétrographie et géochronologie Rb–Sr de la transition calco-alcalin – alcalin fini-panafricaine dans l'Adrar des Iforas (Mali): accrétion crustale au Précambrien supérieur. In: Klerkx, J. and Michot, J. (Eds), *African Geology.* Tervuren. 115–146.

—— **and Black, R. 1987.** Alkaline magmatism subsequent to collision in the Pan-African belt of the Adrar des Iforas (Mali). In: Fitton, J. G. and Upton, B. G. J. (Eds). Alkaline Igneous Rocks, Geological Society, Special Publication 30, 381–401.

Ly, S., Lesquer, A., Ba, H. and Black, R. 1984. Structure profonde du batholite occidental de l'Adrar des Iforas (Mali): une synthèse des données gravimétriques et géologiques. *Rev. Géol. Dyn. Géogr. Phys.,* **25,** 33–44.

Miyashiro, A. 1974. Volcanic rock series in island arcs and active continental margins. *American Journal of Science,* **274,** 321–355.

Pagel, M. and Leterrier, J. 1980. The subalkaline potassic magmatism of the Ballons massif (Southern Vosges): shoshonitic affinity. *Lithos,* **13,** 1–10.

Petterson, M. G. and Windley, B. F. 1985. Rb–Sr dating of the Kohistan arc–batholith in the Trans-Himalaya of north Pakistan, and tectonic implications. *Earth and Planetary Science Letters,* **74,** 45–57.

Radier, H. 1957. Contribution à l'étude géologique du Soudan oriental (A.O.F.). I. Le Précambrien saharien du sud de l'Adrar des Iforas. *Thèse Strasbourg. Bull. Serv. Géol. et Prosp. Min. A.O.F.*, **26**, 1331 pp.
Thorpe, R. S., Potts, P. J., and Francis, P. W. 1976. Rare earth data and petrogenesis of andesite from the north Chilean Andes. *Contr. Min. Petrol.*, **54**, 65–78.
Weis, D., Liégeois, J.P. and Black, R. 1986a. Whole rock U–Pb isochrons and deep mantle origin for a continental alkaline ring-complex (Tadhak, Mali). *Earth and Planetary Science Letters*, submitted.
——, **Liégeois, J. P. and Javoy, M. 1986b.** The Timedjelalen alkaline ring-complex and related N–S dyke swarms (Adrar des Iforas, Mali). A Pb–Sr–O isotopic study. *Chem. Geol.*, in press.
Williamson, J. H. 1968. Least square fitting of a straight line. *Canadian Journal of Physics*, **46**, 1845–1847.

Appendix: analytical techniques

The Sr isotopic compositions, after separation on ion-exchange columns, were measured on Re double filament on a FINNIGAN–MAT 260 mass spectrometer (Belgian Centre for Geochronology). The NBS 987 standard has given a value of 0·710235 ±0·000026 (2 σ_m) during previous measurements (Liégeois and Black 1984) and 0·710215 ±0.000015 during the measurement of the new data (Ibdeken, Tin Seyed, IGU and Kidal Assemblage). The ages were calculated following Williamson (1968) and all the errors are quoted at the 2σ level. $^{87}Rb = 1.42\ 10^{-11}$ a^{-1}. The Rb and Sr concentrations were measured by XRF when above 30 ppm and by isotope dilution when <30 ppm. The errors on the Rb/Sr ratios are estimated at 2 per cent.

The chemical analyses can be obtained from the authors on request.

Geochemistry of the Tioueine Pan-African granite complex (Hoggar, Algeria)

Abla Azzouni-Sekkal

Institut des Sciences de la Terre, U.S.T.H.B., BP 9, Dar-el-Beida, Alger

and

Jean Boissonnas

Commission of the European Communities, Directorate-General XII, 200 rue de la loi, B-1049 Brussels

---

The Tioueine complex belongs to a group of concentrically zoned late Pan-African granites intruding the eastern branch of the Pharusian belt in the Hoggar. It comprises an outer mass of alkaline granites, both hypersolvus and subsolvus, cross-cut by arcuate dykes of hypersolvus ferroedenite-bearing syenite and granite, and a later central core of monzonitic granite.

Tioueine granites are A-type, alkaline, and display only moderate postmagmatic alteration. As a whole, the complex appears to illustrate the transition from orogenic calc-alkaline (monzonitic) to anorogenic alkaline magmatism.

KEY WORDS   Algeria   Hoggar   Pan-African   granite   Transitional monzonitic-alkaline   magmatism

1. Introduction

The Tioueine complex (14 × 9 km) is the southernmost representative of a conspicuous group of granite plutons, sometimes referred to as the Taourirt granites, which intrude the Pharusian (= Upper Proterozoic) belt of the Hoggar, in its northeastern part (Figure 1; bibliography on the Pharusian in Bertrand *et al.* 1983). These plutons are notable for their concentric structures; their emplacement appears to have been controlled by late Pan-African movements along N–S transcurrent faults (Boissonnas 1973).

The Tioueine massif was emplaced into metavolcanic rocks and diorites of lower Pharusian age. Rb–Sr dating has yielded a whole-rock isochron age of 560 ± 40 Ma ($Sr_i = 0.705$) using the constant $\lambda = 1.47$ (Boissonnas *et al.* 1969). Values recalculated to 575 ± 12 Ma and $Sr_i = 0.7073$ by Cahen *et al.* 1984). The need to refine these early results is acknowledged, particularly in view of the zircon age of 583 ± 7 Ma obtained recently by Bertrand *et al.* (1986) on a late tectonic pluton of the Pharusian belt.

0072–1050/87/TI0213–12$06.00
© 1987 by John Wiley & Sons, Ltd.

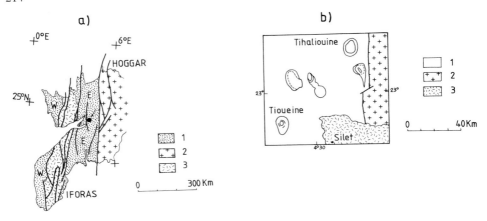

Figure 1. Situation of the Tioueine complex. (a) Structural sketch of the western and central part of the Touareg shield (simplified from Bertrand et al. 1983). 1, Pharusian belt (W= western branch, E= eastern branch); 2, Central Hoggar (gneiss reactivated in the Pan-African orogeny); 3, In Ouzzal–Iforas block (Eburnean granulites and reactivated gneiss). Arrow indicates the position of Tioueine Massif. (b) The southern "Taourirt" granites of the Pharusian belt (after Boissonnas 1973). 1, Pharusian belt; 2, Central Hoggar; 3, basalts.

2. Major units of the complex

Two sharply contrasting units are present in the Tioueine complex: a composite mass of alkali granites and a central core of monzonitic granite (Figure 2.)

2a. The outer unit

1. The largest part of the unit, and indeed of the whole complex, consists of a pink coarse-grained granite of distinctly hypersolvus affinity. Perthite is by far the dominant feldspar. Minor plagioclase occurs as altered oligocase (+ sericite and fluorite). Subsolidus albite is well developed, although not all pervasive. Mafic minerals (ferroedenite, see Figure 3, annite-rich biotite,

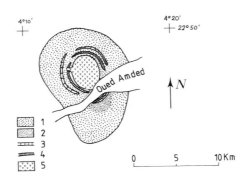

Figure 2. Sketch map of the Tioueine complex (simplified from Boissonnas 1973). 1, main (hypersolvus) granite of the, peripheral unit 2, subsolvus granite of the peripheral unit. 3 and 4, granite and syenite of the arcuate dykes. 5, central granite.

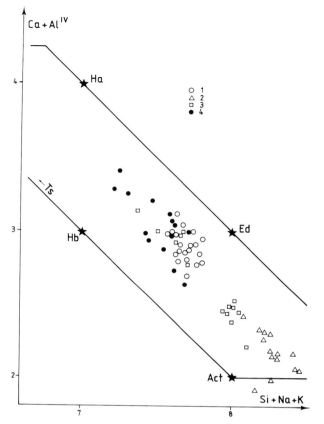

Figure 3. Amphiboles from individual rock samples plotted in the diagram $Ca+Al^{IV} - Si + Na + K$ (Giret et al. 1980). Symbols: 1, hypersolvus granite from the peripheral unit; 2 and 3, granite and syenite from the arcuate dykes; 4, central granite. Act= actinolite, Ed= edenite, Ha= hastingsite, Hb= hornblende, Ts= tschermakite.

see Figure 4) are scarce and are often broken down to iron oxides. In practice the rock appears to fluctuate somewhat between a hypersolvus and subsolvus type throughout the unit.

2. A subsolvus finer grained variety occurs in the interior of the unit. It contains both oligoclase and weakly perthitic K-feldspar, rather more biotite and less amphibole than the main granite.

3. Both granites are cross-cut by arcuate dykes of medium grained syenite, quartz syenite and granite. One can broadly define an external system, complex and predominantly granitic, and an inner dyke of syenite. To the south, the dyke system is reduced to a single composite body, granitic on the convex (external) side and grading northwards to syenite on the concave side.

Locally, particularly in the north, the inner dyke is the most clearly intrusive of the two, with knife-edge contacts and chilled margins, whereas the external system appears more difficult to delineate precisely. This may be accounted for by supposing a slightly younger age of the inner dyke.

Most of the dyke rocks are hypersolvus, essentially composed of perthite.

Ferrohedenbergite (Figure 5), ferroedenite (Figure 3) and very minor amounts of annite (Figure 4) exist in the syenites. The amphibole has two habits: discrete grains, and rims around pyroxene crystals. It sometimes displays small rims of riebeckite. Accessories (zircon, allanite, magnetite) are conspicuously idiomorphic. Acicular secondary actinolite/ferroactinolite (Figure 3) and pockets of reddish-brown amorphous material provide evidence for limited postmagmatic alteration.

Granitic dyke rocks are pyroxene free and contain much less amphibole and accessories than the syenites.

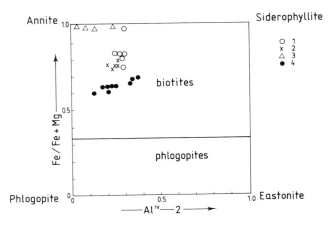

Figure 4. Biotites in the diagram $Fe/Fe + Mg - (Al^{IV} - 2)$. Symbols: 1 and 2, hypersolvus and subsolvus granites from the peripheral unit; 3, dykes; 4, central granite.

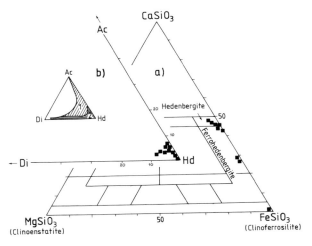

Figure 5. Pyroxenes from syenite and quartz-syenite. (a) Diagram Ca–Mg–Fe (Poldervaart and Hess 1951). The three samples clinopyroxene from qtz-syenite plot closest to $FeSiO_3$ apex of diagram. (b) Diagram Ac–Di–Hd, $Hd = Fe + Mn - (Na + K)$, $Ac = Na + K$, $Di = Mg$. Clinopyroxene compositional range (after Bonin and Giret 1985): 1, peralkaline silica saturated subseries; 2, metaluminous silica saturated subseries.

2b. The central granite

This is a grey, medium grained 2-feldspar granite, containing oligoclase (subidiomorphic and zoned), perthite, ferroedenite, ferrohornblende and biotite (Figures 3 and 4), plus subordinate amounts of secondary chlorite and epidote. The central granite is clearly intrusive, with sharp contacts, in all rocks of the outer unit.

On a Streckeisen QAP diagram (Figure 6a), the Tioueine points are scattered from the alkaline and peralkaline field (syenite) to the calc-alkaline-high K field (central granite and subsolvus granite of the outer unit). The hypersolvus granite of the periphery plots in the field of aluminous granitoids of alkaline provinces. An A-type affinity of the complex is also clear on the diagram of Figure 6b.

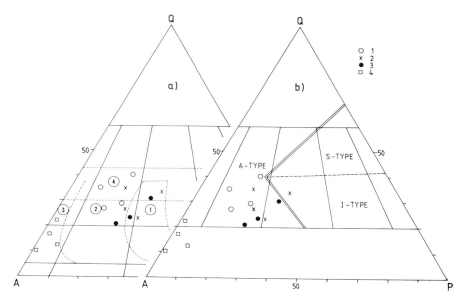

Figure 6. QAP modal diagrams for Tioueine samples. (a) Fields of selected global granitoid series (after Lameyre and Bowden 1982): 1, calc-alkaline — high K; 2, aluminous in alkaline provinces; 3, alkaline and peralkaline; 4, overlapping field of granitoids formed by crustal fusion. (b) after Bowden (1985). Symbols: 1 and 2, hypersolvus and subsolvus granites of the peripheral unit; 3, syenite; 4, central granite.

3. Major element geochemistry

The analytical data are presented in Table 1. As expected, MgO and CaO are low in the peripheral unit, higher in the central granite. The difference for MgO parallels that found in the biotites (Figure 4). Within the outer unit, the hypersolvus–subsolvus granite is well reflected in some CaO values.

On the alkalis-silica diagram (Figure 7), Tioueine samples plot within the alkaline field close to the boundary with the subalkaline field as defined by Miyashiro (1978), and also on the trend indicated for monzonitic series (Lameyre et al. 1982).

Throughout the complex, normative corundum is in the range 0 to 1 per cent. This value, combined with mineralogical data (Bowden 1985), is indicative of metaluminous to weakly peraluminous granites.

Table 1. Analyses of Tioueine samples

	1	2	3	4	5	6	7	8	9	10	11	12	13
SiO_2	75·65	74·88	75·3	75·38	77·14	76·62	76·76	75·81	75·51	73·93	73·9	74·38	68·62
TiO_2	0·13	0·18	0·13	0·10	0·09	0·06	0·14	0·11	0·22	0·19	0·21	0·13	0·26
Al_2O_3	11·51	12·98	12·53	12·85	12·52	12·16	12·37	13·24	12·92	13·67	13·39	13·02	15·05
Fe_2O_3*	1·71	1·83	1·63	1·23	1·17	1·16	1·22	1·27	1·63	1·54	1·83	2·47	4·33
MnO	0·04	0·04	0·04	0·03	0·03	0·03	0·04	0·05	0·06	0·05	0·04	0·05	0·10
MgO	0·27	0·14	0·18	tr	tr	0·08	0·15	tr	0·09	0·15	0·18	0·03	tr
CaO	0·38	0·43	0·45	0·1	0·23	0·10	0·35	0·41	0·46	0·64	0·94	0·42	0·62
Na_2O	3·60	4·03	4·08	4·03	4·38	4·23	3·48	3·80	4·01	4·07	4·16	4·38	4·87
K_2O	4·56	4·63	4·45	5·13	4·37	4·27	4·85	4·97	4·59	4·59	4·17	4·78	5·88
P_2O_5	0·04	tr	tr	tr	tr	tr	tr	tr	tr	tr	tr	tr	tr
L·O·I·	0·66	0·76	0·53	0·62	0·12	0·51	0·29	0·34	0·20	0·72	0·42	0·1	0·12
Total	98·55	99·90	99·32	99·47	100·05	99·22	99·65	100·00	99·69	99·55	99·24	99·76	99·85
Ba	39	159	140	25	<10	38	397	420	309	638	384	113	107
Co	<10	51	34	28	<10	<10	362	286	198	532	43	<10	<10
Cr	<10	<10	<10	<10	<10	16	61	53	66	68	<10	<10	12
Cu	<10	<10	<10	<10	<10	<10	<10	<10	<10	<10	56	<10	<10
Ni	<10	22	15	<10	<10	<10	<10	<10	<10	<10	13	<10	<10
Sr	<10	46	19	<10	<10	<10	194	211	202	293	71	60	53
V	<10	<10	<10	<10	<10	<10	211	185	206	220	<10	<10	23
Rb	136	191	193	160	225	155	224	181	200	209	213	151	193

* Total iron as Fe_2O_3
< Value is the detection limit of the analytical technique.
Analyst: C.R.P.G., Nancy.
Samples: 1 to 6, hypersolvus granite of the outer unit; 7 to 12, subsolvus granite of the outer unit; 13 to 17, syenite; 18, quartz syenite; 19 and 20, granite of the arcuate dykes; 21 to 24, central granite.

	14	15	16	17	18	19	20	21	22	23	24
SiO_2	68·41	66·12	65·22	65·29	69·00	73·67	75·91	72·14	70·02	70·98	71·84
TiO_2	0·16	0·27	0·51	0·22	0·21	0·08	0·15	0·24	0·38	0·29	0·32
Al_2O_3	15·60	16·25	16·77	17·19	14·51	12·82	11·72	13·78	14·56	14·28	14·00
$Fe_2O_3^*$	2·94	4·49	4·14	3·13	3·43	1·78	1·92	2·2	2·50	2·31	2·29
MnO	0·07	0·01	0·08	0·09	0·07	0·04	0·04	0·05	0·05	0·04	0·05
MgO	0·08	tr	0·03	0·03	0·03	0·03	tr	0·05	0·61	0·42	0·65
CaO	0·57	0·83	2·03	0·71	0·5	0·18	0·24	0·66	2·18	1·40	1·34
Na_2O	5·33	5·39	4·85	5·35	4·91	4·50	4·00	1·32	4·06	4·20	3·97
K_2O	5·73	5·69	5·09	6·88	5·36	4·87	4·32	3·87	3·81	4·45	4·59
P_2O_5	tr	tr	0·09	tr	tr	tr	tr	4·58	tr	tr	tr
L.O.I.	0·85	0·62	0·65	0·50	0·69	1·01	0·65	0·62	0·66	0·63	0·85
Total	99·74	99·67	99·46	99·39	98·71	98·98	98·95	99·46	98·83	99·00	100·20
Ba	10	72	1002	109	44	33	21	466	825	408	484
Co	877x	18	<10	<10	22	<10	51	39	37	33	<10
Cr	<10	<10	<10	45	<10	<10	<10	<10	<10	<10	26
Cu	<10	<10	<10	<10	<10	15	<10	<10	<10	<10	<10
Ni	185	<10	<10	<10	<10	<10	63	<10	<10	11	<10
Sr	<10	18	209	<10	<10	<10	<10	126	227	132	154
V	<10	<10	<10	<10	<10	<10	16	<10	11	<10	34
Rb	165	153	148	165	162	152	218	200	134	190	188

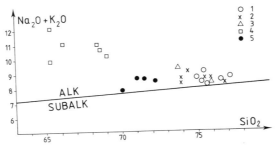

Figure 7. Silica–alkalis diagram. (1), boundary of alkaline and subalkaline fields, after Miyashiro (1978). Symbols: 1 and 2, as in Figure 6; 3 and 4, granites and syenite of the arcuate dykes; 5, central granite.

4. Trace element geochemistry

Perhaps the most striking feature is the abundance of Rb (relatively uniform values of 200 ppm and more). On the Sr/Rb diagram (Figure 8), there is a hint of a negative correlation for the points of the central granite, which suggests magmatic evolution by feldspar fractionation (Arth 1976). All samples plot in the field ascribed by Bonin et al. (1979) to granites which have retained their magmatic features, as opposed to the field of hydrothermal alteration in which Rb is strongly enriched. Similarly, Rb does not vary significantly on the K_2O/Rb diagram (Figure 9). It would seem therefore that postmagmatic fluids have not been very active, a conclusion which appears to be substantiated further by the absence of mineralization.

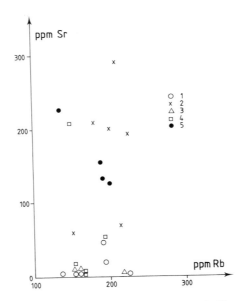

Figure 8. RB/Sr diagram. Same symbols as in Figure 7.

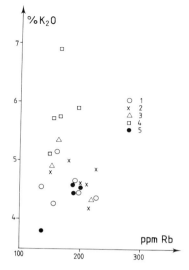

Figure 9. K$_2$O/Rb. Same symbols as in Figure 7.

Ba and Sr are correlated positively in the various granites (Figure 10). This may reflect a genetic link between hypersolvus and subsolvus granites of the main unit. The central granite is clearly distinctive.

Ten samples were analysed for Y and REE by plasma emission spectrometry in the C.R.P.G., Nancy. Values are generally high. Y ranges from 60 to 133 ppm. A chondrite-normalized REE spectrum (Figure 11) shows the following features:

1. All samples from the peripheral unit display strong Eu anomalies similar to aluminous and peralkaline anorogenic granites (Bowden et al. 1979; Bowden 1985).

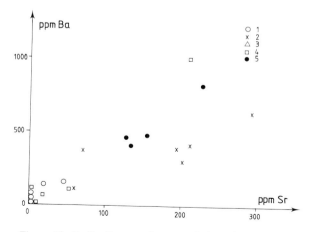

Figure 10. Ba/Sr diagram. Same symbols as in Figure 7.

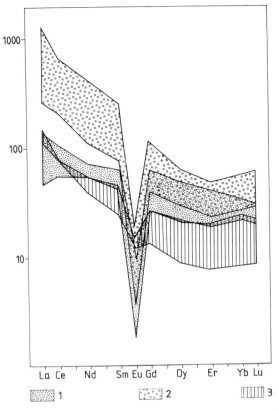

Figure 11. Chondrite–normalized REE Patterns. Symbols: 1, both granites of the peripheral unit; 2, dykes ; 3, central granite.

2. LRee are strongly enriched and fractionated in the arcuate dykes $(La/Yb)_N \simeq 9$ to $33 \cdot 5$.
3. Rare-earth patterns are similar for the hypersolvus and subsolvus granites of the main unit. $(La/Yb)_N$ remains constant at $\simeq 4 \cdot 5$. In general, except for one sample of hypersolvus granite, there is no marked relative enrichment in the HRee, as would be the case if postmagmatic fluids had been very active (Le Bel et al. 1984; Bowden 1985). This seems consistent with the suggestion derived above from the K_2O/Rb diagram.
4. Samples from the central granite show weak Eu anomalies and strong Ree fractionation, $(La/Yb)_N$ varies between $6 \cdot 5$ and $14 \cdot 5$.

Taking into account the above geochemical data, attention should be drawn to two problems concerning the evolution of the complex:

1. The central granite cannot be derived in situ from the peripheral unit. A new stage of melting at depth is needed. If one postulates that surface volcanism may have accompanied the emplacement of the hypersolvus granites, a resulting pressure drop could trigger remelting (Harris and Marriner 1980). Indeed there is no proof of volcanic activity in the Tioueine complex, but the suggestion seems plausible for this type of complex.

2. How are the hypersolvus and subsolvus granites of the peripheral unit related? This appears to be an intriguing problem, still unresolved. Two alternative explanations come to mind: (i) progressive inwards crystallization of the peripheral unit and related water enrichment in the magma, ultimately leading to a 2-feldspar granite, or (ii) a process involving the remelting of hypersolvus granite as proposed by Martin and Bonin (1976).

Both processes imply a slightly younger crystallization age for the subsolvus granite. However, this hypothesis seems at variance with the geochemical observed trends.

5. Conclusions

Based on the evidence presented in the paper it is concluded that most granites of the Tioueine complex belong to the A-type spectrum. The characteristics of these A-type granites are shown by the similarity to the complexes of the Iforas alkaline province (Ba et al. 1985). The aluminous trend of Tioueine granites can be broken down in two components: (i) hypersolvus with perthite, hedenbergite, Ca–Na amphiboles and subordinate biotite; (ii) subsolvus with K feldspar, oligoclase, Ca amphibole, biotite and chlorite. Albitization and the development of secondary minerals such as chlorite, muscovite, epidote and fluorite, although widespread, are not particularly important. This is another feature of the Iforas complexes, contrasting with the greisenization and pervasive albitization observed in the younger granites of Nigeria.

Considered as a whole, with its succession of contrasted granites, the Tioueine complex displays features of both alkaline and calc-alkaline magmatism. Geochemical data appear to substantiate the suggestion made earlier by Boissonnas (1973) that the Tioueine massif and related granites of the Pharusian belt reflect the transition from orogenic to anorogenic magmatism. Similar work is under way on other complexes of the same group, in which subsolvus calc-alkaline high K granites appear to prevail.

Acknowledgements. The authors are indebted to R. Black, B. Bonin and J. Lameyre for constructive discussions during the preparation of the manuscript. The figures were drafted in J. Verkaeren's laboratory at Louvain-la-Neuve. Special thanks are due to P. Bowden for his patience with a delayed manuscript and for his assistance in revising the English.

References

Arth, J.G. 1976. Behaviour of trace elements during magmatic processes: a summary of theoretical models and their applications. *J. Res. U.S. Geol. Surv.*, **4**, 41–47.

Ba, H., Black, R., Benziane, B., Diombana, D., Hascott-Fender, J., Bonin, B., Fabre, J. and Liegeois, J.P. 1985. La province des complexes annulaires alcalins sursaturés de l'Adrar des Iforas, Mali. *Journal of African Earth Science* **3**, 123–142.

Bertrand, J.M.L., Caby, R. and Leblanc, M. 1983. La zone mobile pan-africaine de l'Afrique de l'Ouest. In: Fabre, J. (Ed.), *Afrique de l'Ouest, Introduction Géologique et Termes Stratigraphiques*, Pergamon Press, 35–45.

———, **Meriem, D., Lapique, F., Michard, A., Dautel, D. and Gravelle, M. 1986.** Nouvelles données sur l'age de la tectonique pan-africaine dans le rameau oriental de la chaîne pharusienne (région de Timgaouine, Hoggar, Algérie). *C.R. Acad. Sc. Paris, D.* **302**, 437–440.

Boissonnas, J. 1973. Les granites à structures concentriques et quelques autres granites tardifs de la chaîne pan-africaine en Ahaggar (Sahara central, Algérie). *Thèse Doct. Etat, Ed C.N.R.S. — C.R.Z.A., série géologie*, No. 16, 2 vol., 662 pp.

———, **Borsi, S., Ferrara, G., Fabre, J., Fabries, J. and Gravelle, M. 1969**. On the early cambrian age of two late orogenic granites from west-central Ahaggar (Algerian Sahara). *Canadian Journal of Earth Science*, **6**, 25–37.

Bonin, B., Bowden, P., Vialette, Y. 1979. Le comportement des éléments Rb et Sr au cours des phases de minéralisation: l'exemple de Ririwai (Liruei), Nigéria. *C.R. Acad. Sc. Paris, sér. D*, **289**, 707–710.

——— **and Giret, A. 1985.** Clinopyroxene compositional trends in oversaturated and undersaturated alkaline ring complexes. *Journal of African Earth Sci.*, **3**, 175–183.

Bowden, P. 1985. The geochemistry and mineralization of alkaline ring complexes in Africa (a review). *Journal of African Earth Sci.*, **3**, 17–39.

———, **Bennett, J.M. Whitley, J.E. and Moyes, A.B. 1979.** Rare earths in Nigerian Mesozoic granites and related rocks. In: Ahrens, L.E. (Ed.), *Origin and Distribution of the Elements (Second Symposium). Phys. Chem. Earth*, **11**, 479–491.

Cahen, L., Snelling, N.J., Delhal, J. and Vail, J.R. 1984. *Geochronology and Evolution of Africa.* 512 pp., Clarendon Press, Oxford.

Giret, A., Bonin, B. and Leger, J. M. 1980. Amphibole compositional trends in oversaturated and undersaturated alkaline plutonic ring complexes. *Canadian Mineralogy*, **18**, 481–495.

Harris, N.B.W. and Marriner, G.F. 1980. Geochemistry and petrogenesis of a peralkaline granite complex from the Midian Mountains, Saudi Arabia. *Lithos*, **13**, 325–337.

Lameyre, J. and Bowden, P. 1982. Plutonic rock types series: discrimination of various granitoid series and related rocks. *Journal of Volcanol. Geotherm. Research*, **14**, 169–186.

———, **Black, R., Bonin, B., Bowden, P. and Giret, A. 1982.** The granitic terms of converging plutonic type series and associated mineralisations. In: *Geology of Granites and their Metallogenic Relations*, Proc. Int. Symposium Nanjing, Oct. 26–30, 1982.

Le Bel, L., Li Yi-dou, and Sheng Ji-Fou 1984. Granite evolution of the Xihuashan–Dangping (Jiangxi, China) tungsten-bearing system. *Tschermaks Min. Petr. Mitt.*, **33**, 149–167.

Martin, R.F. and Bonin, R. 1976. Water and magma genesis: the association hypersolvus granite – subsolvus granite. *Canadian Mineralogy*, **14**, 228–237.

Miyashiro, A. 1978. Nature of alkalic volcanic rock series. *Contr. Miner. Petrol*, **66**, 91–104.

Poldervaart, A. and Hess, H.H. 1951. Pyroxenes in the crystallization of basaltic magmas. *Journal of Geology*, **59**, 472–489.

Reconnaissance geological mapping and mineral exploration in northern Sudan using satellite remote sensing

P. S. Griffiths
Griffiths Remote Sensing, 7e Vera Road, London SW6 6RW, U.K.

P. A. S. Curtis
Geosurvey International Ltd., Orchard Lane, East Molesey, Surrey KT8 0BY, U.K.

S. E. A. Fadul
Geological and Mineral Resources Department, P.O. Box 410, Khartoum, Sudan

and

P. D. Scholes
Geosurvey International Ltd., Orchard Lane, East Molesey, Surrey KT8 0BY, U.K.

In suitable terrain, Landsat and other satellite imagery can be used for the production of reconnaissance geological maps which contain a comprehensive range of lithological and structural information. The method described here is based on aerial photogeological interpretation. This approach does not require extremely sophisticated image processing and is most effective when supported by a limited, controlled field check. Maps produced in this way can be called 'image-geological' maps to distinguish them from (aerial) photogeological and conventional geological maps.

Using this procedure image-geological maps have been completed at 1:250 000 scale of an area of 92 000 km^2 of the Sudanese Nubian Desert, based on interpretation of Landsat MSS and RBV imagery and one swath of SIR–A radar imagery as well as on a rapid field check. The work was undertaken to provide a basis for detailed planning and execution of a regional mineral exploration programme that is being conducted by the Egypt–Sudan Integration Project and was initially assisted by the UNDP. This paper describes the main geological and mineral exploration results and the interpretability of the satellite imagery. The complementary nature of methods of geological mapping based on satellite imagery, aerial photography and fieldwork is discussed briefly. It is shown that the absolute reliability and accuracy of each method is a less important issue than an understanding of their different potential uses in a regional mineral exploration programme.

KEY WORDS   Satellite remote sensing   Regional mineral exploration   Reconnaissance geological mapping   Sudan   Image-geological map

0072–1050/87/TI0225–25$12.50
© 1987 by John Wiley & Sons, Ltd.

1. Introduction

This is a case history of recent satellite remote sensing geological work carried out by Geosurvey International on behalf of the Egypt–Sudan Integration Project and financed by the United Nations Development Programme (UNDP). Our study area consists of the Sudanese sector of the Egypt–Sudan Area of Integration (Figure 1). This isolated, very thinly populated desert region is attracting new interest because of its proximity to Lake Nasser (Nuba Lake) and because it is believed to have potential for gold mineralization as well as a wide range of other natural resource commodities. Since the Sudanese sector of the Area of Integration had not previously been systematically geologically studied and the actual potential of the region, therefore, is not known, the UNDP designed a programme of regional mineral exploration which would begin with reconnaissance geological mapping by satellite remote sensing methods and then use the resulting maps for the planning and execution of subsequent stages of the project.

The role of the present satellite image mapping study in Sudan makes it a very suitable case history with which to demonstrate three general and interrelated facts. Firstly, reconnaissance geological mapping based primarily on satellite imagery can be shown to have considerable scientific merit when the method of interpretation and its relationship to more conventional geological mapping is understood. Secondly, satellite imagery mapping is more cost-effective than many other forms of geological mapping. This is possibly because satellite image mapping may provide answers sufficient to the purpose of mineral exploration or, more likely, because the satellite image mapping can be used to identify and reduce in size the areas of interest which require more detailed and expensive investigation. Finally and given the first two factors, optimum use is made of the satellite image mapping work, by both planners and scientists, if it is carried out at the right stage of the total exploration programme. We hope it is clear from these remarks that this case history is primarily concerned with a pragmatic approach to remote sensing geology, carried out under commercial, rather than purely academic conditions.

We are particularly concerned with the use of Landsat multispectral scanner (MSS) and return beam vidicon (RBV) imagery for reasons which will be made clear below. In this paper, the term 'image-geological map' is used for a map produced mainly by the interpretation of Landsat MSS and/or similar imagery and containing a comprehensive range of lithological and structural information (much as would be expected from a conventional geological map). We find this term useful because it expresses a family relationship with the photogeological map and photogeological mapping techniques (described by for example, Allum 1961; Allum 1966; Miller 1961), while it also expresses the differences in map content which may arise as a result of the different parameters of aerial photography and satellite imagery (discussed, for example, by Nash *et al.* 1980).

2. General account of the work in Sudan

2a. General statement

Our study area (Figure 1) consists of 92 000 km^2 of extremely arid, mountainous, and sandy desert terrain cut into two unequal parts by the River Nile and the southern end of Lake Nasser (Nuba Lake).

RECONNAISSANCE MAPPING AND MINERAL EXPLORATION

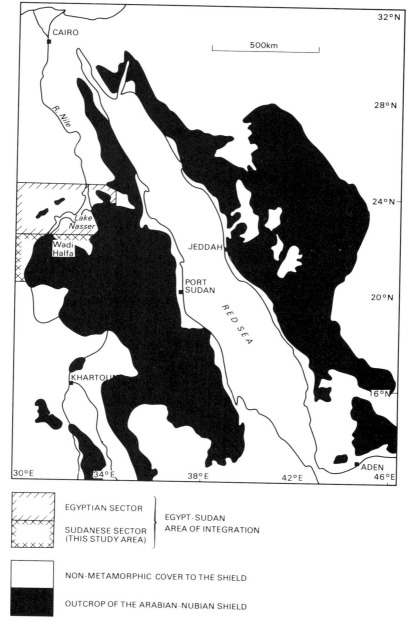

Figure 1. The location of the study area in relationship to the outcrop of the Arabian–Nubian Shield.

The work was completed within the period late 1983–late 1984 in a total of 30 man-months of image interpretation and a further 7 man-months of fieldwork. The fieldwork was carried out by both Geosurvey and GMRD (Geological and

Mineral Resources Department of Sudan) personnel and was supported periodically by the UNDP Project Coordinator. Work planning and most of the image interpretation was carried out by the two senior authors (PSG and PASC), while all authors were involved in the field activities.

2b. Imagery used

Most of the interpretation was carried out on 1:250 000 scale hardcopy Landsat 1 MSS imagery which was selectively processed in-house using Geosurvey's digital image processing facility. The coverage of the study area by eight MSS scenes is shown by Figure 2. Two MSS image types were produced for each scene. These were, firstly, a linearly contrast-stretched false colour composite (FCC) image of bands 4, 5 and 7 and, secondly, a logarithmically contrast-stretched, black-and-white band 7 image (e.g. Figure 3). The linearly stretched FCC imagery provided a broad range of information in all terrain and geological conditions, while the logarithmically stretched band 7 imagery preferentially emphasized tonal variation in areas of outcrop darkened by desert weathering conditions. In addition, band 7 imagery was produced in a rectified form, formatted into four sheets, each with a $2^0 \times 1^0$ area, which cover the whole study area. The imagery was geometrically corrected to a standard map projection using control from the best available topographical maps. The corrected and formatted band 7 imagery was used as a compilation base for the interpretation and fieldwork.

The choice of MSS imagery was restricted to a single-band and FCC types because these represent the widest possible utility relative to cost. This applies to the present use of the imagery for reconnaissance geological mapping. Also, an important aspect of a development project is that such imagery will definitely continue to be of value subsequent to our work, when they are passed to the client and are more likely to be used by non-specialists in remote sensing.

Most of the study area is covered by good quality Landsat 3 return beam vidicon (RBV) imagery. Also, about 15 per cent of the study area is covered by part of one swath of Shuttle Imaging Radar-A (SIR-A) imagery (Figure 2). Both the RBV and SIR-A imagery were used as 1:250 000 scale hardcopy in support of the MSS imagery. Because of their high ground resolution relative to the MSS data, the RBV and SIR-A imagery were a useful complement to the interpretation of detailed structure. For the same reason, the RBV images (enlarged to 1:125 000 scale) were particularly useful for navigation in the field. (We found no positive evidence that any of the information content of the SIR-A imagery is due to imaging of the shallow subsurface: our limited comparisons of the MSS, RBV and SIR-A image types with actual ground conditions in low relief, sand-and-gravel covered areas suggest that the detailed information content of the SIR-A imagery is due primarily to other factors, such as its high ground resolution.)

At the time when the present study was carried out, neither the Landat thematic mapper (TM) nor the SPOT satellite imaging systems were operational. Also, no adequate coverage of the study area by aerial photographs was available. Such image types would theoretically provide, in comparison with Landsat MSS data, greater spectral resolution (TM data), greater spatial resolution (TM, SPOT and, especially, aerial photographic imagery) and systematic stereoscopic coverage (aerial photography and, potentially, SPOT). We do not believe that the unavailability of any of these image types was detrimental to our objective of reconnaissance, regional geological mapping. On the contrary, Landsat MSS imagery frequently offers sufficient spectral and spatial resolution for such a purpose, at manageable data processing and interpretation costs.

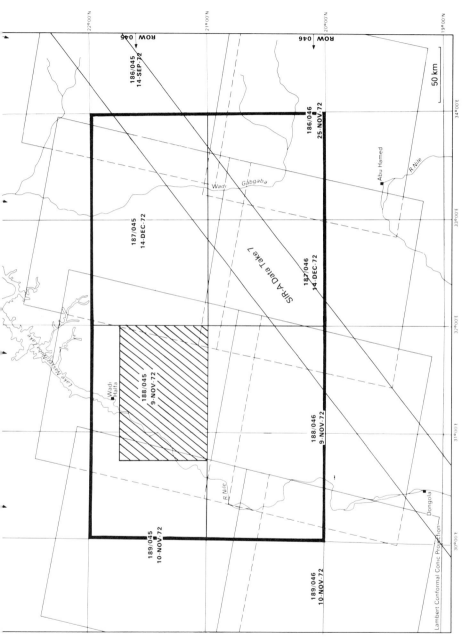

Figure 2. Coverage of the study area by Landsat MSS and SIR-A imagery used in this work. The boundary of the study area is indicated by the heavy line. Each MSS scene is a parallelogram identified by its path/row number and date of acquisition. The SIR-A imagery consists of a single, 50 km-wide strip (Date Take 7) crossing the study area from ENE to WSW. Landsat MSS scene 188/045 (Wadi Halfa) is reproduced as Figure 3. The area of the extract from an image-geological map (Figure 4) is indicated by hatching.

PATH/ROW 188/045:9-NOV-72　　　　　　　　　　LANDSAT MSS BAND 7
GEOSURVEY INTERNATIONAL

Figure 3. Landsat 1 MSS scene 188/045:9-NOV-72 centred near Wadi Halfa (Figure 2). This is a logarithmically contrast stretched image of band 7, printed in black and white. The scene size is 185 × 185 km.

2c. Method of geological study

The approach to the work was determined by several factors, including our previous experience of remote sensing work in Sudan and similar African terrains, the imagery available, the quality of the existing geological information and the contracted time budget. Our approach also took into account the requirement that the resulting image-geological maps should be useful in planning the subsequent steps of the mineral exploration programme. Consequently, we were particularly concerned with the locations and setting of known mineral occurrences during both map compilation and fieldwork. However, it is worth emphasizing that it was not part of our immediate task to use either the imagery or the fieldwork to try to locate major new mineral occurrences. We believe that the use of Landsat MSS and similar image data for such 'anomaly picking' in poorly known regions is usually a much less effective use of the technology than the production of image-geological maps.

Subsequent to the acquisition and processing of imagery, the work was carried out in three distinct stages, as follows:
1. Preliminary interpretation; including literature review and compilation of available information on mineralization, culminating in the production of 1:250 000 scale provisional image-geological maps and legend.
2. Field checking of the provisional maps, including brief study of as many mineral occurrences as possible.
3. Final interpretation; including revision of the provisional maps and preparation of a technical report which includes detailed recommendations for subsequent work in the mineral exploration programme.

Figure 4 is an example of part of a completed image-geological map. Figures 5, 6, and 8 are all directly based on the final results of the mapping.

2d. Previous knowledge of the study area

The main features of the geology of the study area and its environs are a Precambrian to early Palaeozoic 'Basement' of metamorphic rocks (the Arabian–Nubian Shield) and intrusive complexes, all overlain by a 'Cover' of relatively undeformed Phanerozoic sedimentary and minor volcanic units. This general situation appears to be true of much of northeast Africa, including Egypt and Sudan (Cahen et al. 1984) and the present study area (Figure 1). Prior to our work, the geological setting of the study area was known only from very localized observations and by extrapolation from adjacent, better known areas.

The region has been visited by several geologists throughout this century but their observations have all been of restricted application to our study area. For example, Dunn (1911) was concerned only with some of the larger gold workings in the study area. Hume (1934, 1935, 1937) traversed the Nile route throughout what is now the Area of Integration. Sandford (1935) produced a reconnaissance geological map at 1:4 000 000 scale of western and northwestern Sudan which is primarily concerned with the Cover rocks. Maley (1970) made the only detailed geological map (at 1:25 000 scale) of any part of our study area but his work was restricted to the former (drowned) site of the Second Cataract at Wadi Halfa (Figure 2). Vail et al. (1973) produced a detailed geological map of the Third Cataract, which lies just to the south of the study area. They also carried out a reconnaissance traverse northwards into the study area and made a sketch map (which proves to be broadly correct) of the distribution of metavolcanic rocks between the Second and Third Cataracts.

Apart from such selective visits, the study area was only known more indirectly in the form of summary accounts of the geology of all of Sudan (Whiteman 1971; Vail 1978) and of 1:2 000 000 scale geological maps of the whole country (Vail 1978; GMRD 1981). For much of northern Sudan, such overviews rely greatly on extrapolation from better known areas, such as the Red Sea Hills of Egypt and Sudan and the Bayuda Desert of Sudan. Later work by Klitzsch (1983) and his team in Egypt and northwestern Sudan suggests that, although this large region contains Cover rocks of diverse types and of from Silurian to end-Mesozoic age, the Cover rocks in the study area are most likely to be Mesozoic, continental sandstones (the Nubian Sandstone Formation *sensu stricto*).

The existing knowledge of gold mineralization in and around the study area requires particular mention. Gold has been mined in this region from Pharaonic until modern times (Dunn 1911; Graham 1929; Vercoutter 1959). Six small mines which were working during the early part of this century but are now closed are restricted mostly to the east bank of the Nile between the Second and Third Cataract. The only mine located some distance east of the Nile is at Umm

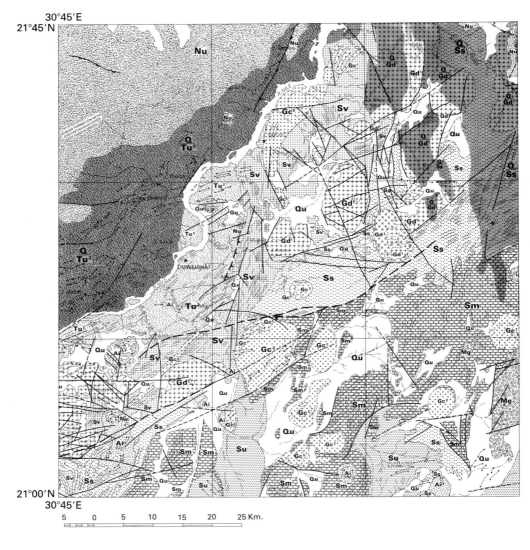

Figure 4. Extract from original image-geological map produced at 1:250 000 scale. The extract is located within the central part of the Landsat MSS scene reproduced in Figure 3 (compare locations on Figure 2). The main geological and structural features of the map extract are as follows. Thick lines are various categories of fracture lineament. Fine lines include boundaries between lithological associations ('image-geological units'), strike ridges and (with dumbbell ends) dykes and pegmatites. The dark toned area labelled Q indicates aeolian sand-cover to pre-Nubian rocks (*cf.* Figures 3 and 5). Qu indicates undifferentiated alluvial and aeolian deposits. Nu and Nh indicate different surface expressions of the Nubian Formation. Non-metamorphic alkali syenite–granite bodies are indicated

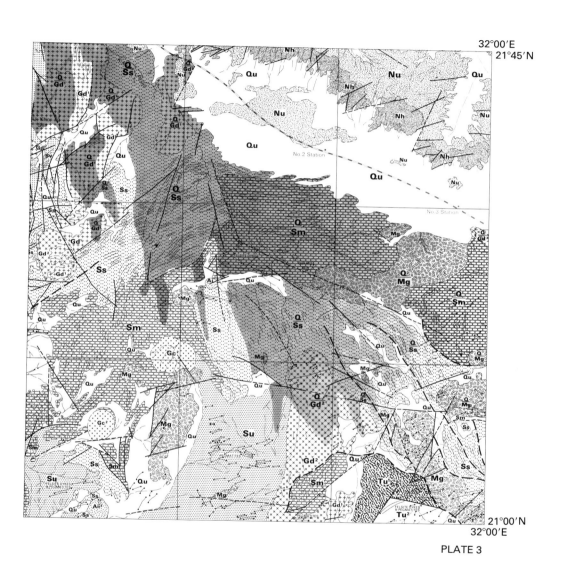

PLATE 3

by Ai. Variably deformed and metamorphosed igneous bodies are indicated by Mg (gabbros, diorites), Gd and Gc (discordant and concordant granodiorite–granite bodies, respectively). Low metamorphic grade rocks include Ss (dominantly pelitic–psammitic metasediments), Sv (dominantly mafic metavolcanic rocks), Sm (metasediment–metavolcanic assemblages with a high proportion of marbles and calcargillites) and Su (undifferentiated metasediments–metavolcanics). Paragneiss–orthogneiss 'terrains' are indicated by Tu (dominantly quartzofeldspathic and banded gneisses and schists). Solid triangles indicate abandoned gold mines and ancient small gold workings located by the field party.

Nabari (about 50 km NNE of Station Six, see Figure 6). Very recently, Minex Developments PLC (personal communication) have visited and relocated many gold occurrences, mostly of very small size, within the study area. This information was kindly made available to us early in our work and incorporated into the preliminary interpretations. Almond et al. (1984) have also recently visited and reported on a small number of gold occurrences immediately to the east of our study area.

3. Results

3a. General statement

Landsat MSS and RBV imagery, like aerial photography, only contain direct information on phenomena at the earth's surface. (In the present study area, this is also effectively true of the SIR–A imagery. See comments in section 2b.) It follows that the geology, which is essentially three-dimensional in expression, can only be inferred indirectly and (since surface conditions vary independently of the geology) incompletely from the imaged characteristics of the earth's surface. The method, results, and — ultimately — the uses to which the interpretation may be put, are all constrained (but not crippled) by this fact.

The remainder of this part of the paper summarizes and discusses the results of the satellite mapping study. It covers all stages of the work, beginning with a short discussion of how surface conditions (particularly the geomorphology) have affected the image interpretation (section 3b) and then proceeding in as logical a sequence as possible through stages which are at a successively distant remove from the actual information content of the imagery. The main part of this account is concerned with the final mapping results (sections 3c, 3d). This leads to a discussion of the geological setting of gold and other mineralization (section 3e). Finally, the account closes with a review of new exploration work which can be planned on the basis of the mapping results and appraisal of mineralization (section 3f).

3b. Surface conditions and geological interpretation

This section is supported by Figures 3, 4, and 5. Because of the extremely arid environments of northern Sudan, surface conditions are determined predominantly by geomorphological processes. In this region, aeolian and alluvial transport and deposition have the greatest overall effect on degree of exposure. Although alluvial processes are presently virtually inactive except along the Nile (Butzer and Hansen 1968), their effect on existing landforms is still strong. Also, where rocks are exposed, desert weathering is the strongest single control on their image colour/tone: we found that the effect of such weathering is often not directly related to the colour index of the fresh rock.

As may be predicted (Allum 1961; Sparks 1971), none of the surface processes effective in northern Sudan has a uniform effect on the surface expression of particular lithologies. Many problems of interpretation arise from this simple fact. However, the following examples illustrate how attention to detail in image interpretation, with some support from field observations, usually results in basically reliable geological mapping.

Figure 5 shows the relative distribution in the work area of the Basement and Cover rocks and of some major landforms which affect geological interpretation (see also Figure 3). Prevailing southerly winds originating in eastern Libya/western Egypt move large quantities of sand into Sudan, most of the sand probably being derived from Nubian Sandstone and similar Cover rock outcrops. Figure 5 illustrates how these factors affect local variations in the extent of inundation of rock outcrop by sand. Note that the River Nile also forms a local but efficient barrier to sand movement. The sand cover causes two particular interpretation problems, with different solutions; the location of the Basement/Cover rock boundary and the mapping of lithologies within the Basement. These are discussed below.

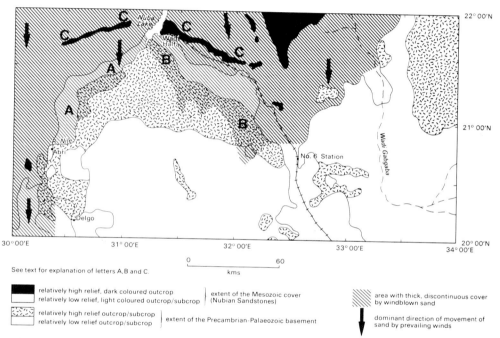

Figure 5. The relationship between some major landforms and the distribution of Basement and Cover rocks in the study area.

To the west of the Nile (Figure 5, area A), light image colour/tone due to heavy sand cover effectively obscures any differences in actual colour of Basement and Cover rocks. However, small scale differences in relief expression of the Basement and Cover at their contact results in subtly different textural expressions on the imagery which could be exploited by the interpreters. When checked in the field, the interpreted boundary proved to be very accurately located.

Southeast of Wadi Halfa (Figure 5, area B), sand inundation of the Basement effectively obscures any image colour/tonal variation which could be reliably attributed to lithological variation. Consequently, this area was initially interpreted only in terms of structural variations (fractures, rock-layering) expressed as patterns of relief (i.e. changes in hill form seen in plan). However, it took very little field traversing to establish that, for example, sprawling massifs of rounded hills are late-tectonic gabbros/diorites, while long, very sinuous ridges and valleys are well layered, intensely folded marbles, calc-argillites, and subordinate metabasites. Once a general but clear connection had been established in the field between a particular pattern of relief, a structural (tectonic) style and a lithological assemblage, it was possible to return to the imagery and reinterpret lithological variation in area B (Figure 5) with a high degree of confidence that was substantiated by later traverse results. The final results can be judged from a comparison of Figure 3 with Figure 4.

Variations in relief and image colour/tone proved misleading in one outstanding case. Areas marked C on Figure 5 are strike ridges within the Nubian Sandstones which are outstandingly darker than adjacent, low relief areas of Nubian outcrop. Such ridges were initially interpreted as distinctive lithological horizons within the Nubian Sandtones. However, the fieldwork revealed no obvious, qualitative lithological difference between the ridges and the adjacent terrain. It seems that both relief types consist of cross-bedded, light coloured, immature arenites, with a small proportion of thin, discontinuous arenite bands which are dusky brown in colour due to a haematitic matrix and cement. On the ridges, the dark colour of the subordinate, haematitic colour dominates the image appearance, because this rock tends to form slabs of debris which cover most of the hill surface. On gentler ground, the debris from the haematitic rock is more thoroughly broken and dispersed, so that the image appearance more closely reflects the true proportions of the rock types. There is no absolute safeguard against such an interpretation error other than field check.

3c. The Basement

Division of the Basement rocks. The distribution of the most important groups of rocks defined in this study is shown in Figure 6. The Basement consists of metamorphic and igneous rocks of presumed Precambrian to Palaeozoic age (Cahen *et al.* 1984). Prior to fieldwork, units were distinguished on the imagery on the basis of structural style, structural relationships, and indications of lithological variation such as image colour/tone and patterns of relief. Taken overall, the single great contribution of the fieldwork to the production of the image-geological maps was to show what are the actual dominant lithologies and their variations from unit to unit. It proved more difficult to establish consistent relationships between units across the study area.

Paragneiss–orthogneiss terrain. This unit is characterized by extensive areas of generally low, undulating relief, broken by occasional larger ridges and inselbergs. The majority of the rocks are quartzofeldspathic (with or without micas and amphiboles) gneisses and schists. Flaggy augen gneisses are common in zones of intense shear. There are local occurrences in these rocks of sillimanite, staurolite, and garnet but more detailed study is required to establish the actual metamorphic grade.

The stratigraphic relationships between this 'terrain' and most of the other units of the Basement are not certain. Two broad possibilities are illustrated in Figure 7. This problem has already attracted attention in the literature on Sudan (e.g. Almond 1982; Shackleton 1979; Vail 1983). We are unable to resolve the problem because, firstly, outcrop patterns visible on the imagery can often be interpreted

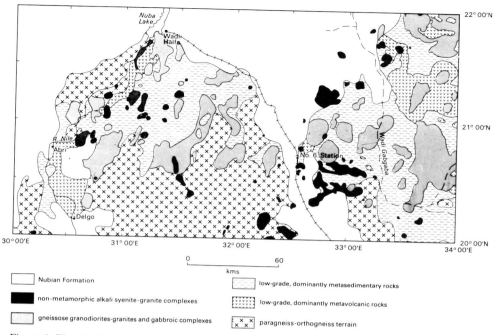

Figure 6. The distribution of major rock groups in the study area. The Mesozoic cover is represented by the Nubian Formation. The Precambrian–Palaeozoic basement is represented by the remaining groups.

in opposing ways (is the dome-like structure related to an inlier or an intrusion?) and, secondly, most rock unit boundaries are insufficiently well exposed to be explained by rapid fieldwork.

Low-grade metasedimentary and metavolcanic rocks. Such rocks are typified, on the imagery and on the ground, by abundant strike ridges and a strong planar fabric which is often due to a schistosity parallel or subparallel to a surviving stratification. The main metasedimentary rock types are pelitic to psammitic schists, marbles, calc-argillites, slates, and phyllites. Metamorphic rocks of demonstrable or presumed volcanic origin include mafic to intermediate lavas and pyroclastic rocks, metabasites, amphibolites, chloritic schists, and subordinate to rare felsic volcanic rocks.

There appear to be three main types of low-grade metamorphic rock assemblage, as follows (see also Figures 6 and 8):

Dominantly mafic metavolcanic rocks,
Dominantly pelitic–psammitic metasedimentary rocks,
Mixed sequences of marbles, calc-argillites, mafic metavolcanic and pelitic–psammitic metasedimentary rocks.

This division into assemblages is made partly on the basis of variations of structural style and colour/tone visible on the imagery (mentioned in section 3b). However, the fieldwork showed that almost all the metasedimentary and metavolcanic lithologies listed above are represented in each mappable assemblage. Also, no regionally applicable stratigraphy could be discerned in the field.

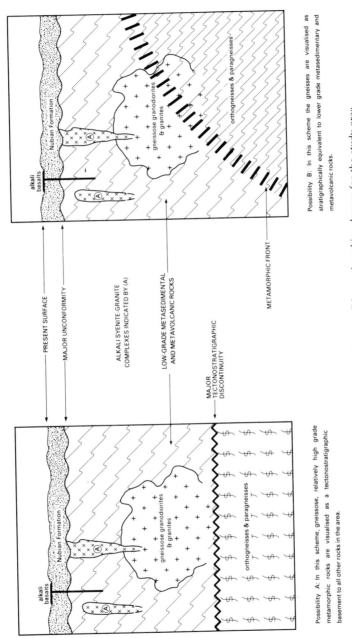

Figure 7. Diagrams illustrating two contrasting possible stratigraphic schemes for the study area.

The basic differences between the units are in the relative proportions of common lithologies and in the influence such lithological variations have upon the general style of deformation (and, hence, image appearance).

Gneissose granodiorite–granite and gabbro complexes. These variable metamorphosed and deformed igneous complexes have no single characteristic feature on the imagery. Some have the recognizable form of discordant intrusions but these are difficult to separate on the imagery from the alkaline complexes described in the next section. Other dome-like, usually strongly gneissose occurrences are difficult to distinguish (and perhaps are not distinct) from the gneisses of the 'orthogneiss–paragneiss terrains'. Finally, some elongate occurrences, again of more gneissose rocks, are difficult to distinguish from metasedimentary and metavolcanic units. The actual distribution (Figure 6) and general characteristics of these complexes only became clear as a result of the fieldwork.

The lithological variation within the igneous complexes ranges from granite, through adamellite, granodiorite, diorite, and gabbro to one localized occurrence of a layered pyroxenite–gabbro (there are no indications that any of these rocks are related to ophiolites). Rocks in the granite–granodiorite range usually but not always form complexes discrete from those of mafic composition. All rocks in all occurrences show some evidence of metamorphism, usually in the form of a gneissosity associated with shearing: such effects are slight and very localized in the case of the mafic bodies and are more commonly intense and pervasive in the case of the granitic–granodioritic bodies.

Non-metamorphic alkali syenite–granite complexes. These complexes generally form distinct hill masses which display circular to ovoid or, occasionally, irregular forms. Some of the elongate masses shown on Figure 6 actually consist of a succession of intersecting ring structures. Dyke swarms are sometimes associated with the main intrusive masses. There is little doubt from the field relationships that these complexes are the youngest element of the Basement and are also older than the Nubian Sandstones.

The most common rock in these complexes is a coarse grained, feldspar-phyric to equigranular subalkaline syenite. Less common plutonic rocks are alkaline syenite (including some nepheline syenite) and alkaline to peralkaline granite. Volcanic rocks, usually pyroclastic deposits, occur in some of the complexes.

3d. The Cover

Nubian Sandstone Formation. This unit of sedimentary rocks covers a large part of the study area (Figure 6). Recent work by Klitzsch (1983), outside the present study area, suggests that these rocks are all of Mesozoic age. We have already shown in section 3b that it does not appear possible to subdivide this rather uniform sedimentary sequence. The dominant lithology is a moderately to poorly sorted arenite, consisting mostly of subangular quartz grains with minor constituents of clay and mafic mineral fragments. Such rocks usually exhibit moderately to well developed cross-bedding in units 1–5 m in thickness. Some arenites throughout the sequence are conglomeratic. The basal conglomerate is usually no more than 25 cm. in thickness. There are also occasional thin, usually discontinuous bands of claystone and siltstone as well as the already described haematite-rich horizons.

These sedimentary rocks are well fractured but rarely exhibit other than minor normal faults. They dip gently and uniformly away from the Basement outcrops which occupy the central and eastern parts of the study area. Probably, these dips are related to the post-Nubian development of a large, gentle dome which occupies the second big bend in the course of the Nile downstream of Khartoum (Figure 1).

Alkali basalts. Numerous very small, dark hills have proved to be occurrences of basaltic rocks. These are located throughout the central and western parts of the study area, with the greatest concentration around the Nile in the area of Delgo (Figure 6). Almost all known occurrences rest on an eroded surface of Nubian Sandstones. Some of these occurrences may be plugs but it is likely that many are the remnants of flows which were erupted onto a well-dissected, post-Nubian landscape.

Most of the rocks are fresh, aphyric, or sparsely olivine-phyric basalts. Some more densely olivine and augite-phyric rocks occur. It is tentatively suggested that these are all alkali basaltic rocks of Caenozoic age.

3e. Mineralization

The setting of gold mineralization. At the beginning of the study, the locations of about 90 gold occurrences, including a few recently operative small mines, were known with varying precision. However, there was little systematic knowledge of the setting of gold mineralization. Since the gold occurrences have no image expression (except that the extensive slimes ponds at Umm Nabari mine appear as a bright spot on the FCC imagery), part of the fieldwork was concerned with direct, if rapid observation of as many gold occurrences as possible.

A total of 108 gold occurrences were located in the field, including lode and placer types. Most of these occurrences are concentrated in the eastern part of the study area (Figure 8). In the whole study area, almost 50 per cent of all occurrences lie within the outcrop of low metamorphic grade mafic metavolcanic rocks. Also, the great majority of lode occurrences are related to quartz veins in these rocks or, less frequently but still significantly, in ferruginous marbles and calc-silicate rocks. Some occurrences fall within the area of outcrop of granodioritic–granitic bodies (Figure 8) but not many of these are of lode type.

The results of the combined image interpretation and fieldwork show that there is a positive relationship between the distribution of gold occurrences and large zones of intense, brittle–ductile shear, both the quartz veins and the gold being concentrated in dilational parts of the zones. Almond *et al.* (1984) examined the setting of a small number of gold occurrences immediately to the east of the present study area and concluded that there is a significant relationship between gold mineralization and sheared granodioritic rocks. However, the evidence from the present study, based on observation of many more occurrences, suggests that the factor common to all lode occurrences, is a shear zone environment in weakly metamorphosed rocks, the most likely host being the mafic metavolcanics, the least likely host being the granodiorites–granites.

Many of the observed ancient gold workings appear to be associated with placer deposits. These include both eluvial and alluvial types, the latter associated with the relic wadi system. The discovery in the field of some 'palaeodrainage' channels exposed at the stratigraphic base of the Nubian Sandstones suggests the possibility

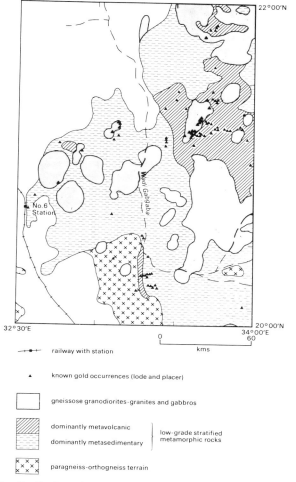

Figure 8. The distribution of gold occurrences in the eastern part of the study area.

of locating palaeoplacer gold deposits in such a situation. (An interesting investigation in the future would consist of a study of the relationships between pre-Nubian palaeodrainage channels and the present relic wadi system.)

Other mineralizations. During the course of fieldwork some other, usually very minor mineral occurrences were observed. Those which are possibly most significant are widespead occurrences of disseminated pyrite in low-grade metasedimentary and metavolcanic rocks. The general association of both gold and such mineralization with metasedimentary–metavolcanic rocks suggests a tentative comparison with the 'island-arc type assemblages' identified in the Basement of Egypt and Saudi Arabia. Such comparisons make it possible that the study area has significant potential not only for gold but also for base metals.

Possible uranium mineralization associated with one alkali syenite–granite body is indicated by an anomalously high spectrometer reading. No indications of uranium mineralization have been found in the Nubian Sandstones during this study. This does not, however, entirely preclude the longer term chances of finding roll-front or similar deposits. The general structural disposition of the Nubian Sandstones means that it is unlikely that these rocks contain hydrocarbon reservoirs within the study area.

3f. Selecting exploration targets

Our study for the UNDP was completed by making technical recommendations for the succeeding stages of the exploration programme. In part, such recommendations amount to a 'suggested reading' of the image-geological maps made by those geologists most intimately involved in their production and, therefore, most aware of their strengths and limitations.

Ideally, selection of mineral prospection target areas is based on sound theoretical models of mineralization in the context of a known geological setting. In practice, however, reconnaissance geological mapping of an unknown area does not provide sufficient information for refined modelling and it is, therefore, necessary to proceed at a more pragmatic level.

We selected five subareas suitable for detailed geological and structural mapping (covering 22 per cent of the total original study area). The most important areas for further study are along the Nile, between Wadi Halfa and Abri (included in the area of Figure 4), and to the east of Wadi Gabgaba (within the area of Figure 6). Detailed mapping and comparison of these areas, supported by radiometric age dating and other laboratory studies, would greatly help resolve many of the complex problems of Basement geology which have been outlined in this paper (cf. our discussion of Figure 7 in section 3c).

We have shown above that the study area has most potential for gold and (less certainly) base metal mineralization. Sufficient information has been gathered from a large number of small gold occurrences to infer the general setting of gold mineralization with reasonable certainty. Consequently, it is possible to identify ten target areas suitable for detailed gold and base metal exploration. These selected target areas range in size from 125 to 1730 km^2 and constitute about 10 per cent of the total Basement outcrop of the original study area. Each target area contains at least some known mineral occurrences, while all such areas are distinguished by an association of weakly metamorphosed (especially metavolcanic) rocks which have been deformed by brittle–ductile shear.

Judging from the observed variation and structural dispositions of rock-types in the study area, an equally wide range of types of mineralization is possible. Since no significant mineral showings have yet been found to support any of these possibilities, it is not possible to make detailed recommendations which are not already implied by the contents of the image-geological maps. Again, more detailed or specific forms of investigation are required to investigate the possibilities. Also, airborne geophysical investigations may be suggested as a complement to satellite remote sensing. Ways in which data from airborne magnetic and radiometric surveys may be used as reconnaissance geological mapping tools have been described recently by Batterham et al. (1983). Also, such geophysical data can be used for 'anomaly picking' attempting to identify particular kinds of mineralized body. A combined airborne magnetic, radiometric, and VLF survey, at 1 km line spacing, of the area of continuous outcrop of Basement rocks (Figure 6), covering about 72 000 km^2, can be recommended.

We made an attempt to assess the potential of the study area for ground water. The area has little obvious potential at any distance from the Nile and the associated outcrops of Nubian Sandstones. The possible exception is in large alluvial bodies associated with the dry course of Wadi Gabgaba (Figure 5). Further exploration will be necessary to locate water required in support of any future development of mineralization at any distance from the Nile.

4. Discussion

4a. General statement

This case history is concerned with how satellite imagery, particularly Landsat MSS data, can be used for producing reconnaissance geological maps (image-geological maps) which are a scientifically valid and economical contribution to mineral exploration. The role of satellite remote sensing for geological mapping as a contribution to mineral exploration has been given serious consideration since the beginning of earth resources exploration from space (noteworthy reviews have been presented by Viljoen et al. 1975; Nash et al. 1980). However, the fact that there remains articulate resistance to the use of any remote sensing in geological mapping (concisely stated by, for example, Whittle 1983), shows that the presentation of such remote sensing case histories and general discussion of their implications remain justifiable.

The discussion is concerned with three themes which, as already suggested in the Introduction to this paper, are all of general importance and closely interrelated. These themes are, firstly, the scientific status of (satellite) image-geological mapping, secondly, its economical character compared to other mapping methods and, finally, the optimum role of satellite remote sensing in a 'multistage' approach to mineral exploration planning.

4b. The scientific status of image-geological mapping

Remote sensing, particularly through the use of multispectral scanner data, can be used for direct lithological mapping; that is, through the use of spectral information to classify rock types according to characteristic surface reflectances (Abrams and Siegal 1980; Rothery 1984). However, as shown by Rothery (1984), the method depends on the field collection of quantitative reflectance measurements and is limited in its application by geological factors such as metamorphism, surface factors such as weathering and instrumental factors such as the spectral parameters of the remote sensing system.

Because of such limitations, it is not practicable to attempt to use Landsat MSS data for reconnaissance geological mapping of a large, inaccessible and poorly known region based on computer classification of the spectral signatures of rock types. The remaining approach to geological mapping based on satellite remote sensing makes use of the contextual and textural information in the imagery (especially those features of the imagery related to variations in the patterns of relief) as the basis of combined structural and lithological mapping. The present study in Sudan demonstrates that useful results may be obtained by this method but that fieldwork is still required to verify and improve them.

The outstanding scientific problem at this point is that, whereas mapping based on computer classification of spectral data is quantifiable at both the remote sensing and field stages, this is not true of mapping based on human interpretation of contextual and textural image information, which is essentially qualitative at both the remote sensing and field verification stages. The problem may be at least clarified, if not solved by a comparison with similar difficulties which have been encountered during the history of (aerial) photogeological mapping, the methods of which most closely resemble those employed in the present case history in Sudan.

More than twenty five years ago, J. A. E. Allum felt it necessary to justify the role of aerial photogeological mapping as a geological technique. His comments (Allum 1961, p. 525) are relevant to the present discussion: '(An) apt comparison is between an aerial photograph and a petrological microscope...(and) leads to an understanding of both the value and limitations of photogeology. The petrographer, using his microscope, studies a small piece of rock very closely, and all those characteristics of the rock which require an area larger than that of a thin section to become apparent are unseen by him. The field geologist has not the advantage of seeing the microscopic characteristics which are able to express themselves in the area of a single outcrop. The photogeologist, however, is in effect even further away from the rock than the field geologist. Many of the macroscopic characteristics expressed in single outcrops are lost to him, but in partial compensation for this he sees those rock characteristics which require great areas for their expression'. The same comments can be extended to include the interpreter of Landsat MSS and similar imagery, who can no longer examine directly the contents of an outcrop but can examine instantaneously the distribution and spatial relationships between outcrops (and other geologically determined surface features) over greater areas than those envisaged by Allum.

This extension of Allum's argument to include satellite remote sensing is, in effect, a restatement of the perceived value of the synoptic coverage of satellite imagery in comparison with both aerial photography and fieldwork. The general potential of a simultaneous overview from space of the earth's geological features was grasped during the early manned space flights and has been subsequently discussed more formally by, for example, Viljoen et al. (1975) and Abrams and Siegal (1980). The latter authors expressed the advantages of satellite imagery relative to aerial photography as a 'trade-off' between image area (ground coverage) and image detail (ground resolution).

The general conclusion to be derived from these considerations is that the ultimate role of satellite imagery in geological studies is the provision of a regional, even global context for the appraisal of data obtained as a result of fieldwork.

4c. The economical character of image-geological mapping

There are two aspects to the relative economy of geological mapping by satellite remote sensing: the cost and the time involved. The following discussion does not analyse the Sudan case history in detail, mainly because the detailed logistical and financial structure of one mapping project cannot be projected effectively onto others carried out over different sizes of area at different scales and under varying geological, geographical, economic, and commercial circumstances, as well as under varying objectives as perceived by both the client and the operator.

A simple measure of cost for comparative purposes is in U.S. dollars per square kilometre. The 1983 prices paid for the study at 1 250 000 scale in Sudan have been converted as accurately as possible into prices which would apply nearer the time of writing in 1985/86. At this time, the work described in this paper in Sudan would cost the client between three and four dollars per square kilometre to cover the total project area of 92 000 km^2. Such a price, determined by competitive open tender for the contract, would reflect the difficult market conditions which

RECONNAISSANCE MAPPING AND MINERAL EXPLORATION

have prevailed throughout the 1980s and would allow for little profit by the operator.

The operating costs of an image-geological mapping programme would normally cover the following main technical categories:
— image acquisition,
— image processing and production,
— image interpretation and related office studies,
— fieldwork,
— cartographic draughting and map production.

As a proportion of the whole, most of the cost is related to the last three categories; the most expensive single item is the fieldwork (even though, as in the Sudan study, the proportion of office to field-based geologists' time is more than 4:1; see section 2a), while the expenses of the office interpretation work and fair-drawn map production are similar. If the fieldwork component of the Sudan study were excluded, it would cost the client at present prices betwen two and three dollars per square kilometre. Image acquisition costs are worth considering in more detail, because they are more predictable and also because their low cost per unit area is a significant part of the economy of satellite image-geological mapping.

At 1986 prices, a single Landsat MSS scene in computer compatible tape form costs 660 dollars from EOSAT Data Center. Conversion of this cost into a figure per unit area has to take into account the number and locations of scenes required, compared with the size and location of the project area (see Figure 2). The cost is 5·7 cents per square kilometre in the case of the Sudan study.

The easiest comparison of the above costs is for a photogeological study based entirely on the use of vertical aerial photographs which provide stereoscopic overlap. The costs of such a study vary widely, especially due to variations in the size of the project area and the scale of the photography used. If an area of 10 000 km^2 and a photographic scale of 1:30 000 is selected, the minimum total cost to the client would be six dollars per square kilometre. A reduction in the size of the project area and a parallel increase in photographic scale would noticeable increase the cost per unit area. An important point is that such cost estimates exclude both fieldwork and image acquisition. Fieldwork costs need not be discussed further.

Image (aerial photograph) acquisition for a photogeological study may represent an important problem in terms of both logistics and costs. Whereas Landsat MSS imagery of acceptable quality is now available for most of the world's continental surface, there are many isolated regions (including the present study area in Sudan) which are not covered by suitable aerial photographs or by any aerial photographs at all. Using the example of an area of 10 000 km^2 to be covered by 1:30 000 scale black-and-white photography, the approximate acquisition costs at present prices would be five dollars per square kilometre (nearly a hundred times greater than the satellite image acquisition costs quoted above).

In summary, while a satellite image-geological mapping study of the Sudan area at 1:250 000 scale would now cost the client between two and three dollars per square kilometre (excluding fieldwork), an aerial photogeological study of part of the area at 1:30 000 scale would cost at least four times as much.

One aspect of the relative time requirements of image-geological and other kinds of mapping study has been alluded to already: that is, the greater likelihood of having to mobilize an aerial photographic survey than of having to make special arrangements to acquire satellite imagery. More generally, the office and fieldwork components of geologists' time in different kinds of mapping study can be broadly compared. The present case history in Sudan required 30 man-months of image interpretation and a further seven man-months of fieldwork, carried out within the period of one year, to map 92 000 km^2 at 1:250 000 scale. The office-based component of the work was, therefore, completed at a rate of about 3000 km^2 per man-month. In the more hypothetical case of aerial photogeological mapping of 10 000 km^2 of the study area at 1:30 000 scale, a comparable rate for the office-based component of the study would be approximately 1000 km^2 per man-month. There is no fixed way of calculating the amount of fieldwork that would be needed to verify such an aerial photogeological study, but it is reasonable to suggest that the same 10 000 km^2 would also need as much as 6–8 man-months. Finally, the time requirements of a conventional geological mapping programme based primarily on fieldwork are difficult to quantify. However, basic geological mapping carried out by Geological Surveys in many African countries has often been planned on the principle of covering a single Quarter Degree sheet (approximately 3000 km^2 in area) during one year by one geologist, including a single field season of about five to six months duration.

This comparison of time requirements is crude but does illustrate the relatively high 'turn-around' time of satellite image-geological mapping compared with other mapping methods.

4d. The multistage approach to mineral exploration

The merits of geological mapping based on satellite remote sensing, taken separately, include the technical possibilities of the method, its relatively low cost per unit area and the relatively rapid completion times required for a study. Considered collectively, such merits indicate that, especially when large regions are to be investigated, the method is ideally suited as a reconnaissance tool, producing results whic are suitable for use in subsequent planning of the investigation. The case history in a virtually unknown region of northern Sudan illustrates how much information can be obtained, as well as how coherent a picture of both lithological variation and structural relationships may be gained from application of the qualitative interpretation methods described.

Exploration targets may be defined as an immediate result of the image-geological mapping and followed-up accordingly without further use of remote sensing methods. The available exploration budget often dictates such a step. However, we wish to conclude with remarks on a more theoretical organization of a major investigation of an unexplored region.

Remote sensing techniques applied to any natural or human science are frequently regarded as part of a 'multistage' approach to problem solving, in which decisions on successive stages of effort and expenditure are based on a progression from relatively generalized, space- or airborne studies towards relatively specialized and/or intensive ground based studies. The application of this concept to mineral exploration is illustrated schematically by Figure 9.

In the case of the work described in this paper in Sudan, which can be regarded as the initial stage in the investigation (Figure 9), appropriate activities at the intermediate stage would include the use of Landsat TM data, airborne geophysical data and radar, aerial photography and — at particularly high relative cost — airborne multispectral scanners. An important point is that all such surveys, with the general exception of airborne geophysics, must be supported by field verification if the best use is to be made of the data. The ideas of iterative image

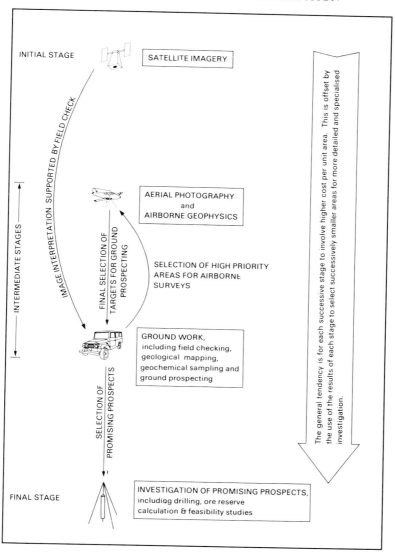

Figure 9. The concept of multistage mineral exploration.

interpretation and of integration of interpretation of different data sets are also very important (see also Nash *et al.* 1980).

Such a multistage approach, leading finally (one may hope) to detailed evaluation of identified prospects, requires time and considerable effort in both planning and execution. However, it is reasonable to argue, as new economic mineral deposits become increasingly difficult to find, that such a thorough approach, based on the best use of both modern technology and human interpretation abilities, is cost-effective in the end.

Acknowledgements. We are most grateful to Abdullah Hassan Ishag, Chairman of the Board of Directors of the Egypt–Sudan Area of Integration for his kind permission to publish this summary of the results of our work in Sudan. We also extend thanks to the UNDP for their supervisory role and practical assistance in this study, especially to Warren L. Anderson, the Project Coordinator, and to Garth ap Rees, UNDP Resident Representative in Khartoum, and all his very cooperative staff. Similar thanks for practical assistance in Sudan are extended to the staff of the GMRD in Khartoum. We are grateful to many colleagues, especially Barrie Hamilton, Martin Leese, Adrian Lloyd-Lawrence, Lynda Taylor and Colin Willers, for advice and assistance. We are also grateful to two anonymous referees for comments which greatly helped us clarify the objectives of this paper. Finally, particularly warm thanks are extended to Charles F. Withington, Advisor in Remote Sensing to the UNDP, for his encouragement to publish the results of our satellite mapping studies.

References

Abrams, M. J. and Siegal, B. S. 1980. Lithologic mapping. In: Siegal, B. S. and Gillespie, A. R. (Eds), *Remote Sensing in Geology*. John Wiley, New York. 381–418.
Allum, J. A. E. 1961. Photogeological interpretation of areas of regional metamorphism. *Trans. Inst. Min. Metall.*, **70**, 521–543.
—— **1966**. *Photogeology and Regional Mapping*. Pergamon Press, London.
Almond, D. C. 1982. New ideas on the geological history of the Basement complex of northeast Sudan. *Sudan Notes Rec.*, **59**, 106–136.
——, **Ahmed, F., and Shaddad, M. Z. 1984**. Setting of gold mineralization in the northern Red Sea Hills of Sudan. *Economic Geology*, **79**, 389–392.
Batterham, P. M., Bullock, S. J., and Hopgood, D. N. 1983. Tanzania: integrated interpretation of aeromagnetic and radiometric maps for mineral exploration. *Trans. Inst. Min. Metall.*, **90**, B83–B92.
Butzer, K. W. and Hansen, C. L. 1968. *Desert and River in Nubia: Geomorphology and Prehistoric Environments at the Aswan Reservoir*. The University of Wisconsin Press, Madison.
Cahen, L., Snelling, N. J., Delhal, J., Vail, J. R. Bonhomme, M. and Ledent, D. 1984. *The Geochronology and Evolution of Africa*. Clarendon Press, Oxford. 512pp.
Dunn, S. C. 1911. Ancient gold mining in the Sudan. *Rep. Wellcome. Trop. Res. Lab. Khartoum*, **4**, 207–215.
GMRD 1981. *Geological Map of the Sudan*, 1:2 000 000 Scale, Geological and Mineral Resources Department, Sudan.
Graham, G. W. 1929. Gold in the Anglo-Egyptian Sudan. *Report 15th Session International Geological Congress*, **2**, 79–280.
Hume, W. G. 1934, 1935, 1937. *Geology of Egypt, Vol. II: The Fundamental Precambrian Rocks of Egypt and The Sudan. Part I: The Metamorphic Rocks, Part II: The Later Plutonic and Minor Intrusive Rocks, Part III: The Minerals of Economic Value*. Survey Department, Cairo.
Klitzsch, E. H. 1983. Geological research in and around Nubia. *Episodes*, **3**, 15–19.
Maley, J. 1970. Introduction à la géologie des environs de la deuxième cataracte du Nil au Soudan. In: Vercoutter, J. (Ed.), *Migrissa I. Dir. Gen. Relat. Cult. Scient. et Techn., Min. Etrang.*, 122–157.
Miller, V. C. 1961. *Photogeology*. McGraw-Hill, New York.
Minex Developments PLC personal communication. Unpublished prospecting data, from northern Sudan.
Nash, C. R., Boshier, P. R., Coupard, M. M., Theron, A. C., and Wilson, J. G. 1980. Photogeology and satellite image intepretation in mineral exploration. *Miner. Sci. Eng.*, **12**, 216–244.
Rothery, D. A. 1984. Reflectances of ophiolite rocks in the Landsat MSS bands: relevance to lithological mapping by remote sensing. *Journal of the Geological Society of London*, **141**, 933–939.
Sandford, K. S. 1935. Geological observations on the north-west frontiers of the Anglo-Egyptian Sudan and adjoining part of the southern Libyan Desert. *Quarterly Journal of the Geological Society of London*, **91**, 323–381.
Shackleton, R. M. 1979. Precambrian tectonics of north-east Africa, In: Tahoun, S. A. (Ed.), *Evolution and Mineralization of the Arabian–Nubian Shield*, 2. Pergamon Press, Oxford. 1–6.
Sparks, B. W. 1971. *Rocks and Relief*. Longmans, London.
Vail, J. R. 1978. Outline of the geology and mineral deposits of the Democratic Republic of the Sudan and adjacent areas, including 1:2 000 000 scale Geological Map of Sudan. *Overseas Geological and Mineralogical Resources*, **49**.
—— **1983**. The Arabian–Nubian Shield: Proterozoic continental accretion. *Report Act. Department of Geology Portsmouth Polytechnic*, **9**, 7–10.

——,Dawoud, A. S., and Ahmed, F. 1973. Geology of the Third Cataract, Halfa District, Northern Province, Sudan. *Bulletin of the Geological Survey of the Sudan*, **22**.
Vercoutter, J. 1959. The gold of Kush. *Kush*, **7**, 120–153.
Viljoen, R. P., Viljoen, M. J., Grootenboer, J., and Longshaw, T. G. 1975. ERTS–1 imagery: an appraisal of applications in geology and mineral exploration. *Mineralogical Science Engineering*, **7**, 132–168.
Whiteman, A. J. 1971. *The Geology of the Sudan Republic*. Clarendon Press, Oxford.
Whittle, R. 1983. So what's wrong with remote sensing? *World Mining*, **September 1983**, 104–105.

Section 3
Rare-metal pegmatites and their mineralization

This chapter consists of two papers, both reviews, concerning pegmatites in Africa. The first paper by von Knorring and Concliffe discuss potential rare-element pegmatites which are widespread on the African continent, especially south of the equator, along the Kibaran orogenic belt extending from Uganda, through Rwanda and Kivu, and Maniema and Shaba provinces of Zaire. They are also prominent within the Mozambiquian fold belt of Pan-African age in East Africa, Mozambique, Madagascar, and within the Damaran fold belt in Namibia. Other areas include the Namaqualand metamorphic complex along the Orange River belt, and the basement rocks in Zimbabwe. In their paper, von Knorring and Concliffe concentrate on the mineralogy and geochemistry of the pegmatite deposits in Africa, paying particular attention to the occurrences of tin and tantalum mineralization.

There are two distinctive periods of pegmatite formation in Africa developed at end-orogeny times during the closing stages of the Kibaran and the Pan-African orogeny. These periods correspond to approximately 900 Ma and 500 Ma. According to Matheis the pegmatites in Nigeria, like those in other parts of Africa, are also Pan-African in age. Indeed most of Africa's rare-metal pegmatite potential is linked to the closing stages of the Pan-African orogeny. In his paper Matheis concentrates on the Nigerian tin province in terms of its economic rare-metal mineralization paying particular attention to occurrences of mineralized pegmatites in the Nigerian Precambrian basement. Matheis concludes that reactivation of old tectonic lineaments during the Pan-African orogeny provided excess fluid and heat to form late Proterozoic and early Palaeozoic rare-metal pegmatites by partial melting and selective leaching. Later plate tectonic movements and anorogenic activity caused partial resetting of the isotopic ages of the pegmatitic minerals.

© 1987 by John Wiley & Sons, Ltd.

Mineralized pegmatites in Africa

O. von Knorring and E. Condliffe
Department of Earth Sciences, University of Leeds, U.K.

Potential rare-element pegmatites are widespread on the African continent, especially south of the equator, along the Kibaran orogenic belt extending from Uganda, through Rwanda and Kivu, Maniema and Shaba provinces of Zaire. They are also prominent within the Mozambique belt in East Africa, Mozambique and Madagascar and within the Damara sequence in Namibia and moreover, within the Namaqualand metamorphic complex along the Orange River belt. Another characteristic, lithium-bearing pegmatite province is found in the basement rocks of Zimbabwe. Although beryl has been the most important mineral mined in the past, lithium, rare-earths, tin, niobium–tantalum, and uranium minerals are also recovered.

Furthermore, some pegmatites in Africa constitute the main source of attractive gemstones, in addition to ceramic feldspar and mica.

KEY WORDS   Pegmatites   Mineralization   Kibaran   Pan-African   Beryl
Amblygonite–montebrasite   Pollucite   Xenotime   Cassiterite
Columbite–tantalite   Equatorial central and south Africa

1. Introduction

Although the major granite pegmatite regions of Africa have been known for a long time (Landes 1935), large-scale exploration of the deposits only began during and after the Second World War, when certain strategic minerals were in great demand.

The present study reviews the principal pegmatite regions in Africa, where extensive mining of minerals has taken place. Special attention has been paid to the mineralogy and geochemistry of the pegmatite deposits, and in particular to tin and tantalum mineralization.

Pegmatites have come to occupy a place of increasing significance, since they constitute the source of many important minor and trace elements used by present-day technology, e.g. lithium, caesium, beryllium, scandium, yttrium, rare-earths, tin, zirconium, hafnium, niobium, tantalum, and uranium. They also provide industrial minerals like ceramic feldspar, mica, and high purity quartz, and a large variety of attractive and highly priced coloured gem minerals and ornamental

0072–1050/87/TI0253–18$08.00
© 1987 by John Wiley & Sons, Ltd.

stones used in jewellery. Much of the information in this study comes from mineralized pegmatites in the equatorial, central and southern Africa, including Madagascar, where the senior author has investigated numerous occurrences in the course of some thirty years.

Mineralized pegmatites are widespread in Africa and occur in the older cratonic areas which have remained stable over the past 1500 Ma — as well as in the comparatively younger orogens, consisting of mobile zones which have undergone a number of orogenic deformations during the past 1200 Ma (Clifford 1966).

It is interesting to note that most of the extensive mineralized pegmatite areas so far known are situated south of the equator and are associated with the younger orogens in the Kibaran belt of Uganda, Rwanda, eastern Zaire, and in the Mozambique belt of East Africa, Mozambique, and Madagascar, and in the Damaran of Namibia (Figure 1).

One of the largest, continuous regions of mineralized pegmatites in Africa, is the Namaqualand–Gordonia–Kenhardt region or the Orange River pegmatite belt which is also of Kibaran age.

In contrast, the important lithium pegmatites of Zimbabwe are found within much older basement rocks of the Zimbabwe craton.

2. The economically important minerals in African pegmatites

Economic interest in pegmatites began during the early part of this century and expanded rapidly in the 1950s. In the following account an outline of some of the more important rare-element minerals will be given.

2a. Lithium

Lithium is one of the characteristic elements of granite pegmatites, where it is concentrated to varying degrees and pegmatites, containing appreciable amounts of lithium minerals are termed lithium pegmatites. Normally, within an extensive pegmatite field only a few distinct lithium pegmatites may occur but in some pegmatitic terrains, especially in parts of central and southern Africa and Madagascar distinct geochemical provinces of lithium can be established. The lithium minerals found in these pegmatites vary to a great extent from one pegmatite region to another. In some areas lithium silicates are the dominant minerals, in others, lithium phosphates may be the only lithium minerals found. Among independent lithium minerals found in typical lithium pegmatites the following may be mentioned: amblygonite–montebrasite and lithiophilite–triphylite series of lithium phosphates and the silicates eucryptite, spodumene, petalite, lithian micas, and holmquistite.

Members of the amblygonite–montebrasite series $LiAl(PO_4)(F,OH)$ are perhaps the most widespread of all lithium minerals. Although the name amblygonite has generally been applied to the series, analyses show that montebrasite is the dominant member in almost all the lithium pegmatites in Africa and elsewhere.

Closely associated with montebrasite are the less common minerals of the lithiophilite–triphylite series $Li(Mn,Fe)PO_4$. These two groups of lithium-phosphates are characteristic of mineralized pegmatites; they occur in nodular form or in aggregates of varying size; occasionally gigantic nodules up to several tons in weight have been observed.

It has been noted in many regions that the lithium-phosphates are important indicators of economic beryllium pegmatites.

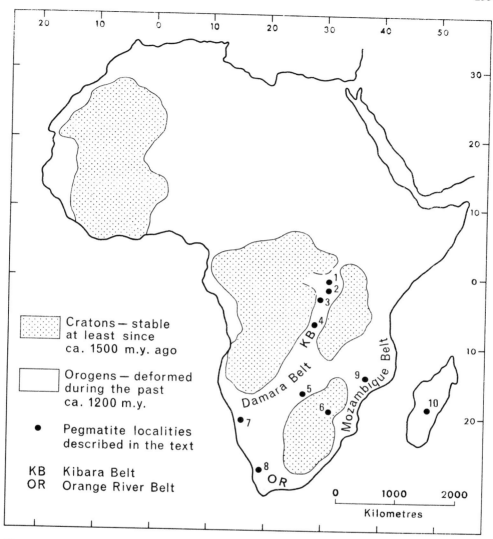

Figure 1. The distribution of the principal economic pegmatites (pegmatite fields) in equatorial, central and southern Africa in relation to the major structural units. 1. Southwestern Uganda; 2. Gatumba area, Rwanda; 3. Kobokobo, Kamituga area, Kivu, Zaire; 4. Manono, Shaba, Zaire; 5. Kamativi, Zimbabwe. 6. Bikita, Zimbabwe; 7. Karibib area, Namibia; 8. Noumas, Namaqualand, South Africa; 9. Muiane area, Alto Ligonha, Mozambique; 10. Antsirabe area, Madagascar. Structural units from Clifford (1966).

Spodumene, $LiAlSi_2O_6$ is the more abundant lithium mineral and has been noted in all the major pegmatite regions of Africa. It is, however, commonly kaolinized or replaced in many pegmatites, particularly in the Kibaran belt of Uganda, Rwanda, and Zaire. A characteristic feature of many spodumene pegmatites is their close association with tin mineralization, and many deposits in Rwanda, Zaire, Zimbabwe, and Namibia have been mined for tin on a large scale. In spite of some very large deposits of spodumene, e.g. Manono in Zaire, the mineral itself has been rarely mined for lithium.

Petalite $LiAlSi_4O_{10}$ is generally much rarer than spodumene but in some African pegmatites it has been found in extremely large concentrations e.g. Bikita in Zimbabwe and a great number of pegmatites in Namibia. Eucryptite $LiAlSiO_4$, is the rarest of the lithium–aluminium-silicates and has usually been observed in a number of petalite-rich pegmatites, although it also occurs in some dominantly spodumene-bearing pegmatites.

Although pure, glassy eucryptite has been found in some localities in Zimbabwe, e.g. Bikita, the majority of finds consists of fine-grained, almost porcellaneous intergrowths of eucryptite and albite that replace spodumene or as eucrytite-quartz intergrowths after petalite.

Lithium micas, pink to purple and occasionally grey to brown in colour are a group of striking materials typically associated with lithium pegmatites. They commonly form separate core units and are intergrown with variable amounts of quartz and albite (cleavelandite).

There is an apparent correlation between lithium and fluorine, moreover, the fluorine-rich lithium micas are commonly associated with other fluorine-bearing minerals such as topaz and fluorite.

In some fractionated lithium pegmatites the lithian mica units are of great economic importance since they carry a variety of rare-element accessories, e.g. hafnian zircon, bismuth and especially tantalum minerals.

2b. Caesium

Pollucite $(Cs,Na)Al_2Si_4O_{12}.H_2O$, the only independent mineral of caesium, is a characteristic, although a rare constituent of highly fractionated or complex lithium pegmatites. It is commonly found in pod-like bodies varying in size and is closely associated with petalite, spodumene, and cleavelandite. It resembles quartz, and smaller nodules may be easily overlooked.

Larger concentrations of pollucite have been observed at Bikita, Zimbabwe, Morrua, and Muiane areas in Mozambique and at Karibib in Namibia.

Additional caesium-bearing minerals found in pollucite pegmatites are caesian analcite, caesian beryl and the rare rhodizite from Madagascar. It has been also noted that microline feldspars and lithian micas have enhanced rubidium and caesium contents in pollucite-bearing pegmatites.

2c. Beryllium

Beryllium is one of the major trace elements associated with most grantie pegmatites and beryl is by far the commonest and economically the most important beryllium mineral.

Among the rarest beryllium minerals found in African pegmatites are bavenite, bertrandite, bityite, chrysoberyl, euclase, faheyite, gadolinite, genthelvite, hambergite, helvite, herderite, hurlbutite, milarite, moraesite, phenakite, and rhodizite. Apart from chrysoberyl, bertrandite, and gadolinite which may be relatively common in some areas, most of the other minerals are rare or very rare and have been recorded from a few localities.

The typical beryl-bearing pegmatites vary in internal structure and composition, but most have prominent quartz cores, distinct zoning, and well-developed lithium and/or sodium phases.

The position of beryl within a pegmatite is variable, but commonly it is located at the quartz core margin, together with microline, albite (cleavelandite), muscovite, lithian mica, and lithium-phosphates. Sometimes exceptionally large masses of beryl may be confined to the quartz core. In highly fractionated lithium pegmatites excellent crystals of the caesian beryl morganite have been noted in pockets of the quartz core.

Another and most important form of beryl and chrysoberyl mineralization is their sporadic occurrence in some pegmatitic granites and migmatites. Among the many varieties of beryl the greenish-yellow or common beryl is found in almost all mineralized pegmatites. Most of the glassy and coloured varieties tend to be associated with lithium pegmatites. Within this group of glassy beryls, the important gem varieties are to be found — the blue to blue-green aquamarine, pink morganite, the golden yellow heliodor, moreover the colourless goshenite and the unusual, black opalescent beryl from the Alto Ligonha pegmatite field in Mozambique. These beryls are chemically distinct in that they contain significant amounts of alkalies and water. Some of the morganites from Alto Ligonha, in particular, are rich in caesium.

2d. Yttrium and rare-earths

The larger part of yttrium and rare-earths in granite pegmatites is confined to a number of minerals e.g. xenotime, monazite, allanite, and a variety of complex uraniferous rare-earth niobotantalites.

Pegmatites within the Mozambique belt, in particular, are known for their high content of rare-earth minerals (von Knorring 1970).

Another important area is the eastern section of the Orange River pegmatite belt, where numerous zoned pegmatites carry xenotime, monazite, gadolinite, allanite, euxenite, and fergusonite (Hugo 1969). Although xenotime and monazite are seldom seen in large amounts in pegmatites, they may be economically recovered from the eluvium surrounding the pegmatite bodies. In some monazites from Gile and Muiane areas of the Alto Ligonha region in Mozambique the following order of abundance of the REE has been observed:

1. Ce>Nd>La>Pr>Sm>Gd>Dy
2. Ce>Nd>La>Sm>Pr>Gd>Dy
3. Sm>Ce>Nd>Gd>La>Pr>Dy

The samarium maximum obtained in some thorium-rich monazites is unusual and they may be the first monazites on record, where Sm>Ce and Gd>La. There are some similarities between the monazites of the Alto Ligonha region and those studied by Heinrich et al. (1960), particularly the fractionation of samarium in the monazites from the Muiane area of Alto Ligonha and those from Brown Derby in Colorado.

Hugo (1969) has recorded monazite analyses from the Orange River pegmatites belt, showing the following order of abundance of the REE:

$$Ce>Nd>La>Gd>Pr>Sm>Dy$$

Xenotime from the Morrua area, Mozambique, has been investigated by Sahama et al. (1973) and the following order of abundance of the REE was noted:

$$Gd>Dy>Sm>Yb>Er>Nd$$

The Morrua xenotime appears to be unique in exhibiting a strongly pronounced Gd maximum. Generally Yb and Dy maxima are observed in xenotimes. The monazites from Alto Ligonha have been analysed by Professor A. Willgallis and his team at the Free University of Berlin, with whom a study of monazites and other REE minerals from African localities is currently in progress.

2e. Tin

Long before pegmatites were mined for mica, beryl and lithium minerals, eluvial and alluvial cassiterite together with some niobium–tantalum derived from pegmatites and closely related mineralized quartz veins, were recovered in many parts of Africa. Cassiterite SnO_2, is by far the most common mineral of tin. It is heavy, hard, extremely resistant to weathering, and the bulk of the world's tin is mined from alluvial deposits. In pegmatites tin is commonly associated with typical lithophile elements, but according to its position in the periodic system, tin is also chalcophile in character and may be found in sulphide ore deposits in the form of stannite or some other sulphide.

Occasionally minerals of the stannite group may be present in pegmatites. During the present study the rare tin mineral nigerite was found in a tin pegmatite in Namibia. Although cassiterite is often regarded as a relatively pure mineral corresponding to the chemical formula SnO_2, earlier spectographic and recent microprobe analyses have shown that cassiterite can accommodate a variety of elements e.g. Ti, Mn, Fe, Nb, Ta, and W in trace and minor amounts in its structure (Dunn et al. 1978; Hassan and von Knorring 1984; Černý and Ercit 1985).

It must be emphasized, that the tantalum content in cassiterite is of great economic importance, since the largest supply of tantalum in recent years has been obtained from tantalum-bearing slags, derived from tin smelting. Cassiterites from a number of pegmatite occurrences have been analysed by microprobe, in order to determine how niobium–tantalum minerals may be intergrown with cassiterite or occurring as inclusions or exsolutions in the cassiterite host. Moreover, variable amounts of these elements have been noted in the cassiterite matrix in solid solution, together with smaller contents of titanium, manganese, and iron.

The mineral inclusions or exsolutions in pegmatitic cassiterites may consist of the following phases: ferrocolumbite, by far the commonest phase, tantalite, manganotantalite, wodginite, ferrowodginite, and tapiolite. They are usually found as irregular grains or blebs some 20–50 μ across, dispersed in the cassiterite host.

Although these inclusions may be quite common in cassiterites from certain deposits, quantitatively the bulk of niobium–tantalum is contained in the cassiterite matrix.

The cassiterite from the mineralized quartz veins showed the lowest contents of niobium–tantalum, usually below one per cent ($Nb_2O_5 + Ta_2O_5$), whereas in the pegmatitic cassiterites the amount varied from some one per cent of combined oxides, up to 15·5 per cent ($Nb_2O_5 + Ta_2O_5$). The highest niobium content was 2·85 per cent Nb_2O_5 and the maximum tantalum content observed was 13 per cent Ta_2O_5.

Generally the cassiterites from lithium pegmatites gave distinctly higher values and the highest average contents of ($Nb_2O_5 + Ta_2O_5$) were noted in some highly fractionated lithium pegmatites and in cassiterites closely associated with tapiolite, wodginite, or some other high-tantalum minerals. The minor elements, titanium, manganese, and iron were also enhanced in high-tantalum cassiterites.

It has been suggested by Černý and Ercit (1985), that cassiterite may incorporate a 'monotapiolite' component in the same manner as rutile, and in the majority of cases this may be true. However, the present study has also shown that high-tantalum cassiterites may be devoid of iron and manganese or contain only traces of these elements.

2f. Niobium and tantalum

Niobium–tantalum minerals are present in almost all granite pegmatites, although in most cases in rather small amounts. Larger concentrations of the more important tantalum-rich minerals, e.g. manganotantalite and microlite, may be found in highly fractionated lithium pegmatites. In an earlier paper, von Knorring and Fadipe (1981) have described the mineralogy and geochemistry of niobium–tantalum in various African pegmatites and outlined their occurrence and distribution.

On the basis of their structure and chemical composition the niobium–tantalum minerals may be grouped as follows:
1. Columbite–tantalite
2. Ixiolite
3. Wodginite
4. Tapiolite
5. Stibiotantalite–bismutotantalite
6. Simpsonite, thoreaulite, etc.
7. Pyrochlore–microlite
8. Uranium-rare earth-bearing niobium–tantalum minerals

Columbite–tantalite $(Fe,Mn)(Nb,Ta)_2O_6$, is the commonest and most widespread group of niobium–tantalum minerals found in granite pegmatites of all kinds. Of the four end-members in this series, ferrocolumbite with some manganese and tantalum is the most common mineral. Ferrotantalites, on the other hand are relatively rare and the end-member has not been observed. The manganiferous end-members, mangano-columbite and mangano-tantalite are frequently found, although restricted to fractionated lithium pegmatites (von Knorring and Condliffe unpublished data).

3. Pegmatites from equatorial, central and southern Africa

3a. Uganda

The mineralized pegmatites of southwest Uganda are related both temporally and spatially to post-tectonic, often pegmatitic granites within the Kibaran (Karagwe–Ankolean) rocks. The granites are very coarse-grained, partly albitized, quartz–microcline–muscovite granites which have been intruded into syntectonic gneissic granites, migmatites, and sedimentary rocks of the Kibaran belt. The pegmatites transect both granites and schists and the majority are found close to the contact of the syntectonic granitoid rocks with either the schists or post-tectonic pegmatitic granites (Reedman and Lowenstein 1971). In southwest Uganda some 150 pegmatites have been worked for beryl and 15 of these have each yielded between 100 and 500 tons of this mineral. Reedman and Lowenstein (1971) distinguished five major groups of pegmatites, on the basis of mineralogical and textural considerations, as follows:
1. Microline–muscovite primary pegmatites
2. Albite–muscovite replaced pegmatites
3. Spodumene–albite replaced pegmatites
4. Lithian muscovite–albite replaced pegmatites
5. Quartz–mica–kaolin pegmatites

The pegmatites of group 1 are the products of primary crystallization with little or no replacement. Pegmatites of groups 2–4 are characterized by zones rich in sodium- or lithium-bearing minerals which have replaced pre-existing minerals formed during the primary crystallization phases. The group 5 pegmatites are highly kaolinized.

As the rare-element mineralization constitutes the most important economic aspect of pegmatites their classification may be conveniently based on certain type minerals (von Knorring 1970), and thus the pegmatites of Uganda may be grouped as follows:
1. Beryl + columbite–tantalite +/− cassiterite
2. Beryl + columbite–tantalite +/− cassiterite + montebrasite +/− lithium–manganese–iron phosphates
3. Beryl + columbite–tantalite +/− cassiterite + lithium silicates (e.g. spodumene and lithian micas)
4. Cassiterite +/− columbite–tantalite +/− beryl

Southwest Uganda is also an important tin province and cassiterite has been found in numerous pegmatites and in specific quartz–muscovite veins, related to the pegmatite mineralization in the area and represented by group 4.

The cassiterite deposits are mainly found in the southern parts of the Ankole and Kigesi districts and in the adjoining areas of Tanzania and Rwanda.

In the older basement rocks of Uganda, within the Tanzania craton, numerous pegmatites are also known but they are devoid of tin and have not been investigated to the same extent. However, three of the larger lithium pegmatites, namely, Mbale estate, Wampewo, and Lunya have been mined in the past for amblygonite, beryl, feldspar, mica, and tantalite — the rare bismutotantalite was first recorded from the Wampewo pegmatite.

3b. Rwanda

Within the Kibaran rocks of Rwanda, mineralized pegmatites and cassiterite-bearing quartz veins are of common occurrence. Here, as in the adjoining areas of Uganda, tin is the dominant element.

The major mineralized pegmatites are situated around Gatumba in the north western part of the country where many lithium pegmatites have been mined for cassiterite, beryl, amblygonite–montebrasite, and niobium–tantalum minerals, for a considerable time.

Lithium is a characteristic element of the pegmatites in this general area, and in addition to spodumene and amblygonite group of minerals, large amounts of the rarer lithium minerals, eucryptite and holmquisite have been noted. The latter mineral forms extensive, exomorphic haloes around some of the pegmatites (von Knorring and Hornung 1965). The Buranga pegmatite at Gatumba is particularly noted for its rich variety of rare phosphate minerals (von Knorring 1970; von Knorring and Fransolet 1975, 1977; von Knorring et al. 1977; von Knorring and Sahama 1982).

3c. Zaire

Mineralized pegmatites and associated tin-bearing quartz veins are found in many areas of eastern Zaire. The most prominent of these, are situated in the Kivu and Shaba provinces. Because cassiterite was the main mineral mined in the past, the various pegmatite fields are generally known as tin-bearing regions. The pegmatites are associated with post-tectonic granitic intrusions of Kibaran age and as in southwestern Uganda they are found within the granitic batholiths or more commonly in the aureole of mica schists surrounding the intrusives.

In the Maniema region numerous smaller mineralized interior pegmatites and quartz veins are known to carry cassiterite, niobium–tantalum minerals, some wolframite, and beryl. Most of the tin mining has been confined to rich eluvial and alluvial deposits.

In the eastern part of the Kivu province, e.g. Kabunga and the Mumba–Numbi areas, lithium pegmatites are of common occurrence, carrying spodumene, amblygonite–montebrasite, lithian mica, beryl, cassiterite, and various niobium–tantalum minerals. The largest beryllium pegmatite known in Africa, is Kobokobo, in the Kamituga area, some 100 km southwest of Bukavu. Over 5000 tons of beryl have been mined from this single pegmatite, in addition to large amounts of amblygonite, cassiterite, and columbite–tantalite (Safiannikoff and van Wambeke 1967).

Within the Kibarides of Shaba province many tin pegmatites are known but the most important of these is the spectacular lithium pegmatite at Manono. The combined length of the two pegmatites en echelon, Manono and Kitololo is some 14 km and the width is in places up to 700 m. Spodumene is a common mineral, associated with microcline, albite, quartz, and muscovite. There is a marked enrichment of cassiterite in the albitized zones and in pockets of greisen along the contacts with the mica schist.

In addition to cassiterite, columbite–tantalite and some thoreaulite are also recovered. The rare, tantalum–niobium–tin mineral, thoreaulite was first found at Manono.

A recent examination of the thoreaulite revealed two other rare tantalum minerals, namely, rankamaite and lithiotantite $Li(Ta,Nb)_3O_8$, of which the latter mineral is new for Africa.

3d. Zimbabwe

Pegmatites are widespread in Zimbabwe and hundreds have been mined for mica, beryl, cassiterite, niobium–tantalum, and lithium minerals during the past fifty years. Zimbabwe has been one of the largest pegmatite mineral producers in Africa and at one time the largest lithium-ore producer in the world.

Detailed investigations, especially on beryllium mineralization were carried out by the U.K. Atomic Energy Authority in 1959–1963 and the results were documented by Ackerman et al. (1966). Subsequently a simple classification of the Zimbabwe beryl pegmatites was presented by Gallagher (1975) on the basis of the presence or absence of lithium minerals and the relative abundance of albite and potassium feldspar as follows:
1. Perthite-rich pegmatites
2. Albite-rich pegmatites
3. Albite-perthite pegmatites
4. Lithium pegmatites

Pegmatites of types 1 and 3 are the most abundant and collectively contain large amounts of beryl, whereas the albite pegmatites are relatively rare. The lithium pegmatites often display complex internal zoning and may be subdivided into those containing lithium mica with or without other lithium minerals, and those in which spodumene, amblygonite–montebrasite or petalite form the only significant minerals.

Three distinct pegmatite regions can be distinguished in Zimbabwe:
1. Miami–Urungwe region
2. Central region
3. Kamativi region

The Miami–Urungwe pegmatite region is situated in the northern part of the country, east of Lake Kariba. Here, industrial sheet mica and beryl have been mined from numerous pegmatites emplaced in schists, gneisses, and amphibolites.

According to Wiles and Tatham (1963) the economic muscovite pegmatites of Zimbabwe are always hosted in already highly micaceous rocks. The beryl pegmatites are chiefly of the well-zoned, potassium-rich type, lithium minerals being very rare or limited to a few occurrences of montebrasite.

This general area is also known for a variety of excellent gemstones, e.g. aquamarine, golden beryl, blue topaz, and euclase.

The major pegmatite fields of Zimbabwe are found within the extensive central pegmatite region in the central and eastern part of the country and are associated with the older rocks of the Zimbabwe Craton. The pegmatites are typically marginal or exterior in relation to the granitic batholiths, being emplaced in the surrounding amphibolitic rocks (greenstones) or, in some cases, within the marginal parts of the gneissose granitic intrusions.

Bikita, in the southeastern part of the central pegmatite region is one of the most interesting pegmatites in the world, with perhaps the highest concentration of lithium within a single pegmatite deposit known. Bikita is unique in that all the major lithium minerals are represented, and moreover, occur in exceptionally large quantities, e.g. amblygonite–montebrasite, eucryptite, spodumene, petalite, and lepidolite, in addition, to large units of pollucite, caesian beryl, cassiterite, and a variety of niobium–tantalum minerals, a characteristic feature of highly fractionated granite pegmatite. Other extensive, mineralized lithium pegmatites are found in the Harare and Mtoko–Makaha areas.

The Benson pegmatites some 30 km north of Mtoko in the northeastern part of Zimbabwe are emplaced in amphibolites. They are well zoned and exhibit extensive lithium–sodium replacement units containing unusually large concentrations of tantalum minerals. The Benson claims enclose 40 pegmatites of which 12 are known to contain lithium minerals. The metasomatic alteration of the wall rock amphibolites is especially pronounced, where the lithium amphibole holmquistite has been found in the aureoles of eight lithium pegmatites.

The major lithium minerals here are montebrasite, spodumene, and lepidolite. Economic accessories are pollucite, caesian beryl, hafnian zircon, cassiterite, columbite–tantalite, manganotantalite, wodginite, several varieties of microlite, stibiotantalite, and simpsonite, a varied mineralization, indicative of highly fractionated lithium pegmatites (Hornung and von Knorring 1962).

The third pegmatite region of great economic importance is the Kamativi tin belt in northwestern Zimbabwe, where according to Rijks and van der Veen (1972), extensive, flat-dipping cassiterite pegmatites are emplaced as dome-like structures in schists, gneisses, and granites.

Three distinct phases of mineralization have been noted:
1. Microcline–quartz–spodumene
2. Quartz–albite
3. Muscovite–quartz

carrying disseminated cassiterite with some columbite–tantalite.

Large bodies of spodumene-rich units have been observed in addition to some accessory montebrasite and beryl.

Age determinations have shown that the emplacement of the cassiterite pegmatites most probably took place some 1000 Ma ago.

Zimbabwe is also well known for its excellent emeralds and alexandrites in the

Belingwe, Filabusi, Novello, and Chikwanda pegmatite occurrences, associated with serpentinites in the southern part of the country.

3e. South Africa

The majority of mineralized pegmatites in South Africa are found in the northwestern Cape Province, connected with rocks of the Namaqualand metamorphic complex of Kibaran age — a smaller number of pegmatites occur within rocks of Kaapvaal craton in Eastern Transvaal.

The Orange River pegmatite belt in northwestern Cape, which runs from the Kenhardt area in the east to Richtersveld in Namaqualand in the west, a distance of some 450 km is perhaps the most extensive, continuous pegmatite belt in Africa.

In the eastern sector of the belt Hugo (1969) has distinguished between homogenous and inhomogenous pegmatites, the former being poorly zoned and rarely mineralized, whereas the latter are generally well zoned and may contain mineable concentrations of rare-element minerals, industrial mica, and ceramic feldspar.

The inhomogeneous or zoned pegmatites may be further divided according to their mineral content and complexity of the internal structure into two groups:
1. Simple pegmatites, that contain either andalusite, corundum and apatite, and rare-earth minerals.
2. Complex pegmatites, that contain beryl, lithium minerals, columbite–tantalite, and cassiterite.

Beryl is one of the commonest minerals found in most pegmatites, except in the adalusite pegmatites. In addition, chrysoberyl has been noted in a few simple and complex pegmatites, and commonly rare-earth minerals, e.g. allanite, monazite, xenotime, gadolinite, euxenite, and fergusonite. Typical lithium pegmatites are surprisingly rare in the eastern sector of this large pegmatite region, and only a few carry spodumene, lithian mica and some lithium–manganese–iron phosphates. In Namaqualand or the western sector of the Orange River pegmatite belt, lithium pegmatites are well represented and are highly mineralized (Schutte 1972).

These pegmatites constitute the main source of industrial minerals, e.g. amblygonite–montebrasite, lithian mica, spodumene, pollucite, beryl, niobium–tantalum minerals, bismuth minerals, and in some pegmatites, large amounts of muscovite and ceramic feldspar have been mined.

The most important lithium pegmatite in this area is the gigantic Noumas or Blesberg, a dyke-like body, which is over 100 m long and up to 40 m wide, discordantly emplaced in a granodiorite. The pegmatite is well zoned, containing a dominant quartz core with a large amount of blocky microcline, in addition to mineralized replacement units of cleavelandite, some lithian mica, and muscovite. At the core margin large crystals of spodumene are seen, extending partly into the quartz core and the intermediate zone, also known as the spodumene zone which carries the bulk of spodumene together with cleavelandite, some microcline, and quartz. The wall zone is rich in muscovite and beryl, the main minerals mined at the present time. Beryl is also found in the spodumene zone and in the quartz core.

Tantalite and bismuth minerals occur sporadically throughout the pegmatite and at one time high concentrations of uraniferous microlite were seen in some replacement units.

The phosphates apatite and lithiophilite are of common occurrence in the Noumas pegmatite and large nodules and aggregates of lithiophilite, often displaying numerous secondary phosphates, some of them of great rarity, e.g. faheyite, the second occurrence in the world, have been noted in the quartz core. Amblygonite–montebrasite, on the other hand, has not been observed, which may be due to the apparent paucity of fluorine in this particular pegmatite. In addition to the mineralized pegmatites, numerous scheelite-bearing quartz veins, carrying some apatite and tourmaline, are found throughout the Orange River pegmatite belt. In places they grade into the pegmatites and appear to be related to the pegmatite mineralization, as in other parts of Africa. There is a marked paucity of tin in this extensive pegmatite belt, and apart from a few occurrences in the Kenhardt area, where cassiterite is associated with some spodumene pegmatites, larger amounts of cassiterite have only been found in the Upington area. Here, a large tourmaline pegmatite is being mined for cassiterite, wolframite, and scheelite.

3f. Namibia

The mineralized pegmatites of Namibia are found in two major regions — a northern region, roughly bounded by the Ugab River in the north and the Swakop River in the south — and a southern region, comprising pegmatites south of Keetmanshoop and in the Warmbad area, close to the Orange River.

The former pegmatites are associated with granitoid rocks of the Damara orogen (470–550 Ma), and are commonly emplaced in metasediments as extensive injection dykes, whereas the pegmatites in the south, are found in mica schists, gneisses, and gabbroic rocks of the Namaqualand metamorphic complex, which is considered to be of Kibaran age.

Some of the major mineralized pegmatites are found in the Karibib–Usakos area, where various rare-element minerals have been mined for the last fifty years, e.g. amblygonite–montebrasite, lepidolite, petalite, pollucite, beryllium–niobium–tantalum, and bismuth minerals, in addition to industrial mica, ceramic feldspar, quartz, and a large variety of attractive gemstones. Two of the larger and most productive lithium pegmatites, Rubicon and Helicon, where extensive mining of lithium minerals and beryl has taken place, contain comparatively small amounts of niobium–tantalum minerals. Even the gigantic lepidolite units are almost devoid of microlite or columbite–tantalite. However, an examination of some extensive, rather narrow pegmatitic satellite dykes bordering the main Rubicon deposit, has shown considerable amounts of tantalum mineralization along the strike, and usually associated with quartz-rich swellings that have developed into highly fractionated pegmatite units, rich in lepidolite, petalite, cleavelandite, pollucite, caesian beryl, topaz, tourmaline, and bismuth minerals.

A distinct lepidolite–cleavelandite–topaz unit, marginal to the quartz core is highly mineralized with colourless to yellow and black microlite, commonly associated with secondary bismuth minerals.

Generally, columbite–tantalite, manganotantalite, microlite, tapiolite, and wodginite have been recorded from a number of smaller lithium pegmatites in the Karibib area (von Knorring and Fadipe 1981).

The well-known gem tourmaline deposits in the Karibib–Usakos area are usually associated with large, zoned pegmatites, frequently with dominant quartz cores and characteristic intermediate and wall zones, consisting of muscovite, quartz and feldspar in graphic intergrowth. Lithium mineralization is always present in the form of montebrasite, some lithian mica and zoned lithian tourmalines, in addition to some cassiterite, columbite–tantalite, topaz, and large amounts of black tourmaline particularly in the marginal zones.

Most of the pegmatitic tin and tantalum in Namibia is confined to three, well-defined, NE–SW trending so-called tin belts, each over 100 km long.
1. The Northern Tin Belt — follows the Uis lineament, from the Uis mining area towards Cape Cross.
2. The Central Tin Belt — trends from the Nainais tin field along the Omaruru River to the Tjirundu mountains, northwest of Omaruru town.
3. The Southern Tin Belt — extends from the Arandis area, south of Spitzkoppe to the Otjimbojo tin mining field east of Karibib.

At the present time the main production of tin comes from the Uis tin mining area, where gigantic pegmatites up to 1000 m long and some 100 m wide are exploited. Columbite–tantalite is recovered as a by-product. The pegmatites appear to be unzoned, coarse to very coarse-grained bodies containing microline, variable amounts of albite, muscovite, and quartz. Occasionally, some lithium minerals are present, such as, montebrasite, lithian mica or petalite, and some beryl.

The cassiterite is closely associated with albite and muscovite and shows marked concentration in veinlets and patches of muscovite and sugary albite, and in sporadic greisenized parts of the pegmatite.

Numerous pegmatites, often highly mineralized are found in the desert area along the northern tin belt southward from Uis and south of Brandberg, the so-called Desert Mines. They frequently contain cassiterite, columbite–tantalite, wodginite, tapiolite, montebrasite, spodumene, beryl, and tourmaline as accessory minerals.

In the central tin belt, the major mineralized pegmatites are confined to the northern bank of the Omaruru River, extending over a distance of some 30 km.

They all carry cassiterite as the major ore constituent with some columbite–tantalite, wodginite, and beryl.

At Tsomtsaub, an interesting occurrence of the rare tin mineral nigerite was noted together with cassiterite, chrysoberyl, and columbite–tantalite in a sillimanite-bearing quartz rock, confined to the marginal parts of a pegmatite, emplaced in cordierite-bearing schists.

The southern tin belt follows a distinct structural feature, the Omaruru lineament, from the Arandis tin mining area up to the Erongo mountains. This tin belt is characterized by the wall-like pegmatite dykes, which are intrusive into the Damara metasediments and are particularly well developed here. Due to their higher resistance to weathering, as compared with the schistose host rock, the pegmatites stand out in the desert terrain as prominent wall-like features and can be traced for great distances.

By far the larger number of dykes are conformable injection dykes, or sheets, of variable magnitude. They are often parallel, multiple dykes with chilled micaceous margins and consist of microcline albite, muscovite and quartz and may occasionally show a crude zoning. Muscovite is invariably enriched along the margins of the dykes and among accessories, cassiterite, blue–green apatite and variable amounts of black tourmaline are often noted.

Although the majority of dykes appear to be regular bodies, here and there, pinch and swell structures are observed. The dilated parts are often enriched in quartz and may occasionally develop into highly mineralized zoned pegmatites.

Perhaps the best known tantalum pegmatites in Namibia are those from the Tantalite valley, some 30 km south of Warmbad in the southern part of the country. Geologically, these pegmatites belong to the much larger pegmatite region of the Orange River belt, within the Namaqualand metamorphic complex.

The tantalum mineralization is particularly well developed in three of the larger lithium pegmatites, Homestead, White City, and Lepidolite, which are emplaced into gabbroic rocks. Tantalite–microlite, montebrasite, spodumene, caesian beryl, common beryl and bismuth minerals have been recovered.

3g. Mozambique

The large pegmatite region of Mozambique is situated in the northeastern part of the country and most of the important pegmatite fields fall within the triangle Quelimane–Cuamba–Nampula. In this vast area, hundreds of pegmatites have been mined in the past for a variety of minerals, of which in recent years beryl, tantalum minerals, and some attractive gemstones have been the most important.

Mozambique produced close to 13000 tons of beryl during the period 1950–1964, and was the leading beryl supplier in Africa.

Lithium pegmatites are well represented in the southern part of the pegmatite region, they are highly mineralized and are commonly of large size. Internal structures are well developed, with dominant quartz cores, large intermediate zones of blocky microcline with muscovite and extensive lithian mica, spodumene, and cleavelandite replacement units. Frequently weathering has kaolinized the feldspars and spodumene, with the result that pegmatites with low rare-element content can be economically mined. Especially large amounts of beryl and biobium–tantalum minerals may be found in the eluvium and recovered on a large scale.

According to Bettencourt Dias (1961), the economically important pegmatites are genetically related to a group of late equigranular granitic intrusions and the pegmatites are emplaced in schistose rocks surrounding the intrusions. These late Pan-African granites have an age of around 500 Ma.

The lithium pegmatite area, often referred to as the Alto Ligonha region, extends from Mocubela in the south to Alto Ligonha in the north, a distance of some 170 km, and comprises the highly mineralized lithium pegmatites of Muiane, Mutala, Morrua, and Marropino. Apart from lithium, exceptional concentrations of other rare-elements have been noted in this general area, e.g. caesium, beryllium, scandium, yttrium, rare-earths, zirconium, hafnium, niobium, tantalum, antimony, and bismuth (von Knorring et al. 1969; von Knorring 1970 and Correia Neves et al. 1974).

Tantalum is a characteristic element of this region, and at one time, in addition to beryl, large quantities of tantalum minerals, e.g. columbite–tantalite, manganotantalite and microlite were mined in the Morrua, Marropino and Mutala areas.

Mozambique is also well known for its great variety of attractive pegmatitic gemstones. The pink caesian beryl or morganite, the tourmalines rubellite, verdelite together with the polychrome varieties, are of the highest quality (Sahama et al. 1979). Important emerald occurrences are also known in the Morrua–Gile areas. The mineral is found in pegmatitic dykes, lenses or irregular bodies composed of plagioclase feldspar, quartz, and biotite in variable proportions and associated with an amphibolitic rock. Molybdenite, scheelite, pyrite, apatite, and calcite are characteristic accessories in these emerald occurrences.

3h. Madagascar

Madagascar is well known for its great variety of pegmatites that have been admirably documented by Lacroix (1922). Similarities in mineralization, geochemistry, and age between the pegmatites of Madagascar and Mozambique indicate, that these regions were affected by the same orogenic events and Cahen and Snelling (1966) regard Madagascar as the southernmost province of the Mozambique orogenic belt. Although pegmatites are practically found along the whole length of the island, the most important may be grouped into four regions — from north to south as follows:

1. Tsaratanana region
2. Andriamena–Ankazobe region
3. Antsirabé–Sahatany–Ambatofinandrahana region
4. Ampandramaika–Malakialina region

In general, the potassium dominant pegmatites, with microcline and muscovite, show, according to Giraud (1957b), distinct zoning and localized mineralization, and they contain larger amounts of beryl, including gem varieties, and columbite–tantalite.

Uraniferous niobotantalites may be rather common in some fields, e.g. Andriamena, Ankazobe, and Antsirabe. The 'sodolithic' types of pegmatites, which are the principal carriers of a large variety of gemstones in the Antsirabe and Sahatany areas, are numerous, but generally rather small in size. It must be emphasized, however, that lithium pegmatites, comparable in size to those of Mozambique, do not exist.

It appears, that tin is also scarce and only traces of cassiterite have been noted.

Tsaratanana region. In this region, some 250–300 km north of Antananarivo a number of highly mineralized pegmatites are known of which Betanimena, Ampandrakely, and Berere are the most important. Here, large amounts of beryl, columbite–tantalite, various uraniferous rare-earth–niobotantalites, monazite, xenotime, and mica were mined.

The pegmatites in this general area are emplaced in amphibolites and migmatites and in the extensive Berere field, Giraud (1957a) distinguished three main types: muscovite pegmatites, biotite–muscovite or two mica pegmatites and biotite pegmatites.

The muscovite pegmatites have dominant quartz cores, generally regular zoning and they are the main carriers of beryl and niobium–tantalum minerals. In the two mica pegmatites the internal structure is well developed and characterized by a distinct graphic zone. In general these pegmatites carry less beryl and columbite than the muscovite pegmatites and, moreover columbite is frequently replaced by euxenite and ampangabeite and other uraniferous and rare-earth minerals. In the biotite pegmatites, beryl is seldom present but fergusonite and samarskite may be locally important. A geochemical characteristic, common to many pegmatites in this area is the presence of large concentrations of scandium, Phan *et al.* (1967), von Knorring (1970).

Andriamena–Ankazobe region. This large pegmatite region is situated some 170 km north of Antananarivo and extends south towards Befanamo, the locality of the rare scandium–silicate, thortveitite, and southwest toward Ankazobe. The pegmatites are generally smaller in size than those of Berere and they are emplaced in amphibolites and gneisses. The pegmatites here are highly weathered, leaving the white quartz cores as markers — the resistant, economic minerals are commonly recovered from the eluvium. At the present time, beryl (sometimes of gem quality), columbite–tantalite, and the uraniferous rare-earth–niobo-tantalites are principally recovered.

Antsirabé-Sahatany-Ambatofinandrahana region Within this area northwest and south of Antsirabé, there are many distinct pegmatite fields and individual pegamatites with a unique mineralization. Some pegmatites are closely associated with the Vavavato granitic intrusion others are emplaced in schists, quartzites, migmatites and gneisses. A large number of lithium pegmatites, generally of small dimensions, are found within dolomitic marbles, south of Antsirabé in the Sahatany valley and farther south in the Ambatofinandrahana region. Besairie (1962, 1965) distinguished the following major types of pegmatites in this general region:
1. Uraniferous pegmatites, containing a variety of niobotantalites.
2. Uraniferous pegmatites with beryl and gem beryls.
3. Beryl-bearing pegmatites, including the major pegmatites with larger amounts of industrial beryl.
4. Pegmatites carrying varieties of gem beryl.
5. Lithium pegmatites (pegmatites 'sodolithique') southwest of Betafo and in the Sahatany valley, which are largely mineralized with multicoloured beryls, tourmalines, spodumene and spessartine garnets.

The pegmatites to the northwest of Antsirabé are commonly uraniferous carrying a variety of yttrium-rare-earth niobotantalites e.g., betafite, euxenite, fergusonite and samarskite. In addition they are the main producers of industrial beryl. One of the most outstanding mineralized pegmatites is Anjanabonoina, situated some 37 km ESE of Betafo. It is best known for its variety of gemstones, e.g. kunzite (spodumene), morganite (Cs-beryl), coloured tourmalines, phenakite, danburite and the exceptionally rare, beryllium-boron mineral hambergite.

Ampandramaika-Malakialina region This extensive pegmatite region is situated some 30 km SSE of Mandrosonoro, on the main Ambositra-Morondava road in central Madagascar. Originally the pegmatites in this area were exploited for industrial mica, gem beryl and tourmaline but during the 1950's the main beryl and columbite-tantalite mining in the country appears to have been concentrated here.

The pegmatites are emplaced in a variety of rocks including schists, amphibolites and migmatitic gneisses. Several large granitic intrusions occur east of the north-south trending pegmatite region. The mineralized pegmatites are of varied size and composition and often exhibit well developed internal structures. A common mineralogical feature of many occurrences is a dominant quartz core which in places may be rose-coloured, and an intermediate microcline zone, with variable amounts of muscovite, some biotite and frequently a great abundance of marginal tourmaline. Industrial beryl, often in giant crystals, occasionally up to 30 tons in weight is commonly found in the core margin, smaller crystals are confined to the wall zone. Niobium-tantalum minerals are found in the core zone or in albitic replacement units. A number of lithium pegmatites carrying montebrasite, lithium-manganese-iron phosphates, beryl and variable amounts of lithian mica are found in widely scattered areas.

4. Conclusions

The mineralized pegmatites described in this review are from the younger orogens in equatorial central and southern Africa, namely the Kibaran-, Mozambique-,

Damaran and Orange River belts, and from the Tanzanian and Zimbabwe-Kaapvaal cratons. A great number of extensive pegmatites have developed in these major structural units and many deposits have been mined for a great variety of industrial minerals.

The mineralized pegmatites are not evenly distributed along the structural configurations but form localized concentrations of pegmatite dykes (swarms) often closely associated with major granitic intrusions, that are frequently controlled by large-scale tectonic features. It has been suggested, that many of the present mineralized pegmatites are genetically connected with post-tectonic granites, and although geochemical data are lacking, there are clear indications that many important pegmatites are extensions of late-stage pegmatitic granites. Frequently, the pegmatites have invaded the country rock to such extent that pegmatitic gneisses have been formed. In this case, large areas may be mineralized, carrying the typical pegmatite minerals as accessories, e.g. beryl, chrysoberyl, rare-earth and uranium minerals. Economically, pegmatite mineralization of this kind can be very important but is generally less known and commonly overlooked.

Although the lithium pegmatites constitute only a small proportion of all the pegmatites in most areas, economically they are by far the most important in all the pegmatite regions of Africa. Apart from large concentrations of lithium, these pegmatites are also the main carriers of tin, tantalum and gemstones. Of particular interest is the frequent presence of tin mineralization along the entire length of the Kibaran belt continuing into the Kamativi region of Zimbabwe and the Damara sequence in Namibia. In the lithium pegmatites of the Mozambique belt, however, tin is rather scarce or absent (von Knorring, 1970).

References

Ackermann, K. J., Branscombe, K. C., Hawkes, J. R., and Tidy, A. J. L. 1966. The geology of some beryl pegmatites in southern Rhodesia. *Transactions of the Geological Society of South Africa*, **69**, 1–38.

Besairie, H. 1962. Contribution a la minéralogie de Madagascar. *Ann. Géol. Madagascar*, **29**.
—— 1965. Géologie économique de la sous-préfecture d'Ambatofinandrahana. *Doc. Serv. Géol. Madagascar*, **170**.

Bettencourt Dias, M. 1961. Os pegmatitos do Alto Ligonha. *Bol. Serv. Geol. Min. Mozambique*, **27**, 17–36.

Cahen, L. and Snelling, N. J. 1966. The geochronology of equatorial Africa, Amsterdam.

Černý, P. and Ercit, T. S. 1985. Some recent advances in the mineralogy and geochemistry of Nb and Ta in rare-element granitic pegmatites. *Bull. Minéral*, **108**, 499–532.

Clifford, T. N. 1966. Tectono-metallogenic units and metallogenic provinces of Africa. *Earth and Planetary Science Letters*, **1**, 421.

Correia Neves, J. M., Lopes Nunes, J. E., Sahama, Th. G., Lehtinen, M. and von Knorring, O. 1974. Bismuth and antimony minerals in the granite pegmatites of northern Mozambique. *Rev. Cienc. Geol. Lourenço Marques*, **7**, 1–37.

Dunn, P. J., Gaines, R. V., Wolfe, C. W. and Barbosa, C. do P. 1978. Epitaxial wodginite and cassiterite from Lavra Jabuti, Baixio, Galilea, Minas Gerais, Brazil. *Mineral Record*, **9**, 14–18.

Gallagher, M. J. 1975. Composition of some Rhodesian lithium–beryllium pegmatites. *Transactions of the Geological Society of South Africa*, **78**, 35–41.

Giraud, P. 1957a. Le champ pegmatitique de Berere à Madagascar. C.C.T.A. Conférence de Tananarive Compt. Rend. Serv. Géol. Madagascar, **1**, 125–133.

—— 1957b. Les principaux champs pegmatitique de Madagascar. C.C.T.A. Conférence de Tananarive Compt. Rend. Serv. Géol. Madagascar, **1**, 139–150.

Hassan, W. F. and von Knorring, O. 1984. Niobium–tantalum minerals from Peninsular Malaysia. *Geological Society Malaysia, Bulletin*, **17**, 33–47.

Heinrich, E. Wm., Borup, R. A. and Levinson, A. A. 1960. Relationships between geology and composition of some pegmatitic monazites. *Geochim. Cosmochim. Acta*, **19**, 222–231.

Hornung, G. and von Knorring, O. 1962. The pegmatites of the North Mtoko region, Southern Rhodesia. *Transactions of the Geological Society of South Africa*, **65**, 153–180.

Hugo, P. J. 1969. The pegmatites of the Kenhardt and Gordonia districts, Cape Province. *Memoirs of the Geological Survey of South Africa*, **58**.

von Knorring, O. and Hornung, G. 1965. Pegmatite investigations in East Africa. *9th ann. Rep. res. Inst. afr. Geol., Univ. Leeds*, 44–45.

——, Sahama, Th. G., and Lehtinen, M. 1969. Scandian ixiolite from Mozambique and Madagascar. *Bulletin of the Geological Society of Finland*, **41**, 75–77.

—— 1970. Mineralogical and geochemical aspects of pegmatites from orogenic belts of equatorial and southern Africa. In: Clifford, T. N. and Gass, I. G. (Eds), *African Magmatism and Tectonics*, Oliver and Boyd, Edinburgh, 157–184.

—— and Fransolet, A. M. 1975. An occurrence of bjarebyite in the Buranga pegmatite, Rwanda. *Schweiz. Mineral. petrogr. Mitt*, **55**, 9–18.

—— and Fransolet, A. M. 1977. Gatumbaite, a new species from Buranga pegmatite, Rwanda. *N. Jb. Miner. Mh.*, H. **12**, 561–568.

——, Lehtinen, M. and Sahama, Th. G. 1977. Burangaite, a new phosphate mineral from Rwanda. *Bulletin of the Geological Society of Finland*, **49**, 33–36.

—— and Fadipe, A. 1981. On the mineralogy and geochemistry of niobium and tantalum in some granite pegmatites and alkali granites of Africa. *Bull. Minéral.*, **104**, 496–507.

—— and Sahama, Th. G. 1982. Some FeMn phosphates from the Buranga pegmatite, Rwanda. *Schweiz. Mineral. Petrogr. Mitt.* **62**, 343–352.

Lacroix, A. 1922. *Minéralogie de Madagascar*, Paris.

Landes, K. K. 1935. Age and distribution of pegmatites. *American Mineralogy*, **20**, 81–105, 153–175.

Phan, K. D., Foissy, B., Kerjean, M., Moatti, J. and Schiltz, J. C. 1967. Le scandium dans les minéraux et les roches encaissantes des certaines pegmatites malgaches. *Bull. B.R.G.M.* **3**, 77.

Reedman, A. J. and Lowenstein, P. L. 1971. Economic geology of the beryl-bearing pegmatites of southwest Uganda. *Trans. Instn. Min. Metall.*, **80**, B4–17.

Rijks, H. R. P. and van der Veen, A. H. 1972. The geology of the tin-bearing pegmatites in the eastern part of the Kamativi district, Rhodesia. *Mineral Deposita*, **7**, 383–395.

Safiannikoff, A. and Van Wambeke, L. 1967. Le pegmatite radioactive à béryl de Kobokobo et les autres venues pegmatitiques et filoniennes de la région de Kamituga, Kivu, Rép. du Congo. *Mineral Deposita*, **2**, 119–130.

Sahama, Th. G., von Knorring, O. and Rehtijärvai, P. 1973. Xenotime from Morrua, Mozambique. *Bulletin of the Geological Society of Finland*, **45**, 67–71.

——, —— and Törnroos, R. 1979. On tourmaline, *Lithos*, **12**, 109–114.

Schutte, I. C. 1972. The main pegmatites of the area between Steinkopf, Vioolsdrif and Goodhouse, Namaqualand. *Memoir of the Geological Survey of South Africa*, **60**.

Wiles, J. W. and Tatham, N. A. 1963. The muscovite pegmatites of southern Rhodesia and their exploitation. Pegmatites in Southern Rhodesia, a Symposium. *Instn. Min. Metall. S. Rhodesia*, 10–18.

Nigerian rare-metal pegmatites and their lithological framework

G. Matheis

Technical University Berlin (West), Special Research Project Arid Areas, Berlin, West Germany

Most of Africa's rare-metal potential is linked to Proterozoic pegmatite fields except for the Jurassic tin-bearing ring complexes of the Jos Plateau Nigeria. In order to establish the position of the Nigerian tin province in terms of economic rare-metal mineralization, occurrences of mineralized pegmatitic phases of the Nigerian basement complex, including the so-called 'Older Tin-Fields', have been investigated. Rb/Sr geochronology confirms the end Pan-African ages for the pegmatites in the range 580–530 Ma. Although emplaced within the same time span, the mineralogy, geochemistry, and mineralization of these pegmatites differ according to the lithology of their host-rock. Pegmatite occurrences from southwest Nigeria, which are emplaced into Proterozoic meta-sedimentary–metavolcanic sequences, are enriched in tantalum relative to niobium. These geochemical characteristics are also displayed by the composition of eluvial heavy-mineral concentrates as well as by the trace-element distribution in cassiterites.

Although Pan-African granites with an age range of 700–520 Ma are dispersed throughout Nigeria, associated rare-metal mineralized pegmatites are known almost exclusively along a southwest–northeast striking belt about 400 km long, which intersects the Jos Plateau tin-fields. A direct genetic link between the rare-metal bearing pegmatites and proximal granite occurrences was never observed. It is suggested that reactivation of old tectonic lineaments during the Pan-African orogeny provided excess heat and fluid to concentrate rare-metal pegmatites by partial melting with selective leaching from the country rocks, i.e. their lithological framework.

KEY WORDS   Pan-African   Pegmatites   Tin   Mineralization   Nigeria

1. Introduction

Rare-metal mineralization in general and tin mineralization in particular display a time-related development during geological evolution: from high-temperature ore phases (oxides only) in pegmatites during the Precambrian to low-temperature ore associations (oxides + sulphides) in greisens and vein-complexes of Phanerozoic ages. Thus, nearly all of the Ta, Be, Cs, Li, and a considerable amount of the Sn, W, Th, U, Nb, and REE resources are linked to pegmatoid phases of mainly Precambrian age. Typically, such ore-potential pegmatites occur in Proterozoic amphibolite-grade metamorphic terrains of mixed volcanosedimentary sequences with fluid (F, Cl) generating potentials, which may include partly reworked Archaean greenstone associations. Mining is mostly from eluvial placers and/or from deeply weathered primary deposits. The typical pegmatoid rare-metal occurrences of the shield areas in western, central and southern Africa, in Brazil, in Canada, in western Australia, in India, and the U.S.S.R. have mineralogical indicators in common. In decreasing order of significance, these are: fluorine-rich muscovite, Li-silicates, tourmaline, and beryl — the main ore association of cassiterite, tantalite, and columbite — the lack of large mineral growth — occurrences of gold and minor scheelite in the same heavy-mineral concentrates together with the Sn–Nb–Ta ore minerals.

Outstanding in their tin-mining potential, the Jurassic ring complex province of Nigeria is unique among the widespread occurrences of petrologically similar anorogenic ring complexes elsewhere in Africa. These Younger Tin-Fields of Nigeria are preceded by the rare-metal pegmatites of end Pan-African age, which are confined to a southwest–northeast trending belt of about 400 km length extending from southwestern to central Nigeria. The recognition of two different tin sources was first mentioned by Falconer (1924). However, Jacobson and Webb (1946) pointed out the considerable time gap between both mineralization phases on the basis of detailed petrographic studies of Sn–Nb–Ta bearing pegmatites of central Nigeria. Wright (1970) was the first to suggest a genetic link between Pan-African and Jurassic rare-metal mineralization in Nigeria, based on geological and mineralogical field observations. Detailed geochemical and geochronological data were not available at the time and the question had to be left open.

About 95 per cent of the Nigerian tin and columbite output has been produced from the anorogenic Jurassic ring complexes of the Jos Plateau, however, most of the tantalum has been mined from pegmatitic sources. Detailed production figures are difficult to obtain for the small scale mining operations in the Older Tin-Fields. Available information, which shows a considerable variation within the pegmatite province, is summarized in Table 1.

Alluvial and eluvial placer concentrates, which form the basis of typical small scale mining operations, are situated close to the pegmatitic source rocks due to the high degree of physical and chemical weathering under the prevailing tropical conditions. This rapid disintegration of the coarse-grained source rocks coupled with short distances of transportation by seasonal streams have resulted in a very inconsistent pattern of heavy mineral concentrations which are reflected in the fluctuations in the mining history of the Older Tin-Fields.

Table 1. Production estimates from the Nigerian Older Tin-Fields (Sources: Jacobson *et al* 1951; Dempster 1959; Schätzl 1971)

Locality*	Timespan	Cassiterite (in tons)	Columbite (in tons)	Tantalite (in tons)
Ijero (2)	1942–59	156	7·4	3·4
Ijero (2)	1944–70	247	13·0	5·0
Kabba (3)	1942–50	?	16·0	17·5
Egbe (3)	1950–70	117	19·0	20·0
Niger (*)	1942–50	?	11·0	7·2

* locality-numbers refer to Figure 1; Niger province mining sites are located between River Niger and sites 4+5

The present contribution aims to summarize the results obtained (Matheis 1979; Matheis et al. 1984; Matheis and Caen-Vachette 1983) and to add additional geochemical and geochronological data, indicating the close genetic relationship between the types of rare-metal mineralization and their lithological framework.

2. Geological setting

Located between the West Africa Craton and the Congo Craton, most of the Nigerian basement has been reactivated by the Pan-African orogeny although still older tectonometamorphic events such as the Liberian (2·8–2·2 Ga), the Eburnean (2·1–1·6 Ga), and the Kibaran (1·4–0·9 Ga) are preserved (Odeyemi 1979; Rahaman 1978). Based on the Geological Map of Nigeria 1: 2,000,000 (1974), the basement complex is divided into three lithological units, namely, the undifferentiated migmatite complex of Proterozoic to Archaean origin, the metavolcano-sedimentary belts of late Proterozoic age and the Older Granite complex of late Precambrian–lower Palaeozoic age (700–480 Ma).

Rare-metal pegmatites which contributed to the mining output given in Table 1, are known to occur in two major regions, namely, in southwestern and central

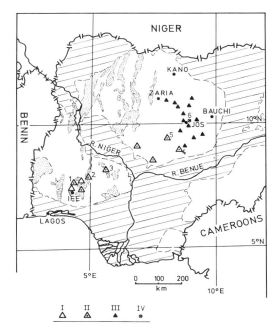

Figure 1. Distribution of pegmatites in the Nigerian Tin Province, including major geological units: Precambrian basement complex undifferentiated (blank), late Proterozoic meta–volcano–sedimentary belts (stippled), Cretaceous and younger sedimentary rocks (hatched).
Main components of the Nigerian Tin Province:
I   barren pegmatite fields of Pan-African age
II   rare-metal mineralized pegmatites of Pan-African age (Older Tin-Fields)
III   rare-metal bearing ring complexes of Jurassic age (Younger Tin-Fields)
IV   rare-metal bearing basement units
Localities of investigated productive pegmatite fields: 1, Iregun; 2, Ijero/Ikoro; 3, Egbe (Kabba); 4, Wamba; 5, Jemaa; 6, Gurum.

Nigeria (Figure 1), although other pegmatites of no economic value have been reported from other areas (e.g. Calabar district, southeastern Nigeria). The two major regions provide different lithological host rocks to the pegmatites. Rare-metal pegmatites in southwestern Nigeria are exclusively confined to the metavolcanosedimentary belts while those in central Nigeria are found only in gneissic basement environments.

2a. Pegmatites of southwestern Nigeria

The geological setting of southwestern Nigeria is characterized by the occurrence of late Proterozoic metavolcano–sedimentary belts of mafic composition which have been compared with typical greenstone associations (Olade and Elueze 1979; Wagner 1980; Klemm et al. 1984). In addition, many of these metabasic rock units display cataclastic and mylonitic textures indicating their association with major shear zones (Elueze 1984) of continental dimensions (Guiraud and Alidou 1981). There are three productive rare-metal mining areas in southwestern Nigeria, namely Iregun, Ijero, and Egbe which are detailed below.

Near the present mining site of Iregun, strongly weathered quartz–muscovite pegmatites are enclosed by mafic to ultramafic amphibolite rocks of mainly tholeiitic affinity (Wagner 1980) which forms part of the so-called Ilesha goldfield. The mining concentrates of cassiterite, columbite, tantalite, and gold barely exceeds one ton annually.

At Ijero, two types of pegmatite are emplaced into a series of biotite-schists and gneisses with intercalated amphibolites (Table 2).

Tabular pegmatites are steeply dipping, sharply bordered bodies, tens of metres in extent, as well as a massive cupola-shaped pegmatite body covering an area of about 7 km^2 which shows concentric zoning (Figure 2). The dominant feldspar of the tabular pegmatites is microcline, while the massive pegmatite carries mainly oligoclase–albite. Compared with the other pegmatite fields, the Ijero area is

Table 2. Geochemical and geochronological data of the Ijero/Ikoro pegmatite field, SW Nigeria

Sample number	Rock type	Na/K	K/Rb	Rb/Sr	Li	Sn	Nb	Rb/Sr model-age
46b	gneiss	0·19	187	4·8	64	5	6	
74	gneiss	0·16	150	8·0	66	5	–	
46a	pegm. mobilizate	0·11	210	8·2	51	13	45	
74a	pegm. mobilizate	0·05	254	8·4	68	5	61	674±23 ma
43	qu–mica–schist	2·33	19	0·4	23	5	17	
5-point isochron: 683±10 ma; Io=0·7169; MSWD = 0·97								
22	muscovite–schist	0·05	78	47	195	47	25	560± 8 ma
62	tab. pegm. miner.	0·04	51	47	614	83	163	549±11 ma
76	tab. pegm. miner.	0·06	146	15	58	19	159	568±22 ma
7	mass. pegm. barren	1·35	148	7·0	41	5	41	528±34 ma
34	mass. pegm. barren	0·72	149	27	53	8	51	
14b	mass. pegm. miner.	2·73	72	17	85	29	95	
19b	mass. pegm. miner.	4·96	104	4·6	70	15	51	
3-point isochron: 545±10 ma; Io=0·7155; MSWD=0·4								

For sample locations refer to Figure 2; model-ages based on Io=0·712

particularly rich in schorl which forms large clusters of quartz–tourmaline-aggregates at the contact between biotite-schists and massive pegmatite (Figure 2). Former mining sites of eluvial accumulations are known to be close to both types of pegmatite. Mining output in Table 1 indicates the dominance of columbite over tantalite.

The Egbe pegmatites occur close to porphyritic granite complexes (Figure 3) in contrast with the first two localities where comparable granites are at least 15 km away from the pegmatites. The ore-bearing pegmatites are massive, lensoid bodies, partly albitized feldspar–quartz–mica to quartz–mica varieties, which are densely intercalated with amphibolites originally mafic in composition (Ohiwerei 1978) and conformably emplaced along the regional NNE trend (*cf.* Figure 7). Egbe is the oldest mining area in southwestern Nigeria, well known for its high tantalite output (Table 1), although present mining does not exceed three tons of concentrate per year. Egbe is also the type locality for 'nigerite', a Sn-rich gahnite first

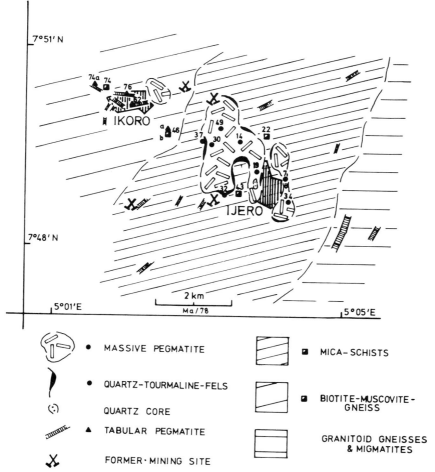

Figure 2. Geological map and sample locations of the Ijero/Ikoro pegmatite field (after Emofurieta 1977).

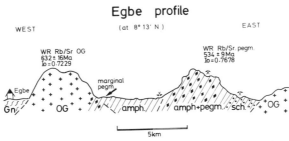

Figure 3. Geological profile and geochronological relation between Older Granite complex and the rare-metal productive amphibolite–pegmatite complex around Egbe (after Ohiwerei 1978). Gn = gneiss, sch = schist, pegm. = pegmatite, OG = older granite, amph. = amphibolite.

described by Jacobson and Webb 1947, and Bannister *et al.* 1947) associated with quartz–sillimanite rocks. Quartz–sillimanite nodules, intergrown with cassiterite, were reported from the massive Ijero pegmatites by Kayode (1972).

2b. Pegmatites from central Nigeria

These occurrences extend from Abuja–Keffi towards the Jos Plateau (Figure 1) and are emplaced mainly into gneissic basement. Detailed geological mapping around these pegmatites has been completed by Kuester (1985) in the Jemaa–Wamba area. Geological sketches of the most important former mining sites are also given in Jacobson and Webb (1946). Albitization is a dominant feature of this rare-metal mineralization which carries tin almost exclusively. 'Columbite in any appreciable quantity is practically unknown in Nassarawa' (Raeburn 1924), in contrast to the ore associations in southwestern Nigeria. The NNE trend is the dominant structural setting for the tabular pegmatites.

Old pegmatitic tin workings are reported also from areas adjacent to the ring complexes in the Zaria and Bauchi provinces (Figure 1), but their mining output has been rather limited (Jacobson and Webb 1946). It should be noted that investigations carried out by Pastor and Ogezi (1986) in the Gurum area (Locality 6 in Figure 1) indicate stanniferous mineralization in pre-Pan-African basement.

3. Geochronological aspects

Four localities of rare-metal mining, i.e. the three main occurrences in southwestern Nigeria and one close to Wamba in central Nigeria, were investigated by Rb/Sr studies for whole rock and mineral samples, respectively. For comparison, two barren pegmatite bodies in southwestern Nigeria (Figure 1) as well as some Pan-African granite occurrences were included in these investigations. Analytical details, individual sample data, and isochrons for 27 whole rock and seven mineral analysis are presented in Matheis and Caen-Vachette (1983). Additional geochronological results from basement gneisses and pegmatites (11 whole rock and 13 mineral Rb/Sr analyses) are summarised in Table 2 and Table 3.

3a. Wamba area

In the central Nigerian pegmatite field, only the Gwon–Gwon locality near Wamba has been sampled for Rb/Sr dating. A 3-point isochron age gave 555+/−5 Ma and three muscovite model ages range from 537 to 522 Ma (Table 3).

Table 3. Rb/Sr model-ages of pegmatoid minerals (Io=0·712)

Location–mineral	Rb ppm	Sr ppm	Radiogenic Sr	Age (Ma)
Iregun (mineralized)				
muscovite	5·060	21	93·7%	513±13
Aramoko (barren)				
muscovite	971	6·5	86·0%	490±14
biotite	1·883	16	58·0%	185± 6
Ijero (mineralized)				
muscovite	4·931	18	96·6%	594±14
oligoclase	1·872	9·0	91·0%	523±16
muscovite	2·378	8·0	95·0%	487±14
biotite	1·959	16	59·0%	183± 5
Ikoro (barren)				
muscovite	1·516	25	66·0%	482±17
Egbe (mineralized)				
muscovite	3·002	15	94·5%	537±20
muscovite	4·924	29	90·7%	536±17
muscovite	5·329	13	99·0%	524±19
Wamba/Gwon–Gwon (mineralized)				
muscovite	2·868	11	95·3%	537±13
muscovite	2·775	13	92·0%	537±12
muscovite	3·028	14	92·0%	522±11

Whole rock Rb/Sr dating from the Nassarawa Eggon granite complex, which is about 30 km south of Wamba, indicate an emplacement age of 535+/−8Ma (Umeji and Caen-Vachette 1984). In contrast to the Jurassic biotite ages obtained in southwestern Nigeria (Table 3), the Rb/Sr model-age of biotite from Nassarawa was 512+/−10 Ma.

3b. Polyphase end Pan-African events

The Pan-African emplacement and related cooling history of both Older Granites and pegmatitic complexes varied to a considerable extent within Nigeria. If published data are compiled as a histogram (Figure 4) at least three main episodes over the period 700–480 Ma can be recognized. Although the dates of the episodes were determined by different analytical methods, indicating different closing temperatures and thus reflecting thermotectonic events of varying degrees, the clustering of ages into these three main episodes is regarded as a reflection of distinctive geological processes related to Pan-African orogeny. Additional results by Egbuniwe et al. (1985) indicate that peralkaline A-type granitoids were emplaced in northwestern Nigeria around 510 Ma at the closing stages of the Pan-African.

Maximum ages calculated on the basis of Io = 0·7000 from the Rb/Sr isochrons obtained in our studies, indicate crustal residence times from 1200 Ma for the Wamba and Egbe Older granites to about 1075 Ma for the Aramoko Older granite suggesting the onset of magma generation in the Nigerian domain during the peak of Kibaran activity in Central Africa. The massive, cupola-type mineralized pegmatites at Wamba and barren pegmatites at Osu, range in age from 750–700 Ma which coincides with a thermotectonic event preserved in the basement gneisses

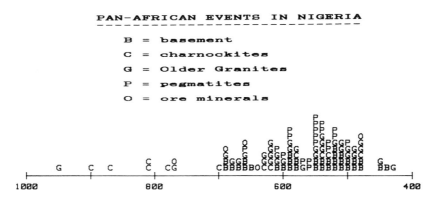

Figure 4. Distribution of published geochronological data in Ma (various analytical methods) for the Pan-African province of Nigeria; included are whole rock ages for silicate rocks according to references discussed in the text and mineral ages for galena after Tugarinov et al. (1968).

at Ijero and Egbe. The mineralized massive Ijero pegmatite indicates a crustal residence time from 600 Ma to 545 Ma. For all the tabular pegmatites, the maximum ages are identical with the upper error limits suggesting a 'fast' solidification. This sequence of varying crustal residence times supports the palingenetic origin of the Older Granites, the partly palingenetic components of the massive pegmatites, and the hyrothermal conditions of tabular pegmatite emplacement.

4. Geochemical indicators

The geochemical investigations were aimed at identifying element indicators for the 'degree of specialization', according to Tischendorf (1977), of the Nigerian rare-metal pegmatites and to test their use as pathfinders for exploration purposes (Matheis 1979; Matheis et al. 1982). In addition to whole rock major and trace element analyses for all investigated pegmatite occurrences, specific rare-metal indicators in rocks and their minerals have been analysed for selected localities (Table 4 and Table 5). More detailed work in the central Nigerian pegmatite field is presently in progress (Kuester 1985). To compare the mineralization characteristic of the Older and Younger Nigerian Tin provinces, the rare-metal concentrations of heavy mineral concentrates, and cassiterites from active mining sites were studied (Matheis et al. 1984).

To obtain representative data for pegmatites, precautions were taken to avoid material of excessive grain size by using chip sampling. The average sample weights of pegmatite were about 5 kg. The major and trace element data were obtained by a fully automatic energy-dispersive X-ray fluorescence spectrometer (XRFS) from fused beads, and Nb and Sn by wavelength-dispersive XRFS from pressed powder pellets (Ta was below the detection limit). Li, Cs, and Be were determined by atomic absorption spectrometry after hydrofluoric acid dissolution and F by ion-sensitive electrodes after NaOH fusion. All geochemical data were calibrated against international standards. The analyses of the mineral concentrates were made by XRFS and ICP-Emission Spectrometry, and cassiterite by neutron activation analysis at the Hahn-Meitner-Reactor Centre/Berlin.

Table 4. Main element distribution in the Nigerian pegmatite fields

rock type	SiO$_2$	TiO$_2$	Al$_2$O$_3$	Fe$_2$O$_3$	MnO	MgO	CaO	Na$_2$O	K$_2$O	P$_2$O$_5$	Ign. loss
If/mb	72.6	0.14	15.2	1.6	0.04	0.25	0.88	1.7	0.4	0.02	0.58
Os/ma	75.1	0.05	14.1	0.51	0.04	0.16	0.52	3.9	4.5	0.16	0.64
Os/OG	69.1	0.06	17.5	1.2	0.02	0.13	1.2	2.8	7.9	0.43	0.35
Ar/ta	74.4	0.07	15.1	0.89	0.04	0.20	0.79	1.4	7.9	0.17	1.3
Ar/OG	65.3	0.86	16.0	6.6	0.01	1.0	2.4	2.5	3.6	0.40	0.83
Ij/gn	74.0	0.22	14.1	2.2	0.05	0.66	0.91	1.1	6.0	0.18	0.67
Ij/mb	73.5	0.02	16.3	1.2	0.08	0.30	0.61	0.74	8.9	0.16	1.2
Ij/ms	63.7	1.3	18.4	8.8	0.08	1.9	0.18	0.29	4.9	0.08	4.3
Ij/ta	76.1	0.05	15.3	1.5	0.08	0.35	0.34	0.48	6.7	0.12	1.8
Ij/ba	76.2	0.04	14.3	0.50	0.02	0.18	0.44	3.7	3.6	0.14	0.77
Ij/ma	76.2	0.03	14.1	0.81	0.09	0.19	0.36	5.0	1.4	0.14	1.1
Eg/gn	75.3	0.01	14.5	1.2	0.35	0.04	0.21	3.9	3.6	0.08	0.72
Eg/OG	71.3	0.38	14.5	3.2	0.08	0.48	1.8	2.6	5.3	0.07	0.52
Eg/mp	74.3	0.06	14.2	1.4	0.24	0.11	0.41	3.8	4.9	0.11	0.61
Eg/cpf	73.5	0.04	15.4	0.60	0.15	0.23	0.28	4.6	4.9	0.19	0.74
Eg/cpm	81.1	0.04	12.4	0.74	0.10	0.20	0.18	1.5	3.1	0.10	1.7
Pa/ma	76.0	0.01	14.6	0.57	0.14	0.02	0.50	3.1	6.0	0.01	0.30
Pa/OG	74.4	0.02	15.7	0.57	0.06	0.15	1.4	1.5	7.7	0.01	0.39
Wa/ma	73.5	0.03	15.4	0.95	0.12	0.12	0.50	3.0	4.1	0.26	1.27
Wa/OG	67.8	0.04	16.1	1.4	0.18	0.06	0.38	7.1	0.92	0.52	1.45

Total iron as Fe$_2$O$_3$: number of samples analysed indicated in brackets
Sample description (for localities cf. Figure 1):

If/mb   = Ife quarry, fsp–qu–mobilizates in bi-gneiss (2)
Os/ma   = Osu massive pegmatites, barren (7)
Os/OG   = Osu Older Granite (1)
Ar/ta   = Aramoko tabular pegmatites, barren (8)
Ar/OG   = Aramoko Older Granite (1)
Ij/gn   = Ijero bi–musc-gneiss (2)
Ij/mb   = Ijero fsp–qu-mobilizates in Ij/gn (2)
Ij/ms   = Ijero mica-schist (1)
Ij/ta   = Ijero tabular pegmatites, productive (4)
Ij/ba   = Ijero massive pegmatites, barren (2)
Ij/ma   = Ijero massive pegmatites, productive (5)
Eg/gn   = Egbe bi-gneiss (1)
Eg/OG   = Egbe Older Granite (3)
Eg/mp   = Egbe marginal pegmatites, barren (3)
Eg/cpf = Egbe complex, fsp–qu–mica pegmatites, productive (6)
Eg/cpm = Egbe complex, mica–qu–fsp pegmatites, productive (8)
Pa/ma   = Pategi massive pegmatite, barren (1)
Pa/OG   = Pategi Older Granite (1)
Wa/ma   = Wamba massive pegmatites, productive (3)
Wa/OG   = Wamba Older Granite (1)

4a. Petrogenetic element patterns

Rare-metal bearing potentials of acidic source rocks can be ascertained from geological, geochemical, and petrological ore-controlling factors referred to as the 'degree of specialization' (Tischendorf 1977). Such metallogenetically specialized

Table 5. Element ratios and trace element distribution in the Nigerian pegmatite fields

Rock type	Na/K	K/Rb	Ba/Rb	Ba	Rb	Sr	Y	Zr
If/mb	0·18	254	2·4	741	309	112	55	238
Os/ma	0·77	200	2·7	433	166	145	47	211
Os/OG	0·63	280	5·6	971	172	759	52	245
Ar/ta	0·18	243	2·0	539	270	69	51	165
Ar/OG	0·69	113	4·7	1230	264	257	43	481
Ij/gn	0·19	169	2·1	645	300	55	54	195
Ij/mb	0·08	232	1·7	540	317	56	50	157
Ij/ms	0·06	78	1·4	750	525	11	48	209
Ij/ta	0·06	145	1·4	525	383	38	53	158
Ij/ba	0·91	147	2·0	407	204	14	47	183
Ij/ma	3·2	48	2·1	515	240	16	52	203
Eg/gn	0·98	162	8·0	2300	287	875	45	557
Eg/OG	0·47	203	5·1	1030	201	429	47	427
Eg/mp	0·69	98	1·4	570	414	16	58	216
Eg/cpf	0·83	33	0·32	395	1225	25	113	189
Eg/cpm	0·43	19	0·37	485	1315	52	121	196
Pa/ma	0·52	89	0·97	546	559	12	83	195
Pa/OG	0·19	166	1·1	425	385	67	70	207
Wa/ma	0·86	52	1·2	685	585	120	76	210
Wa/OG	7·7	37	4·7	980	210	355	46	266

Sample identification as Table 4

granitoids have been considered to be products of either extreme magmatic fractionation or substantial hydrothermal alteration, and are characterized by low levels of Fe, Ca, Mg, Ti, Ba, Sr, and Zr whereas Rb, Li, Cs, F, Be, Y, and the ore-forming elements Sn, Nb, and Ta are often enriched.

Another important factor related to rare-metal mineralization is the degree of albitization which is reflected in the Na/K ratio (Figure 5). While Jacobson and Webb (1946) emphasized the association of the tin mineralization with intensive albitization in the Older Tin-Fields of Nigeria, de Kun (1965), based on his experience in central Africa, stated that '. . .in pegmatites tin does not favour albitized zones to the same extent as niobium'. A detailed discussion on the tin-bearing pegmatites of southwestern Nigeria (Matheis et al. 1982) demonstrated the heterogeneous nature of feldspar composition in the different varieties of mineralized pegmatite, a feature which is reflected in the element compositions (Tables 4 and 5). Potash feldspar is most dominant in the tabular pegmatites and the feldspar–quartz–mobilizates in the basement gneisses, while albitization is only found in the cupola-type massive pegmatites from the Ijero and Wamba areas, respectively. There is no linear relationship between Na/K ratios and rare-metal mineralization — except for a weak Nb correlation with albitization. But it seems that either high K or Na concentration correlates with rare-metal enrichment. In the Egbe tin-field, all ores are related to micaceous pegmatites. One of the most

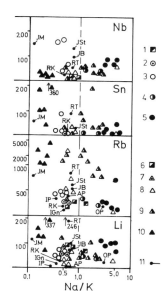

Figure 5. The degree of albitization (Na/K) vs. ore indicators Sn and Nb and petrogenetic indicators Li and Rb. Rock symbols (detailed description cf. Table 4):
IJERO district = 1, Ar/OG; 2, Ij/gn; 3, Ij/ta; 4, Ij/ba; 5, Ij/ma;
EGBE district = 6, Eg/OG; 7, Eg/gn; 8, Eg/mp; 9, Eg/cpf; 10, Eg/cpm;
11 = reference data as follows:

IGn	(9)	basement gneisses, Ife quarry	Mbata (1978)
IP	(11)	concordant pegmatites in IGn	
OP	(9)	barren oligoclase pegmatites, Osu	Emofurieta
AP	(9)	barren microcline pegmatites, Aramoko	(1977)
JB	(43)	barren biotite-granite, Jos ring complexes	
JSt	(73)	stanniferous bi-granite, Jos RC	Olade (1980)
JM	(67)	mineralized bi-granite, Jos RC	
RK	(19)	barren rapakivi, Tarkki/Finland	Haapala
RT	(28)	tin-bearing rapakivi, Väkkärä/Finland	(1977)

useful plots to highlight increasing degrees of 'specialization' is K/Rb vs. Rb (Figure 6) which separates barren pegmatites from productive pegmatites. The barren varieties from all the investigated localities are found to plot close to the differentiation trend of granitoid complexes (Stavrov et al., 1969) at K/Rb ratios above 100 and Rb values below 500. Mobilizates, marginal pegmatites, and barren tabular pegmatites (Table 4) follow the barren 'pegmatite trend' in Figure 6 at K/Rb values above 100. With regard to the mineralized pegmatites, however, there is a distinctive break from the 'normal' differentiation trend and it is difficult to reconcile their formation to 'advanced' fractionation of Older Granite complexes, none of which shows any geochemical tin-specialization. The three pegmatite fields Ijero, Egbe, and Wamba occupy distinct fields in Figure 6 which reflect a development due to the controlling factors of their respective lithological frameworks.

4b. Rare-metal indicators

Element ratios are useful for discriminating between barren and mineralized pegmatites. K/Rb and Rb/Sr ratios identify both fractional crystallization processes (Stavrov et al. 1969; El Bouseily and El Sokkary 1975) and tin-bearing source rocks (Tischendorf 1977). Ba/Rb links tin-bearing granitoids with different tectonic settings (Groves and McCarthy 1978) and the Mg/Li ratio, introduced by Beus and Sitnin (1968), is an indicator of rare-metal potential in apogranites and pegmatites. The latter ratio was successfully applied to delineate buried pegmatites by geochemical patterns in residual lateritic soil (Matheis 1979). Still the best discriminants are the pegmatoid indicator elements Rb, Li, Cs, F, and to a lesser extent Be. Again, the various pegmatite fields display distinct geochemical features which separate them according to the geological setting.

At Ijero, the pegmatoid mobilizates are almost identical in composition with their gneissic host rocks (Table 2), especially as expressed by the K/Rb and Rb/Sr ratios, in contrast with the younger rare-metal pegmatites. Although there are few data, it is interesting to note the relative Nb enrichment for both the pegmatoid mobilisates and the tabular pegmatite. These pegmatites are distinctly different in age (650–550Ma) but may have been formed by similar processes within the biotite–muscovite–gneiss (Figure 2). In contrast, the massive Ijero pegmatites and

Table 6. Trace elements in pegmatites and their minerals from the Egbe area, SW Nigeria

Sample type	Sn	Nb	Li	Rb	Cs	Be	F	K/Rb
			Whole rock data					
Eg/OG (2)	<10	16	82	302	36	2·3	570	135
Eg/mp (4)	12	37	16	1000	37	49	150	78
Eg/cpf (5)	42	41	71	1050	45	209	1050	28
Eg/cpm (5)	158	26	167	1730	99	50	3900	16
			Mineral data					
Eg/OG–fsp (2)			33	385	19	0·2	–	214
Eg/mp–fsp (4)			13	1135	37	4·1	–	46
Eg/cpf–fsp (1)			40	3350	90	14	–	26
Eg/cpf–mica (5)			275	2830	120	30	–	22
Eg/cpm–mica (5)			306	2850	144	24	–	21

Mineral data from Kempf 1981; concentrations in ppm; sample identification as in Table 4.
Correlation Coefficients:
Sn(rock)/Rb(rock) = +0·61
Sn (rock)/F (rock) = +0·70
Sn(rock)/Li(rock) = +0·81
Sn(rock)/Be(rock) = +0·08
Sn(rock)/Cs(rock) = +0·80
Sn(rock)/Li(mica) = +0·89
Sn(rock)/Cs(mica) = +0·75

Eg/OG–fsp feldspar from Egbe older granite
Eg/mp–fsp feldspar from Egbe marginal pegmatites (barren)
Eg/cpf–fsp feldspar from Egbe complex productive pegmatites
Eg/cpf–mica mica from Egbe complex productive pegmatites
Eg/cpm–mica mica from Egbe complex productive pegmatites
See Table 4 for more detailed descriptions

the Egbe pegmatites are significantly poorer in Nb (Figure 5). Sn values in the productive pegmatites are above 10 ppm but rather erratically distributed in the Ijero occurrences which, as a whole, have Sn concentrations well above those of the Egbe pegmatites. This trend of maximum values in the Egbe pegmatites holds true for all rare-metal indicators except Nb.

The Egbe pegmatites have been studied in greater geochemical detail than the other locality. The regional distribution pattern of the most important indicator elements in relation to their geological setting is outlined in Figure 7.

The central pegmatite–amphibolite complex has the maximum values for Sn, Rb, Li, F, and extremely low K/Rb, while Nb concentration is nearly equal in all the investigated pegmatites (Table 6). Although reliable Ta data on whole rock are not available, semi-quantitative XRFS measurements indicate a correlation with Sn. Similarly, fluorine is particularly high in the central pegmatite–amphibolite complex but has only background levels in the marginal pegmatites close to the Older Granite complex. On the basis of the geochronological evidence, a genetic link between the proximal Older Granite complex and the rare-metal pegmatites has to be ruled out. This is supported by the geochemical evidence which indicates that there are two separate rock formations which have been generated within different crustal levels from different source material.

Reference data for barren mobilizates and pegmatites from non-productive areas in southwestern Nigeria (Figure 5) plot close to the Older Granites and barren pegmatites from mining areas, pointing to a common petrogenetic origin. Geochemical indicators in feldspar and white mica have been used in various attempts to identify pegmatoid rare-metal potentials Gaupp et al. 1983, 1984). A limited number of feldspar and muscovite samples of the Egbe pegmatites analysed by Kempf are presented in Table 6. So far, contrasts between barren and productive pegmatites are found to be quite significant for Rb and Cs. The method

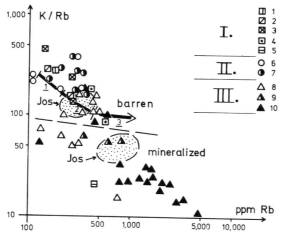

Figure 6. K/Rb vs. Rb distribution pattern of the Older Tin-Fields. Average distribution fields (stippled) of the Younger Tin-Fields are based on data of Olade (1980); the arrow indicates the normal differentiation trend from granodiorite (1) to granite (2) to pegmatite (3) after Stavrov et al. (1969). Rock symbols:
I   Older Granite complexes of Osu (1), Aramoko (2), Egbe (3), Pategi (4), Wamba (5);
II   barren pegmatite complexes of Osu (6), Aramoko (7);
III   productive pegmatite complexes of Ijero (8), Wamba (9), Egbe (10).

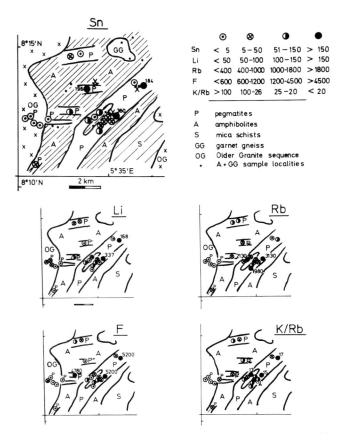

Figure 7. Distribution of Sn and rare-metal path-finders in granitoid–pegmatoid rocks of the Egbe mining area. Concentration-ranges are indicated by symbols, anomalous values are given in numbers (*cf.* cross-section of Figure 3 for geological situation).

is applicable even in muscovite. In the metabasalt-hosted mineralized pegmatites at Iregun and Egbe there are the highest Rb concentrations in muscovites, while the Ijero and Wamba muscovites, which are derived from pegmatites emplaced into gneissic basement units, have more intermediate concentrations.

4c. Ore minerals

The ore concentrate in the Nigerian pegmatite fields is recovered from alluvial placers close to their source rocks. In general then the compositions can be accepted as representative of the primary mineralization. Within the framework of investigations on mining concentrates of the Nigerian tin province (Matheis *et al.* 1984), a total of thirteen concentrates from eight pegmatite localities have been studied out of which six concentrates from five localities have been simultaneously analysed by three methods (Möller and Dulski 1983). There is a marked difference between the occurrences in central and southwestern Nigeria, with considerable variations even within these regions (Table 7). Evidently, the element patterns of the handpicked cassiterites give the most representative picture of the primary mineralization of each mining site which is also reflected in the total concentrate compositions and in the selectively HF-leached distribution patterns.

Table 7. Geochemical indicators in mining concentrates of the pegmatitic tin-fields of Nigeria

(A) Tin concentrates

No.	Locality	SnO_2	Nb_2O_5	Ta_2O_5	Fe_2O_3	Sn/Nb	Sn/Ta	Nb/Ta
	CENTRAL NIGERIA:							
25	Gurum Bmnt	89.0	4.5	0.3	6.5	19.8	297	15.0
26	Jemaa Peg	94.0	4.0	0.3	0.5	23.5	313	13.3
29	Wamba Peg	88.0	3.0	1.0	8.0	29.3	88	3.0
	SOUTHWESTERN NIGERIA:							
30	Egbe Peg	78.0	5.5	7.5	9.0	14.2	10.4	0.7
31	Egbe Peg	80.0	6.0	9.5	4.5	13.3	8.4	0.6
32	Ijero Peg	80.0	6.5	4.5	9.0	12.3	17.8	1.4

(XRFS Analysis; K. Weber-Diefenbach, Univ. München; oxide concentrations in %)

(B) HF-Leached concentrates

No.	Locality	Nb	Ta	W	Fe	Ti	Mn	Hf	La	Zr	Ce
		ppm									
	CENTRAL NIGERIA:										
25	Gurum Bmnt	1400	20	50	1.0%	2400	290	11	178	70	316
26	Jemaa Peg	110	<10	<10	1500	270	130	2.2	21	51	6.5
29	Wamba Peg	2900	470	130	1.5%	1.1%	1100	20	135	144	388
	SOUTHWESTERN NIGERIA:										
30	Egbe Peg	8900	2.4%	180	4.8%	3.6%	3400	58	32	164	192
31	Egbe Peg	1.2%	3.3%	160	2.1%	1.2%	3600	58	72	161	243
32	Ijero Peg	1.1%	8100	250	4.2%	3.6%	2200	47	47	503	156

(ICP+AAS Analysis: W. Schreiber/W. Becker, TU Berlin)

(C) Trace Element ratios in HF-leached concentrates

No.	Locality	Nb/Ta	Fe/Ti	Ta/W	Zr/Hf	Ta/Hf	La/Ce
	CENTRAL NIGERIA:						
25	Gurum Bmnt	70	4.3	0.4	6.4	0.02	0.56
26	Jemaa Peg	>11	5.6	–	3.2	0.05	3.2
29	Wamba Peg	6.2	1.4	3.6	7.2	0.2	0.35
	SOUTHWESTERN NIGERIA:						
30	Egbe Peg	0.4	1.3	135	2.8	896	0.21
31	Egbe Peg	0.4	1.8	208	2.8	112	0.3
32	Ijero Peg	12	1.2	800	11	42	0.3

Table 7. (contd)

(D) Cassiterites

No.	Locality	Sn %	Ta	W	Zr ppm	Hf	Sc	Th	La
	CENTRAL NIGERIA:								
25	Gurum Bmnt	64	8450	3100	1200	130	125	178	410
26	Jemaa Peg	74	2400	1800	270	71	36	26	22
29	Wamba Peg	74	8950	32	1200	210	3	315	695
	SOUTHWESTERN NIGERIA:								
30	Egbe Peg	32	6·57%	190	340	190	24	nd	130
31	Egbe Peg	56	7·86%	510	810	250	7	nd	40
32	Ijero Peg	74	1·75%	80	420	160	7	nd	14

(NAS-Analysis; P. Dulski, H.M.I. Berlin; nd: below detection limit)

No.	Locality	Sn/Ta	Sn/W	Ta/W	Ta/Hf	Ta/Sc	Zr/Hf
	CENTRAL NIGERIA:						
25	Gurum Bmnt	76	206	2·7	65	77	9·2
26	Jemaa Peg	308	411	1·3	33	66	3·8
29	Wamba Peg	83	23125	279	43	2980	5·7
	SOUTHWESTERN NIGERIA:						
30	Egbe Peg	4·9	1684	345	346	2740	1·8
31	Egbe Peg	7·1	1098	154	314	11200	3·2
32	Ijero Peg	42	9250	218	109	2500	2·6

Table 7 (A) shows the tin dominance in the Pan-African mineralization as already seen from the mining output in Table 1. This tin dominance is valid for the whole 'tin-province' of Nigeria. The relative concentrations of Sn–Nb–Ta in concentrates from both the Older and Younger Tin-fields of Nigeria and two concentrates from the Air Mountains in Niger are displayed in Figure 8.

The absolute values in Table 7 show that the Nb concentrations are nearly always constant while Ta enrichment in southwestern Nigeria occurs at the expense of Sn. As a general trend in minor and trace element distributions, the central Nigerian tin-fields have higher concentrations of W, Zr, and Sc while the occurrences in southwestern Nigeria are dominated by Ta, Hf, Ti, Fe, and Mn. There is a tendency for a simultaneous increase in Ta and Ti to its maximum in the Egbe pegmatites which are emplaced into a series of tholeiitic amphibolites. These distribution patterns support the conclusion that there are no uniform geochemical characteristics for the pegmatoid rare-metal mineralization and that the host-rock lithology seems to have made a considerable imprint on the individual mineralization event.

Möller and Dulski (1983) included Nigeria cassiterites supplied by the author (cf. Table 7D) in their investigations of worldwide tin deposits. Using Zr and Hf contents as genetic indicators, they found distinct clustering at Zr/Hf values around 28 for the Andes and Sardinia tin provinces compared with Zr/Hf ratios around 4 for the various African rare-metal pegmatites. In contrast to Möller and Dulski's conclusions, who considered that the two ratios represent pneumatolytic (+/−4) and hydrothermal (+/−28) ore formation, respectively, Zr/Hf is considered in this

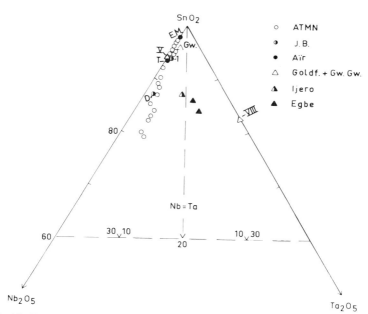

Figure 8. Sn–Nb–Ta distribution in mining concentrates (crude) from the Nigerian Tin Province and from the Aïr Mountains (Niger). Triangles = pegmatitic source rocks (Goldf. + Gw.Gw. = central Nigeria; Ijero + Egbe = SW Nigeria), circles = ring-complex derived concentrates (ATMN + J.B. = Jos Plateau).

work to indicate different crustal evolutions of the regional lithological framework in which the tin mineralization occurred. The lower Zr/Hf ratio, is found exclusively in deposits emplaced into stable cratonic areas, while higher Zr/Hf ratios are typical for young orogenic zones. This latter environment, is characterized by low-temperature ore associations including stannite which is also found in the Jos Plateau ring complexes (Kinnaird 1984) but with Zr/Hf ratios in cassiterites of 2·7 to 5·7.

5. Conclusions

The economically most important rare-metal producing pegmatoid source rocks are of end Pan-African age located in shield areas. They are typically products of crustal recycling and related to major tectonic lineaments (Groves 1982; Kazansky 1982; Beus 1982). The emplacement of the Nigerian rare-metal bearing pegmatites clearly occurred within the last of three recognizable events of the Pan-African in Nigeria, contemporaneous with the formation of the widespread Older Granite complexes. The geological, geochronological, and geochemical evidence presented does not confirm a genetic link between these two rock units but suggests an origin independent of each other. Similar conclusions were obtained by Beckinsale (1979) from studies in the tin-belt of southeast Asia 'In general, it has not been established that particular cassiterite-bearing pegmatites which intrude a granite actually fractioned from the same magma that produced the granite host rock'.

In contrast to the juvenile Pan-African crustal formation of the Arabian–Nubian Shield of northeastern Africa (Duyverman *et al*. 1982), the Nigerian basement complex shows Archaean relict ages and distinct tectonometamorpic overprints of Eburnean and Pan-African ages. The Pan-African covers the time span 700 Ma to 500 Ma with three periods of major crustal activity being preserved at 675 +/−15 Ma, 620+/−25 Ma. Similar age sequences are reported from the neighbouring basement areas of Togo/Benin and the Cameroons (Lasserre and Soba 1979), marking a common influence by the tectonic regime between West Africa and Congo cratons. The resulting repeating patterns of crustal reactivation are typical for the whole of West Africa and have 'probably been a fundamental factor in determining the occurrence and distribution of economic mineral deposits.' (Wright 1985). The lack of such repeated crustal reactivation is considered by Harris (1985) to be one of the main reasons for the relatively poor mineralization of the Arabian shield.

Field evidence as well as geochemical and geochronological data for the 'Older Tin-Fields' of Nigeria reveal that rare-metal pegmatites are not the final fractionation phases of contemporaneous Older Granite complexes but products of high grade metamorphic conditions which were enhanced along a deep-seated, reactivated continental lineaments by higher heat flow and addition of fluid phases. The host-rock lithology as contributed significantly to the individual characteristics of each occurence as demonstrated by the marked differences between the pegmatite fields of southwestern and central Nigeria.

Obviously, rare-metal mineralization at the earlier stage of geological evolution had a distinct affinity to mantle-derived mafic or ultramafic volcanic sequences rather than to acidic intrusive complexes which are dominant in the late Phanerozoic.

Acknowledgements. Field work and an important part of the geochemical studies were carried out in co-operation with postgraduate students at the University of Ife (Nigeria) whose efforts contributed substantially to the results presented. The geochronological part would not have been possible without the efforts of Madame M. Caen-Vachette and the laboratory in Clermont-Ferrand.

Additional analytical assistance was provided by K. Weber-Diefenbach/Munich, P. Dulski/Berlin and postgraduate students at the University of Giessen/FRG. For very helpful discussions on the subject over the past years, I would like to thank in particular P. Bowden of the University of St. Andrews and J. B. Wright of the Open University.

Financial support through grants and scholarships was provided by the German Academic Exchange Service (DAAD), the University of Ife and the German Research Foundation (DFG).

References

Ajakaiye, D. E. 1977. A gravity survey over the Nigeria younger granite province. In: Kogbe, C. A. (Ed.) *Geology of Nigeria*. Elizabethan Publ. Comp., Lagos, 207–224.
——, **Hall, D. H. and Millar, T. 1983**. Æromagnetic anomalies across the Nigerian Younger Granite Province. Oral Presentation *IAGA Session IUGG General Assembley*, Hamburg, August 1983.
Bannister, F. A., Hey, M. H. and Stadler, H. P. 1947. Nigerite, a new tin minreal. *Mineralogical Magazine*, **28**, 129–136.

Barsukov, V. L. 1975. The source of ore material. In: Tugarinov, A. I. (Ed.), *Recent Contributions to Geochemistry and Analytical Chemistry* Wiley, New York & Toronto, 303–310.

Beckinsale, R. D. 1979. Granite magmatism in the tin belt of South-East Asia. In: Atherton, M. P. and Tarney, J. (Eds), *Origin of Granite Batholiths, Geochemical Evidence*. Shiva Publ., Nantwich/UK, 34–44.

Beus, A. A. 1982. Metallogeny of Precambrian rare-metal granitoids. *Proc. Int. Symp. Archean Early Proterozoic Geol. Evol. Metallog., Rev. Brasil. Geosci.*, **12**, 1–3, 410–413.

—— 1984. On metallogeny of Precambrian granitoids. *27th Int. Geol. Congr. Moscow, Abstr. Vol. IX, Part 1*, 147–148.

—— and Sitnin, A. A. (1968). Geochemical specialization of magmatic complexes as criteria for the exploration of hidden deposits. *Proc. XXIII Intern. Geol. Congr. Prague*, **6**, 101–105.

De Booder, H. 1982. Deep-reaching fracture zones in the crystaline basement surrounding the West Congo System and their control of mineralization in Angola and Gabon. *Geoexploration*, **20**, 259–273.

Bowden, P. 1970. Origin of the Younger Granites of Northern Nigeria. *Contr. Mineral. Petrol.*, **25**, 153–162.

—— and Jones, J. A. 1978. Mineralization in the younger granite province of northern Nigeria. In: Stemprok, M., Burnol, L. and Tischendorf, G. (Eds) *Metallization Associated with Acid Magmatism*, Vol. 3. Ustredni Ustav Geologicky, Prague, 179–180.

Correia, N. J. 1984. Metallogenesis of granite pegmatites. *27th Int. Geol. Congr. Moscow, Abstr. Vol. VI, Sect. 12*, 225 pp.

Dempster, A. N. 1959. Report on mining in Ijero district, Ekiti division. *Geol. Surv. Nigeria Rep.*, 1386, Kaduna.

Dulski, P., Möller, P., Villalpando, A., and Schneider, H.-J. 1982. Correlation of trace element fractionation in cassiterites with the genesis of the Bolivian metallotect. In: Evans, A. M. (Ed.) *Metallization Associated with Acid Magmatism*. Wiley, London, 71–83.

Duyverman, H. J., Harris, N. B., and Hawkesworth, C. 1982. Crustal accretion in the Pan-African: Nd and Sr isotope evidence from the Arabian Shield. *Earth and Planetary Science Letters*, **59**, 315–326.

Egbuniwe, I. G., Fitches, W. R., Bentley, M., and Snelling, N. J. 1985. Late Pan-African syenite-granite plutons in NW Nigeria. *Journal of African Earth Science*, **3**, 427–435.

El Bouseily, A. M. and El Sokkary, A. A. 1975. The relation between Rb, Ba, and Sr in granitic rocks, *Chemical Geology*, **16**, 207–219.

Elueze, A. A. 1984. Mineralogical and chemical variations in ore minerals-bearing metabasites in the Precambrian Basement Complex of Nigeria, in relation to cataclasis. *Natural Res. Develop.*, **19**, 63–70. Tübingen, F. R. Germany.

Emofurieta, W. O. 1977. *Geochemical Studies of Pegmatites around Ijero, Ikoro, Aramoko and Osu in South-Western Nigeria*. Unpublished M. Sc. Thesis, University of Ife, Nigeria.

Falconer, J. D. 1924. *Geological Survey of Nigeria Bulletin*, No. 5, Preface, Kaduna.

Galdeano, A. 1981. Les mesures magnetiques du satellite Magsat et la derive des continents. *C. R. Acad. Sci. Paris*, **293**, Ser. II, 161–164.

Gaupp, R., Morteani, G., and Möller, P. 1983. Die Aufsuchung und Bewertung Tantal-führender Pegmatite. *Erzmetall*, **36**, 244–250 und 294–300.

Gaupp, R, Morteani, G., and Möller, P. 1984. Tantal-Pegmatite: Geologische, petrologische und geochemische Untersuchungen. *Monograph Series on Mineral Deposits*, **23**, 124 pp., Berlin–Stuttgart.

Grant, N. K. 1971. A compilation of radiometric ages from Nigeria. *Journal Mining Geology, Nigeria*, **6**, 37–54.

—— 1978. Structural distinction between a metasedimentary cover and an underlying basement in the 600-m.y.old Pan-African domain of northwestern Nigeria, West Africa. *Geological Society of America Bulletin*, **89**, 50–58.

Groves, D. I. 1982. The Archean and Earliest Proterozoic Evolution and Metallogeny of Austalia. *Proc. Int. Symp. Archean Early Proterozoic Evol. Metallog., Rev. Brasil Geosci.*, **12**, no. 1–3, 135–148.

—— and McCarthy, T. S. 1978. Fractional crystallization and the origin of tin deposits in granitoids. *Mineral. Deposita*, **13**, 11–26.

Guiraud, R. and Alidou, S. 1981. La faille du Kandi (Benin), temoin du rejeu fini-crétace d'un accident majeur à l'échelle de la plaque africaine. *C. R. Acad. Sc. Paris*, Ser. II, **29**, 779–782.

Haapala, I. 1977. Petrography and geochemistry of the Eurajoki stock, a rapakivi-granit complex with greisen-type mineralisation in southwestern Finland. *Geological Survey of Finland Bulletin*, no. **286**.

Harper, C. T., Sherrer, G., McCurry, P., and Wright, J. B. 1973. K–Ar retention ages from the Pan-African of northern Nigeria. *Geological Society of American Bulletin*, **84**, 919–926.

Harris, N. B. W. 1985. Alkaline complexes from the Arabian Shield. *Journal of African Earth Science*, **3**, 83–88.

Imeokparia, E. G. 1984/1985. Geochemistry of intrusive rocks associated with molybdenite mineralization (Kigom Complex, northern Nigeria). *Chemical Geology*, **47**, 261–283.

Jacobson, R. R. E. 1944. Columbite and tantalite. *Geological Survey of Nigeria Report*, 711, Kaduna.

Jacobson, R. and Webb, J. S. 1946. The pegmatites of central Nigeria. *Geological Survey of Nigeria Bulletin*, 17, Kaduna.

—— and Webb, J. S. 1947. The occurrence of Nigerite, a new tin mineral in quartz–sillimanite-rocks from Nigeria. *Mineralogical Magazine*, **28**, 118–128.

Jacobson, R. E., Caley, A., and Macleod, W. N. 1951. The occurrence of columbite in Nigeria. *Geological Survey of Nigeria Occasional*, **9**, Kaduna.

Jacobson, R., Snelling, N. J., and Truswell, J. F. 1963. Age determinations in the geology of Nigeria, with special reference to the Older and Younger Granites. *Overs. Geol. Mineral. Res.*, **9**, 168–182, London.

Jaques, E. H. 1945. Reconnaissance of the Kabba–Ilorin pegmatite area. *Geol. Surv. Nigeria Ann. Rep.*, **1944**, Kaduna.

Katz, M. B. 1985. East African Rift and Northeast Lineaments: continental spreading — transform system? *7th Conf. African Geol., Botswana Nov. 1985*, Abstr. Vol., 32–33, Geol. Soc. Africa.

Kayode, A. A. 1972. Quartz–sillimanite nodules from the Ijero–Ekiti cassiterite-bearing pegmatite. *Journal Mining Geology Nigeria*, **7**, 13–17.

Kazansky, V. I. 1982. Metallogenic processes in the early history of the Earth. *Proc. Intern. Symp. Archean Early Proteroz. Geol. Evol. Metallog., Rev. Brasil. Geosci.*, **12**, no. 1–3, 476–483.

Kempf, W. D. 1981. *Mineralogisch–geochemische Untersuchungen zur Prospektion zinnführener Pegmatite von Südwest-Nigeria*. Unpubl. M. Sc. Thesis, University of Giessen, F. R. Germany.

Kinniard, J. A. 1984. Contrasting styles of Sn–Nb–Ta–Zn mineralization in Nigeria. *Journal African Earth Science*, **2**, 81–90.

Klemm, D. D., Schneider, W., and Wagner, B. 1984. The Precambrian metavolcano-sedimentary sequence east of Ife and Ilesha/SW-Nigeria. A Nigerian greenstone belt? *Journal African Earth Science*, **2**, 161–176.

Kogbe, C. A., Ajakaiye, D. E., and Matheis, G. 1983. Confirmation of a rift structure along the Mid-Niger Valley, Nigeria. *Journal African Earth Science*, **1**, 127–131.

Kuester, D. 1985. Trace element distribution in pegmatitic muscovites from Central Nigeria. *13th Coll. African Geol.*, Sept. 1985, St. Andrews, Scotland.

de Kun, N. 1965. *Mineral Resources of Africa*. Elsevier, Amsterdam.

Lasserre, M. and Soba, D. 1979. Migmatisation d'âge panafricain au sein des formations camerounaises apartenant à la zone mobile de l'Afrique centrale. *C. R. Somm. Soc. geol. France*, **1979**, fasc. 2, 64–68.

Lesquer, A., Betrao, J. F., and De Abreu, F. A. M. 1984. Proterozoic links between northeastern Brazil and West Africa: a plate tectonic model based on gravity data. *Tectonophysics*, **110**, 9–26.

Manning, D. A. 1982. An experimental study of the effects of fluorine on the crystallization of granitic melts. In: Evans, A. M. (Ed.), *Metallization Associated with Acid Magmatism*. Wiley, London. 191–203.

Matheis, G. 1979. Geochemical exploration around the pegmatitic Sn–Nb–Ta mineralization of SW-Nigeria. *Geological Society of Malaysia Bulletin*, **11**, 333–351.

—— 1981. Trace-element patterns in lateritic soils applied to geochemical exploration. *Journal of Geochemical Exploration*, **15**, 471–480.

—— and Caen-Vachette, M. 1983. Rb–Sr isotopic study of rare-metal bearing and barren pegmatites in the Pan-African reactivation zone of Nigeria. *Journal of African Earth Science*, **1**, 35–40.

——, Dulski, P., and Schreiber, W. 1984. Geochemische Charakterisierung von Schwermineralseifen und Kassiteriten der nigerianischen Zinnprovinz. *Fortschr. Mineral.*, **62**, Bh. 1, 149–150, Stuttgart.

——, Emofurieta, W. O., and Ohiwerei, S. F. 1982. Trace element distribution in tin-bearing pegmatites of Southwestern Nigeria. In: Evans, A. M. (Ed.), *Metallization Associated with Acid Magmatism*. Wiley, London, 205–220.

Mbata, A. 1978. *Trace Elements and Major Elements of Pegmatites and their Surrounding Host Rock in the Oke D. O. Quarry, Ile-Ife*. Unpublished B. Sc. Thesis, University of Ife, Nigeria.

Möller, P. and Dulski, P. 1983. Fractionation of Zr and Hf in cassiterites. *Chemical Geology*, **40**, 1–12.

Neiva, A. M. R. 1982. Geochemistry of muscovites and some Physico-chemical conditions of the formation of some tin–tungsten deposits in Portugal. In: Evans, A. M. (Ed.), *Metallization Associated with Acid Magmatism*, Wiley, London. 243–259.

Odeyemi, I. B. 1979. Orogenic events in the Precambrian basement of Nigeria, West Africa. *Newsl. IGCP 108-144*, **3**, 18–23.

O'Driscoll, E. S. T. 1985. Observations of the lineament–ore relation. *Meeting Royal Soc. London 'Major Crustal Lineaments and their Influence on the Geological History of the Continental Lithosphere'*, London, March 1985, Abstr. Vol., pp. 8–9.

Ohiwerei, S. F. 1978. *Geochemical Indicators of Pegmatites in the Egbe Mining Area of Kwara State of Nigeria*. Unpublished M. Sc. Thesis, University of Ife, Nigeria.

Olade, M. A. 1980. Geochemical characteristics of tin-bearing and tin-barren granites. *Economic Geology*, **75**, 71–82.

Olade, M. A. and Elueze, A. A. 1979. Petrochemistry of the Ilesha amphibolites and Precambrian crustal evolution in the Pan-African domain of SW Nigeria. *Precambrian Research*, **8**, 303–318.

Pastor, J. and Ogezi, A. E. 1986. New evidence of cassiterite-bearing Precambrian Basement rocks of the Jos Plateau, Nigeria — the Gurum case study. *Mineral. Deposita*, **21**, 81–83.

Plimer, I. R. 1980. Exhalative Sn and W deposits associated with mafic volcanism as precursors to Sn and W deposits associated with granites. *Mineral. Deposita*, **15**, 275–289.

—— 1984. The role of fluorine in submarine exhalative systems with special reference to Broken Hill, Australia. *Mineral. Deposita*, **19**, 19–25.

Priem, H. N. A., Boelrijk, N. A. I. M., Hebeda, E. H., Verdurmen, E. A. Th., Verschure, R. H., and Bon, E. H. 1971. Granitic complexes and associated tin mineralizations of 'Grenville' Age in Rondonia, Western Brazil. *Geological Society of America Bulletin*, **82**, 1095–1102.

Raeburn, C. 1924. The tinfields of Nassarawa and Ilorin Provinces. *Geological Survey of Nigeria Bulletin*, **5**, Kaduna.

——, Bain, A. D. N., and Russ, W. 1927. The tinfields of Zaria and Kano Provinces; tinstone in Calabar District. *Geology Survey Nigeria Bulletin*, **11**, Kaduna.

Rahaman, M. A. 1978. Review of the basement geology of southwestern Nigeria, In: Kogbe, C. A. (Ed.), *Geology of Nigeria*, Elizabethan Publ. Comp., Lagos, 41–58.

——, Emofurieta, W. O., and Caen-Vachette, M. 1983. The potassic-granites of the Igbeti Area: further evidence of the poly-cyclic evolution of the Pan-African Belt in Southwestern Nigeria. *Precambrian Research*, **22**, 75–92.

——, Van Breemen, O., Bowden, P., and Bennett, J. N. 1984. Age migrations of anorogenic ring complexes in Northern Nigeria. *Journal Geology*, **92**, 173–184.

Routhier, P. 1980. *Ou sont les metaux pour l'avenir?* Memoir BRGM no. 105, Orleans.

Schätzl, L. 1971. *The Nigerian Tin Industry*. Unpublished Rep. Nigerian Inst. Soc. Econ. Res., Ibadan, Nigeria.

Scheidegger, A. E. and Ajakaiye, D. E. 1985. Geodynamics of Nigerian shield areas, *Journal African Earth Science*, **3**, 461–470.

Setter, J. R. D. and Adams, J. A. S. 1985. Rare-element mineralization in the Quitman Mountains — Sierra Blanca Igneous Complex, Trans-Pecos Texas. *6th Int. Conf. Basement Tectonics Santa Fe/ N. Mexico, Abstr. Vol.*, 33–34.

Sibuet, J.-C. and Mascle, J. 1978. Plate kinematic implications of Atlantic Equatorial Fracture Zone Trends. *Journal of Geophysical Research*, **83**, B7, 3401–3421.

Simpson, P. R. and Hurdley, J. 1985. Relationship between metalliferous mineralization and Sn–U–F-rich mildly alkaline high heat production (HHP) granites in the Bushveld Complex, South Africa. *Proc. IMM Conf. 'High heat production (HHP) granites, hydrothermal circulation and ore genesis', St. Austell, Sept. 1985*, Inst. Mining Metall., London, 365–382.

Snelling, N. J. 1964. *Overs. Geol. Surv. Ann. Rep.* **1963**, 113 pp.

Stavrov, O. D., Stolyarov, I. S., and Iocheva, E. I. 1969. Geochemistry and origin of the Verkh–Iset granitoid massif in central Ural. *Geochem. Intern.*, **6**, 1138–1146.

Tattam, C. M. 1944. The occurrence of cassiterite in Ife–Ilesha Division. *Geological Survey of Nigeria Annual Report*, **1943**, Kaduna.

Tauson, L. V. 1984. The geochemistry of Precambrian bedrocks, *Journal of Geochemical Exploration*, **21**, 487–501.

Taylor, B. E. and Friedrichsen, H. 1983. Light stable isotope systematics of granitic pegmatites from North America and Norway. *Isotope Geoscience*, **1**, 127–167.

Tischendorf, G. 1977. Geochemical and petrographic characteristics of silicic magmatic rocks associated with rare-metal mineralization. In Stemprok, M., Burnol, L. and Tischendorf, G. (Eds.), *Metallization Associated with Acid Magmatism*, vol. 2. Ustredni Ustav Geologicky, Prague. 41–98.

Tubosun, I. A. 1983. *Geochronologie U/Pb du Socle Precambrien du Nigeria*. Unpubl. Ph. D. Thesis, Universite du Languedoc, Montpellier, France.

Tugarinov, A. I., Knorre, K. G., Shanin, L. L., and Prokofieva, L. N. 1968. The geochronology of some Precambrian rocks of southern West Africa. *Canadian Journal Earth Science*, **5**, 639–642.

Turner, D. C. 1986. Magma distribution and crustal extension in the Nigerian Younger Granite province: evidence from the Wase area. *Journal African Earth Science*, **5**, 243–247.

Umeji, A. C. and Caen-Vachette, M. 1984. Geochronology of Pan African Nassarawa Eggon and Mkar–Gboko granites, S. E. Nigeria. *Precambrian Research*, **23**, 317–324.

Van Breemen, O., Pidgeon, R. T., and Bowden, P. 1977. Age and isotopic studies of some Pan-African granites from north-central Nigeria. *Precambrian Research*, **4**, 307–319.

Varlamoff, N. 1972. Central and West African rare metal granitic pegmatites, related aplites, quartz veins and mineral deposits. *Mineral. Deposita*, **7**, 202–216.

Wagner, B. M. 1980. *Zur Geochemie des Amphibolitkomplexes im präkambrischen Basement von SW-Nigeria, östlich von Ile–Ife*. Unpubl. Ph. D. Thesis, Universität München, F. R. Germany.

Wedepohl, K. H. (Ed.) 1969–1978. *Handbook of Geochemistry*. Springer-Verlag, Heidelberg.

Wright, J. B. 1970. Controls of mineralization in the Older and Younger Tin Fields of Nigeria. *Economic Geology*, **6**, 945–951.

—— 1985. *Geology and Mineral Resources of West Africa*. G. Allen and Unwin, London.

Section 4
Phanerozoic anorogenic magmatism: plate tectonic implications and mineralization

Phanerozoic alkaline magmatism on the African continent is considered in this part of the book. The scene is set by Kinnaird and Bowden who consider the implications of the Pan-African orogeny to Phanerozoic subvolcanic complexes for providing source materials for magmas and mineralization by assimilation and/or melting, and assess the isotopic constraints on magma generation and the source of mineralizing fluids throughout the African plate.

The African continent consists of a number of terranes accreted during major orogenies. Phanerozoic fold belts in Africa are limited to the northwest and southwest fringes. Over the African shield as a whole therefore, the Phanerozoic plutonism is anorogenic in nature and dominated by the formation of alkaline ring complexes. Prior to 180 Ma ago, Africa formed part of Gondwanaland. It was the exploitation and reactivation of the Pan-African shear zones and transcurrent faults during the period leading up to and following the fragmentation of Gondwana, that controlled the locations of Phanerozoic intraplate magmatism in Africa. Many of these intersecting shear zones and transcurrent fault systems formed a network of NE-SW and NW-SE lineaments which transect Africa. These structures have been reactivated at different times throughout the Phanerozoic and dictate the location of the alkaline igneous complexes and their mineralization.

The earliest anorogenic activity which began at the end of the Pan-African orogeny, was confined to a relatively short period of geological time and shows no significant widespread mineralization throughout the African plate. Where reactivation occurred in the Palaeozoic the alkaline ring-complexes are more economically mineralized. It is within the group of alkaline ring-complexes of Mesozoic age that major mineralization is found. The initiation and location of these centres is related to the separation of South America and Africa. There is widespread anorogenic activity centred along major lineaments and reactivated shear zones throughout Africa. A group of late Mesozoic to Cainozoic volcanic and subvolcanic activity is mainly connected with re-exploitation of zones of faulting and rifting. Some of these subvolcanic centres are also mineralized. This magmatism is almost exclusively restricted to areas of the convection-generated

© 1987 by John Wiley & Sons, Ltd.

stress fields and much of it is controlled by sites of former Pan-African tectonothermal activity.

From the economic viewpoint it is the granite/syenite complexes and carbonatite centres that are the most important. Economic elements in carbonatite complexes include Nb, F, P, Mn, Sr, Ba, and LRee. In certain cases there are also enrichments in V, Cu, Zn, Mo, Pb, Th, and U. Within the granite/syenite complexes elements of economic concentration are Nb, Sn, W, Zn, with Mo, Cu, Pb, Ree, U, and Th enrichment. In terms of the geochemistry and mineralization, there are many parallels between such alkaline granite/syenite complexes and carbonite centres, particularly with the cause and effects of alkali metasomatism and especially with the abundance of columbite, pyrochlore, zircon and complex Ti–Zr silicates, uranium/thorium abundances, rare earths, and oxide–sulphide assemblages.

Orogeny in Africa has provided a variety of source materials for Phanerozoic magmatism and mineralization. There is a considerable range in age, chemical composition of the plutonism, and variety in the styles of mineralization and ore deposition.

The movement of the African plate in Palaeozoic times based partly on studies completed in the Palaeozoic alkaline ring complexes from Aïr (Niger) coupled with palaeomagnetic data from elsewhere, is considered by Van Houten and Hargraves. Currently available Palaeozoic palaeomagnetic data from Gondwanaland can be interpreted in terms of either a migration of the pole from northern Africa to southern Africa between Ordovician and late Palaeozoic times, or a rapid excursion of the pole from northern Africa to southern Africa during Upper Ordovician to Lower Silurian times, followed by a return to central Africa in late Devonian times, thereafter continuing southwards again. To support the second hypothesis, the authors discuss pertinent stratigraphical evidence from western Gondwanaland including the distribution of glacial deposits and cold and warm-water faunas. This record, although meagre and to some extent contradictory, appears to favour a drift history consistent with a rapid southerly excursion of the pole by Lower Silurian times.

This section continues with a detailed discussion by Bonin, Platevoet and Vialette on alkaline magmatic provinces which occur globally in regions subjected to cratonization. The authors show that at the end of any major orogeny alkaline magmatism developed when relative plate motions became stationary. They illustrate their arguments with examples from the Corsica alkaline province as part of a larger Western Mediterranean province, emplaced at the end of the Hercynian orogeny. The author's conclusions are then applied to West Africa. Bonin *et al.* conclude that Pan-African domains have been the sites of later alkaline provinces, during various Phanerozoic time periods since the end of the Pan-African orogeny in which subsolvus granites initially predominated then hypersolvus granites superseded them during various Phanerozoic periods of magmatism. Finally with thickening of the continental lithosphere mixed and undersaturated complexes developed particularly along the rift zones, with increased proportions of carbonatites.

The Nigerian anorogenic province is recognized as one of the well researched areas for classical A-type granitic ring complexes with clear examples of hydrothermal alteration and mineralization. There are numerous geochemical studies which have been completed in Nigeria over the past decade. Some of the implications of the geochemical data are reviewed in this volume by Batchelor and co.

The section concludes with an ore petrology and microprobe study of mineralization in Nigeria associated with sodic, potassic, and acid metasomatism specifically

SECTION FOUR PHANEROZOIC MAGMATISM

in the Ririwai complex. According to Ixer and his coworkers, the biotite granites at Ririwai have undergone considerable post-crystallization alteration and mineralization. Investigations have revealed that the opaque and accessory minerals of Zr, Hf, U, Th, Nb, Ta, Ti, Sn, and the REE recrystallized in various forms at different stages during metasomatic changes in biotite granites. Ixer *et al.* also describe late stage changes which have precipitated Zr, U, Pb, Nb, Ta, and Ti as complex intergrowths of uranium-bearing plumbopyrochlore, columbite, ilmenite, rutile, anatase, and zircon. The textural and chemical evidence suggests that each metasomatic alteration process essentially dissolved or replaced earlier formed mineral phases and then reprecipitated them with new compositions compatible with the evolving fluids at lower temperatures.

African anorogenic alkaline magmatism and mineralization — a discussion with reference to the Niger-Nigerian Province

Judith Kinnaird and Peter Bowden
Department of Geology, University of St Andrews, Fife, Scotland

---

The close of the Pan-African was characterized by a change from subduction and arc related magmatism to post tectonic alkaline magmatism in Saudi Arabia and West Africa. For the remainder of the Phanerozoic, folding and tectonism was more restricted in Africa than on any other continent and was limited to the Caledonian, Hercynian and Alpine orogenies in the extreme north, and the Cape fold belt at the southern tip of the continent, with an equivalent orogenic belt on the extreme east of the Arabian peninsula. Thus Phanerozoic magmatism was dominantly anorogenic in nature and alkaline in character throughout the African Plate. Such magmatism was initiated by reactivation of deep-seated lineaments with the peak of activity during the Mesozoic, which can be correlated with fragmentation of Gondwanaland. The Phanerozoic alkaline magmatism is characterised by small centres of subvolcanic to plutonic nature often in the form of ring complexes. These can be subdivided into two separate associations consisting of Provinces dominated by centres of oversaturated magmatism such as Nigeria and Sudan. There are also provinces with undersaturated complexes and carbonatites with oversaturated magmatism such as in Namibia, Angola, and the East African rift system. Despite the contrasting nature of undersaturated and oversaturated magmatism many comparisons can be drawn between carbonatites and granitic complexes. Such comparisons include the source of the magmas, the mechanisms of emplacement, the cause and effect of alkali metasomatism and processes of hydrothermal alteration, the trace element geochemistry, the abundance of zircon, uranium, thorium and complex Ti-Zr silicates, the importance of columbite, pyrochlore, fluorine, and rare earth enrichment in mineralization, and in their current and future economic potential. The common feature is the development and reactivation of deep-seated lineaments which were not only controlled the emplacement of anorogenic complexes but also acted as channelways for hydrothermal mineralizing fluids. The development of a fluid phase during crystallisation occurred in both undersaturated and oversaturated alkaline rocks. Subsequent hydrothermal modifications, subsolidus reactions, geochemical variations and mineralization were dependent upon whether this fluid phase was retained or expelled.

KEY WORDS   Alkaline granites Carbonatites Africa Structural setting Geochemistry Mineralization Phanerozoic Cameron Niger Nigeria Sudan Saudi Arabia Namibia

1. Introduction

Most of the African continent consists of various terranes of different Precambrian ages accreted during major orogenies. The Pan-African orogeny in late Proterozoic, played a major role in re-assembling old cratons and terranes to form the

0072–1050/87/TI0297–44$22.00
© 1987 by John Wiley & Sons, Ltd.

supercontinent Gondwanaland. The close of the Pan-African was characterized by a change from subduction and arc related magmatism to post tectonic alkaline magmatism in early Phanerozoic in Saudi Arabia (Drysdall et al. 1986) and West Africa (Liegeois et al, this volume).

Recent detailed investigations in Saudi Arabia (Drysdall et al. 1986) has shown that the oversaturated province of end-Pan African age consists of a number of accreted terranes. These terranes can be modelled from their granitic types and from the basement into which they were emplaced. During the first stage of evolution, 900–680 ma ago, magmatism was dominated by intermediate plutonic rocks of tonalitic and trondhjemitic affinities. Approximately 680 to 600 Ma ago there was a switch from arc magmatism to collision related magmatism resulting in the formation of granodiorites and monzogranitic compositions up to 610 Ma ago. There was then a rapid shift from collision related magmatism around 610 Ma to the formation of anorogenic peraluminous to peralkaline alkali-feldspar granites throughout the Saudi Province in the period from 610–510 Ma (Stoeser 1986). Thus the most important feature of this region is the sequential development of granitic rock types ranging in composition from calcic-alkali through alkali-calcic (calc-alkaline) to alkaline. The model for the temporal evolution of the Saudi Arabian region during Upper Proterozoic to early Phanerozoic times is similar to the formation of the Iforas Province in West Africa.

In West Africa (Liegeois et al. this volume) have shown how magmatism changed with time and source during the late Proterozoic. A large composite batholith was emplaced >620 Ma ago close to the Pan-African suture between the 2000 ma old West African craton and the Trans-Saharan mobile belt (Liegeois et al, this volume). The collision which occurred between 620 and 580 Ma was accompanied by the intrusion of abundant granitoids. The post tectonic stages between 580 and 540 Ma are characterised by strike-slip movements, reversals in the stress field and a rapid switch from calc-alkaline to alkaline magmatism. According to Liegeois et al. the calc-alkaline magmas originated from the classical subduction source of depleted upper mantle, which was modified by hydrous fluids of the subducted ocean plate. In contrast the post-collision alkaline magmatism was produced from an asthenospheric source.

Thus the studies in Saudi Arabia and West Africa have monitored the change in magmatism related to changes in tectonism and shifting sources of magma origin. At the close of the Pan-African there was a change to anorogenic magmatism and small alkaline complexes were widely emplaced in both Saudi Arabia and West Africa. These complexes have been termed A-type granites of short time duration by Bowden (1985) to distinguish them from later A-type granites which were related to reactivation along deep seated lineaments.

A-type granites were emplaced throughout the Phanerozoic over the whole African Plate (Figure 1). Within the same time span — and often within the same province — undersaturated magmatic centres, sometimes with carbonatite complexes were also emplaced. The Phanerozoic alkaline magmatism is characterised by small centres of subvolcanic to plutonic nature often in the form of ring complexes. These can be subdivided into two separate associations:
  (i) Provinces dominated by centres of oversaturated magmatism
  (ii) Provinces with undersaturated complexes and carbonatites with oversaturated magmatism.

The anorogenic centres consist of multiple intrusions of different petrological facies and often occur as ring complexes. Such complexes include ring dykes,

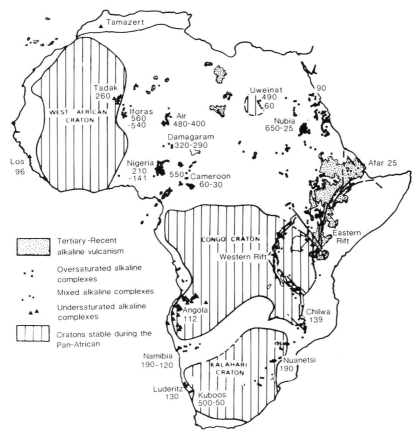

Figure 1. Distribution of alkaline magmatism in Africa (after Black, 1984). For details on Arabia see Figure 6.

partial ring dykes and cone sheets. Outcrop patterns are arcuate, annular, rectilinear and elliptical with varying diameters ranging from less than 1 km to 30 km or greater. The majority of ring complexes represent the eroded roots of volcanoes and their calderas. They have structural aspects which change with the depth of erosion so that depending on the age of the various massifs different structural features such as sheeted units, migrating cauldrons, resurgent calderas, and bell-jar intrusions can be observed. Even within one province the alkaline ring complexes can show a wide variety of structural forms.

It is now widely accepted that carbonatite magmas have originated in the mantle and rock types found associated with some alkaline complexes such as kimberlites point to a deep source within the asthenosphere. In contrast the source of the A-type granites has been attributed to both mantle or crustal origins. Recent oxygen, sulphur, neodymium and samarium studies combined with Sr and Pb isotopic data indicate that the Nigerian A-type granites have both a mantle, lower and upper crustal component. If this is the case for other alkaline centres, then the consistency in the rock types, the mineralogy, geochemistry, ore distribution, type and abundance between A-type complexes in Africa seems all the more remarkable.

2. Sequential development of alkaline magmatism in Africa

Alkaline magmatism has occurred over the whole African Plate throughout the Phanerozoic. There were several major periods of activity:
  (i) Palaeozoic
  (ii) Mesozoic
  (iii) Mesozoic-Cainozoic

(i) The earliest Phanerozoic anorogenic activity began at the end of the Pan-African orogeny. Magmatism, which was confined to a relatively short period of geological time, resulted in the formation of high-level ring complexes which are predominantly alkali in nature. These intrusives were linked in time and space to the closing phases of the Pan-African event and accompanied the uplift which followed it. In Niger-Nigeria and also Sudan a more or less continuous sequence of magmatism and mineralization continues into the Mesozoic. The Trans-Saharan belt granites range from 585 to ca 505 Ma and a similar age range is observed in northeast Africa (Cahen et al. 1984).

(ii) The second group is of Mesozoic age. The initiation and location of these centres was related to the fragmentation of Gondwanaland and the separation of South America from Africa. There was widespread anorogenic activity centred along major lineaments and reactivated shear zones throughout Africa.

(iii) A third group of late Mesozoic and Cainozoic volcanic and subvolcanic activity was mainly connected with reactivated zones of faulting, uplift and rifting. The alkaline magmatism predates the basaltic and related lavas associated with the evolution of rift valleys from Cretaceous times onwards.

There is no distinction in age between the two types of provinces, thus oversaturated and mixed provinces can be found within any of the above age ranges. For example, early Phanerozoic undersaturated magmatism can be found in the reactivated Western rift within the same age span as the oversaturated magmatism of the Iforas oversaturated province in West Africa. Similarly the Mesozoic oversaturated magmatism of Nigeria has an almost identical age range to the mixed province in Namibia, with oversaturated magmatism close to the coast and undersaturated magmatism up to 100 km inland. More recent magmatism of late Mesozoic to Cainozoic age shows evidence of mixed province development in the Red Sea Hills in contrast to solely oversaturated magmatism in Cameroon.

It is also notable when examining the age ranges within the differing provinces that magmatism may have occurred during more than one period in any one province. For example, in the Nubian Province of Sudan, the Uweinat Province of Libya and the Cameroon Province of West Africa, magmatism can have occurred at widely spaced intervals. Also in some of the provinces, e.g. Sudan, there was a trend from oversaturated towards undersaturated magmatism with time. What is notable however, is that the Palaeozoic-Mesozoic range of ages in the oversaturated provinces of northern Africa, is not extensively developed south of the Sahara.

Figure 1 shows that the Phanerozoic oversaturated magmatism occurs exclusively in Pan-African domains whereas undersaturated magmatism, ignoring Tertiary to Recent volcanism, tend to be located on the cratons or in the southern prolongation of the Mozambique belt.

3. The structural setting of alkaline anorogenic magmatism

The cause of the Phanerozoic intraplate alkaline magmatism throughout Africa within the different time periods discussed above has been actively debated in the literature (Figure 2). The plume hypothesis first advocated by Morgan (1971) was used initially by Duncan et al. (1972) and later by numerous other workers to explain the linear arrays of alkaline igneous activity. However Bailey (1977) pointed out that the chemistry volume relationships as well as the recurrence of magmatic activity in the same regions at different time periods was not consistent with the plume concept. Instead he preferred that arching, volatile fluxing related to continental uplift was responsible for alkaline magmatism. More recently the idea that mantle metasomatism has greatly enriched the mantle in fugative constituents, alkalis, and the incompatible elements provides an ideal source for some of the alkaline magmas.

The Phanerozoic plutonic alkaline complexes which were emplaced within the areas of regional swell structures were not randomly distributed but their locations were controlled by the exploitation and reactivation of Pan-African shear zones and transcurrent faults during the period leading up to and after fragmentation of Gondwanaland. Many of these intersecting shear zones and transcurrent fault systems formed a network of NE–SW and NW–SE lineaments which transect Africa (Furon 1963). Such deep seated lineaments provided channelways not only for the rise of magma but they also acted as pathways for the movement of ore-rich hydrothermal fluids. These structures have been reactivated at different times throughout the Phanerozoic and dictated the location of the alkaline igneous complexes and their mineralization. In some cases discussed later, reactivation during the Mesozoic and Tertiary resulted in mineralization of complexes emplaced in late Proterozoic and early Phanerozoic.

Therefore, there is an overall structural relationship between the structural pattern established during the Pan-African and most of the Phanerozoic evolution of the continent. The reasons for reactivation changed with time and before considering individual provinces in detail it is essential to consider the structural constraints on magmatism during the different Phanerozoic periods.

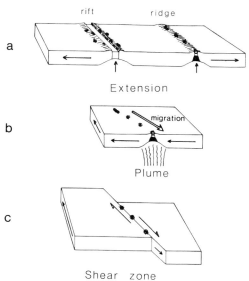

Figure 2. Schematic controls of alkaline magmatism (after Bonin, 1982). (a) effects of extensional magmatism, causing linear doming expressed as a ridge or ultimately as rifting; (b) hot spot model generated by a mantle plume; (c) alkaline magmatism located along shear zones subjected to reverse movement (harpoon effect, Black et al. 1985).

(i) Early Phanerozoic

The Pan-African orogeny is regarded as a major event in the reconstruction of Gondwanaland and marked the close of the Precambrian and the beginning of the Phanerozoic. The earliest anorogenic activity began therefore at the end of the Pan-African orogeny in West Africa and Saudi Arabia reactivating major lineaments.

Recent models for the evolution of the Arabian Shield indicate its development through a series of orogenies associated with collisional events along suture zones between five microplates (Stoeser and Camp 1984). The fundamental control of major tectonic lineaments and fractures on localising magma generation, ascent and emplacement can be well established. Belts of plutonic rocks and linear arrays of individual plutons occur but their structural relationships remain largely unexplained. According to Agar (1986) the late-stage alkaline magmatism, is post-tectonic relative to regional deformation but syn-tectonic relative to shear zones. Rectilinear contacts are common, further suggesting fracture control of emplacement. Agar (loc. cit) also notes that in Saudi Arabia, plutons in the accreted terranes may be emplaced in one orogeny and affected by a subsequent event.

In the Iforas Province of West Africa, Liegeois et al. (this volume) have been able to relate changing stress fields to successive stages of magmatism. The oblique collision between the stable West African craton and the Trans-Saharan mobile belt in West Africa, which occurred between 620 and 580 Ma, was marked by granitoid emplacement at the end of a D2 east-west compressional event. The post collission D3 deformation 580-540 Ma ago marked a reversal in the stress field and now corresponds to a northeast-southwest shortening direction. After uplift, the D3-D4 event resulted in a complete reversal in direction of maximum compressive stress and a rapid switch from calc-alkaline to alkaline magmatism. (Black et al. 1985, Liegeois et al, this volume.)

(ii) Mesozoic

Prior to 180 ma ago, Africa formed part of Gondwanaland. It was the fragmentation of Gondwanaland in Mesozoic times that was responsible not only for the reactivation of lineaments in former Pan-African mobile terranes but also for the formation of oceanic transform systems.

Along the continental margin of the western edge of Africa there are a series of northeast-southwest lineaments which may continue into oceanic transform systems (Sykes 1978). It was Marsh (1973) who first proposed that the southwest African and south American alkaline igneous provinces can be correlated with the initial break-up of Gondwanaland. This proposal was based on the fact that the lineaments lie on small circles centred on a Cretaceous pole of rotation which gave rise to the transform fault systems defined by Francheteau and Le Pichon (1972). An example of one of these lineament systems is the Namibian Province of Jurassic to Cretaceous age. It spans a similar age range to the oversaturated complexes of Nigeria.

The Namibian Province (190-120 ma) comprises of a series of oversaturated and undersaturated complexes aligned along northeast trending lineaments coinciding with the axis of the central zone of the Damaran orogenic belt of late Pan-African age. The separation of southern Africa from South America resulted in the reactivation of pre-existing northeast trending lineaments with the formation of incipient grabens. These tectonic disturbances appear to be important not only for the emplacement of anorogenic complexes but also because they have provided active channels for metasomatic mineralizing fluids (Pirajno and Jacob 1987). The ring complexes have been grouped by Prins (1978) into:

(i) granitic types (Brandberg, Erongo, Gross and Klein Spitzkoppe)
(ii) differentiated basic complexes (Cape Cross, Doros, Messum and Okonjeje)
(iii) peralkaline complexes (Paresio and Etaneno)
(iv) carbonatite complexes (Ondurakorume, Osongombo, Okorusu and Kalkteld)

The oversaturated complexes occur towards the western coast in a zone characterised by abundant Pan-African granites, whilst the more undersaturated alkaline centres developed up to 100 km inland. Similar associations are also noted in the Angolan province and the Luderitz province but the proportions of oversaturated alkaline complexes is extremely limited. Nevertheless the majority of the igneous complexes near the Atlantic coast of southwestern Africa show age ranges consistent with the sequential opening of the south Atlantic. Thus the ring complex provinces of Angola, Namibia and Luderitz (Figure 1) all lie on narrow zones extending from the coast for several hundred kilometres inland. Their development, distribution and petrographic features differ from the complexes associated with rifting. The compositional variation is different for each lineament array in the proportion of oversaturated to undersaturated complexes, although kimberlites and carbonatites occur in each zone thus increasing the economic potential of the region. Similar associations are recorded along the major northwest-southeast lineaments in South America.

(iii) Mesozoic-Cainozoic

In contrast to the continental margin magmatism linked to transform systems, intraplate magmatism within the African continent is concentrated along zones of rifting. The orientation of these rift systems were themselves controlled either by ancient structures within the Mozambique and Limpopo fold belts or by lineaments within cratons.

Magmatism is almost exclusively restricted to areas of the convection generated stress fields (Liu 1980) and much of it is controlled by sites of former Pan-African tectono-thermal activity, one important exception being the eastern rift valley and associated volcanics, which is essentially superposed on an area cratonic in relation to the Pan-African activity, another exception being the Jebel Uweinat Massif (Libya). The other cratons are devoid of Cainozoic magmatism.

There is thus an overall relationship between the structural pattern established during the Pan-African and most of the Phanerozoic evolution of the continent. Kennedy (1965) pointed out this relation in connection with the Mesozoic-Cainozoic coastal basins of Africa, formed by faulting and rifting which developed into the breaking up of Gondwana. This relationship can now be extended to the western rift so that most of the swells of present day Africa are on sites of former Pan-African activity whereas most of the basins are situated on cratons of that time. It is suggested that post Pan-African uplift produced doming and tension which in turn produced faulting and rifting, favourable to the rising and extrusion of lavas. The East African rift system follows Pan-African lineaments as well as fracturing across some craton (Chorowicz et al., this volume). When the rift system crosses cratons the rift features are not always obvious.

Thus alkaline magmatism may or may not be associated with rifts but is often accompanied by crustal doming in an extensional regime (Figure 2).

Strike-slip movement and reactivation of transcurrent faults plays a major role as well as transformed faults whose position were determined by older lineaments on the continents (Sykes 1978). It can be concluded that pre-existing faults determine the location of alkaline magmatism, and that their reactivation is caused by lithospheric stress related to changes in plate motion, orogenic events along plate margins, and fragmentation of Gondwanaland. The variations in the degree of alkalinity and silica saturation of alkaline provinces may be related to the age and thickness of the continental lithosphere. The composition of this lithosphere also plays a role in determining the economic potential of the alkaline provinces.

4. Structural setting of the alkaline complexes

In Africa, dominantly oversaturated alkaline provinces have been identified in Niger, Nigeria, Cameroon, Arabia, Sudan, Egypt and Ethiopia (Figure 1). These oversaturated complexes, which range in age from the Precambrian-Cambrian boundary to the Tertiary period, occur exclusively in Pan-African domains. In most of these oversaturated alkaline provinces the subvolcanic intrusions are dominantly syenitic to granitic in composition with basic rocks occupying 5 per cent or less of the total area. Small proportions of basic rocks are found in Niger (Aïr), and to a limited extent in the Cameroon (Mboutou) where gabbros are associated with layered sequences including leucogabbros, anorthosites and monzo-anorthosites (Husch and Moreau 1982, Jacquemin et al. 1982).

There are similarities in the structural features of all the anorogenic alkaline centres whether they are undersaturated or oversaturated. They are exposed at all levels of erosion from volcanoes to the root zones. The plutons are characterised by their discordant nature and contacts that are sharply defined. The intrusives are generally homogeneous, unfoliated and porphyritic facies are common. The subvolcanic nature of many centres is demonstrated by their circular outcrop and association with nested ring complexes and ring dykes which are often polygonal in nature.

Most of these oversaturated plutonic provinces are composed of chains of complexes often arranged concentrically as roots of volcanoes. In some instances the outer limits of the complexes are defined partly or completely by a ring dyke. This intrusion, generally composed of granite porphyry, is the chief structural element of the complex controlling the distribution of both volcanic and subvolcanic magmatism at high levels in the crust. A typical cross-section through an alkaline ring complex is shown in Figure 3.

It illustrates the form of the volcanic and subvolcanic products found at different erosional levels as well as the types and styles of extrusion and intrusion. At certain erosional levels the volcanic products are preserved through caldera collapse following the classical caldera model based on Glen Coe. However at some centres there is abundant evidence for resurgent caldera formation (Valles caldera model) so that the updomed volcanic cover has been eroded leaving a boss-like promontory of biotite granite and its associated mineralization. In certain cases the granitic magma has not erupted onto the land surface but has been confined to a series of sheets and ring fracture systems often referred to as underground cauldron formation. Where this has occurred trapped fluids can be retained within the complex possibly leading to enhanced mineralization potential.

Where the volcanic rocks have been preserved by downfaulting of the caldera the magmatic evolution of anorogenic complexes can be clearly established from

AFRICAN ANOROGENIC ALKALINE MAGMATISM

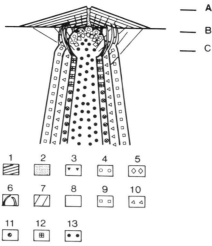

Figure 3. Cross-section of an oversaturated alkaline complex. African alkaline centres can be exposed to different levels shown by A, B, and C. (1) volcanic pile consisting of minor basalts, with comenditic and pantelleritic trachytes and rhyolites; (2) quartz-porphyry volcanic feeder; (3) greisen zone in biotite granite cupola; (4) albite zone; (5) microclinite zone; (6) ring dykes; (7) cone sheets; (8) country rock; (9) syenite; (10) fayalite granite; (11) arfvedsonite granite; (12) arfvedsonite albite apogranite; (13) biotite granite.

preserved volcanic sequences. Although the majority of volcanics are dominantly rhyolitic, some alkaline centres have preserved occasional successions of hawaiites, mugearites and trachytes confirming the dominantly alkaline trend from transitional ne-normative or hy-normative basalts. Igneous breccias containing a melange of acid and basic rocks indicate that both types of magma were available for mixing and eruption. In some instances mixing has occurred to produce volcanic suites of andesitic compositions with fragmented and broken feldspar and quartz phenocrysts.

For the undersaturated complexes along the major rift zones, all levels of erosion can be studied from the alkalic volcanoes (e.g. Oldoinyo Lengai) to the partially eroded centres (e.g. Napak in Uganda) to the well exposed root zones at Chilwa Island and elsewhere in Malawi (Woolley and Garson 1970). Each locality represents a chronostratigraphic horizon in Phanerozoic alkaline undersaturated magmatism (Figure 4) but the three geographically separated centres can be combined to examine different structural levels from volcanic edifice to eroded core. Based upon these three examples it is possible to speculate that modern volcanic products of carbonatite volcanoes are nephelinite, agpaitic phonolite, and natrocarbonatite: the corresponding subvolcanic equivalents in the cores and roots are ijolite, miaskitic nepheline syenite, and sovitic or alvikitic carbonatite. The carbonatites display all the principal modes of emplacement of subvolcanic rocks as well as occurring as lavas and pyroclastic deposits. Many carbonatites occur as igneous bosses and plug-like bodies with margins disturbed by attendant fenitisation and carrying xenoliths, often flow oriented, from the surrounding country rocks. Many other carbonatites occur as dykes, veins and cone-sheets with well-marked cross-cutting relationships and chilled margins. A typical cross section of a carbonatite-bearing alkaline ring complex is provided in Figure 4.

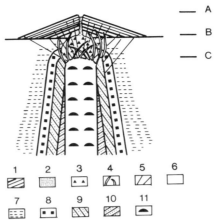

Figure 4. Cross-section through a carbonatite complex. African carbonatites represented by (A) Oldoinyo Lengai; (B) Napak; (C) Chilwa Island. (1) alkalic stratovolcano with phonolite, nephilenite; (2) natrocarbonatite; (3) breccia zone; (4) carbonatite ring dykes; (5) carbonatite cone sheets; (6) country rock; (7) fenitized country rock; (8) syenite fenite; (9) nepheline syenite; (10) ijolite; (11) carbonatite core.

The diagram provides an indication of the form of the volcanic and subvolcanic products found at different erosional levels, as well as the types and styles of extrusion and intrusion. The relationship between volcanism and subvolcanic intrusions can be observed in East and Central Africa. For instance carbonatites, ijolites and associated nepheline syenites are found in both the Western and Eastern Rift. In the Western rift and its immediate surroundings nepheline syenites, with or without carbonatites, are found from Kivu Province through Tanzania to Malawi (Figure 5).

5. Age variations in different oversaturated provinces

The anorogenic alkaline oversaturated plutonism occurred in the three distinct periods discussed earlier. Thus end-Pan African magmatism is characterised by A-type granites of short time duration in Arabia and the Iforas province of Mali; late Palaeozoic-Mesozoic anorogenic centres occur in Sudan, Niger and Nigeria; with Tertiary magmatism in the Cameroon.

The Arabian Shield consists of one of the largest Provinces of alkali granites in the world. Forty-nine major and more than a dozen minor alkali granite plutons are concentrated in the Midyan and Hijaz terranes and the Nabitah orogenic belt, alkali granites do not occur in the southwestern and easternmost parts of the Shield (Figure 6).

Where they occur the alkali granite pluton is the last major intrusive phase. Radiometric dating (Stoeser 1986) has shown that the alkali granites were emplaced within a time span of about 180 Ma (686–518) and all but four formed between 630 and 565 Ma. In addition to the alkali granite plutons, major peralkaline rhyolite dyke swarms occur in the northeastern Shield. Although there is no obvious distribution pattern, there seems to be a slight younging trend eastwards across the northeastern part of the Shield (Stoeser 1986). The alkali granites are spatially associated with other granites. They commonly occur in plutons and

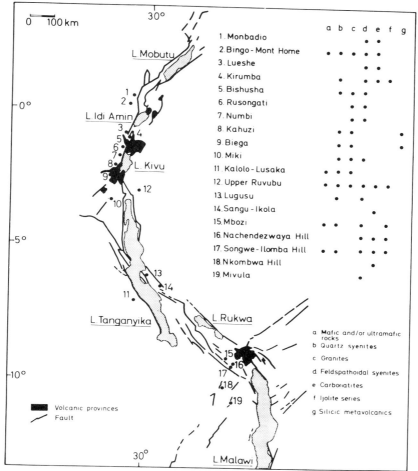

Figure 5. Geological sketch map of the complexes located along the same lineaments as the Western rift (after Tack et al. 1984).

complexes in which alkali granite forms an outer rim and metaluminous hypersolvus or subsolvus granite forms the core (Harris and Marriner 1980, Radain 1981, Stoeser and Elliot 1980).

The Iforas province of Mali in West Africa is composed of nordmarkites, peralkaline and metaluminous granites and granite porphyries and has been dated by Liegeois and Black (1983, 1984) and Ba et al.(1985). Liegeois et al, this volume describe the alkaline complexes to be about 15 in number typically varying in size from 20 to 30 km in diameter. Whole-rock Rb/Sr analyses have yielded Cambrian ages between 560 and 540 ma. Alkaline massifs of similar age have been recorded in northwest Hoggar and in southern Benin in the Pan-African belt close to the West African craton.

Both the Niger and Nigeria Province (Figure 7), has similar rock types although the relative proportion changed with time from north to south. Thus peralkaline granites and quartz syenites predominate in the Aïr where they may be associated with anorthosites, leucogabbros and lenses of iron-titanium oxides often as layered, funnel-shaped intrusions (Black 1965, Black et al. 1967, Husch and Moreau 1982, Leger 1985) whereas aluminous granites predominate in Nigeria. Even within Nigeria there is an overall change in lithological dominance from north to south with volcanics and peralkaline granites dominant in the north and biotite granites dominant in the south. The Niger-Nigeria Province ranges from lower Palaeozoic in Air (480–400 Ma) through Carboniferous in the Damagaram (320–290 Ma) to Mesozoic in Nigeria (215–140 Ma). During this period there has been a southerly shift of centres of magmatic activity recording a period of practically continuous within-plate activity.

The Nubian Province (Figure 8) comprises over 130 alkaline complexes which are predominantly in Sudan but which extend into Egypt and Libya in the north and into Ethiopia and Uganda in the southeast and south. The province includes alkali granites and syenites, rarer nepheline syenites and associated extrusive trachytes, rhyolites and ignimbrites (Vail 1985). As in the Niger-Nigeria province there is a petrological change with time. This is reflected in the degree of alkalinity and silica-saturation since the Palaeozoic complexes are generally oversaturated whereas undersaturated rocks and a carbonatite appear in the Mesozoic (Black et al. 1985). The Nubian Province comprises complexes which range in age from 650 to 25 Ma, the most recent being in the Afar (Vail 1985). There are five peaks of plutonism within this period (Black et al. 1985) and the time range is comparable with that of oversaturated alkaline magmatism in West Africa. However, there is no progressive change in age or distribution pattern of the episodic magmatism as in Niger-Nigeria, so that alkaline complexes of different ages occur within the same region. The random distribution of ages of the complexes is a result of reactivation of the regions over various periods of time possibly related to the ultimate opening of the Red Sea rift system.

The Cameroon oversaturated province is dominated by a number of Tertiary alkaline granite ring complexes (Lasserre 1978) emplaced into end-Pan African basement, and orientated along the north-northeasterly Cameroon line. Some of these centres like Poli, Mayo Darlé, are mineralized, while in others like Mboutou, and Goldes Velda (Jacquemin et al. 1982) there are rock associations reminiscent of the Air gabbro-anorthosite-syenite association. However, in contrast to the Niger-Nigeria province there is no sequential age progression of the ring structures but there is an apparent random range in age between 60 and 30 ma.

6. Petrological variatons in oversaturated complexes

Volcanic rocks have been preserved by downfaulting of the caldera in many centres. Although ignimbritic rhyolites dominated vulcanism, some centres erupted minor quantities of mildly alkaline basalt lavas which are intercalated with the acid pyroclastics. Some alkaline centres have preserved occasional successions of hawaiites, mugearites and trachytes confirming the dominantly alkaline trend from transitional ne-normative or hy-normative basalts. Igneous breccias containing a melange of acid and basic rocks indicate that both types of magma were available for mixing and eruption. In some instances mixing has occurred to produce volcanic suites of andesitic compositions with fragmented and broken feldspar and quartz phenocrysts.

The subvolcanic assemblages include fayalite hedenbergite granites, amphibole granites, albite rich- and ablite-poor aegirine arfvedsonite granites, and biotite

AFRICAN ANOROGENIC ALKALINE MAGMATISM

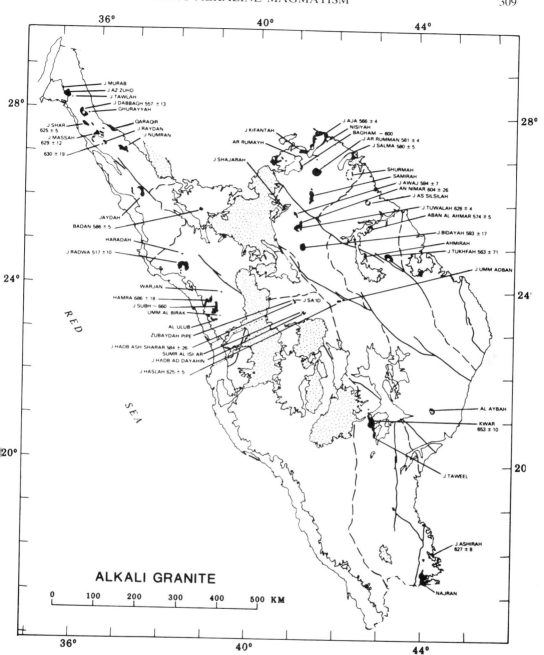

Figure 6. Geological sketch map of Arabia with location of the various alkaline granite complexes (taken from Stoeser 1986).

granites with minor gabbros, monzogabbros, anorthosite, syenites and nordmarkites.

As far as the granites are concerned there is a natural progression from volcanic feeder intrusions to subvolcanic intrusions with fayalite and hedenbergite. The volcanic feeder intrusions are a minor but important link during the caldera-forming stage between the subvolcanic roots and the overlying volcanic pile. They are represented by quartz porphyries and granite porphyries. Quartz porphyry

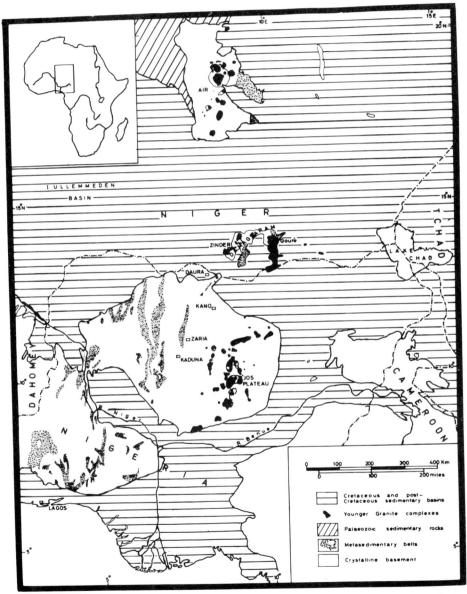

Figure 7. Geological sketch map of the Niger-Nigeria province showing the distribution and ages of the ring complex province.

Table 1. Ferromagnesian minerals in alkaline anorogenic granites

olivine	pyroxene	amphibole	mica
Fayalite Fa_{92-99}	ferro-augite hedenbergite ferrohedenbergite aegirine-hedenbergite aegirine	edenite ferro-actinolite ferrowinchite ferrorichterite arfvedsonite	biotite annitic biotite ferrous siderophyllite lithian siderophyllite protolithionite zinnwaldite cryophyllite lepidolite

often has ignimbritic textures but intrusive forms which are welded and consolidated in ring fissure/and or cone sheets. Arrested growth forms of hedenbergite, feldspar and quartz are often visible in thin section. Granite porphyry represents the vesiculated magma that followed behind the quartz porphyries emplaced into the ring fracture. In these rocks, variations in the composition and activity of the alkaline residual phase is observed by the appearance of Ca-Na amphibole replacing hedenbergite and fayalite, or a more sodic assemblage with arfvedsonite, aenigmatite, and riebeckite together with annitic mica. Often with more continued reactions with residual fluids, the porphyritic texture is lost, the groundmass mineralogy coarsens resulting in the formation of an equigranular texture of biotite granites and peralkaline granites as the principal rock types. Subdivisions of these alkaline granites and syenites can be made according to the plotting position of the salic constituents in the Streckeisen QAP modal diagram and the dominant but variable range of the ferromagnesian mineral types (Table 1).

There are three types of syenite which can be petrologically and geochemically distinguished: (i) Syenites with plagioclase phenocrysts (varying from An_{60} in the core to An_{30} at the rim) are recognised chemically by the presence of normative anorthite (>3 per cent). They also contain iron-rich olivine (Fa_{90}) and hedenbergite, and alkali feldspar either as separate grains or rimming the plagioclase phenocrysts. These syenites are often of minor occurrence but provide the link to alkali feldspar syenites from monzonitic compositions. (ii) Syenites with alkali feldspar phenocrysts contain interstitial sodic ferrohedenbergite, with fayalitic olivine partly destabilised due to the build up in alkalinity. (iii) There are also metasomatic syenites containing important concentrations of ordered microcline and/or albite. They are often referred to as microclinites or albitites. A characteristic feature is the presence of partially filled miarolitic cavities with spheres of calcite, fluorite and cassiterite. The occurrence of these metasomatic syenites is a good indicator of fluid activity and potential mineralization.

7. The effect of late stage fluids

The mineralogical assemblages of many of these complexes is often the result of reactions with residual fluids (Table 2). Fluids affect the late magmatic and particularly the postmagmatic (subsolidus) crystallization history of a cooling subvolcanic pluton and to some extent the overlying volcanic pile. Such hydrothermal alteration has undoubtedly modified the original granite chemistry and has been responsible for the introduction and dispersion of ore minerals which bear some similarities to economic concentrations in carbonatites. The sequence of

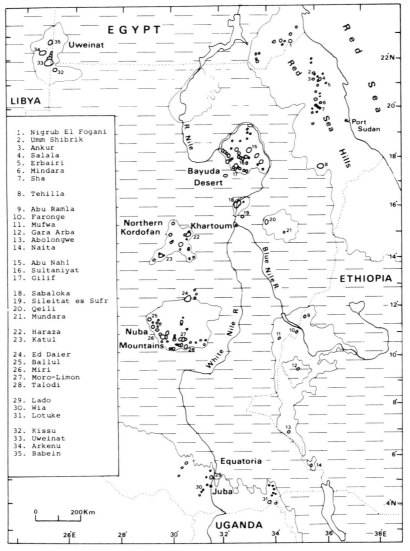

Figure 8. Alkaline complexes in the Nubian province which extends into Libya, Egypt, Ethiopia and Uganda (taken from Vail 1985).

subsolidus hydrothermal alteration processes that have modified the original mineralogy of the granites can be recognised by structural changes in alkali feldspar, by the growth of new micas, by local textural variations, by changing compositions of opaque minerals and by the assemblage of accessory minerals. Such changes have been described in detail for the Nigerian anorogenic granites by (Kinnaird 1985, Kinnaird *et al.* 1985). Those granites emplaced as ring dykes in polygonal or circular fractures with steep-sided contacts show limited hydrothermal alteration. It is only in the granite cupolas with shallow outward dipping contacts where fluids

Table 2. Metasomatic mineral assemblages associated with late stage fluids in oversaturated complexes.

Sodic peralkaline	Potash peraluminous	Acid	silica	late-stage
albite	albite	microcline	quartz	quartz	kaolinite
ferrorichterite-	zinnwaldite-	chlorite	zinnwaldite-	siderite	montmorillonite
Li-arfvedsonite	lepidolite	annite-	siderophyllite-	haematite	epidote
aegirine	zircon	siderophyllite-	annite		chlorite
aenigmatite	cryolite	zinnwaldite	sericite		siderite
amblygonite	epidote	lepidolite	chlorite		
astrophyllite	chlorite		topaz		
chevkinite	siderite		zircon		
narsarsukite			fluorite		

were retained, that metasomatic reactions were most intense (Bowden and Kinnaird 1984a and b).

In the Nigerian province there were four periods of hydrothermal alteration that were important economically; sodic metasomatism (albitization), potassic metasomatism (microclinization), acid metasomatism (greisenization) and silica metasomatism (silicification).

Similar processes of alteration have been described in Niger (Perez 1985), Cameroon (Nguere and Norman 1985), Sudan (Almond 1967, 1979; el Samani 1985), Arabia (Jackson 1986) and Egypt (el Ramly et al. 1971) although no detailed work on the order of alteration processes has been carried out in these provinces. Later lower temperature process of alteration such as hematization, chloritic, zeolitic or argillic alteration also occurred but they were more restricted in occurrence and limited in economic importance since no new ores were added at these stages.

In Nigeria, on textural and field evidence the earliest process of alteration was sodic metasomatism. This alteration process occurred before jointing or faulting of the granites and therefore its effects are disseminated through a granite roof zone. The actual mineral assemblages resulting from sodic metasomatism depended on the intensity of rock-fluid interaction and on the initial mineralogy of the original rock-type. Sodic metasomatism has occurred in both peralkaline and biotite granite cupolas.

In the peralkaline granites, which are are characterised by normative acmite and petrographically by blue sodic amphibole, a range of subsolidus modifications during sodic metasomatism can be inferred. The deep blue sodic amphibole, occasionally with green-blue cores together with aegirine, exhibit a wide variation in colour and texture. Arfvedsonite granites range from equigranular to porphyritic full of drusy cavities, and in colour from pink to white. With increasing modal content of albite they grade into arfvedsonite albite granites. The alkali amphiboles therefore belong to a continuously varying series, ranging from ferrorichterite to arfvedsonite (Figure 10).

Often amphiboles may show two distinct forms, one as euhedral prisms of ferrorichterite, the other as irregular interstitial masses of Li-arfvedsonite of late subsolidus recrystallization. Aegirine may replace or be replaced by arfvedsonite suggesting that there were fluctuating physico-chemical compositions in the subsolidus. Arfvedsonite and aegirine are accompanied by accessory aenigmatite and

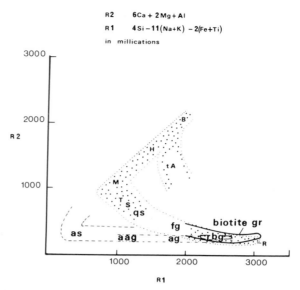

Figure 9. Cationic diagram R1-R2 for various African alkaline complexes. Volcanic trend B: transitional basalt, H: hawaiite, tA: trachyandesite, M: mugearite, T: trachyte, R: rhyolite; subvolcanic trends, S: syenite, qs: quartz syenite, fg: fayalite granite, rbg: riebeckite biotite granite, as: alkaline syenite, aag: arfvedsonite albite granite, ag: arfvedsonite granite.

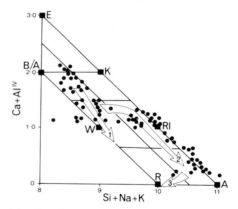

Figure 10. Amphibole compositional variations for alkaline granites and syenites. Adapted from O'Halloran 1985. W: winchite, Rl: richterite, A: arfvedsonite, R: riebeckite.

traces of annitic mica with secondary iron-titanium oxides. In addition to changes in amphibole composition during sodic metasomatism the original perthitic feldspars were progressively altered, first by the gradual coarsening of the albite and microcline domains within the perthite and ultimately by the growth of new albite laths.

In the biotite granites slight sodic metasomatism resulted in coarsening domains in perthite as described above, together with changing mica compositions. With increasing fluid reaction, a zinnwaldite albite granite was produced, or more rarely a lepidolite albite apogranite. In these extensively modified facies, thin sections show a 'snowball' texture in which abundant laths of ordered albite are surrounding and enclosing larger anhedral crystals of untwinned turbid intermediate microcline and unstrained alpha quartz. The zinnwaldite mica shows ragged skeletal prisms which vary in colour from pale honey through shades of pale blue or green to grey. Remnant cores of annitic or siderophyllite mica may be overgrown by zinnwaldite mica but it is not unusual to find the original mica has been completely replaced. Accessory minerals include topaz as a common accessory together with zircon and cryolite with the development of epidote, chlorite and siderite when alteration continued to low temperatures.

Sodic metasomatism was economically important for the formation of the niobium-bearing ore minerals — pyrochlore and columbite — together with accessory minerals enriched in rare-earths, thorium and uranium. Such mineralization occurs in Saudi Arabia and Nigeria.

Potash metasomatism which followed sodic metasomatism in Nigeria, may also be a disseminated process. In other instances it is restricted to localised alteration pods in the granite margins and immediately beneath the volcanic cover, whilst in the Ririwai and Tibchi complexes of Nigeria potash metasomatism affected fissure wallrocks with progressively less replacement of sodium by potash outwards into the granite. Mineralogically the process was characterised by microclinization of feldspar due to the partial or complete replacement of sodium by potassium in the feldspars of the biotite granites, often accompanied by conspicuous haematitic reddening. Such alteration resulted in a turbidity of the feldspars which resembles kaolinisation, but can be recognised as microcline on XRD studies. Feldspar alteration is linked to a growth of new mica in the compositional range annite through siderophyllite and zinnwaldite to lepidolite. Potash metasomatism is characterised by an oxide assemblage of ores dominated by cassiterite and wolframite, with accessory sulphides (Table 3). This process of alteration has not been described in detail for other provinces although as indicated later, the process may have occurred elsewhere but remain unrecognised. For example Almond (1967) describes feldspars that have been completely 'kaolinized' and then replaced marginally by zinnwaldite. Such kaolinization may actually be microclinization.

During acid metasomatism there was a progressive breakdown of 'granitic' minerals in response to changing K^+/H^+ ratios in the fluid. Perthite or microcline feldspar was destroyed and replaced by mica in the compositional range from zinnwaldite through siderophyllite and may be accompanied by sericite ± chlorite ± topaz ± fluorite ± cryolite. This mica-quartz assemblage development is often referred to as greisenization. Such greisenization may be locally pervasive within a granite roof zone e.g. in Niger, Nigeria and Saudi Arabia or may be concentrated along fissures and fractures in Niger, Nigeria, Cameroon, Sudan, Egypt and the Red Sea Hills. The petrological characteristics and the extent of the greisen development may depend upon the intensity of earlier stages of sodic and/or potash metasomatism. Thus in Nigeria it has been possible to recognise greisenized perthite granite, greisenized sodic ± potash altered granite, greisenized basement or volcanic rocks. The associated mineralogical assemblages for these possible variants are discussed in Bowden and Kinnaird (1984b) and Kinnaird (1985). The greisenisation process is economically important because it was a major episode of disseminated oxide mineralisation dominated by the crystallization of abundant fine-grained cassiterite and wolframite with other ores (Table 3).

In Saudi Arabia the greisenization process was responsible for the formation of wolframite (with scheelite) and molybdenite deposition (with powellite) together with pyrite, arsenopyrite, stannite, chalcopyrite, pyrrhotite, bismuthinite, galena,

Table 3. Ore minerals characteristic of mineralised anorogenic complexes.

Sodic	Potash	acid	silica	late-stage
pyrochlore	cassiterite	cassiterite	cassiterite	smithsonite
columbite-tantalite	wolframite	wolframite	wolframite	goethite
cassiterite	zircon	sphalerite	sphalerite	limonite
zircon	columbite	molybdenite	TiO_2 minerals	jarosite
galena	monazite	monazite	bismuth	haematite
thorite	ilmenite	pyrite	bismuthinite	cerussite
xenotime	rutile	arsenopyrite	bismuthite	pyromorphite
ilmenite	molybdenite	genthelvite	haematite	bornite
magnetite	powellite	phenakite	chalcocite	azurite
cryolite	sphalerite	stannite	covellite	malachite
bastnaesite	stannite	chalcopyrite	galena	chalcanthite
synchesite	chalcopyrite	pyrrhotite	stannite	pyrolusite
genthelvite	arsenopyrite	bismuth	pyrite	
gargarinite	genthelvite	bismuthinite	marcasite	
apatite		galena	chalcopyrite	
chevkinite		siderite	cubanite	
wiikite		zircon	pyrrhotite	
uraninite		marcasite	mackinawite	
		greenockite	terahedrite	
		gold		
		argentite	bastnaesite	
			zircon	
			uraninite	
			monazite	

sphalerite, cassiterite, gold and argentite. In Sudan, the Saboloka complex has accessory cassiterite and wolframite, altering to haematite and manganese oxides with goethite, limonite, jarosite, molybdenite, scheelite, powellite, galena, fluorite and malachite.

The final major process of ore formation and hydrothermal alteration is silica metasomatism. Like the potash and acid metasomatism, the process may be pervasive or vein controlled. Quartz may be pervasively deposited into vugs in a cupola created by earlier alteration, or it may replace all earlier formed minerals. Even more common are the veinlets which are found in virtually all biotite alkali-feldspar granite masses. There is a major deposition of ores dominated by cassiterite, wolframite and sphalerite (Table 1) associated with quartz vein development, particularly in the lodes of the Ririwai and Tibchi complexes of the Nigerian Province. In Saudi Arabia Jackson (1986) also records bastnaesite, zircon, uraninite and monazite associated with silicification at Jebel Hamra.

Argillic alteration, which involves the formation of clays (kaolinite and montmorillonite groups) at the expense of feldspar, is a late-stage process. It is apparent in only a few complexes in Nigeria although it is intense around some complexes in other provinces, e.g. Mayo Darlé in Cameroon. Usually such argillic alteration is superimposed on an earlier alteration of the feldspars particularly following intense albitisation.

It is important to emphasise that there is considerable mineralogical variation from complex to complex in the subsolidus assemblages and the sequential changes related to the different processes of alteration. There is as much variation within a Province as between provinces. It is important to emphasise that for Nigeria where the whole range of processes has been defined, that in most complexes only one or two stages of alteration are developed to a small degree.

For Niger, the sequence of alteration stages for the Taghouaji complex is similar to that of Nigeria. Perez (1985) gives the succession as albitization, greisenization, then 'muscovitization', silicification and finally kaolinization. In fact however, there is little difference between the greisenization and 'muscovitization' except in the degree of mica readjustment to hydrothermal fluids. Although Perez states that a microclinization process has not yet been identified, he does describe amazonite veins in the biotite granite. Since the microcline produced during hydrothermal alteration is turbid and does not have the characteristic tartan twinning and can only be identified with certainty by XRD, it is possible that the process may have occurred, if not at Taghouaji, in some of the other Niger complexes. At Taghouaji, the occurrence of greisen veins and the absence of a greisenised cupola is interpreted as due to the level of erosion. In summary, for Taghouaji, microclinization has not been observed: albitization and kaolinization are weakly developed.

In Sudan, Almond (1979) has described hydrothermal alteration and mineralization in the Abu Dom area of the Sabaloka Complex. An early feldspar-phyric greisen — which may be equivalent to the potash metasomatism phase in West Africa, is followed and brecciated by a later greisenization phase. This was followed by the emplacement and mineralization of the stockwork quartz veins. The order of processes is similar to that in West Africa, although no albitization process is described. Unlike many of the West African complexes however, the dominant ore mineral in the stockworks is wolframite and topaz is very scarce.

Hydrothermal alteration in the tin-mineralized Mayo Darle complex of the Cameroon line shows greisenization, silicification, chloritization and haematization with zones of intense kaolinization near the tin deposits.

8. The nature of the late-stage fluids

The PVTX nature of the fluids which were responsible for hydrothermal alteration can be monitored by a study of fluid inclusion characteristics. A study of fluid inclusions within the volcanic rocks of the Saiya Shokobo Complex in Nigeria has shown that at the magmatic stage the silicate melt, which was trapped as globules of silicate melt, co-existed with an immiscible saline fluid. The fluid phase itself separated into two immiscible phases possibly as a result of pressure release on eruption, and resulted in a CO_2-rich gas phase and a saline fluid phase. The initial fluid was over 600°C with a salinity equivalent to >60 wt per cent NaCl and a density as high as 1.4 gm/cm^3. The fluid inclusion population within the granites often do not contain melt inclusions suggesting that quartz has recrystallised at postmagmatic temperatures. The inclusions within the granites indicate a range of fluids trapped over a period when fluids cooled from >600 to <100°C.

During sodic metasomatism loss of CO_2 appears to have been important for the deposition of uranium-enriched pyrochlore or thorite, and zircon which is also Hf-rich. Fluids were saline but ranged in temperature from >600°C to less than 300°C. Potash metasomatism covers a similar temperature range although much of the CO_2 has been lost from the fluid at this stage. The process cannot be entirely temperature dependent and the Na^+/K^+ ratio in the fluid must have been largely responsible for determining which process operated. Pervasive acid metasomatism (greisenization) results from fluids that were in the temperature range 420–220°C and which were subject to periodic boiling particularly in the temperature range 350–380°C. A minimum depth of 0.1 km has been assessed for the pervasive acid metasomatism of the Ririwai Complex. Vein-controlled acid metasomatism generally took place in the temperature range 270–380°C from Na-Cl-F fluids that boiled as pressure was released in fissures and along incipient fractures. Major cassiterite deposition took place between 300 and 380°C in greisens often from fluids that were boiling. Silicification took place at less than 380°C with deposition down to less than 70°C from fluids that were gradually decreasing in salinity from ~ 10 eq wt per cent NaCl to almost zero. Major sulphide deposition was in the range 240–380°C.

Broadly similar observations were made for Mayo Darlé in Cameroon. Nguere and Norman (1985) suggest that there was a saline fluid phase in the granite during or soon after crystallization with a salinity up to 60 eq wt per cent NaCl. Fluid inclusion studies on the quartz-tin veins indicate Na-Cl-F fluids at temperatures between 200 and 600°C and salinities up to 65 eq wt per cent NaCl which were subject to periodic boiling at temperatures above 385°C. Pressure fluctuations from hydrostatic to lithostatic are indicated during mineralization.

For the Jebel Eyob tungsten deposit of Sudan, El Samani (1985) has shown that the stockwork of greisen and quartz veins contain a generation of H_2O-CO_2 bearing inclusions that were trapped between 320 and 340°C at a pressure of less than 0.4 kbar.

The processes of hydrothermal alteration established within the Niger/Nigerian province have equivalents in all tectonic settings. There are, however, several significant difference between A-type altered granites from I- and S-type hydrothermally altered granites. Firstly, in the African anorogenic granite ring complexes, there is no phase of tourmalinization and secondly, in no other tectonic setting is there the same sequence of alteration processes. Thus in Nigeria, where the alteration processes have been best documented, the sodic or albitization process is the earliest, followed by potash metasomatism. This does not seem to have occurred in any other tin provinces such as Bolivia, Czechoslovakia, Cornwall or Australia. It is more usual for potash metasomatism to preceed the sodic process and this is related to the chemical evolution of the granites and their residual fluids.

9. Geochemistry of the alteration processes

The major element data showing the various mineralogical effects of sodic metasomatism, potassic metasomatism, and acid metasomatism linked to silica metasomatism, can be displayed in the cationic Q-F diagram (Figure 11).

This approach first devised by De la Roche (1964) to study the various Hercynian granitic rocks in France, was adapted for the display of greisenization by Charoy (1979). The diagram is effectively a display of the dominant mineralogical components based on the relative cationic equivalents of quartz, albite, and microcline in the analysed rock. Not only can the salic components be displayed but mica analyses can be plotted into the diagram to define the greisenisation processes between quartz-rich greisens and mica-rich greisens. By calculating the millicationic components in terms of the parameters $Q = \frac{1}{3} Si - (K + Na + \frac{2}{3}Ca)$ and $F = (K - (Na + Ca))$ for any granitic rock suites which have undergone various degrees of subsolidus reactions, the degree of disturbance from granite minimum compositions can be clearly demonstrated. Furthermore the individual dominant process of either sodic metasomatism, potash metasomatism, or acid metasomatism can be defined. Also the paragenetic mineralisation associations occupy specific areas in the Q-F diagram.

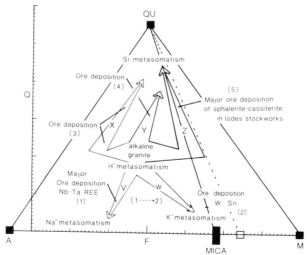

Figure 11. Q-F diagram showing principal stages of metasomatism. Trends V: important Nb-Ta REE mineralisation; W: superimposed K metasomatism; X: Na^+H^+Si metasomatism; Y: either an absence of major Na-K metasomatism, or the result of K overprinting Na metasomatism with substantial ore deposition; Z: K+H+Si metasomatism with major deposition of sphalerite and cassiterite; mineral associations 1, 2, 3, 4, 5 are different depending on the composition of the ore forming fluids, the composition of the host rocks and the PT conditions of ore formation. Mineral associations discussed in the text and presented in Tables 2 and 3.

The geochemical data indicate that each of the alteration processes is characterized by a change in alkali element ratios. These changes are accompanied by an enrichments and depletions in specific trace elements. The chondrite-normalised plots of trace element abundances (Figure 12) illustrate the chemical effects associated with the alteration processes. There is clearly REE mobility during alteration, although the pattern of depletion or enrichment varies from that described for S.W. England (Alderton et al. 1980). The absolute REE concentrations and the chondrite-normalized patterns for the unmineralised amphibole or biotite granites shows limited variation. When compared with the oversaturated syenite from the Kila Warji complex (Figure 12a), the chondrite-normalised curves

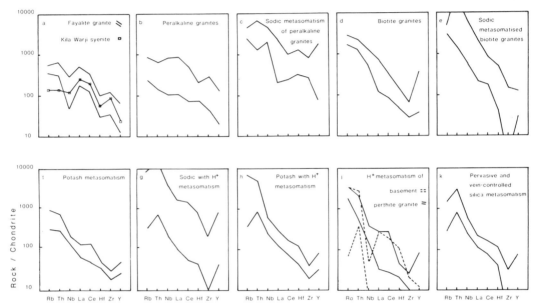

Figure 12. Chondrite normalised plots of trace element abundances showing mobility related to metasomatism (after Kinnaird et al. 1985, Figure 6).

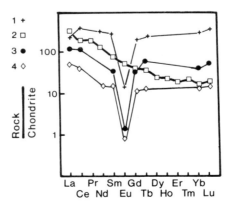

Figure 13. Chondrite-normalised rare-earth patterns for (1) arfvedsonite albite apogranite, (2) augite syenite, (3) biotite granite, (4) microcline-rich borders to greisen veins in mineral biotite granite (adapted from Bowden, Figure 14, 1985).

are distinctly flat, with significant negative Eu anomalies in all the granites, the magnitude of which varies with the alteration processes.

Major element variations during sodic metasomatism show an increase in Na and Fe and a decrease in K. There is a substantial enrichment in all REE's, particularly in the peralkaline facies (Bowden *et al.* 1979) which displays an increase in all trace elements including the high field strength elements Nb, Zr and Hf (Figure 12c and e). The biotite granites show a substantial enrichment in Rb, Th, Nb, La, Ce and Hf but sometimes a depletion in Zr and Y.

During potash metasomatism there is an increase in K and a concomitant loss of Na. There is a decrease in abundance of most trace elements when compared with the average for the initial rock type, and particularly when compared with elevated values resulting from sodic metasomatism. Some trace elements however like Rb and Sr are enhanced, reflecting the increased potash content. There may be a slight increase in elements like Ni and Ba also combined with enhanced Sn and Zn levels. There is a depletion in the whole rare earth spectrum (Figure 13), on chondrite-normalised curves potash-metasomatised wallrocks of the Ririwai lode show the greatest REE depletion (Kinnaird *et al.* 1985).

Chemically, hydrogen ion metasomatism is characterised by a marked decrease in potash and alumina due to feldspar breakdown and sometimes by increases in Fe_2O_3 and SiO_2. Figure 12 shows a depletion in Zr relative to unaltered biotite granite and an increase in some of the trace element populations compared with the potash metasomatism pattern, particularly in Th, Ce and Y. However, trace element patterns in greisens are very variable, since they are affected by earlier processes. Thus where H^+ metasomatism is superimposed on an earlier sodic effect, the chondrite-normalised pattern (Figure 12g), shows a close similarity to that for sodic metasomatism and where superimposed on a potash effect (Figure 12h), the potash pattern is reflected. Where H^+ metasomatism has been superimposed on an unaltered biotite perthite granite, the trace element pattern (Figure 12j), shows a close similarity to the original granite, whereas there is a notable depletion in Nb for the basement. The rare-earth patterns (Figure 13), show that there is partitioning of the LREE to the greisen mineral assemblage, a slight enrichment of Eu coupled with increasing Yb and Lu and with the interpolated HREE assemblage partitioned slightly into the fluid. There is a marked increase in Li, which is accommodated in the new mica of the greisen, and there is a substantial increase in the ore-forming elements, notably Sn, Pb, Zn, W and Cu.

The silicification process shows the obvious increase in silica, balanced by a decrease in all the other major elements except iron in some cases. No detailed data exist for the rare-earth behaviour during the silification process. The chondrite-normalised plots (Figure 12k), shows that while there is an overall decrease in the spectrum of trace elements compared with biotite granite, there is a selective dilution, most notably in Zr. There is an increase in the ore-forming elements during silicification particularly in Sn, Zn, W, Bi, Cu, Mo and Pb, sometimes in appreciable quantitites.

10. Mineralization and economic aspects of the oversaturated ring complexes

Over the continent of Africa as a whole, the major potential of the oversaturated complexes is for niobium, tin and tungsten mineralization. Apart from limited tin-tungsten mineralization in central Sudan (Almond 1967, Vail 1978, 1979), Arabia

(Jackson, 1986) and the Red Sea Hills (el Ramly *et al*. 1971, Vail 1979) most cassiterite and wolframite production has been related to the Nigerian tin province. The Nigerian province has been a major world producer of tin, and alluvial production of cassiterite in Nigeria is recorded from early this century. Since 1905 when records began, more than 700,000 tonnes of cassiterite concentrate have been produced from the alluvial deposits.

11. Styles of mineralization produced by hydrothermal alteration

Mineralization of a pluton generally occurs in the apical or marginal zones or in satellite dykes. Different styles of mineralization tend to characterise different parts of a granite pluton and five separate zones have been recognised: the roof, marginal and contact zones of a pluton, the country rock and surrounding ring dyke (Kinnaird 1985, Jackson 1986).
    (i) pegmatitic pods in roof and marginal zones with columbite, genthelvite or beryl, uraninite, topaz, thorite as at Jabal Tarban, Saudi Arabia, Harwell in Nigeria etc
    (ii) pervasive metasomatic disseminations in cupola zones of columbite, xenotime, pyrochlore, thorite, rare earths, monazite, Hf-rich zircon + cassiterite
    (iii) stockworks and sheeted veins with cassiterite, wolframite, sulphides and associated minerals
    (iv) mineralized ring dykes with cassiterite, wolframite, sphalerite, molybdenite and galena
    (v) brecciated greisen deposits of sphalerite, wolframite, cassiterite
    (vi) fissure-filling lodes with cassiterite, wolframite and various sulphides
    (vii) late replacement bodies with cassiterite and sulphides
    (viii) quartz rafts and veins with wolframite scheelite, bismuth minerals, sometimes abundant cassiterite and/or sulphides
    (ix) late-stage veins of fluorite with sphalerite, galena and other sulphides
    (x) alluvial deposits of cassiterite, columbite, xenotime, monazite, fergusonite, ilmenite, magnetite, thorite, zircon and pyrochlore
    (xi) residual soils enriched in wolframite and cassiterite

No particular type of hydrothermal alteration is restricted to one particular style of mineralization. Thus it is possible to have disseminated sodic, potassic, acid or silica metasomatism; altered ring dykes may show the effects of all the processes; pegmatitic pods may be generated at any hydrothermal alteration stage and fissure filling lodes show all types of hydrothermal alteration with earlier alteration assemblages overprinted by later ones.

12. Economic potential

The economic potential varies according to the style of mineralization and the mineralogical association. Two distinctive mineralogical associations can be recognised in all the oversaturated provinces: a pyrochlore-columbite type with rare-earth enrichment and an oxide-sulphide assemblage with cassiterite, wolframite and sulphides (Table 3).

(i) The pyrochlore/columbite association is related to sodic metasomatism of granite cupola zones. Usually such mineralization occurs disseminated throughout a granite cupola although concentrations may occur in pegmatitic sheets and veins.

In Saudi Arabia, both pyrochlore and columbite occur together in late Pan-African alkaline anorogenic granites (610–510 Ma; Stoeser 1986). Three different styles of Nb-Zr-REE mineralization have been distinguished and summarised in Jackson (1986); disseminations in porphyritic alkali microgranite plutons and stocks such as Ghurayyah, Jabal Tawlah, and Umm al Birak; disseminations in layered aplite-pegmatite sheets in the apical or marginal zones of alkali granite porphyry plutons such as Jabal Sa'id (Habd ash Sharar pluton) and Sumr al Ishar (Habd ad Dayahin pluton); and disseminations in discordant felsic veins and pegmatites in Umm al Birak, Habd ad Dayahin, Babal Awja, Jabal Tuwalah and Jabal Kuara. The principal accessory minerals with the pyrochlore and columbite-tantalite are monazite, cassiterite, bastnaesite, synchesite, gargarinite, thorite and zircon. In addition accessory ixiolite and betafite have been recognised at Jabal Umm al Suqian (Bokhari et al. 1986).

In Nigeria pyrochlore and columbite occur in different facies, pyrochlore characterises sodic metasomatised peralkaline granites whereas columbite occurs in zinnwaldite albite granites resulting from sodic metasomatism of biotite perthite granite.

The pyrochlore-bearing granites occur in six localities in Nigeria in the Ririwai, Dutsen Wai, Shere, Buji, Ropp and Kigom Complexes. The uranium-enriched pyrochlore which forms distinct irregularly-distributed, honey coloured octahedra is accompanied by Th-rich monazite, cloudy prisms of uranium-, thorium- and hafnium-enriched zircon, amblygonite, stellate clusters of astrophyllite, cryolite, thomsenolite, villiaumite and sometimes by narsarsukite and chevkinite.

Columbite is known to occur in varying quantities in all complexes where fine-grained granites occur since many of the alkali biotite granites of Nigeria show slight sodic metasomatism. However, the most intense albitization and highest primary enrichment occurs in localised parts of the Jos Bukuru complex and the Odegi area of the Afu complex. In these areas the granite has been decomposed to the consistency of clay by late-stage argillic alteration. This allows the extraction of ore minerals by the use of monitor and gravel pumps. There is a large variation in columbite content from <30 to >2220 ppm Nb_2O_5. There is an associated substantial enrichment in heavy rare-earth elements and also uranium in the ores, particularly in the thorite, xenotime, monazite and zircon.

In Niger, a similar mineralogy to the albitized peralkaline granites of Nigeria has been noted by Perez (1985). Pyrochlore and columbite also occur separately in different facies, with pyrochlore in peralkaline granites and columbite in peraluminous granites. In the peralkaline granite, pyrochlore is accompanied by bastnaesite-synchesite, zircon, monazite, apatite, chevkinite, wiikite, Fe-, Ca- and REE-silicates, sphene rich in niobium and/or rare earths, oxy-fluorides of titanium and rare-earths and more rarely, fergusonite, fersmite, niobotitanates, xenotime, fluorite, and fluocerite in the Taghouaji complex. The uranium mineralization at Bilete and Goundai belong to this mineral association. In the biotite granites disseminated columbite is accompanied by thorite, zircon, anatase, fluorite, fluocerite, topaz and more rarely by xenotime, monazite, ilmenite, haematite, magnetite, iron-titanium oxides, cheralite, wiikite, niobo-titanates and bastnaesite.

(ii) The oxide-sulphide assemblage of ores is related to potassic, acid and silica metasomatism and occurs in Niger, Cameroon, Sudan, Saudi Arabia and Nigeria. It is a later mineral association than the previous rare-earth pyrochlore-columbite assemblage. In some provinces cassiterite is the dominant ore of this type, in others cassiterite is subordinate to wolframite. In some localities sulphides, such as sphalerite or galena, are the major ore mineral and this is especially the case in late-stage veins. The mineralization type occurs in a wider style of deposits than the earlier type and whilst minerals of the association may be disseminated in granite cupola zones they may also occur in sheeted veins, stockworks, replacement bodies in brecciated deposits and late-stage veins.

In Niger cassiterite is disseminated in biotite granites at Tin Tajet, Sirret, Agalak and Bagazeuan. The most important mineralization however, occurs in the El Meki, Guissat and Taghouaji complexes. In these complexes, in addition to disseminated cassiterite in the biotite granite, there are cassiterite-bearing greisens with wolframite, columbite, chalcopyrite and more rarely beryl, small veins of galena with or without a gangue mineral, veins of quartz with wolframite predominating and accompanied by cassiterite and chalcopyrite and late-stage veins of fluorite ± barytes. Small veinlets carrying sphalerite and chalcopyrite occur in the volcanics. In the Taghouaji complex the greisens veins with cassiterite, also contain TiO_2 minerals, zircon, columbite, sphalerite, pyrite and pyrrhotite; and the quartz veins contain cassiterite, wolframite, pyrrhotite, fluorite, beryl, pyrite, sphalerite, molybdenite, arsenopyrite, galena and native bismuth. The late-stage barytes veins contain pods of fluorite, galena and beryl (Perez 1985).

In Cameroon, the Mayo Darlé complex of the Cameroon line consists of rhyolitic volcanics with clasts of benmoreite, granite porphyry quartz-bearing syenite, biotite and riebeckite granites. All the mineralization occurs in biotite granite and most occurs in conjugate fractures. Tin mineralization occurs in a stockwork of veinlets 2.3 km^2 in areal extent with grades of 0.3% Sn. There are vertical and horizontal lode veins with 2 to 20% SnO_2 which occur near several highly silicified breccia pipes which themselves are barren. The greisen veins consist of quartz and zinnwaldite with accessory topaz and cassiterite.

In Sudan a wolframite- and cassiterite-bearing stockwork has been described from Abu Dom, in the Saboloka complex, 88 km north of Khartoum (Almond 1967). The mineralized stockwork centres around a small lens-shaped mass of greisen 300 m long, which lies on the contact of a porphyritic microgranite ring dyke. Almond believes that the hydrothermal fluids have originated from a nearby mass of biotite-muscovite granite. The veins forming the stockwork are commonly less than 5 cm wide and are composed largely of translucent white quartz with translucent to transparent quartz lining the numerous drusy cavities. These contain wolframite which has oxidized at the surface to form a mixture of manganese and iron oxides and cassiterite which is much sparser and more patchily distributed. These are accompanied by goethite, limonite, jarosite, powellite, galena, fluorite, molybdenite, scheelite, calcite and malachite in approximate order of abundance. The greisen lens in the centre of the stockwork is composed of quartz-zinnwaldite with sparse accessory magnetite, haematite, cassiterite, zircon and rare topaz. The greisen contains numerous miarolitic cavities, up to 10 cm in diameter, lined with quartz and/or cassiterite.

In Saudi Arabia, in addition to the polymetallic rare-earth pyrochlore-columbite mineralization in the alkali granites described above, vein or pegmatite deposits of fluorite are also commonly associated with alkali granites for example at Habd ad Dayahin, Jabal Habd ash Sharar, Jabal at Tuwalah and Umm al Birak. Galena-bearing veins also occur at Jabal Habd ad Dayahin and Jabal Habd ash Sharar. Elsewhere in Saudi Arabia there are polymetallic veins, stockworks and replacement bodies with deposits of Ag, As, Au, Be, Bi, Cu, F, Fe, Mo, Pb, Sn, Te,

Th, U, W, Zn minerals. However these occur in a Ca granites, granodiorites, tonalites and diorites which belong to the period from 900–610 ma.

In Nigeria, the cassiterite-sulphide assemblage of ores can be found in any of the styles of mineralization listed above, although wolframite and sulphides do not survive in alluvial deposits. Thus cassiterite occurs as an accessory disseminated in many granites that have been affected by hydrothermal alteration throughout the Nigerian province. The values in a particular granite may vary from zero to over 400 ppm SnO_2. Disseminated cassiterite introduced during potassic and acid metasomatism is accompanied by monazite, zircon, cryolite, rutile, ilmenite, columbite, genthelvite or beryl, molybdenite, sphalerite and chalcopyrite. In addition to disseminated mineralization, abundant cassiterite, varying from small to large anhedral or euhedral crystals, which may exceed 1 cm in size, are found in mineralized ring dykes, veins, sheeted veins, replacement bodies, and stockworks throughout the Nigerian Province. In greisen veins the accessory assemblage listed above is supplemented by stannite, pyrite, arsenopyrite, siderite, bismuth minerals, marcasite, galena, cubanite, pyrrhotite, powellite, greenockite and mackinawite. Silica metasomatism resulted in quartz veins with cassiterite, wolframite, sphalerite, native bismuth, bismuthinite, bismutite, haematite, chalcocite, covellite and tetrahedrite. Late fluorite-rich veins carry a mixed assemblage of the same oxide and sulphide ores dominated by cassiterite, wolframite, sphalerite and chalcopyrite. Supergene alteration of the primary ores had led to the secondary formation of smithsonite, cerussite and pyromorphite, bornite, azurite, malachite, chalcanthite, haematite, limonite and jarosite.

Major primary ore deposits occur in lode systems in the Ririwai and Tibchi complexes (Figure 14). The lodes are aligned along fractures formed during the updoming of the underlying central biotite granites. In the Tibchi complex the elliptical intrusion has a long axis orientated northwest-southeast whereas in Ririwai the axis of the ellipsoid intrusion lies east-west with an east-west orientation to the lode. In both complexes the lodes are the product of several alteration processes with fluids channelled in enlarged steeply dipping tectonic master joints, which opened during the continued uplift stage. In the Tibchi complex (Ike 1979, 1983) the mineralized veins extend out into the basement which overlies the biotite granite but in both complexes the lode system is confined within the outer ring dyke.

The Ririwai lode has been described in detail in Kinnaird et al. (1985). It extends for a distance of 5 km in an east-west direction and to over 400 m depth and dips to the south at 85°. The maximum surface width of the lode system is 8 m. The lode,which is extensively mineralized, consists of a series of parallel to sub-parallel or braided quartz veins enclosed by zones of grey greisen grading outwards through reddened wallrock into pale pink equigranular biotite perthite granite. Hydrothermal alteration in the lode zone began with potash metasomatism and perthitic feldspar adjacent to the fissure was microclinised. Early monazite, zircon and ilmenite deposition was followed by the formation of cassiterite, wolframite and rutile and finally by the introduction of molybdenite (Kinnaird et al. 1985). During subsequent acid metasomatism microcline was altered to a greisen consisting of green coloured lithium siderophyllite or grey zinnwaldite and quartz. The sequence of oxide ore deposition, which began with early monazite, is similar to that associated with potash metasomatism. However, sphalerite with stannite, pyrite and marcasite and finally chalcopyrite followed molybdenite. During silica metasomatism, in addition to the deposition of quartz into vugs created by earlier processes major fissure-filling quartz veins up to 75 cms formed.

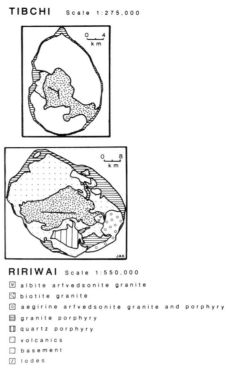

Figure 14. Geological sketch map of the Ririwai and Tibchi lode systems (Kinnaird 1981).

Cassiterite was the first ore to be deposited at this stage followed by a major deposition of sphalerite with traces of stannite, pyrite and marcasite, abundant chalcopyrite, and other minor copper and bismuth ores as exsolution blebs in the main sulphides. Abundant galena was the last major ore. Large cavities (30–100 cm in size) occur within the quartz veins and are infilled with kaolinite. Supergene alteration of the ore minerals is limited.

The Tibchi lode occurs within a biotite granite which has long been recognised as one of the most intensely mineralized in the province (Falconer and Raeburn 1923). Within the granite, there are two lode systems forming a letter Y. One orientated north-south, between 3 and 15 m wide and strike length of at least 1 km (Rockingham 1951); the other (labelled (b) on Figure 14), is orientated northwest-southeast, with similar dimensions and a strike length of over 2 km. This is believed by Turaki (pers comm) to be the earlier of the 2 systems since it does not cross the north-south lode. The north-south lode consists in the south of reddened quartz veins rich in cassiterite and wolframite with an almost complete absence of sulphides. In the north it is poorly mineralized and is often characterised by the development of mica-rich pods. The lode branches in the south with one branch cutting alkali feldspar granite the other cutting micro–syenogranite (Ike 1979). The northwest-southeast lode follows the main axis of the elliptical biotite perthite granite. It is similar to Ririwai with a reddened microcline rich outer facies grading through a greenish grey greisen to fissure-filling quartz, which is sometimes massive and milky, sometimes well crystallized. Oxides dominated by

Oxides dominated by cassiterite are disseminated through the greisen and red quartz-microcline wallrock. Sphalerite and chalcopyrite accompanied by pyrite, molybdenite, arsenopyrite and galena with traces of stannite and other minor sulphides are probably related to silicification. The lode is well exposed on Kogo hill which rises 100 m above the surrounding biotite granite. In both the Tibchi and Ririwai complexes therefore, it is apparent that there are several phases of deposition of the major ore minerals and that mineralization was repeatedly emplaced in the same lode system.

13. Constraints on the source of the magma and the source of the fluids

Different researchers have conflicting opinions as to the source of the A-type granites, and the petrogenesis of the Nigerian intraplate alkaline magmatism is therefore of worldwide as well as regional significance. The Nigerian granites could have been derived from (i) the mantle, (ii) 2,000 Ma mid Proterozoic (Eburnean) cratonic granites and gneisses representative of the lower crust or (iii) late-Proterozoic Pan-African gneisses and granites representative of the upper crust.

13a. Plate tectonic constraints

From the plate tectonic viewpoint there is a correlation between alkaline magmatism and changes in the direction of plate movements provoking reactivation of lithospheric shear zones and rifting within plates (Bowden et al. 1987). If this reactivation caused a reversal in the sense of movement along the north-south megashear zones in the Pan-African domains then the associated oblique sets of transcurrent faults, originally under compression would open and propagate as tensional faults. This would allow fracturing through the continental lithosphere causing pressure release, channelling of volatiles, partial melting and generation of magma from the aesthenosphere. Thus the overall plate tectonic approach suggests that it is the within-plate stress fields and fault reactivation which controlled the sites of alkaline magmatism in the continential lithosphere, and provided the triggering mechanism for diapiric processes of magma generation in the aesthenosphere (Bowden et al. 1987). The regions of crustal doming into which the anorogenic granites were emplaced are similar to those associated with rift valley formation and related magmatism, of which the best examples are in the East African Rift. Pressure release may help towards generating magmas of the olivine basalt-trachyte-phonolite association with geochemically alkaline to peralkaline affinities. This kind of magmatism characterises the non-orogenic igneous activity of continental rifts and swells all over the world.

13b. Petrological and geochemical constraints

The magmatic evolution of the anorogenic ring complexes in Niger and Nigeria is comparable to other alkaline provinces. It can be established from preserved volcanic sequences of ignimbrites and rhyolitic lavas with minor amounts of intercalated intermediate and basic lavas (Turner and Bowden 1979). The subvolcanic assemblages include minor gabbros, monzogabbros, nordmarkite, syenites, albite-rich and aegirine arfvedsonite granites, fayalite-hedenbergite granites, amphibole biotite granites and biotite granites. In the Niger-Nigeria Province, the relative proportions of these subvolcanic rock types changes from north to south. Peralkaline granites and quartz syenites predominate in the Air where they may be associated with anorthosites and leucogabbro, whereas minor proportions of peralkaline granites and major proportions of biotite granites are the most prevalent rock type in Nigeria.

The Nigerian anorogenic granite province however, differs from most other alkaline-peralkaline associations in the dominance of acid ignimbrites and granitic intrusions over basic and intermediate compositions, whereas they are normally subordinate. Also where basic and certain syenitic intrusions occur they do not show strong alkaline tendencies and undersaturated trends (phonolites and nepheline syenites) do not occur. The occurrence of A-type granites, as in Nigeria, without abundant associated basic or intermediate rocks does not favour an origin by fractional crystallization. The crustal uplift, in addition to the regional and local negative gravity anomalies associated with the complexes, preclude the existence of the huge volumes of basic cumulates that would be required to provide the acid differentiates.

It is widely believed that the unusual chemical composition of A-type granitoids results from high F and/or Cl and low H_2O activities. One hypothesis (White and Chappell 1983) involves production from an F or Cl-rich source that is essentially anhydrous. The source rocks have already been depleted in water by the extraction of minimum-melt I-type magmas. This hypothesis explains why A-type granites are always late in the magmatic history of a region and why granitoids are more abundant than other igneous rocks in A-type provinces. For the late orogenic A-type granites of the Lachlan fold belt in south east Australia, partial melting of felsic granulite is the preferred genetic model (Collins et al. 1982). High temperature, vapour-absent melting of the granulitic source generates a low-viscosity, relatively anhydrous melt containing F and possibly Cl. The framework structure of this melt is considerably distorted by the presence of these dissolved halides allowing highly charged cations such as Nb, Ta, Sn, Ga and Zr and probably other elements such as REE to form stable high co-ordination structures, particularly with F. The melt structure determines the trace element composition of the granite. Separation of a fluid phase from A-type magma results in destablization of co-ordination complexes and in the formation of rare-metal deposits commonly associated with fluorite.

13c. Radiogenic isotopic constraints

A considerable amount of isotopic information has become available within the last 10 years on anorogenic alkaline magmatism. In particular stable isotopic data is now available for several A-type granites from Arabia (Radain et al. 1981, Jackson 1986), Cameroon (Jacquemin et al. 1982) and Nigeria (Borley et al. 1976; Sheppard and Bowden in Sheppard 1986). Stable isotopic studies coupled with analyses of radiogenic isotopes of Sr, Nd and Pb have considerably added to our understanding of the source of anorogenic alkaline magmas.

13d. Rb/Sr isotopic systematics

A detailed study of the Rb/Sr systematics of Nigerian granites from six complexes was undertaken by van Breemen et al. (1975). This work was supplemented by a further study on 11 complexes by Rahaman et al. (1984) in which 40 Rb-Sr isotopic measurements were made on mineral and whole rock samples. Most of the rocks analysed have high Rb/Sr ratios, resulting in high precision radiometric ages. Low initial ratios support a model of fractional crystallization from mafic to intermediate sources and the general observation that low ratios characterise A-type granites that are associated with syenites and more mafic rocks, provide additional evidence for this model. However, using a fractionation model, high initial ratios such as those of the Ririwai complex cannot be explained without appealing to mixing or contamination of mantle and crustal material. If a lower crustal source is invoked, then isotopic variation may result from different source rocks and from the time lapse between production of the source and the generation of the A-type granites. Also it is possible that metasomatic processes have a greater effect on changing initial $^{87}Sr/^{86}Sr$ ratios in A-type granites relative to other granites (van Breemen et al. 1975). Other units in the Ririwai complex such as the biotite granite display less variations in initial $^{87}Sr/^{86}Sr$. However, the interpretation of correlated Rb and Sr concentrations as a hydrothermal imprint may not be a unique solution. At very late stages in magmatic differentiation Rb as well as Sr might become a compatible phase in the crystal fractionation process due to simultaneous biotite and alkali feldspar separation, thus also yielding a positive correlation between Rb and Sr. (Dickin et al. 1987). Collins et al. (1982) conclude for the Australian A-type granites that although isotopic evidence suggests that some A-type granites may be produced by fractional crystallisation from a syenitic or even more basic source magma, such a mechanism is not a general explanation for their origin.

13e. Common Pb isotope systematics

Bowden (1970) and Bowden and van Breemen (1972) discussed the isotopic composition of lead in some Nigerian Proterozoic, Palaeozoic and Mesozoic rocks and minerals and concluded that Jurassic granites with 'old lead ages' cannot be assumed to have differentiated from upper mantle but have been derived by fusion of mineralised crustal rocks. Additional data by Tugarinov (1968) obtained on three complexes indicated that most of the lead values lie reasonably close to the 500 ma isochron, while grouping near two distinct growth lines. This suggested that there were isotopic variations in the source region and indicated a crustal origin at least for some of the lead mineralisation.

More recent data by Dickin et al. (1988), yield initial Pb isotopic compositions confirming that lead was largely inherited from mid-Proterozoic (Eburnean) crust and Pan-African crust. This does not however preclude an original mantle input but does suggest that interaction of magma and/or its fluid with crustal rocks has taken place.

13f. Combined Nd/Sm and Sr Isotopic data

In order to determine more about the contribution from possible crustal or mantle sources Dickin et al. (1988) considered the evolution of $^{87}Sr/^{86}Sr$ in each of the possible source materials and combined with Nd/Sm studies.

(i) Evidence for mantle signatures

The composition of the mantle beneath the Jos Plateau cannot be determined directly since the Mesozoic complexes are the only significant magmatic products of the Jos Plateau during the Phanerozoic. Dickin et al. (1987) therefore sampled mantle-derived basic magmas from different ages and from different areas, both the north and south of the Plateau. Cameroon line volcanics and micrograbbos from the Adrar Bous complex of Niger were chosen for study. The micrograbbos, dated at 470 Ma had high initial ratios despite their basic composition indicating that they have been significantly contaminated by crustal material. The composition of the Cameroon Line volcanics (Halliday et al. 1983) provided a much more satisfactory indication of the nature of the Nigerian sub-lithospheric mantle.

For the Cameroon line volcanics, basalts from the continental sector display a wider range of isotopic compositions than those from the oceanic sector, which can be partly attributed to crustal contamination.

(ii) Evidence for mid-Proterozoic crustal signatures

In southwest Nigeria the basement consists of charnockitic rocks of unknown age and a gneissic complex comprising banded gneisses, migmatites, quartzites schists, biotite and amphibole gneisses, amphibolites and marbles metamorphosed in the almandine-amphibolite facies. They are cut by foliated granites which have yielded Proterozoic ages (Rb-Sr WR isochrons of 2270 + 30 Ma and 2280 + 70 Ma and by porphyritic granites, subordinate quartz diorites and two mica granites which are considered to be Pan-African in age. The bulk of the metasedimentary rocks are probably of lower Proterozoic age but an age of 2750 Ma indicates the presence of some remnant Archaen rocks. Rejuvenation during the Pan-African thermotectonic event produced biotite Rb-Sr and K-Ar ages around 500 Ma.

In a further effort to constrain the composition of the deep crust beneath the Jos Plateau a selection of metamorphic rocks were collected from the Zaria area of north central Nigeria, Rb-Sr isotopic studies have been undertaken by van Breeman et al. (1977) on a small selection of samples from the Pan-African basement host of the Jos Plateau granite complexes. Recalculating the data to the new decay constants, these plutons range in age from 596 ± 23 Ma for foliated granites between Bauchi and Gongola River and 654 ± 128 Ma for monzodiorites and a hornblende biotite granite at Rahama to 677 ± 161 Ma for foliated granites at Panyam. When Nd isotopic compositions are corrected using the Rb/Sr dates the initial $143Nd/144Nd$ ratios, expressed in epsilon units (parts per ten thousand deviation from the bulk earth evolution line) are -1.6, -10.5 and -13.1 for the Panyam, Rahama and Bauchi samples respectively. These very enriched source compositions are indicative of a large component of crustal basement in the Pan African granites and in fact the Rahama and Bauchi samples can be little more than remelted crust. When Nd model ages are calculated, they yield 810, 1590 and 1672 Ma ages respectively for the three plutons. The last two ages are at the lower end of the 1600–2400 Ma age range of the Eburnean orogeny which has been recognised in rocks from the Kaduna and Ibadan areas. Initial Sr isotope ratios of these granites (0.710 for Bauchi and 0.708 for Rahama, van Breemen et al. 1977), are consistent with this interpretation of the Nd data, and indicate that the Pan-African crust was reworked from Eburnean basement, which may be present at depth over large areas (Dickin et al. 1987). Similar age data can be obtained from granulite facies xenoliths within the Cameroon line volcanics. These are believed by Dickin et al. (1987) to represent the crustal basement of the Jos Plateau. In general, it would appear that the Amo, Jos and Pankshin magmas may have been contaminated in the Eburnean (?) lower crust.

(iii) Evidence for crustal signatures

Precise identification of the crustal components in the Nigerian granites is difficult, not least because the Pan-African crust largely consists of reworked older basement. However, it would appear that the Shere Hills and Ririwai intrusions contain a large fraction of a Pan-African component. This is compatible with a magmatic process, but with extensive hydrothermal activity and REE mobility, contamination by Pan-African material at the hydrothermal stage must be a possibility. An important consideration therefore is whether contamination by Pan-African crust was introduced at the magmatic stage, or by meteoric fluids.

14. Origin of the fluids

The origin of fluids in hydrothermal systems has been actively debated in the past. Historically, one group regarded such fluid to be dominantly meteoric, surface derived waters, whilst a second group believed that the fluids were juvenile in origin. The question has been partially resolved in recent years for many ore deposits, with the advent of stable isotopic studies. This is possible because the four major types of water — magmatic, metamorphic, seawater and meteoric water — have characteristic stable isotopic signatures. Applied to the analysis of isotopic data on hydrothermal minerals, fluid inclusions and waters from active geothermal systems, isotopic ratios indicate that waters of several origins are involved in ore deposition in the volcanic and subvolcanic intrusive environment (Sheppard 1977). For example, water of solely magmatic origin dominates main stage mineralization of the Casalpalca Ag-Pb-Zn-Cu deposits of Peru, whilst water solely of meteoric origin is responsible for the Cu-Zn-Mn deposits of Butte Montana. However, solutions of more than one origin are important in both porphyry copper and porphyry molybdenum deposits which have similarities in style, alteration processes and mineralization with the Nigerian tin province. In these porphyry deposits, the initial major ore transportation and alteration processes of K-feldspar and biotite alteration are magmatic-hydrothermal events that occurred between 700 and 500°C. These fluids are typically highly saline, Na-K-Ca-Cl rich brines, of more than 15 eq wt per cent NaCl. The convecting meteoric hydrothermal system that developed in the surrounding country rocks with relatively low integrated water/rock ratios (less than 0.5 at. per cent oxygen) substantially collapsed as temperatures in the dying magmatic hydrothermal system fell below 350–200°C. These fluids generally had moderate to low salinities of less than 15 eq wt per cent NaCl. Differences among these deposits are probably related in part to variations in the relative importance of the meteoric, hydrothermal and magmatic hydrothermal effects.

For the Nigerian province there has been no previous detailed study of the fluids responsible for hydrothermal alteration. There are several possibilities for the origin of the fluids. The heat of the intrusions, plus the high heat producing capacity of these granites, could have been responsible for the creation of hydrothermal cells involving meteoric water as envisaged for north and southwest England by Moore (1982) and elsewhere. Alternatively it is possible that uranium introduction and associated mineralization was a result solely of magmatic processes, with the waters solely of magmatic origin. A further possibility exists of saline magmatic water mixing with dilute meteoric water which would be consistent with some of the features observed in fluid inclusion studies.

Clearly, the petrological and radiogenic isotope data indicates that there has been a mantle input to the ring complexes, with contamination by lower and upper crust. The origin of the ore elements and hydrothermal fluids could similarly have a variety of origins.

14a. Petrological and geochemical constraints on fluid composition

Crystallisation appears to have begun with quartz and feldspar as evidenced by occasional granophyric and myrmekitic intergrowths, followed by late-stage magmatic and postmagmatic modifications in many cases. Experimental work (Tuttle and Bowen, 1958; Luth et al. 1964; Wyllie and Tuttle 1964; Wyllie, 1977; Manning 1981) suggests that the high Li and F content of granites such as these is responsible for shifting the position of the ternary cotectic curve away from the quartz apex, thus enlarging the quartz field, and of the shifting of the minima of this curve towards the albite pole with concomitant lowering of the temperature. Manning (1981), also shows a progressive shift of the minimum with increasing F accompanied by enlargement of the quartz field and a depression of the minimum from 730°C for zero fluorine to about 630°C with 4 per cent fluorine. Since the experimental work of Luth (1976), Wyllie (1977), Brown and Fyfe (1970) suggests that the emplacement of high level granitic magmas from a deeper crustal source is unlikely unless the magmas are water undersaturated, it would seem therefore that most of the water released from the crystallizing magma had been exsolved after emplacement and during differentiation. A further line of evidence to suggest that the fluids are largely magmatic comes from the dating of one such vein relative to the host granite. The Ririwai lode and the enclosing biotite granite have a similar age of 171 Ma. This contrasts markedly with other provinces eg. Cornwall where there is a large time gap between granite emplacement and hydrothermal vein formation (Jackson et al. 1982).

14b. Stable isotope constraints

Sulphur

Sulphur isotope studies on primary sulphides in granitic samples and vein samples from Nigeria can be used to define the source region of sulphur. Considering the radiogenic data already discussed this could mean that sulphur in the granites and mineralized veins could have originated in the mantle, the lower crust or the upper crust or be a mixture.

Twelve different samples were selected for study from five widely spaced complexes. These sulphides from the five complexes included sphalerite, chalcopyrite, galena and molybdenite from mineralised veins and lodes. Where possible co-existing pairs of sulphides were chosen with the idea of determining temperatures of formation. The samples chosen are listed in Table 4. All twelve samples gave $\delta^{34}S$ per cent values between −1.44 and +0.8.

If the interpretations based on the stable isotopic data above are valid, then the $\delta^{34}S$ data indicates that the vein sulphur shows a narrow range of mantle values. It would appear therefore that the sulphur in both granites and mineralised veins is of mantle origin. It is perhaps not surprising that the sulphides in the veins show the same characteristics as those from the granite since the sulphides were disseminated throughout some granite plutons as a result of hydrothermal alteration. What is not clear is how long sulphur can retain its mantle memory.

14c. Oxygen and hydrogen isotopic constraints

Isotopically the $\delta^{18}O$ of samples displayed in Figure 15, have undergone albitisation and are not very different from similar granites that have not been sodic metasomatised to any significant extent.

Table 4. Sulphur isotopic compositions for some Nigerian ore minerals

Complex	Sample no.	mineral	$\delta^{34}S_{cdt}(\%_{oo})$
Afu	AF23/	sp	−1·44
	AFJ2	cp	−0·55
Ririwai	J108	sp	+0·80
	R111	ga	−0·61
Saiya	SS 93	sp	+0·72
Shokobo	SS 110A	cp	+0·17
	SS 6/1	sp	+0·07
Tibchi	TK02	ga	−0·01
	TK03	mo	+0·35
Kigom	KG5	ga	+1·48
	KG3	mo	+1·75
Jos Bukuru	JBJ118A	ga	+0·48

sp = sphalerite
ga = galena
mo = molybdenite
cp = chalcopyrite

The conclusion must be that the fluids which were responsible for sodic metasomatism were products of crystallization of the initial granite magma. Borley (1976) and Borley *et al.* (1976) suggested a temperature of 750–700°C as the maximum temperature for the formation of the Kaffo Valley arfvedsonite albite granite with consolidation and albitization occurring at temperatures down to 600°C. This is supported by fluid inclusion microthermometric data which would extend the lower limit.

The δD values of fluids associated with micas from hydrothermally altered rocks (Table 5) have been calculated according to the data of Taylor (1974). Calculated values to as low as −7 for micas of the Ririwai lode have been interpreted to be dominantly of meteoric origin. Calculated values for the other micas suggest that the influence of meteoric water was significantly less. Calculated values for the fluid that could have been in equilibrium with the biotite granites at late-magmatic temperatures is around −50 implying that for most of the micas the fluid was dominantly of magmatic origin.

15. Summary and conclusions

Although the ideas summarized in this part of the paper are specifically based on two or three anorogenic complexes in the Nigerian province, it is worth considering that the interpretation can be generally applied to oversaturated alkaline magmatism throughout Africa. Some anorogenic provinces are considered to be poorly mineralized thus more careful studies are needed to assess the hydrothermal processes in the rocks.

From both the fluid inclusion homogenization temperatures, and the isotopic geothermometers, it would seem that much of the hydrothermal alteration in the

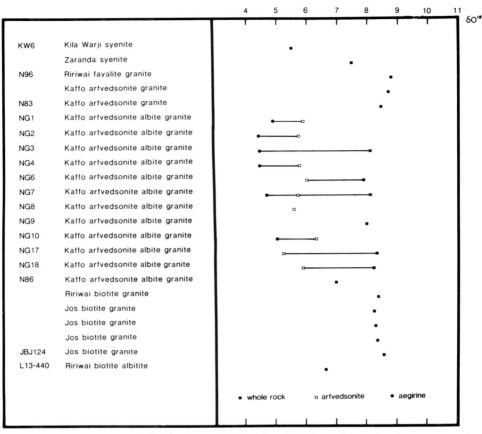

Figure 15. A compilation of $\delta^{18}O$ analyses of mineral and whole rock samples from anorogenic complexes in northern Nigeria. Samples prefixed NG– from Borley et al. (1976); un-numbered samples from Sheppard and Bowden (in Sheppard 1986); remaining samples from Kinnaird 1987.

Nigerian anorogenic province was high temperatures and took place above 300°C. However, the homogenisation temperatures from fluid inclusions in granites indicates that hydrothermal alteration continued to temperatures as low as 200°C. This is supported by Ferry (1985) who demonstrated a clear correlation between the degree of feldspar turbidity and ^{18}O depletion in hydrothermally altered Skye granites. He showed that the temperature of alteration of these granites was about 350–450°C, based on various cation exchange geothermometers and phase equilibria, and that locally temperatures as low as 200°C were recorded. The Nigerian isotopic data is consistent with a fluid which is dominantly magmatic in origin; which derived its ore-forming constituents from the mantle (S + ?), from the old Proterozoic crust (Pb + ?) and from the younger Pan-African crust. Near or at the level of final emplacement of some of the granite magmas, or on eruption of volcanics, the alkaline magma unmixed into two or three phases (silicate melt, saline to hypersaline fluid, CO_2 rich fluid). Meteoric water was only involved where substantial late-stage fracturing occurred as in the Ririwai complex during the formation of the lode.

Table 5. Oxygen and hydrogen isotopic compositions of Nigerian whole rock and mineral separates

Complex	Sample no.	Mineral	$\delta D\permil$	$\delta^{18}O\permil$
Saiya Shokobo	SS9/11	whole rock		7·62
		K feldspar		7·03
		quartz		9·02
		mica	−60	3·51
Saiya Shokobo	SS8/1	microcline		7·64
		mica		6·57
Saiya Shokobo	SS1	whole rock		6·66
		K feldspar		9·22
		quartz		9·03
		mica	−75	3·94
Saiya Shokobo	SS6/1	whole rock		8·54
		quartz		10·28
		mica	−59	4·91
Ririwai	R1/14	whole rock		9·99
		quartz		10·00
		mica	−48	8·10
Ririwai	R2/1C	whole rock		8·22
		quartz		10·70
		mica	−35	5·08

Overall the economic potential varies according to the style of mineralisation and the mineralogical association. There are four economically important phases of hydrothermal alteration recognized from field, textural, and mineralogical evidence: sodic metasomatism (albitization), potassic metasomatism (microclization), acid metasomatism (greisenization) and silica metasomatism (silicification). The residual fluids responsible for hydrothermal alteration carried ore metals which were deposited in response to fluid-rock interaction aided by disequilibrium conditions caused by boiling. Without boiling of the fluid system extensive ore deposition is unlikely to have occurred. Such a simplistic but critical observation explains why some of the African anorogenic centres are relatively unmineralized. The hydrothermal reactions described in this paper can be recognised elsewhere but characteristic textures, mineralogical changes, and fluid evolution related to ore deposition are often absent. More research needs to be completed to confirm these conclusions. However where mineralization has occurred it tends to be concentrated in the exogenetic and endogenetic zones of subvolcanic plutons and the surrounding country rocks. Some marginal mineralization is also recorded in the borders of ring-dykes. Nine different styles of mineralization can be identified including pervasive metasomatic disseminations; pegmatitic pods and lenses; stockworks and lodes. The pyrochlore-columbite association with rare earth enrichment is economically important. However cassiterite dominates the oxide-sulphide assemblage and together with sphalerite makes the anorogenic complexes worthy of commercial exploitation.

Acknowledgements. This paper represents a distillation of experience gained during our research on the ring complexes in Niger and Nigeria, coupled with the opportunity to study the Lueshe carbonatite and other anorogenic centres during field trips organised in conjunction with the first International Congress of IGCP 227 in Zaire during August 1986. We would like to thank all the many people of Nigeria who have helped the project team over the past fourteen years since 1973 for their hospitality, their company, their help or for useful discussion. Accommodation and logistical help was generously donated by the Amalgamated Tin Mines of Nigeria, especially D. Dent Young and his staff at Barakin Ladi. Additional advice and access to property was given by other mining companies including Bisichi-Jantar (Banke, Tibchi) and Gold and Base Ltd (Ririwai). Air photographs and maps were supplied by the Ministry of Lands and Surveys in Jos, Lagos and by the Lands and Survey Division in Tolworth, Surrey. The Minister of Mines and Power, Lagos and his staff in the regional offices are acknowledged for their advice and interest, particularly Bill Ford and Usman Turaki. JAK would like to thank Gerry and Elizabeth Staley for their beer, beryls and preservation of my sanity. Mohammed Zakari's friendship and local knowledge proved invaluable and will always be treasured. The St Andrews Younger Granite Research group consisted of a dozen individuals and all of them helped in some measure but special thanks must go to C. Abernethy, J. Bennett, E. Ike and S. Abaa who collected some of the samples used in the preparation of this work. The University of St Andrews has generously provided considerable help and support for the African research work to continue. Such efforts culminated in the successful 13th Colloquium of African Geology, held in the Purdie Building, the home of the Departments of Chemistry and Geology, during September 1985. Many members of Staff have generously given assistance and advice. Professor E. K. Walton is specifically acknowledged for his encouragement and welcome he gave all members of the research group. We would particularly like to thank Andy Mackie, Jim Allan, Alistair Reid, Kit Finlay, Angus Calder, and Donald Herd for the technical help at the University of St Andrews, they have given over the years. Such help often goes unacknowledged, but without it no research of this nature can proceed. We are also grateful to Richard Batchelor for some of the geochemical data but especially for a sympathetic ear and useful discussion on many geochemical topics. JAK was introduced to the study of fluid inclusions in the laboratory of Bernard Poty of CRPG in Nancy and is grateful to all those people who helped. The assistance of Jacques Leterrier, for guiding us through the intricacies of multicationic parameters was also appreciated. Thanks to the staff at SURRC, East Kilbride, particularly John Whitley, Gus Mackenzie and Tony Fallick for the provision of analytical and isotopic data and their unfailing interest in Nigerian granites. It is also important to acknowledge former members of staff of SURRC including Drs Sheppard, Dickin, Halliday and van Breeman for their involvement. Both PB and JAK have been substantially supported by the Overseas Development Ministry through Research grant R2679. JAK greatly appreciated financial help from: The British Council for the award of a Visiting Fellowship to Ilorin University in 1983; to the University of St Andrews for a research scholarship during the period Oct. 1984 to September 1985; and to the Russell Trust for a travel bursary to visit Zaire. PB is grateful to the Royal Society for their financial assistance to enable him to represent IGCP 227 as the British representative, and to the University of Lubumbashi, Zaire, for their logistic support. PB and JAK also appreciated the opportunity to participate in the excursions to Nyamulagira, Nyiragongo, and Rwanda providing a unique experience to study volcanic features in the Western Rift. Finally, no thanks would be complete without remembering our families who have suffered our absences and

References

Agar, R.A. 1986. Structural geology of felsic rocks in the Arabian Shield; styles, modes and levels of emplacement. *Journ. Afric. Earth Sci.*, **4**, 105-121.

Alderton, D.H.M., Pearce, J.A. and Potts, J.A. 1980. Rare earth element mobility during granite alteration: evidence from southwest England. *Earth Planet. Sci. Lett.*, **49**, 149-165.

Almond, D.C. 1967. Discovery of tin-tungsten mineralization in northern Khartoum province, Sudan. *Geol. Mag.*, **104**, 1-12.

—— 1979. Younger granite complexes of Sudan. In Al-Shanti, A.M. (Ed.), '*Evolution and mineralization of the Arabian-Nubian shield*', *Bull. Inst. Appl. Geol. King Abdulaziz University (Jeddah)*, **1**, pp. 159-176 (published by Pergamon Press, Oxford).

Ba, H., Black, R., Benziane, B., Diombana, D., Hascoet-Fender, J., Bonin, B., Fabre, J. Liegeois, J.P. 1985. La province des complexes annulaires alcalins sursatures de l'Adrar des Iforas, Mali. *Journ. Afric. Earth Sci.*, **3**, 123-142.

Bailey, D.K. 1977. Lithosphere control of continental rift magmatism. *J. Geol. Soc.*, **133**, 103-106.

Black, R. 1965. Sur la signification petrogenetique de la decouverte d'anorthosites associees aux complexes annulaires sub-volcaniques du Niger. *C.R. Acaf. Sci., Paris*, **260**, 5829-5832.

—— 1984. The Pan-African event in the geological framework of Africa. *Pangaea*, **2**, 8-16.

——, Lameyre, J. and Bonin, B. 1985. The structural setting of alkaline complexes. *Journ. Afric. Earth Sci.*, **3**, 5-16.

——, Jajou, M. and Pellaton, C. 1967. Notice explicative dela carte geologique de l'Air a l'echelle 1/500 000. Dir. Mines Geol. Niger.

—— 1984. The Pan-African event in the geological framework of Africa. *Pangaea*, **2**, 8-16.

——, Lameyre, J. and Bonin, B. 1985. The structural setting of alkaline complexes. *Journ. Afric. Earth Sci.*, **3**, 5-16.

Bokhari, M. M., Jackson, N.J. and al Oweidi, K. 1986. Geology and mineralization of the Jabal Umm Al Suqian albitized apogranite, southern Najd region, Kingdom of Saudi Arabia. In Drysdall, A.R., Ramsay, C.R. and Stoeser, D.B. (Eds), '*Felsic plutonic rocks and associated mineralization of the Kingdom of Saudi Arabia*', Sp. Vol. Journ. African Earth Sci., **4**, 189-198.

Bonin, B. 1982. Les granites des complexes annulaires. *Manuels et methodes*, **4**, BRGM, Orleans, 183 pp.

Borley, G.D. 1976. Ferromagnesian mineralogy and temperatures of formation of the younger granites of Nigeria. In Kogbe, C.A. (Ed.), '*Geology of Nigeria*', 159-176. Elizabethan Publishing Co. Lagos.

Borley, G.D., Beckinsale, R.D., Suddaby, P. and Durham, J.J. 1976. Variations in composition and δ^{18} values within the Kaffo albite-riebeckite granite of the Ririwai complex. Younger Granites of Nigeria. *Chem. Geol.*, **18**, 159-176.

Bowden, P. 1970. Origin of the younger granites of northern Nigeria. *Contr. Miner. Petrol.*, **25**, 153-162.

—— and van Breemen, O. 1972. Isotopic and chemical studies on younger granites from northern Nigiera. In Dessauvagie T.F. and Whiteman, A.J. (Eds) '*African Geology*', University of Ibadan Press, Ibadan, 1970.

——, Bennett, J.N., Whitley, J.E. and Moyes, A.B. 1979. Rare earths in Nigerian Mesozoic granites and related rocks. In Ahrens, L.H. (Ed.), '*Origin and distribution of the elements*', 479-491. Pergamon Press.

——, Bennett, J.N., Kinnaird, J.A., Whitley, J.E., Abaa, S.I. and Hadzigeorgiou-Stavrakis, P.K. 1981. Uranium in the Niger-Nigeria younger granite province. *Mineral. Mag.*, **44**, 379-389.

—— and Kinnaird, J.A. 1984a. Petrological and geochemical criteria for the identification of (potential) ore-bearing Nigerian granitoids. In: *Proceedings of the 27th International Geological Congress, Moscow*. VNU Science Press, Utrecht, Vol. 9, 85-119.

—— and Kinnaird, J.A. 1984b. Geology and mineralization of the Nigerian anorogenic ring complexes. *Geol. Jb. (Hannover)*, **B56**, 3-65.

—— 1985. The geochemistry and mineralization of alkaline ring complexes in Africa (a review). *Journ. Afric. Earth Sci.*, **3**, 17-39.

——, Black, R., Martin, R.F., Ike, I.C., Kinnaird, J.A. and Batchelor, R.A. 1987. Niger-Nigeria alkaline ring complexes: a classic example of African Phanerozoic anorogenic mid-plate magmatism. In Fitton J.G and Upton B.J.G. (Eds), '*Alkaline Igneous Rocks*', Geological Society Special Publication (in press).

Breemen, O van, Hutchinson, J. and Bowden, P. 1975. Age and origin of the Nigerian Mesozoic granites: A Rb-Sr isotropic study. *Contr. Mineral. Petrol.*, **50**, 157-172.

——, Pidgeon, R.T. and Bowden, P. 1977. Age and isotopic studies of some Pan-African granites and related rocks, north central Nigeria. *Precamb. Res.*, **4**, 307-319.

Brown, G.C. and Fyfe, W.S. 1970. The production of granitic melts during ultrametamorphism. *Contr. Mineral. Petr.*, **28**, 310–318.
Cahen, L., Snelling, N.J., Delhal, J. and Vail, J.R. 1984. The geochronology and evolution of Africa. Clarendon Press, Oxford, pp. 512.
Chappell, B.W. and White, A.J.R. 1974. Two contrasting granite types. *Pacific Geology*, **8**, 173–174.
Charoy, B. 1979. Definition et l'importance les phenomenes deuteriques et des fluids associes dans les granites. Consequences metallogeniques. *Mem. Sci. de la Terre, Fr.*, **37**, pp. 364.
Chorowicz, J., Le Fournier, J. and Vidal, G. 1987. A model for rift development in East Africa Geological Journal, **22**.
Collins, W.J., Beams, S.D., White, A.J.R. and Chappell, B.W. 1982. Nature and origin of A-type granites with particular reference to southeastern Australia. *Contr. Mineral. Petrol.*, **80**, 189–200.
De la Roche, H. 1964. Sur l'expression graphique des relations entre la composition chimique et la composition mineralogique quantitative des roches cristallines — Presentation d'un diagramme destine a l'etude chimico-mineralogique des massifs granitiques et granodioritiques — application aux Vosges cristallines. *Sci. Terre*, **9(3)**, 293–337.
Dickin, A.P., Halliday, A.N. and Bowden, P. 1987. A lead, strontium and neodymium isotope study of Nigerian Mesozoic ring complexes. (ms in preparation).
Drysdall, A.R., Jackson, N.J., Douch, C.J., Ramsay, C.R. and Hackett, D. 1984. Rare metal mineralization related to Precambrian alkali granites in the northwest Arabian Shield. *Econ. Geol.*, **79**, 1366–1377.
—— and Douch, C.J. 1986. Nb-Th-Zr mineralization in microgranite-microsyenite at Jabal Tawlah, Midyan region, Kingdom of Saudi Arabia. In Drysdall, A.R., Ramsay C.R. and Stoeser, D.B. (Eds), *'Felsic plutonic rocks and associated mineralization of the Kingdom of Saudi Arabia'*, Sp. Vol. Journ. African Earth Sci., **4**, 275–288.
——, Ramsay, C.R. and Stoeser, D.B. 1986. Felsic plutonic rocks and associated mineralization of the Kingdom of Saudi Arabia. *Journ. Afric. Earth Sci.* (special volume) **4**, pp. 291.
Duncan, R.A., Petersen, N. and Hargraves, R.B. 1972. Mantle plumes, movement of the European plate, and polar wandering. *Nature*, **239**, 82–86.
Falconer, J.D. and Raeburn, C. 1923. The northern tinfields of the Bauchi Province. *Bull. Geol. Surv. Nigeria*, **4**.
Faure, G. and Powell, J.L. 1972. *Strontium isotope geology*. Springer-Verlag, New York, pp. 171.
Ferry, J.M. 1985. Hydrothermal alteration of Tertiary igneous rocks from the Isle of Skye, northwest Scotland. II granites. *Contr. Miner. Petrol.*, **91**, 283–304.
Francheteau, J. and Le Pichon, X. 1972. Marginal fracture zones as structural framework of continental margins in the south Atlantic. *Ocean. Bull. Am. Ass. Petrol. Geol.*, **56**, 991–1001.
Furon, R. 1963. *Geology of Africa*. Oliver and Boyd, Edinburgh, pp. 377.
Halliday, A.N., Dickin, A.P. and Fitton, G.E. 1983. Nd-Sr-Pb isotopic evidence for the origins of the Cameroon Line volcanics and the nature of the mantle under West Africa and the eastern Atlantic, *Terra Cognita*, **3**, 122–3.
Harris, N.B.W. and Marriner, G.F. 1980. Geochemistry and petrogenesis of a peralkaline complex from the Midian Mountains, Saudi Arabia. *Lithos*, **13**, 325–336.
Husch, J. and Moreau, C. 1982. Geology and major element geochemistry of anorthositic rocks associated with Palaeozoic hypabyssal ring complexes, Air massif, Niger, West Africa. *J. Volcan. Geotherm. Res.*, **14**, 14–66.
Ike, E.C. 1979. The structure, petrology and geochemistry of the Tibchi younger granite ring complex. Unpubl. PhD thesis. Univ. St Andrews.
—— 1983. The structural evolution of Tibchi ring complex — a cast study for the Nigerian Younger Granite province. *J. Geol. Soc.*, **140**, 781–788.
Jacquemin, H., Sheppard, S.M.F. and Vidal, P. 1982. Isotopic geochemistry (O, Sr, Pb) of the Golda Zuelva and Mboutou anorogenic complexes, North Cameroon: manyle origin with evidence of crustal contamination. *Earth Planet. Sci. Lett.*, **61**, 97–111.
Jackson, N.J., Halliday, A.N., Sheppard, S.M.F. and Mitchell, J.G. 1982. Hydrothermal activity in the St Just Mining District, Cornwall, England. In Evans, A.M. (Ed.), *'Metallization associated with acid magmatism'*, pp. 137–139. John Wiley and Sons.
—— 1986. Mineralisation associated with felsic plutonic rocks in the Arabian Shield. In Drysdall, A.R., Ramsay, C.R. and Stoeser, D.B. (Eds), *'Felsic plutonic rocks and associated mineralisation of the Kingdom of Saudi Arabia'*, Special volume. Journ. African Earth Sci., **4**, 213–227.
Kennedy, W.Q. 1965. The influence of basement structure on the evolution of the coastal (Mesozoic and Tertiary) basins of Africa. In *'Salt basins around Africa'*, pp. 7–16. Inst. Petrol., London.
Kinnaird, J.A. 1985. Hydrothermal alteration and mineralization of the alkaline anorogenic ring complexes of northern Nigeria. *Journ. Afric. Earth Sci.*, **3**, 229–251.
——, Bowden, P., Ixer, R.A., Odling, N.W.A. 1985. Mineralogy, geochemistry and mineralization of the Ririwai complex, northern Nigeria. *Journ. Afric. Earth Sci.*, **3**, 185–222.
Lasserre, M. 1978. Mise au point sur les granitoids dits 'ultimes' du Cameroun. Gisements, petrologie, et geochronologie. *Bull. Bur. Rech. Geol. Minieres*, **4**, 143–159.
Leger, J.M. 1985. Evolution petrologique des magmas basiques et alkalins dans le complexe anorogenique d'Iskou (Air-Niger). *Journ. Afric. Earth Sci.*, **3**, 89–96.
Liegeois, J.-P. and Black, R. 1983. Preliminary results on the geology and geochemistry of the late Pan-African composite batholith of western Iforas (Mali), *Abstracts 12th Colloquim. Afr. geol.*, **62**.
—— and Black, R. 1984. Petrographie et geochronologie Rb-Sr de la transition fini-panafricaine dans

l'Adrar des Iforas (Mali): accretion crustale au Precambrien superieur. In Klerkx J. and Michot, J. (Eds), *'African Geology'*, pp. 115–146.

———, Bertrand, J.M.L. and Black, R. 1987. The subduction and collision related Pan-African composite batholith in the Adrar des Iforas (Mali). A review. *Geological Journal*, **22**.

Liu, H.-S. 1980. Convection generated stress field and intraplate volcanism. *Tectonophysics*, **65**, 225–244.

Luth, W.C. 1976. Granitic rocks. In Bailey, D.K. and Macdonald, R., (Eds.) *'The evolution of the crystalline rocks'*, pp. 335–417. Academic Press, London.

Manning, D.A.C. 1981. The effect of fluorine on liquidus phase relationships in the system Qz-Ab-Or with excess water at 1 kb. *Contr. Miner. Petrol.*, **76**, 206–215.

Marsh, J.S. 1973. Relationships between transform directions and alkaline igneous rock lineaments in Africa and South America. *Earth Planet. Sci. Lett.*, **18**, 317–323.

Moore, J. Mc. 1982. Mineral zonation near the granitic batholiths of south-west and northern England and some geothermal analogues. In Evans, A.M. (Ed.), *'Metallization Associated with Acid Magmatism'*, 229–241. John Wiley & Sons Ltd.

Morgan, W.J. 1971. Convection plumes in the lower mantle. *Nature*, **230**, 42–43.

Nguere, F.R. and Norman, D.I. 1985. The Mayo Darle tin deposit, Cameroon. (Abstract) 302–303; *13th Colloquium of African Geology, St Andrews, Scotland*, September 1985. CIFEG publication occasionelle 1985/3. 397 pp.

O'Halloran, D. 1985. Ras ed Dom migrating ring complex: A-type granites and syenites from the Bayuda desert, Sudan. *Journ. Afric. Earth Sci.*, **3**, 123–142.

Perez, J.B. 1985. Nouvelles donees sur le complexe granitique anorogenique de Taghouaji (Republique du Niger) influence des fluides au cours de la cristallisation. These de l'universite de Nancy.

Pirajno, F. and Jacob, R.E. 1987. Tin-tungsten mineralization in the Brandberg West-Goantagab area of the Damara orogen, Namibia. *Abstr. 851–854. Geocongress, July 86*, Johannesburg, South Africa.

Prins, P. 1978. The geochemical evolution of the alkaline and carbonatite complexes of the Damaraland igneous province, South West Africa. *Ann. Univ. Stellenbosch, Ser. Al.*, **3**, 145–278.

Radain, A.A.M., Fyfe, W.S. and Kerrich, R. 1981. Origin of peralkaline granites in Saudi Arabia. *Contr. Mineral. Petrol.*, **78**, 358–366.

Rahaman, M.A., van Breeman, O., Bowden, P. and Bennett, J.N. 1984. Age migrations of anorogenic ring complexes in northern Nigeria. *Jl. Geol.*, **92**, 173–184.

Ramly, M.F. el, Budanov, V.I. and Hussein, A.A.A. 1971. The alkaline rocks of southeastern Egypt. *Geol. Surv. Egypt*, **53**.

Rockingham, J.E. 1951. Preliminary report on a lode near Kogo, Tibchi. *Geol. Surv. Nigeria*, Unpublished report No. 875.

Samani, Y. el 1985. Microthermometric study of the Tungsten deposit at Jebel Eyob, Red Sea Hills, Sudan. Abstract 306–307; *13th Colloquium of African Geology, St Andrews, Scotland*, September 1985. CIFEG publication occasionelle 1985/3. 397 pp.

Sheppard, S.M.F. 1986. Igneous Rocks: 111. Isotopic case studies of magmatism in Africa, Eurasia and Oceanic Islands. Chapter 10. In Valley, J.W., Taylor, H.P. Jr, and O'Neil, J.R. (Eds), *'Stable isotopes in high temperature geological processes'*, 319–371.

Stoeser, D.B. 1986. Distribution and tectonic setting of plutonic rocks of the Arabian Shield. In Drysdall, A.R., Ramsay, C.R. and Stoeser, D.B. (Eds), *'Felsic plutonic rocks and associated mineralisation of the Kingdom of Saudi Arabia'*, (Special Volume). *Journ. Afric. Earth Sci.*, **4**, 21–46.

——— and Camp, V.E. 1984. Pan African microplate accretion of the Arabian shield. *Bull. Geol. Soc. Amer.*, **96**, 817–826.

——— and Elliott, J.E. 1980. Post-orogenic peralkaline and calc-alkaline granites and associated mineralization of the Arabian Shield, Kingdom of Saudi Arabia. In Al-Shanti, A.M. (Ed.), *'Evolution and Mineralization of the Arabian-Nubian Shield.' Bull. Inst. Appl. Geol. King Addulaziz Univ. (Jeddah)*, **4**, 1–23 (published by Pergamon Press, Oxford).

Sykes, L.R. 1978. Intraplate seismicity reactivation, reactivation of pre-existing zones of weakness, alkaline magmatism and other tectonism post-dating continental fragmentation. *Rev. Geophys. Spac. Phys.*, **16**, 621–688.

Tack, L., de Paepe, P., Deutsch, S. and Liegeois, J.-P. 1984. The alkaline plutonic complex of the Upper Ruvubu (Burundi): geology, age, isotopic geochemistry and implications for the regional geology of the western rift. In Klerkx, J. and Michot, J. (Eds.) *'African Geology'*. J. Musee royal de l'Afrique centrale, Tervuren, 91–114.

Taylor, H.P. Jr. 1974. The application of oxygen and hydrogen isotope studies to problems of hydrothermal alteration and ore deposition. *Econ. Geol.*, **69**, 843–883.

Turner, D.C. and Bowden, P. 1979. The Ningi-Burra complex, Nigeria: dissected calderas and migrating magmatic centres. *Journ. Geol. Soc.*, **136**, 105–119.

Tugarinov, A.I., Kovalenko, V.I., Znamensky, E.B., Legeido, V.A., Sobatovich, E.V., Brandt, S.B. and Tsychansky, V.D. 1968. Distribution of Pb-isotopes, Sn, Nb, Ta, Zr and Hf in granitoids from Nigeria. In Ahrens, L.H. (Ed.), *'Origin and distribution of the elements'*, Monographs in Earth Sciences, **30**, 687–699. Pergamon Press, Oxford.

Tuttle, O.F. and Bowen, N.L. 1958. Origin of granite in the light of experimental studies in the system $NaAlSi_3O_8-KAlSi_3O_8-SiO_2-H_2O$. *Geol. Soc. Am. Mem.*, **74**, 153 pp.

Vail, J.R. 1978. Outline of the geology and mineral deposits of the Democratic Republic of Sudan and adjacent areas. *Overseas Geol. Miner. resourc. Lond.*, No 49.

—— 1979. Outline of geology and mineralization of the Nubian shield east of the Nile Valley, Sudan. In Al-Shanti, A.M. (Ed.), *'Evolution and mineralization of the Arabian-Nubian shield'*. *Bull. Inst. Appl. Geol., King Abdulaziz Univ., (Jeddah)*, **1**, 97–107 (published by Pergamon Press, Oxford).
—— 1985. Alkaline ring complexes in Sudan. *Journ. Afric. Earth Sci.*, **3**, 51–59.

White, A.J.R. and Chappell, B.W. 1983. Granitoid types and their distribution in the Lachlan fold belt, southeastern Australia. In Reddich, J.A. (Ed.), *'Circum-Pacific plutonic terranes'*. *Mem. Geol. Soc. Amer.* **159**, 21–34.

Woolley, A.R. and Garson, M.S. 1970. Petrochemical and tectonic relationship of the Malawi carbonatite-alkaline province and the Luputa-Lebombo volcanics. In Clifford, T.N. and Gass, I.G. (Eds), *'African magmatism and tectonics'*, pp. 237–262. Oliver and Boyd, London.

Wyllie, P.J. and Tuttle, O.F. 1964. Experimental investigations of silicate systems containing two volatile components. Part III. The effects of SO_3, P_2O_5, HCl and Li_2O in addition to H_2O on the melting temperatures of albite and granite. *Am. J. Sci.*, **262**, 930–939.

Wyllie, P.J. 1977. Mantle fluid compositions buffered by carbonates in peridotite-CO_2-H_2O. *J. Geol.*, **85**, 187–208.

Palaeozoic drift of Gondwana: palaeomagnetic and stratigraphic constraints

F. B. Van Houten and R. B. Hargraves
Department of Geological and Geophysical Sciences, Princeton University, Princeton, New Jersey 08544, U.S.A.

Currently available Palaeozoic palaeomagnetic data from Gondwanan continents can be interpreted in terms of either (a) a migration of the pole from northern Africa to southern Africa between Ordovician and late Palaeozoic times, or (b) a rapid excursion of the pole from northern Africa to southwest of South Africa during late Ordovician to early Silurian times, followed by a return to central Africa in late Devonian times, thereafter continuing southward again. With respect to this uncertainty, pertinent stratigraphical evidence from western Gondwana includes the distribution of glacial deposits and cold-water and warm-water faunas. This record, although meagre and to some extent contradictory, appears to favour a drift history consistent with the second (b) of the APW alternatives that involves a rapid southerly excursion of the pole by early Silurian times.

KEY WORDS   Gondwana   Palaeomagnetism   Lower Palaeozoic   Africa   Plate motion   Apparent polar wandering

1. Introduction

Western Gondwana drifted over the south pole during the Palaeozoic Era (Smith *et al*. 1981). During this period western Gondwana was subjected to several episodes of glaciation (Caputo and Crowell 1985). However many of the palaeomagnetic and stratigraphical data supporting this scenario are poorly constrained. This has led to varied interpretations and a situation in need of further critical evaluation. In fact, there are so few mid-Palaeozoic poles from Gondwana, and some of these are discordant, that Scotese *et al*. (1985, p. 63) have asserted that 'the single greatest uncertainty in any Siluro-Devonian (palaeogeographic) reconstruction is the position of the supercontinent of Gondwana'.

Most of the Silurian–Devonian palaeomagnetic poles have been derived from the Tasman (Lachlan) fold belt of southeastern Australia. In as much as the specific palaeogeographic relation of the Tasman terrane to cratonic Australia during mid-Palaeozoic time is unclear (Crook 1980), the interpretation of these

0072–1050/87/TI0341–19$09.50
© 1987 by John Wiley & Sons, Ltd.

palaeomagnetic data with respect to Gondwana as a whole is ambiguous. As a result of this uncertainty several different apparent polar wander (APW) paths for Gondwana have been proposed (Figure 1). If southeastern Gondwana was unconnected to the mainland until late in Devonian time the preferred APW path records a southerly excursion of the pole from northern Africa in Ordovician times to southern Africa by late Palaeozoic times (Figure 1A). On the other hand, if the southeastern Australian poles apply to cratonic Australia, and the rest of Gondwana as well, then a considerably longer APW path, including a relatively rapid south-southwesterly excursion and partial return to the previous track during Silurian–Devonian time, is involved (Figure 1, B, C, D).

In view of this problem a reasonably reliable early Silurian (435 Ma) magnetic pole located just southwest of Cape Town, South Africa, which we have derived from an African source (Hargraves et al. 1987), is pertinent. This new pole position

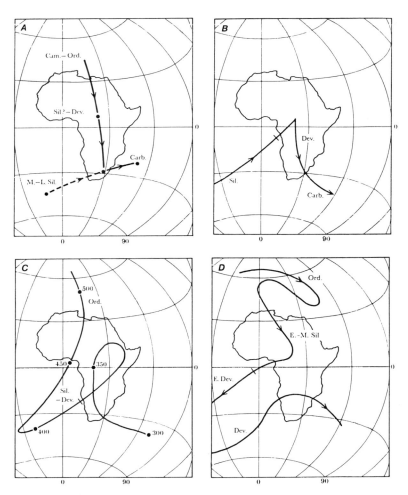

Figure 1. Palaeozoic apparent polar wander paths for Gondwana. (A) McElhinny and Embleton (1974); (B) Schmidt and Morris (1977); (C) Morel and Irving (1978); (D) Goleby (1980). After Livermore et al. (1985)

implies a southerly excursion in the APW path, and it has caused us to review the early and mid-Palaeozoic stratigraphical record of western Gondwana in search of evidence to test this hypothesis, especially in relation to the record of Palaeozoic glaciation. Our concern is restricted to western Gondwana (Figure 2) because the alternative APW paths illustrated in Figure 1 traverse either Africa alone or both Africa and South America, where the distribution of glacial deposits records the passage of the migrating pole.

After surveying the relevant stratigraphy (Figure 3) we discuss the new African palaeomagnetic pole in the light of current compilations of Gondwana data. From this perspective we consider the problems of reconstructing the drift of Gondwana during early and mid-Palaeozoic times. In this review we assume that the Palaeozoic ice caps developed at high latitudes and most readily in the interior of large landmasses. We are aware, however, that the global distribution of the late Proterozoic glacial deposits (Hambrey and Harland 1981) seems to contradict particularly his first assumption. The record of Earth's history may be more complex and obscure than these assumptions suggest.

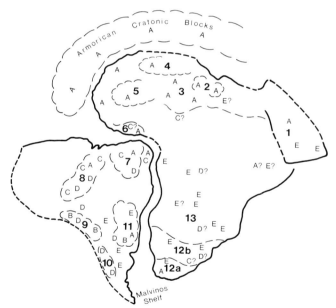

Figure 2. Basins and Palaeozoic glacial deposits of western Gondwana, mainly after Caputo and Crowell (1985) and Hambrey and Harland (1981). Basins: 1. Arabian Basin, NC Saudi Arabia; 2. Kufra and Murzuk basins, S. Libya; 3. Hoggar, S. Algeria; 4. Anti-Atlas and Tindouf Basin, S. Morocco and N. W. Algeria; 5. Taoudeni Basin, Mauritania and Mali; 6. Accra Basin, Ghana; 7. Parnaibo Basin, N. E. Brazil; 8. Amazon–Solimoes Basin, N. C. Brazil; 9. Cancaniri Basin, S. Peru to N. W. Argentina; 10. Argentine Andes, W. C. Argentina; 11. Parana Basin, S. E. Brazil; 12. Cape (a) and Karoo (b) basins, S. Africa; 13. African Platform, S. E. and C. Africa.

Glacial deposits: C? — latest Cambrian to early Ordovician. A — late Ordovician to earliest Silurian. B — early Silurian. C — late Devonian. D — early Carboniferous. E — late Carboniferous

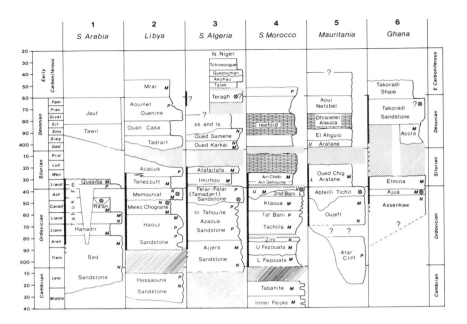
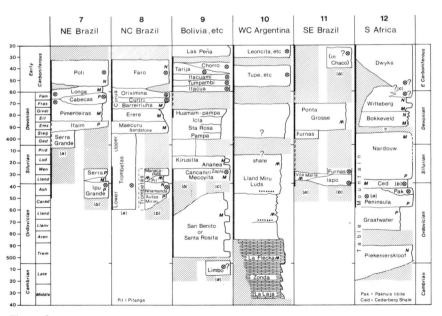

Figure 3.

2. Stratigraphical Record

2a. General statement

A review of the Palaeozoic stratigraphical record of western Gondwana reveals that for the purposes of rigorous reconstruction the information from many localities is sparse, and some of the relevant interpretations appear to be inconsistent or contradictory.

Palaeozoic sedimentary rocks of western Gondwana are preserved along the western margin and in the interior of South America, and on the broad northern and narrow southern margins of Africa (Figure 2). In describing these deposits we refer to basins principally as the geographical location of the record, without implying a particular Palaeozoic tectonic framework.

The Gondwanan deposits (Figure 3) consist mainly of detrital (clastic) successions of quartz-rich sandstones and mudstones (commonly graptolitic shales) in patterns of alternating marine and non-marine facies, and of major transgressions and regressions. They accumulated mostly in a cold-water (Malvinokaffric) realm, although the northern and western cratonic margins lay in warmer water during much of Silurian and Devonian time. Throughout the region three major marine episodes prevailed: late Cambrian–Ordovician sedimentation ending with regression and late Ordovician to early Silurian glaciation; early Silurian sedimentation ending with a widespread late Silurian to earliest Devonian hiatus; and early to late Devonian sedimentation ending with regression and late Devonian to early Carboniferous glaciation (Figures 2, 3, 7), except on the northern margin of western Gondwana where marine sedimentation continued (Wendt 1985).

2b. Cambrian record (540–505 Ma)

Around much of the inner margin of western Gondwana poorly-dated, incomplete, and at least partly non-marine sequences of quartz-rich sandstones and subordinate mudstones were deposited on Precambrian metamorphic rocks (Destombes et al. 1985). Mostly, marine carbonates also accumulated along the fringe of western central South America. Miller (1984) has reported possible Late

---

Figure 3. Correlation chart for western Gondwana. Stratigraphical names in several columns were selected arbitrarily from numerous equivalents (see references below). Alternative intepretations of stratigraphical position are shown in lettered half-width columns. Timescale is after Palmer (1983). M — marine, P — Paralic, N — non-marine; thick left border — cold-water (Malvinokaffric) realm; O — glacial deposit.

1. N. C. Saudi Arabia. McClure 1978; Young 1981.
2. S. Libya. Burollet and Manderscheid 1967; Klitzsch 1981.
3. S. Algeria, N. Niger. Biju-Duval et al. 1981; Hambrey and Kluyver 1981; Legrande 1985.
4. S. Morocco, N. W. Algeria. Destombes 1981; Destombes et al. 1985.
5. Mauritania, Mali. Deynoux and Trompette 1981; Deynoux et al. 1985.
6. Ghana. Talbot 1981; Bär and Riegel 1980; Gray et al. 1985.
7. N. E. Brazil. Caputo and Crowell 1985; Rocha-Campos 1981a. (a) Bär and Riegel 1980; Gray et al. 1985; Soares et al. 1978. (b) Caputo 1985.
8. N. C. Brazil. Caputo and Crowell 1985; Rocha-Campos 1981b, 1981c. (a) Gray et al. 1985. (b) Caputo 1985; Soares et al. 1978.
9. S. Peru to N. W. Argentina. (a) Rocha-Campos 1981e; Miller, 1984. (b) Gray et al. 1985; Laubacher et al. 1982. (c) Caputo and Crowell 1985.
10. W. C. Argentina. Baldis et al. 1984. Frakes and Crowell 1969.
11. S. E. Brazil. (a) Rocha-Campos 1981d; Gray et al. 1985; Copper 1977; Soares et al. 1978. (b) Caputo and Crowell 1985.
12. S. Africa. Rust, 1981a, 1981b. (a) Söhnge 1984. (b) Cocks and Fortey 1986; Tankard et al. 1982. (c) Plumstead 1969; Von Brunn and Stratten 1981.

Cambrian–Early Ordovician glacial deposits in this sequence in Bolivia (see also Rocha–Campos 1981e).

2c. Ordovician record (505–450 Ma)

Mature marine and non-marine sandy deposits with increased amounts of mudstone accumulated around the margin of western Gondwana, in a cold-water realm (Boucot and Gray 1980) that had reached its greatest Palaeozoic extent (Figures 3, 4A). Deposition was essentially continuous in much of northern and southern Africa and western central South America and, as in other parts of the globe, the late Ordovician record was marked by major regression (Berry and Boucot 1973; McKerrow 1979; Soares et al. 1978). Marine flooding apparently did not reach the interior of South America until late Ordovician time when the paralic Autas-Mirim deposits accumulated in the Amazon Basin (Carozzi 1979).

2d. Late Ordovician to early Silurian glaciation (450–435 Ma)

A remarkable record of early Palaeozoic continental glaciation is preserved on western Gondwana (Figures 3, 4A, 4B). Because the age of most of these deposits is not rigorously constrained by dated marine deposits above and below, they have been interpreted in different ways. (1) They may all be essentially coeval records of latest Ordovician to earliest Silurian glaciation (Berry and Boucot 1973; Laubacher et al. 1982; Cooper 1984). (2) They may differ somewhat in age — Caradocian in Saudi Arabia and southern Algeria; Ashgillian in most of northwestern Africa northeastern and southeastern South America and southern Africa; and Llandoverian in western central and southeastern South America (Caputo and Crowell 1985). Söhnge (1984) has found glacial deposits (Hangklip Member) several hundred metres below the top of the Peninsula Formation in South Africa, and suggests that they may be as old as Caradocian because the overlying Pakhuis tillite apparently is Ashgillian in age (Figures 3–12).

2e. Early Silurian marine record (435–425 Ma)

During early Silurian time mudstone and subordinate sandstone, rather like those of the Ordovician, spread widely across northern and southern Africa, western central South America, and into the intracratonic basins of Brazil (Figure 4B). In this transgression northernmost Africa and northwestern South America lay in a warm-water realm whereas the rest of western Gondwana remained in a cold-water zone (Boucot 1985). The precise age of the oldest Silurian marine deposits among the basins varies somewhat or it is poorly constrained. According to published interpretations the transgression could have been essentially synchronous, or diachronously younger from the northern to southern part of western Gondwana.

Fossiliferous marine strata above the late Ordovician to early Silurian glacial deposits in the Brazilian basins, commonly assigned an early Silurian age, may be late Silurian to early Devonian instead (Figures 3–7, 8, 11; Gray et al. 1985).

2f. Late Silurian to earliest Devonian hiatus (425–395 Ma)

Deposition continued locally, but with increasing regression, so that much of western Gondwana was emergent in Late Silurian time (Figure 4C). This regression reached its maximum in latest Silurian–earliest Devonian time (Boucot 1985). The hiatus apparently prevailed longest in northeastern and western Africa, and in the Brazilian basins (Soares et al. 1978), although Gray and others (1985) believe that

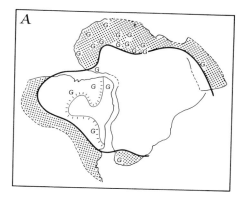
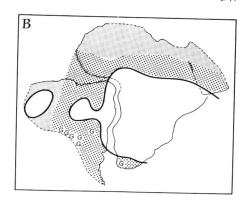
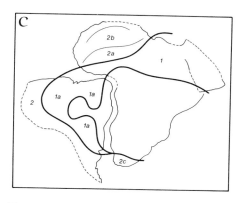
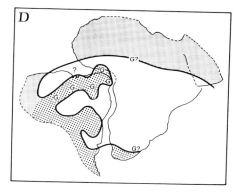

Figure 4. Generalized distribution of Palaeozoic marine deposits and record of glaciation on western Gondwana, modified from Cocks (1972), Gray et al. (1985), and Runnegar (1977). Cold-water (Malvinokaffric) realm — light pattern, warm-water realm — dark pattern; after Copper (1977), Johnson (1979), Boucot and Gray (1980), and Boucot (1985). Dates are from Palmer (1983).
A. Mid-Ordovician shoreline (458–478 Ma). G — late Ordovician – earliest Silurian glacial deposits (435–450 Ma). Position of Brazilian basins outlined for reference.
B. Early Silurian shoreline (425–435 Ma). G — Early Silurian glacial deposits (~430 Ma); G? — probably latest Ordovician (Cocks and Fortey 1986); G' — may be early Devonian glacial deposit.
C. Distribution of late Silurian – earliest Devonian hiatus (395–425 Ma maximum, 400–410 Ma minimum).
  1. Early Silurian directly below early Devonian deposits.
    a. Late Silurian or earliest Devonian may be present.
  2. Essentially complete late Silurian–early Devonian sequence.
    a. Missing locally
    b. Only latest Silurian or earliest Devonian missing
    c. No major hiatus, sequence poorly dated.
D. Mid-Devonian shoreline (375–395 Ma). G — late Devonian glacial deposits (360–370 Ma); G? — may be early Carboniferous glacial deposit.

there are late Silurian deposits in the Amazon and Parnaiba basins (Figures 3–7, 8). The hiatus was briefest along the outer margin of northwestern Africa, in western central South America, and in southern Africa. During much of this episode northern Africa and northwestern and westernmost central South America lay in a warm-water realm (Boucot 1985).

2g. Devonian record (395–360 Ma)

In Early Devonian times western Gondwana was extensively flooded (Boucot 1985). During this transgression only central and southern South America and southern Africa remained in a cold-water zone. By Mid to late Devonian times (Figure 4D) the cold-water realm had spread again to the northern Brazilian basins (Johnson 1979). Near the end of this period, non-marine sediments and glacial deposits developed in the Amazon, Parnaibo, and possibly the Accra basins as well as possibly in Niger and southeastern Africa.

2h. Early Carboniferous record (360–320 Ma)

Paralic to marine facies prevailed in northern Africa while paralic to non-marine sediments, including glacial deposits, accumulated in the Brazilian Basins and along western central South America, as well as in central Africa (McElhinny and Opdyke 1968; Bond 1981) and southeastern Africa (Plumstead 1969; Frakes and Crowell 1970) (Figure 2). This record of an early Carboniferous glaciation reinforces the evidence of glaciation in the northern Brazilian basins in late Devonian times.

3. Deficiencies in the Stratigraphic Record

From our review of the stratigraphic record we now point to significant uncertainties and contradictions that can serve as a focus for future work.

Foremost among these are the major questions: did the tillites now preserved accumulate near the centre of ice caps or toward their fringes? And how far from the pole were the centres? These questions are pertinent because till was deposited as much as 15° from Pleistocene ice centres in the northern hemisphere, and is accumulating 25° from the present Antarctic centre. Furthermore, Pleistocene ice centres were as much as 30–40° from the Earth's rotational axis. Uncertainty about these matters, coupled with ignorance about the size of the early to mid-Palaeozoic ice caps, means that a poorly-constrained pole may have been as much as 40–50° from a particular glacial deposit. It should be noted however, that Pleistocene till only accumulated in low latitudes where an ice cap spread toward the interior of large, northern hemisphere landmasses rather than toward the sea.

Reports of a late Cambrian to early Ordovician glaciation in western central South America and of Cambro-Ordovician glacioeustatic changes in sea level (Miller 1984) contradicts the reasonably reconstructed low latitude position of South America at that time (Smith et al. 1981).

Many of the discussions concerning the late Ordovician to early Silurian glacial records are inconsistent and confused. Some assume that the glaciation was restricted essentially to late Ashgillian times; others emphasize the role of Caradocian and Llandoverian advances. For example, interpretations of the age of glaciation in the Cancaniri Basin (Figure 3–9) range from an erroneous Wenlockian assignment (Hambrey 1985) to Llandoverian or even to late Ashgillian (Laubacher et al. 1982).

Uncertainty about the age of the early Palaeozoic glaciation in South Africa has introduced an additional problem. The older, Hangklip glacial deposit in the Peninsula Formation (Figures 3–12) may be of late Ordovician (Caradocian?) age (Söhnge 1984). The overlying Pakhuis tillite is now considered to be latest Ordovician (Ashgillian; Cocks and Fortey 1986), and not Caradocian as stated by Cooper (1984). Trace fossils from the upper part of the Peninsula Formation, assigned an early Silurian age by Potgieter and Oelofsen (1983, 1984) apparently were misidentified. The Ashgillian Pakhuis glacial deposits, and possibly the Caradocian Hangklip glacial deposits, create an intriguing anomaly because they are so far from the commonly accepted late Ordovician pole position in northern Africa and, in addition, occur along the margin of a flooded shelf, which should have ameliorated the climate. However, if the late Ordovician tillites in both northern and southern Africa accumulated near the fringe of one great ice sheet, they formed within 35° of our postulated late Ordovician geographic pole in central Africa (Hargraves et al. 1987).

If glaciation was limited to latest Ordovician times, for only a few million years, then rapid deglaciation that followed would have produced essentially synchronous regional transgressions. Moreover, the rapid deglaciation must have occurred while the South Pole was still centred on a large landmass; this implies an external, global climate control. If however glaciation persisted from late Ordovician to earliest Silurian times, the ice cap could have waned slowly as it followed the southward path of the pole; this implies a slower, diachronous Silurian transgression. The general geological record indicates important extinctions (Sheenhan 1973; Berry and Boucot 1973) and widespread regression (Brenchley and Newhall 1980) at the end of the Ordovician Period, followed by an extensive early Silurian transgression commonly attributed to glacioeustatic control (Berry and Boucot 1973; McKerrow 1979; Cooper 1984). Bjorlikke (1985), on the other hand, is cautious about ascribing these changes in sea level to the Gondwanan glaciation. He maintains that even if there were eustatic lowering of sea level in northern Europe it coincided with an episode in which tectonic unconformities were developed and therefore is difficult to evaluate. Clarification of these discrepancies is necessary for understanding the role of a melting Gondwanan ice cap in producing the early Silurian transgressions. On present evidence they probably were not synchronous throughout western Gondwana (c.f. Ross 1984).

A major hiatus prevailed throughout much of western Gondwana during late Silurian and earliest Devonian time. Nevertheless none of the reports reassigning late Ordovician to early Silurian sequences a late Silurian to earliest Devonian age (Figure 3) points to the consequences of deposition occurring during the regional hiatus. Confusion is also introduced by assigning an early Devonian age to the Serra Grande Group in the Parnaibo Basin (Bär and Riegel 1980; Soares et al. 1978) and to the Furnas Formation in the Parana Basin (Rocha–Campos 1981d; Copper 1977; Soares et al. 1978) without mentioning the possible presence of glacial deposits (Caputo and Crowell 1985).

By late Devonian times a cold-water regime had spread again to the northern Brazilian Basins, and glacial deposits accumulated there (Caputo 1985) as well as in the Accra Basin. Paradoxically, this glaciation apparently left no record in western central South America or southeastern South America which also lay in a cold climatic zone. Additional complications are introduced by Soares et al. (1978) reporting that the Brazilian sequences probably consist of turbidites, but did not mention the possibility of interbedded glacial deposits. Caputo (1985) reviewed some of this confused situation, but failed to refer to the contradictory interpretation of Soares et al. (1978).

Clearly, these problems require a thorough, critical reappraisal of all the available evidence and differing interpretations.

Table 1. Palaeozoic south poles for Gondwanan continents

Lat	Long	Ref	Symbol	Name and age
AFRICA				
47·0	42·0	1	AA	Anti Atlas 530 Ma
−43·4	08·6	4	AIR	Air intrusives 435 Ma
51·0	353·0	1	BL	Blaubeker fm 471 Ma
13·6	23·5	3	BO	Bokkeveld D
−32·0	64·0	1	CH	Chougrane redbeds Pl
27·5	351·4	2	DAM	Damara belt 450–550 Ma
−26·0	26·0	1	DW	Dwyka varves Cl
−46·0	40·0	1	G	Galula redbeds Pl–Cu
−35·2	43·6	6	GN	Gneiguira redbeds Dm
26·5	10·2	3	GR	Graafwater TM group O
14·0	336·0	1	HK	Hook intrusives Ol–Cau 500±17 Ma
53·0	26·0	1	HM	Hasi-messaud seds O–Ca
05·0	271·0	1	N2	Nama N2 Ca
28·0	345·0	1	NT	Ntonya ring structure 479 Ma
−37·1	42·6	5	SAB	Sabaloka 399±7 Ma
−27·0	89·0	1	SON	Songwe Ketewaka redbeds Pl
−39·0	56·0	1	TAZ	Taztot trachyandesite Pl
50·0	349·0	1	TM	Lower TMS shales O
SOUTH AMERICA				
−17·4	77·6	7	AL	Alcaparrosa Ou
−53·9	53·8	7	AMP	Amana Talampaya Paganzo Pu–Trl
−60·5	95·0	7	AMV	Amana Talampaya villa Union P–Tr
1·8	344·1	9	B1	Bolivian seds O
7·3	346·1	9	B2	Bolivian seds Dl(O?)
83·1	4·5	7	CNT	Campanario north Tilcara Ca
−51·9	71·9	7	CO	Corumbatai Pu–Trl
85·5	337·6	7	CST	Campanario south Tilcara Ca
−44·7	81·6	7	IR	Irati Pm–u
26·4	4·6	9	JU	Jujuy redbeds O
−23·2	61·4	9	LCB	La Colina basalt Pl
−47·0	81·9	7	LCC	La Colina rio Chaschuil P
−45·4	70·2	7	LCH	La Colina los Color highest Cu–Pl
−24·2	58·8	7	LCL	La Colina los Color lower Cu–Pl
−19·4	45·5	7	LCM	La Colina las Mellizas Cu 295±5 Ma
−60·5	64·3	7	LCP	La Colina Paganzo Pl–m≥266·7 Ma
−47·7	54·2	7	LCU	La Colina los Color upper Cu–Pl
−33·1	61·5	9	PA	Paganzo huaco P–C
−63·3	76·0	7	PI	Pillahuinco Pu
−19·6	47·2	9	PIA	Piaui fm Cu
−16·9	58·3	9	PIP	Pipiral fm Cu(–P)
22·6	24·4	8	PIQ	Piquete complex 650–450 Ma
−17·9	14·2	9	PP	Picos Passagem series Dl
−25·1	59·3	9	P1	Paganza 1 los Colorados P–C
−47·7	54·2	9	P2	Paganza 2 los Colorados Pu
24·8	0·5	9	SA	Salta Jujuy redbeds Ca.O
24·9	47·4	7	SU	Suri Ol
−17·1	41·4	9	TA	Taiguati fm C
−52·1	86·3	9	TU	Tupambi fm C
−22·6	57·4	7	TUM	Tubarao Mococa Cu

Table 1. (contd.)

Lat	Long	Ref	Symbol	Name and age
−21·4	52·7	7	TUS	Tubarao Sorocaba itap Cu
39·4	13·2	9	UR	Urucum fm O(–S)
AUSTRALIA				
21·1	19·6	10	AD	Arooma dam seds Cal
−17·4	05·8	10	AV	Ainslie volcanics Dl
25·8	319·3	10	APV	Antrim plat volc Cal
−7·5	24·1	10	BG	Bowning Gp Dl
3·4	22·1	12	CB	Canning Basin reef Du
−31·1	74·1	10	CF	Currabubulla Fm Cu
−25·8	05·4	11	CV	Comerong volcanics Du
31·5	358·3	10	DG	Dundas Gp Cau
71·5	338·7	10	HF	Hudson Fm Cam
−20·5	39·8	10	HG	Housetop granite Dm 375 Ma
67·5	33·1	10	HRS	Hugh river shale Cal–m
43·7	09·7	10	JF	Jinduckin Fm Ol
40·8	10·2	10	LFG	Lake Frome gp Cam–u
−13·2	51·5	10	MD	Mulga Downs Du
−41·4	63·8	10	MG	Main glacial stage Cu
−55·5	98·0	10	MM	Milton monzonite Pu 240 Ma
−26·3	09·3	10	MPl	Mugga porphyry Su 423 Ma
15·7	24·9	10	MS	Mereenie ss S?–D
−29·4	72·2	10	PC	Permo-Carb. volcanics P–C
−35·3	64·2	10	RC	Rocky creek cgl Cu
−48·5	13·5	11	SR	Snowy river volcanics Dl
54·4	51·7	10	SS	Stairway ss Dm
−43·4	333·6	10	SV	Silurian volcanics Sm–u
0·8	343·3	10	TRD	Todd river dolomite Cal
27·1	17·8	10	TS	Tumblagooda ss O?(Cam–Sl)
−32·7	70·9	10	UM	Upper marine latites Pm–u 248 Ma
−49·8	16·7	10	VV	Visean volcanics Cl
−30·7	11·4	10	YA	Yetholme adamellite Cu 320 Ma
ANTARCTICA				
49·2	15·7	10	MS	Mirnyy station charnockite 502 Ma Cau–Ol
17·5	15·3	10	SRM	Sor Rondane intrusives Ol 480 Ma
42·2	18·5	10	TV	Taylor valley lamprophyre Ol 470 Ma

References
1. Brock 1981
2. Corner 1983
3. Bachtadse et al. 1984
4. Hargraves, Dawson, and Van Houten 1987
5. Saradeth et al. 1985
6. Kent et al. 1984
7. Vilas 1981
8. Filho et al. 1985; Filho et al. 1986
9. McElhinny 1973
10. Embleton 1981
11. Schmidt and Embleton 1985; Schmidt et al. 1986
12. Hurley and Van der Voo 1985; Hurley et al. 1985
Note: Poles rotated back to Africa, where appropriate, according to the reconstruction of Smith and Hallam (1970).

4. Review of the Palaeomagnetic record

We have compiled all of those Palaeozoic poles from Gondwanan continents (see Table 1) that are listed in the reviews of McElhinny (1973), Brock (1981), Embleton (1981), Vilas (1981), and Schmidt and Embleton (1985). Several poles published (in paper or abstract) subsequent to the most recent 'official' compilation for a particular continent have been added. All of these data (South Poles) are plotted in Figure 5, and are designated by symbols for each country and letters for abbreviated name of pole, as in the official publications. Rather than reappraise the reliability of each pole, we have included all of those accepted by the regional compilers. In our judgement uncertainty about the age of the palaeomagnetic poles is the most common cause of error and confusion in constructing APW curves. For this reason we have plotted (Figure 5) all of the Palaeozoic data given in Table 1, rotated back to Africa according to the Gondwana reconstruction of Smith and Hallam (1970). Most of the oldest poles are north of Africa or in northern Africa, and Permian poles are concentrated in or southeast of Antarctica. The general direction of the apparent polar wander is clear.

The relative paucity of Silurian–Devonian palaeomagnetic data is emphasized by plotting poles of this time period alone (Figure 6). With the exception of our new Air (Niger) pole, the five poles southwest of southern Africa were obtained from the Tasman foldbelt of southeastern Australia. As pointed out in the introduction, the relevance of data from this terrain to the drift of cratonic Australia, and hence all of Gondwana, is uncertain. Goleby (1980) has presented palaeomagnetic data from southeastern Australia which apparently support a rapid Silurian–Devonian movement of (western) Gondwana and he proposed the APW path illustrated in Figure 1D. In their most recent review Schmidt and Embleton (1985) exlcuded Goleby's results on the grounds that they lacked sufficient sampling density and/or structural control. His study was, however, considered to be a useful reconnaissance. Significantly, Goleby's Cambrian pole and earliest Silurian pole fall broadly in and around northern Africa, as do most Gondwanan poles of the same vintage (compare Figure 5). If his data are at all reliable they suggest that the terranes, including southeastern Australia, were at least close to cratonic Australia in early Palaeozoic time. This regional relation was originally suggested by Morel and Irving (1978), and was the basis for the apparent polar wandering they proposed (Figure 1C). Unfortunately there appears to be no compelling independent geological evidence for either including or excluding southeastern Australia from Gondwana in early and mid-Palaeozoic times (Crook 1980). However, the Air pole, at latitude 43·4 S, longitude 08·6 E ($n=12$ sites, $k=50$, $x95=6·2$), from cratonic Africa, suggests that the movement of Gondwana was similar to that illustrated in Figure 1C, and that the southeastern Australian data are valid for the whole of Gondwana. The drift required by this excursion was 150 mm/yr for 40 Ma, which although high, is not an unreasonable rate. Nevertheless, in view of the uncertain age and structural relation of most of the Palaeozoic poles we must acknowledge that at present there is no unambiguous palaeomagnetic basis for favouring either the straight Ordovician–late Palaeozoic APW path or one which includes a rapid southerly late Ordovician–early Silurian excursion. Confidence in our Niger (Air) pole leads us to favour the latter alternative, but we emphasize that establishing such a history awaits confirmation by finding other similar poles, of a similar age, elsewhere in cratonic Gondwana.

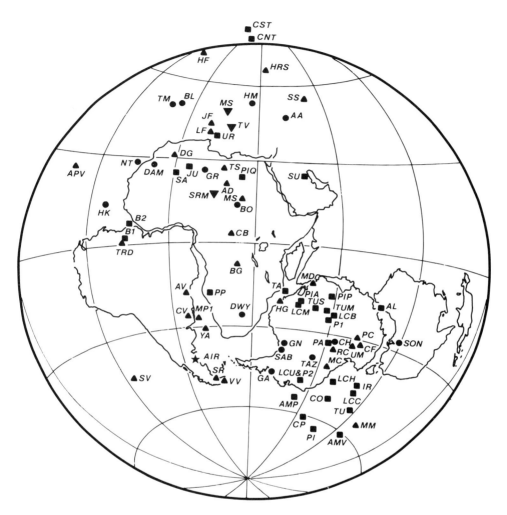

Figure 5. Plot of all published Palaeozoic pole positions for Gondwana (see Table 1 for details and dates), rotated according to fit of Smith and Hallam (1970). Source of data: circles — Africa; squares — South America; upright triangles — Australia; inverted triangles — Antarctica; star — Aïr (Niger) pole. Cambrian and Ordovician poles are mainly north of or in northern Africa; Permian poles lie in and southeast of Antarctica.

5. Preferred reconstruction

On the basis of the available evidence reviewed here we suggest that the following succession of Palaeozoic scenarios occurred as (western) Gondwana drifted toward and oscillated over the South Pole.

Figure 6. Plot of all reported Silurian – Devonian pole positions for Gondwana. Symbols as in Figure 5. Large solid triangle — Canning Basin pole (Hurley *et al.* 1985).

At the beginning of the Palaeozoic Era the South Pole was located just north of Africa in the palaeo–Atlantic ocean, and western Gondwana had no Antarctic ice cap. In Cambrian times quartz-rich sandy deposits began to accumulate around the continental margins which were increasingly transgressed.

During Ordovician times Gondwana drifted over the South Pole which by the end of the Ordovician period was located far inland in northwestern Africa, and all of the supercontinent lay in a cold-water realm. Although glaciation may have begun in Mid-Ordovician times (Spjeldnaes 1981), its oldest record apparently is Caradocian in northern Africa. As the pole continued to shift southwards, widespread glaciation persisted for about 15 Ma, and by early Silurian times reached western central and possibly southeastern South America (Figure 3–11). The general palaeogeographic agreement between the palaeomagnetic pole position and dated glacial deposits, together with faunal evidence of a cold-water realm, is convincing evidence of the usefulness of the palaeomagnetic data for evaluating polar wander paths. Assuming that the dating of relevant deposits is reliable enough to define the ages of the tillites, the persistence of glaciation into the Lower Silurian implies that the widespread very early Silurian transgression (Figure 7) could not have been produced simply by a melting ice cap.

As a result of rapid northward drift during the early Silurian the northern margin of western Gondwana moved into a warm-water realm, and the South Pole had shifted as far southwards as the Malvinos shelf (Figures 2, 6; the location of the Air–Niger pole). In this maritime position the polar ice cap disappeared altogether. A widespread hiatus that developed during this interval (Figure 4C) may have been in part the result of isostatic readjustment but the hiatus also correlates in a general way with a global regression (Boucot 1985). As with the Late Ordovician record of transgressions and regressions, however, there is no simple correlation between waxing and waning of glaciation and global changes in sea level. On the other hand, correlation of widespread transgressions and regressions with changing rates of continental drift (Figure 7) suggests a tectonic control related to fast or slow sea floor spreading (Pitman 1978).

During Devonian times the South Pole shifted rapidly back to central Africa (Murley and Van der Voo, 1985), inducing, at least locally, late Devonian glaciation recorded mainly in northern Brazil (Caputo 1985). This distribution suggests that a barrier north of the Brazilian basins (Johnson 1979) separated them from a warmer water realm along the northern margin of western Gondwana.

As Gondwana began its drift northward again the South Pole shifted southeastward across Africa and the ice cap expanded. In early Carboniferous times glacial deposits accumulated more widely, both in the northern Brazilian basins and

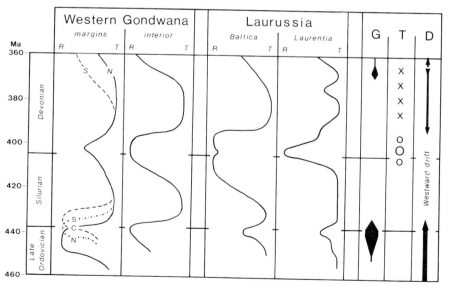

Figure 7. Major Mid-Palaeozoic events. Adapted in part from Boucot (1985), Hallam (1984), and McKerrow (1979). R — regression, T — transgression; N — northern, C — central, S — southern; G — Gondwanan glaciation; T — tectonic activity; X — extensional phase along southern margin of Laurussia and northern margin of Western Gondwana, O — late Caledonian orogeny; D — rate and north and south direction of Gondwana drift, width of arrow is proportional to estimated drift rate.

farther south in western central South America, as well as possibly in central and southeastern Africa.

Although beyond the scope of our review, we note an implied tectonic consequence (Figure 7) of the reconstructed pattern of mid-Palaeozoic drift. The suggested relatively rapid northward drift of Gondwana during Silurian time, accompanied by widespread transgression, may have progressively closed the early Palaeozoic ocean, producing an assemblage of Pangaeic continents essentially like that reconstructed by Boucot and Gray (1980), Boucot (1985) for late Silurian and early Devonian times. The collision of Armorica with Baltica and Laurentia during the late Caledonian–Acadian orogeny (Perroud *et al.* 1984; Dallmeyer and Gee 1986) was accompanied by extensive regression. With a change to rapid southward drift during the Devonian Period an ocean reopened between western Gondwana and those Armorican blocks that remained attached to Laurentia and Baltica, and it was accompanied by a major transgression. This Mid to late Devonian episode of reversed drift was also accompanied by extensional deformation along the southern margin of Laurussia (Leeder 1976; Ziegler 1978) and the northern margin of western Gondwana (Wendt 1985).

References

Bachtadse, V., Van der Voo, R., and Kroner, A. 1984. Palaeomagnetic results from the lower Devonian part of the Table Mountain Group, South Africa. *EOS*, **65**, 863.

Baldis, B., Beresi, M., Bordonaro, O., and Vaca, A. 1984. The Argentine Precordillera as a key to Andean Structure. *Episodes*, **7**, 14–19.

Bär, P. and Riegel, W. 1980. Mikrofloren des hochsten Ordovician bis tiefen Silurian aus dem unteren Sekoni-Serie von Ghana. *Neu. Jb. Geol. Päleon. Abh.*, **160**, 42–60.

Berry, W. B. N. and Boucot, A. J. 1973. Glacio-eustatic control of Late Ordovician–Early Silurian platform sedimentation and faunal changes. *Geological Society of America Bulletin*, **84**, 275–283.

Biju-Duval, B., Deynoux, M., and Rognon, P. 1981. Late Ordovician tillites of the central Sahara. In: Hambrey, M. J. and Harland, W. B. (Eds) *Earth's Pre-Pleistocene Glacial Record* Cambridge, London. 99–107.

Bjorlikke, K. 1985. Glaciations, preservation of their sedimentary record and sea level changes. *Paleogeography, Paleoclimatology, Paleoecology*, **51**, 197–207.

Bond, G. 1981. The Late Paleozoic (Dwyka) glaciation in the Middle Zambezi region. In: Hambrey, M. J. and Harland, W. B. (Eds), *Earth's Pre-Pleistocene Glacial Record*. Cambridge, London. 55–57.

Boucot, A. J. 1985. Late Silurian–Early Devonian biogeography, provincialism, evolution, and extinctions. *Philosophical Transactions of the Royal Society, London*, **B309**, 323–339.

—— and Gray, J. 1980. Cambro-Permian pangaeic models. *Geological Association, Canada, Special Paper*, **20**, 389–419.

Brenchley, P. J. and Newhall, G. 1980. A facies analysis of Upper Ordovician regressive sequences in the Oslo region, Norway — a record of glacio-eustatic changes. *Palaeogeography Palaeoclimatology, Palaeoecology*, **31**, 1–38.

Brock, A. 1981. Paleomagnetism of Africa and Madagascar. In: McElhinny, M. W. and Valencio, D. A., (Eds), *Paleoreconstruction of the Continents*. Geodynamic Series, **2**, American Geophysical, pp. 65–76.

Burollet, P. F. and Manderscheid, G. 1967. Le Devonian en Libye et en Tunisie. *Intern. Symp. Devon. Syst.*, **1**, 285–302.

Caputo, M. V. 1985. Late Devonian glaciation in South America. *Palaeogeography, Palaeoclimatology, Palaeoecology*, **51**, 291–317.

—— and Crowell, J. C. 1985. Migration of glacial centers across Gondwana during Paleozoic Era. *Geological Society of American Bulletin*, **96**, 1020–1036.

Carozzi, A. V. 1979. Petroleum geology in the Paleozoic clastics of the Middle Amazon Basin, Brazil. *Journal of Petroleum Geology*, **2**, 55–74.

Cocks, L. R. M. 1972. The origin of the Silurian *Clarkeia* shelly fauna of South America, and its extension to west Africa. *Palaeontology*, **15**, 623–630.

—— and Fortey, R. A. 1986. New evidence on the South African Lower Palaeozoic: age and fossils reviewed. *Geological Magazine*, **123**, 437–444.

Cooper, M. R. 1984. Discussion of 'Cruziana acacensis — the first Silurian index-trace fossil from South Africa. *Transactions Geological Society of South Africa*, **86**, 53–54.

Copper, P. 1977. Paleolatitudes in the Devonian of Brazil and the Frasnian-Famennian extinction.

Palaeogeography, Palaeoclimatology, Palaeoecology, **21**, 165–207.

Corner, B. 1983. An interpretation of the aeromagnetic data covering the western portion of the Damara Orogen in South West Africa/Namibia. *Special Publication of the Geological Society of South Africa*, **11**, 339–354.

Crook, K. A. W. 1980. Fore-arc evolution in the Tasman Geosyncline: the origin of the Southeast Australian continental crust. *Journal Geological Society of Australia*, **27**, 215–232.

Dallmeyer, R. D. and Gee, D. G. 1986. $^{40}Ar^{39}Ar$ mineral dates from retrogressed ecologites within the Baltoscandian miogeocline: implications for a polyphase Caledonian orogenic evolution. *Geological Society of American Bulletin*, **97**, 26–34.

Destombes, J. 1981. Hirnantian (Upper Ordovician) tillites on the north flank of the Tindouf Basin, Anti-Atlas, Morocco, In: Hambrey, M. J. and Harland, W. B. (Eds) *Earth's Pre-Pleistocene Glacial Record* Cambridge, London. 84–88.

——, **Hollard, H., and Willefert, S. 1985.** Lower Palaeozoic rocks of Morocco. In: Holland, C. H. (Ed.) *Lower Palaeozoic of North-western and West-central Africa* Wiley, Chichester. 91–336.

Deynoux, M., Sougy, V., and Trompette, R. 1985. Lower Palaeozoic rocks of west Africa and the western part of central Africa. In: Holland, C. H. (Ed.) *Lower Palaeozoic of North-western and West-central Africa*. Wiley, Chichester. 337–495.

—— **and Trompette, R. 1981.** Late Ordovician tillites of the Taoudeni Basin, West Africa. In: Hambrey, M. J. and Harland, W. B. (Eds), *Earth's Pre-Pleistocene Glacial Record*. Cambridge, London. 89–91.

Embleton, B. J. J. 1981. A review of the Paleomagnetism of Australia and Antarctica. In: McElhinny, M. W. and Valencio, D. A. (Eds), *Paleoreconstruction of the Continents. Geodynamics Series*, **2**, *American Geophysical Union*, 77–92.

Filho, M. S. D'A., Sato, K., and Pacca, I. G. 1985. Paleomagnetism of metamorphic rocks from the Piquete Complex, Southern Brazil. *Abs. Internat. Assoc. Geomag. Aeron.*, **1**, 225.

——, **Pacca, I. G., and Sato, K. 1986.** Paleomagnetism of metamorphic rocks from the Piquete region — Ribeira Valley, southeastern Brazil (preprint).

Frakes, L. A. and Crowell, J. C. 1969. Paleozoic glaciation — Part I, South America. *Geological Society of America Bulletin*, **80**, 1007–1042.

—— **and Crowell, J. C. 1970.** Late Paleozoic glaciation — Part II, Africa exclusive of the Karroo Basin. *Geological Society of America Bulletin*, **81**, 2261–2286.

Goleby, B. R. 1980. Early Paleozoic paleomagnetism in south east Australia. *J. Geomag. Geoelec.*, **32**, Supl. III, SIII-1 – SIII-21.

Gray, J., Colbath, G. K., deFaria, A., and others 1985. Silurian-age fossils from the Paleozoic Parana Basin, southern Brazil. *Geology*, **13**, 521–525.

Hallam, A. 1984. Pre-Quaternary sea-level changes. *Ann. Rev. Earth Planet. Sci.*, **12**, 205–243.

Hambrey, M. J. 1985. The Late Ordovician–Early Silurian glacial period. *Palaeogeography, Palaeoclimatology, Palaeoecology*, **51**, 273–289.

—— **and Harland, W. B. 1981.** *Earth's Pre-Pleistocene Glacial Record*. Cambridge, London.

—— **and Kluyver, H. M. 1981.** Evidence of Devonian or Early Carboniferous glaciation in the Agades region of Niger. In: Hambrey, M. J. and Harland, W. B. (Eds), *Earth's Pre-Pleistocene Glacial Record*. Cambridge, London. 188–190.

Hargraves, R. B., Dawson, E. M., and Van Houten, F. B. 1987. Palaeomagnetism and age of Mid Paleozoic ring complexes in Niger, and tectonic implications *Geophysical Journal Royal Astronomical Society* **90**. 705–729.

Hurley, N. F. and Van der Voo, R. 1985. Paleomagnetism of Upper Devonian reefal limestones, Canning Basin, Western Australia. *Abs., EOS*, **66**, 259.

——, **Bachtadse, V., Ballard, M. M., and Van der Voo, R. 1985.** Early and mid-Paleozoic paleomagnetism of South Africa and Australia — new constraints for the Gondwana apparent polar wander path. *6th Gondwana Symp.*, Columbus, Ohio, abstr., in press.

Johnson, J. G. 1979. Devonian brachiopod biostratigraphy. In: House, M. R., Scrutton, C. T., and Bassett, M. G. (Eds), *The Devonian System, Spec. Pap. Palaeont.*, **23**, 291–309.

—— **and Boucot, A. V. 1973.** Devonian brachiopods. In: Hallam, A. (Ed.) *Atlas of Palaeobiography* 89–96. Elsevier, Amsterdam.

Kent, D. V., Dia, O., and Sougy, J. M. A. 1984. Paleomagnetism of lower–middle Devonian and upper Proterozoic — Cambrian (?) rocks from Mejeria (Mauritania), West Africa. In: Van der Voo, R., Scotese, C. R., and Bonhommet, N. (Eds), *Plate reconstruction from Paleozoic Paleomagnetism Geodynamics Series*, **12**, *American Geophysical Union*, 99–115.

Klitzsch, E. 1981. Lower Paleozoic rocks of Libya, Egypt, and Sudan. In: Holland, C. H. (Ed.), *Lower Palaeozoic of the Middle East, Eastern and Southern Africa, and Antarctica*. Wiley, New York. 131–163.

Laubacher, G., Boucot, A. J., and Gray, J. 1982. Additions to Silurian stratigraphy, lithofacies, biogeography and paleontology of Bolivia and southern Peru. *Journal of Palaeontology*, **56**, 1138–1170.

Leeder, M. R. 1976. Sedimentary facies and the origins of basin subsidence along the northern margin of the supposed Hercynian ocean. *Tectonophysics*, **36**, 167–179.

Legrande, Ph. 1985. Lower Palaeozoic rocks of Algeria. In: Holland, C. H. (Ed.) *Lower Palaeozoic of North-western and West-central Africa*. Wiley, Chichester. 6–89.

Livermore, R. A., Smith, A. G., and Briden, J. C. 1985. Paleomagnetic constraints on the distribution of continents in the late Silurian and early Devonian. *Philosophical Transactions of the Royal Society, London*, **B309**, 29–56.

McClure, H. A. 1978. Early Palaeozoic glaciation in Arabia. *Palaeogeography, Palaeoclimatology, Palaeoecology*, **25**, 315–326.

McElhinny, M. W. 1973. *Paleomagnetism and Plate-tectonics*. Cambridge Univ. Press, London.

—— and Embleton, B. J. J. 1974. Australian palaeomagnetism and Phanerozoic plate tectonics of eastern Gondwanaland. *Tectonophysics*, **22**, 1–29.

—— and Opdyke, N. D. 1968. Paleomagnetism of some Carboniferous glacial varves in central Africa. *Journal of Geophysical Research*, **73**, 689–696.

McKerrow, W. S. 1979. Ordovician and Silurian changes in sea level. *Journal Geological Society London*, **136**, 137–145.

Miller, J. P. 1984. The Cambro-Ordovician boundary: biomere extinctions, erosion surfaces, and glacio-eustatic(?) fluctuations. *Geological Society of America, Abstracts with Programs*, **16**, 596–597.

Morel, P. and Irving, E. 1978. Tentative palaeocontinental maps for the early Phanerozoic and Proterozoic. *Journal of Geology*, **86**, 535–561.

Palmer, A. R. (compiler), 1983. The Decade of North American Geology 1983 geologic time scale. *Geology*, **11**, 503–504.

Perroud, H., Van der Voo, R., and Bonhommet, N. 1984. Paleozoic evolution of the Armorica plate on the basis of paleomagnetic data. *Geology*, **12**, 579–582.

Pitman, W. C. 1978. Relationship between eustacy and stratigraphic sequences of passive margins. *Geological Society of America Bulletin*, **89**, 1389–1403.

Plumstead, E. P. 1969. Three thousand million years of plant life in Africa. *Geological Society of South Africa, Annex*, **72**, 72 pp.

Potgieter, C. D. and Oelofsen, B. W. 1983. *Cruziana acacensis* — the first Silurian index-trace fossil from southern Africa. *Transactions Geological Society of South Africa*, **86**, 51–54.

—— and Oelofsen, B. W. 1984. Author's reply to discussion of '*Cruziana acacensis* — the first Silurian index-trace fossil from South Africa'. *Transactions of the Geological Society of South Africa*, **87**, 54–55.

Rocha-Campos, A. C. 1981a. Middle–Late Devonian Cabecas Formation, Parnaibo Basin, Brazil. In: Hambrey, M. J. and Harland, W. B. (Eds.), *Earth's Pre-Pleistocene Glacial Record*. Cambridge, London. 892–895.

—— 1981b. Late Ordovician? – early Silurian Trombetas Formation, Axaon Basin, Brazil. In: Hambrey, M. J. and Harland, W. B. (Eds.), *Earth's Pre-Pleistocene Glacial Record*. Cambridge, London. 896–898.

—— 1981c. Late Devonian Curura Formation, Amazon Basin, Brazil. In: Hambrey, M. J. and Harland, W. B. (Eds.), *Earth's Pre-Pleistocene Glacial Record*. Cambridge, London. 881–891.

—— 1981d. Early Paleozoic Iapo Formation of Parana, Brazil. In: Hambrey, M. J. and Harland, W. B. (Eds.), *Earth's Pre-Pleistocene Glacial Record*. Cambridge, London. 908–909.

—— 1981e. The Cambrian (?) Limbo Group of Bolivia. In: Hambrey, M. J. and Harland, W. B. (Eds.), *Earth's Pre-Pleistocene Glacial Record*. Cambridge, London. 910–911.

Ross, R. J. 1984. The Ordovician System, progress and problems. *Ann. Rev. Earth Planet Sci.*, **12**, 307–335.

Runnegar, B. 1977. Marine fossil invertebrates of Gondwanaland: palaeogeographic implications. *3th Gondw. Symp.*, Calcutta, 144–158.

Rust, I. C. 1981a. Lower Palaeozoic rocks of southern Africa. In: Holland, C. H., (Ed.), *Lower Palaeozoic of the Middle East, Eastern and Southern Africa, and Antarctica*. Wiley, New York. 165–187.

—— 1981b. Early Paleozoic Pakhuis tillite, South Africa. In: Hambrey, M. J. and Harland, W. B. (Eds.), *Earth's Pre-Pleistocene Glacial Record*. Cambridge, London. 113–117.

Saradeth, S., Schult, A., and Soffel, H. 1985. Four, new paleomagnetic results from Egypt and Sudan. *Abs. 13th Colloq. of African Geology*, St. Andrews, Scotland, 163 pp.

Schmidth, P. W. and Embleton, B. J. J. 1985. A review of Australian Paleozoic paleomagnetic poles and suspect terranes. Submitted: *Circum-Pacific terrane Conference*.

——, Embleton, B. J. J., Cudahy, T. J., and Powell, C. M. 1986. Prefolding and premegakinking magnetizations from the Devonian Comerong volcanics, New South Wales, Australia, and their bearing on the Gondwana pole path. *Tectonics*, **5**, 135–150.

Scotese, C. R., Van der Voo, R., and Barrett, S. F. 1985. Silurian and Devonian base maps. *Philosophical Transactions of the Royal Society, London*, **B309**, 57–77.

Sheenhan, P. M. 1973. The relation of the Late Ordovician glaciation to the Ordovician–Silurian changeover in North American brachiopod faunas. *Lethaia*, **6**, 147–154.

Smith, A. G. and Hallam, A. 1970. The fit of the southern continents. *Nature*, **225**, 139–144.

——, Briden, J. C., and Hurley, A. M. 1981. *Phanerozoic Paleocontinental World Maps*. Cambridge Univ. Press, New York.

Soares, P. C., Landim, P. M. B., and Fulfaro, V. J. 1978. Tectonic cycles and sedimentary sequences in the Brazilian intracratonic basins. *Geological Society America Bulletin*, **89**, 181–191.

Söhnge, A. P. G. 1984. Glacial diamictite in the Peninsula Formation near Cape Hangklip. *Transactions of the Geological Society of South Africa*, **87**, 199–210.

Spjeldnaes, N. 1981. Lower Palaeozoic palaeoclimatology. In: Holland, C. H. (Ed.), *Lower Palaeozoic of the Middle East, Eastern and Southern Africa, and Antarctica*. Wiley, New York. 199–256.

Talbot, M. R. 1981. Early Palaeozoic? diamictites of southwest Ghana. In: Hambrey, M. J. and Harland, W. R., (Eds.), *Earth's Pre-Pleistocene Glacial Record*. Cambridge, London. 108–112.

Tankard, A. J., Jackson, M. P. A., Eriksson, K. A., Hobday, D. K., Hunter, D. R., and Minter, W.

E. L. **1982**. *Crustal Evolution of Southern Africa, 308 Billion Years of Earth History*. Springer-Verlag, New York.

Vilas, J. F. A. **1981**. Paleomagnetism of South American rocks and the dynamic processes related to the fragmentation of western Gondwana. In: McElhinny, M. W. and Valencio, D. A. (Eds.), *Paleoreconstruction of the Continents. Geodynamics series*, **2**, *American Geophysical Union*, 106–114.

Von Brunn, V. and Stratten, T. **1981**. Late Paleozoic tillites of the Karoo Basin of South Africa. In: Hambrey, M. J. and Harland, W. B. (Eds.) *Earth's Pre-Pleistocene Glacial Record*. Cambridge, London. 71–79.

Wendt, J. **1985**. Disintegration of the continental margin of northwestern Gondwana: Late Devonian of the eastern Anti-Atlas (Morocco). *Geology*, **13**, 815–818.

Young, G. M. **1981**. Early Paleozoic tillites of the northern Arabian Peninsula. In: Hambrey, M. J. and Harland, W. B. (Eds.), *Earth's pre-Pleistocene Glacial Record*. Cambridge, London. 338–340.

Ziegler, P. A. **1978**. North-western Europe: tectonics and basin development. *Geol. Mijnb.* **57**, 589–626.

The geodynamic significance of alkaline magmatism in the western Mediterranean compared with West Africa

Bernard Bonin and Bernard Platevoet
Laboratoire de Pétrologie–Volcanologie, U.A. 728 (C.N.R.S.), Université de Paris-Sud, 91405 Orsay, France

and

Yves Vialette
Departement de Géologie–Minéralogie, L.A. 10 (C.N.R.S.), 5 rue Kessler, 63038 Clermont-Ferrand, France

Alkaline magmatic provinces occur globally in regions subjected to cratonization. Therefore, they may appear just after the end of an orogenic episode, when collision between two continental plates caused rapid uplift and unroofing of orogenic batholiths by isostatic adjustment under intense stress fields. A reversal in the sense of movement in a pre-existing fault system provoked alkaline magmatism. As reversals in the sense of movement are signs of global tectonics, it can be shown that at each orogeny an alkaline magmatism developed, essentially when relative plate motions became stationary. The Corsican alkaline province is a part of a larger western Mediterranean province, emplaced at the end of the Hercynian orogeny. Lithospheric faults along which melts from the mantle have ascended through the continental crust may subsequently act as preferential lines of weakness for oceanic basin development.

In West Africa, alkaline provinces are essentially emplaced in Pan-African domains. A number of provinces are Cambrian in age, forming just after the termination of the Pan-African orogeny and located near major suture zones. Unlike the western Mediterranean province, the lithospheric faults, responsible for the ascent of the alkaline magmas, have not been reactivated in West Africa and no subsequent oceanic opening has occurred. In addition, Pan-African domains have been the sites of later alkaline provinces, during various time periods since the end of the Pan-African orogeny. Subsolvus granites predominate as early as 10 Ma after the last compressive tectonic phase, then hypersolvus granites supersede them during a 500 Ma period. Lastly, mixed and undersaturated complexes developed with increased proportions of carbonatites.

KEY WORDS   Anorogenic magmatism   Phanerozoic   Mediterranean   West Africa

1. Introduction

Alkaline magmatic provinces are classically considered as anorogenic, i.e. 'unrelated to orogenic disturbance' (Bates and Jackson 1980). The anorogenic character

0072–1050/87/TI0361–27$13.50
© 1987 by John Wiley & Sons, Ltd.

has been used to typify A-granites (Loiselle and Wones 1979) and to distinguish them from S- and I-granites of Chappell and White (1974). Bowden (1985) has pointed out that 'A stands for anhydrous, alkaline, anorogenic as well as aluminous' (p. 26).

We shall focus on anorogenic magmatic provinces which display alkaline rocks, in the enlarged meaning proposed by Black et al. (1985) as well as by Bonin and Giret (1985). The adjective 'alkaline' is given a wider sense than by Sorensen (1974) to encompass hypersolvus, metaluminous, and subaluminous granites, and subsolvus, subaluminous, and peraluminous granites, typically associated with peralkaline granites in within-plate environments. The geodynamic significance of these anorogenic provinces will be discussed using two examples: the Permo-Triassic province of Corsica and Western Mediterranean and the Palaeozoic to recent provinces of West Africa. The structural settings at the time of emplacement as well as the rock-type assemblages will be considered as clues to the petrological and geodynamical interpretations of within-plate magmatism.

2. The alkaline province in Corsica

The general features of the geology of Corsican alkaline ring-complexes have been described elsewhere (Bonin 1977, 1980, 1982; Vellutini 1977). In this section, we shall focus on the geodynamic setting of the basement country rocks and on the significance of the ages obtained on Permo–Triassic alkaline ring-complexes.

2a. The geodynamic setting of the basement country rocks

In West Corsica, alkaline ring-complexes cut a large Variscan composite calc-alkaline batholith (Orsini 1980) covered in some places by later calc-alkaline volcanic units (Vellutini 1977). The batholith has yielded ages of 340 to 300 Ma (Bonin et al. 1972; Maluski 1977), clearly related to Variscan orogenic events (Lameyre and Autran 1980). According to the cordilleran-type of magmatic associations, the batholith can be interpreted as the result of an oceanic closure with collision around the Devonian–Carboniferous boundary between a passive margin (part of the Gondwana continent) and the active margin of the southern edge of the North European shield. During and after the collision, the calc-alkaline batholith has been subjected to rapid uplift and unroofing, as in Early Permian times, and calc-alkaline volcanic flows mantled a deeply eroded and peneplaned area.

The calc-alkaline volcanic events are dated as Lower Permian. The basal andesitic flows are lying unconformably on Stephanian coal-bearing layers and some volcanic ash beds which have yielded some Autunian plants (Cordaites) (Vellutini 1977). The complete volcanic series is composed of, from bottom to top: andesitic pyroclastic flows, dacitic pyroclastic flows, ignimbritic dacites and rhyodacites, and rhyolitic ignimbritic flows. Dyke swarms, trending N60–N70, are particularly abundant in the vicinity of the calc-alkaline volcanic plateaus and thus are considered as feeders of the lavas. No clear central volcanic feature has been described and all the units display fissural characters.

A widespread calc-alkaline volcanism occured during the Lower Permian on the southern margin of Europe, from Galicia (Spain) to the Pamirs (U.S.S.R.) (Mossakovsky 1970). The interpretation of this large-scale volcanic event is subject to discussion. According to Mossakovsky (1970), this volcanic event is linked to subduction processes along an active margin, but according to Ziegler (1983), the preferred model, at least for the western part of the volcanic belt, is that of a continental collision between the Gondwana shelf and the Laurasian block. In this scheme, the convergence between the two continental blocks apparently changed from an essentially northwest–southeast direction during the Westphalian to an essentially east–west direction during the Stephanian–Autunian times. This was accompanied by the development of large dextral shear zones between the Appalachians and the Urals (Arthaud and Matte 1977). In the European fold belt, the development of a complex pattern of conjugate shear faults and related pull-apart structures which transected the consolidated Variscan fold belt as well as its foreland (Figure 1). This fault system, often named as 'Tardi–Hercynian' (Arthaud and Matte 1975), remained active from Stephanian to Autunian times.

In western and central Europe, deep crustal fracturing triggered widespread magmatic activity of highly variable chemistry. In the Variscan fold belt itself, magmatic activity is definitely calc-alkaline (e.g. Corsica, Sardinia, Pyrenees, Alps,. . .) whereas the north European foreland is the site of markedly alkaline magmatism (e.g. Scotland, Oslo Graben). According to Ziegler (1983), the so-called Late Hercynian tectonics and their associated igneous activity are essentially anorogenic and are not so much related to the final phase of consolidation of the Variscan fold belt as they are to the first phase of its disintegration.

Note however for the magmatic phases that, in Corsica, situated in the internal part of the Variscan fold belt, if wrench tectonics can be related to pre-Alpine opening, igneous activity is calc-alkaline, i.e. of orogenic affinity. The problem of the relationship between subduction–collision processes and calc-alkaline magmatism is often hidden by the large time interval between the end of tectonic processes and the end of their magmatic responses (e.g. volcanic-plutonic activity in Turkey and Iran, clearly calc-alkaline from Cretaceous to present, whereas subduction processes have ceased probably in Eocene times, Moine Vaziri 1985). During this critical time of about 30–50 Ma, calc-alkaline magmatism continues in the internal parts of fold belts whereas in the external parts and in their foreland, magmatism is alkaline from the outset.

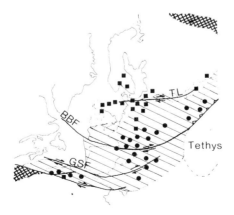

Figure 1. Sketch map of Europe during Stephanian-Autunian times. Symbols: diagonal lines: inactive fold belt, cross-hatched area: active fold belt, squares: *alkaline* igneous centres, circles: *calc-alkaline* igneous centres. Abbreviations: BBF, Bay of Biscay Fault zone, GSF, Gibraltar Strait Fault zone, TL, Tornquist Line.

Table 1a. Ages of emplacement of alkaline ring complexes and associated thermal overprinting

Alkaline complex	Sample	Method	Age (Ma)	References
Monte Cinto	amphibole	K/A isochron	243 ± 15	Bonin (1980)
Porto Scandola–Senino	biotite	$^{39}A/^{40}A$	280 ± 12	
	rhyolites	K/A	208	
Cauro–Bastelica	zircon	WR isochron U/Pb	262 ± 13	
	amphibole	$^{39}A/^{40}A$	310 ± 13	
		K/A isochron	243 ± 15	
	biotite	$^{39}A/^{40}A$	280 ± 9	
		K/A	251 ± 10	
	biotite — WR	Rb/Sr	256 ± 11	
	feldspar — WR	Rb/Sr	201 ± 17	
	hypersolvus	Rb/Sr	244 ± 5	Vialette
	biotite granite	WR isochron	(0·7073)	this paper
Evisa	peralkaline granite	Rb/Sr	246 ± 7	
		WR isochron	(0·7034)	
Cauro–Bastelica	subsolvus	Rb/Sr	238 ± 4	
	biotite granite	WR isochron	(0·7111)	
Tana-Peloso	id.	id.	267 ± 3	
			(0·7132)	
Pastricciola	id.	id.	268 ± 4	
			(0·7222)	

WR Whole rock: A-argon numbers in parentheses refer to initial Sr isotope ratios

Table 1b. Isotopic data on subsolvus biotite granites, Corsica

N°	Rb	Sr	$^{87}Rb/^{86}Sr$	$^{87}Sr/^{86}Sr$
Cauro–Bastelica,				
Aragnasco ring-dyke				
1	348	16·2	63·5	0·9272
2	345	15·2	67·5	0·9414
3	334	16·0	61·8	0·9202
4	310	14·7	62·4	0·9249
5	325	17·9	55·0	0·8942
6	321	12·3	77·8	0·9664
7	351	8·6	123	1·1340
8	(20·54)	473	0·126	0·7120
	(21·16)	473	0·132	0·7118
9	342	44·4	22·5	0·7876
Tana–Peloso,				
Tana centre				
10	387	3·90	322	1·9583
11	291	8·35	107	1·1093
12	400	7·25	170	1·3774
13	234	50·0	13·6	0·7644
14	158	12·17	38·0	0·8530
15	171	58·1	8·55	0·7472
16	269	30·7	25·6	0·8087
17	336	12·1	82·6	1·0302
Pastricciola				
Peraluminous Complex				
19a	488	3·08	555	2·8545
19γ	453	20·0	67	0·9778
26	360	18·4	58	0·9455
27a	767	3·22	929	4·2570
27γ	478	15·0	96	1·0793

2b. Age and significance of the Corsican alkaline province

Cutting across the structures inherited from the Variscan event, about twenty ring-complexes were emplaced from the Middle Permian to the end of Triassic times (Table 1a and Figure 2).

Rb–Sr isotopic data have been obtained on hypersolvus and subsolvus granites, in four ring-complexes: whole-rock isochrons display ages from 268 to 238 Ma. Sr initial ratios are highly variable, from 0·703 in the peralkaline complex of Evisa (Bonin et al. 1978) to 0·722 in the peraluminous complex of Pastricciola. Several processes have been invoked for the high variations of isotopic ratios (van Breemen et al. 1975; Blaxland et al. 1978; Bonin et al. 1979; Vidal et al. 1979) in alkaline rocks: crustal melting, contamination in situ or during the ascent of the magma reworking of previous alkaline granites in the magma chamber (Martin and Bonin 1976), rehomogenization with wall-rocks, life time of the magma. The isotopic data obtained on the batholith have been used to constrain the role of the crust during the emplacement of alkaline magmas (Table 2).

Table 2. Isotopic evolution of the Rb–Sr of the calc-alkaline basement of some alkaline complexes (data from Bonin et al., 1972 and Vialette unpublished)

Alkaline complex	Calc-alkaline rock-type	Rb (ppm)	Sr (ppm)	IIR	Age (Ma)	Δt (Ma)	RIR
Cauro–Bastelica	granodiorite	180	100	0·703	340	100	0·712
Pastricciola	leucocratic granite	250	15	0·707	295	25	0·725

Rb and Sr are mean values, IIR = initial isotopic ratio, RIR = resultant isotopic ratio, Δt = time interval between emplacements of basement and of alkaline complex.
^{87}Rb decay constant = $1·42 \times 10^{-11}$ yr^{-1}.

Crustal derivation or contamination? At 270–230 Ma, the Sr minimum isotopic ratio for rocks belonging to the basement is 0·721–0·725. This implies that the calc-alkaline batholith can be taken as a contaminant but not as the source of Corsican alkaline granites. Moreover, note that no older crust, apart some Palaeozoic sedimentary series and some metamorphic rocks belonging to a post Pan-African lower crust (Vellutini 1977) in a very small volume, has been recognized in this segment of the Variscan fold belt. Thus, the isotopic ratio of 0·703, obtained for the Evisa complex, is considered as a typical mantle signature for the peralkaline acid magmas, without any crustal contamination (Bonin et al. 1978).

The intermediate initial ratio of 0·707 for the hypersolvus biotite granites of Cauro–Bastelica ring-complex (Bonin et al. 1972) can be interpreted in terms of crustal contamination of a mantle-derived magma (Table 3a, Figure 3): a contamination of about 10 per cent is sufficient to enhance Sr initial ratio from its mantle value of 0·703 to its present value of 0·707. But in the Rb–Sr plot (Figure 4), the magmatic trend displayed by the biotite granite makes it unlikely that the contamination has operated *in situ* during the crystallization of the magma. Thus, it is more likely that the contamination has applied to the magma itself, either during its ascent or in the magma chamber (Marsh and Kantha 1978; Bonin 1982).

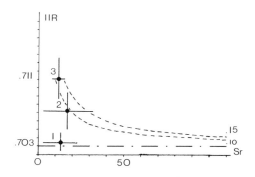

Figure 3. Initial Sr isotopic ratios (IIR) versus Sr concentration (in ppm) 1. Evisa peralkaline complex, 2. hypersolvus biotite granite, Cauro-Bastelica ring-complex, 3. subsolvus biotite granite, Cauro-Bastelica ring-complex. Dashed lines: selective contamination by Sr isotopes from the wall-rocks (0.10 and 0.15, selected ratios of the contaminant) (see text). Note that subsolvus granites have initial Sr isotopic ratios that agree with a contamination process as well as with the remelting process.

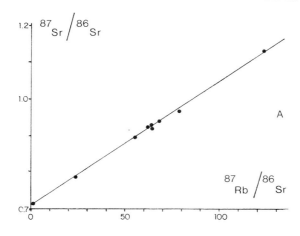
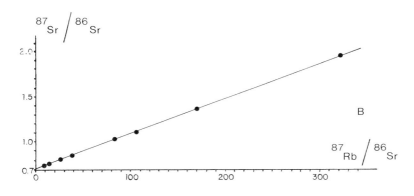
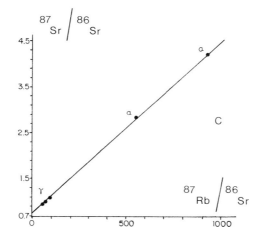

Figure 2. Rb-Sr whole-rock isochrons for biotite subsolvus granites in Corsica. A. Aragnasco ring-dyke, Cauro-Bastelica ring-complex. 8 is an albitite pocket, 9 is sampled at one metre from the albitite pocket, other samples are taken at 100 m at least from the albitite pocket. B. Tana centre, Tana-Peloso complex. C. Pastricciola complex, α is for aplitic dykes, γ is for central granite.

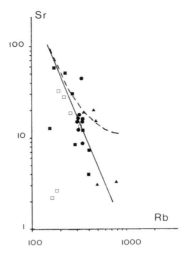

Figure 4. Sr-Rb (in ppm) logarithmic plot. continuous line–magmatic trend without contamination; dashed line–10 percent contamination of consolidated rocks previously following the magmatic trend (see text), open squares: hypersolvus biotite granite, Cauro-Bastelica ring-complex, closed squares: subsolvus biotite granite, Cauro-Bastelica ring-complex, closed circles: subsolvus biotite granite, Tana centre, closed triangles: subsolvus biotite granite, Pastricciola complex.

Table 3. Rb–Sr isotopic data base used to account for variations in isotopic ratios of mantle-derived magmas (original ratio of 0·703)

Crustal contamination by calc-alkaline granodiorites, 245 Ma ago.

Rb (ppm)	Sr (ppm)	RIR
180	100	0·711

Remelting of hypersolvus biotite granites (Cauro–Bastelica).

Rock-type	$^{87}Rb/^{86}Sr$	IIR	Age (Ma)	$\triangle$ t (Ma)	RIR
Biotite granite	50	0·7073	244	6	0·7115

IIR = initial isotopic ratio, RIR = resultant isotopic ratio, $\triangle$ t = time interval between emplacements of hypersolvus granite and subsolvus granite.
^{87}Rb decay constant = $1·42 \times 10^{-11}$ yr^{-1}.
Analytical methods are described in Bonin et al. (1978), the ^{87}Rb decay constant is $1·42 \times 10^{-11}$ yr^{-1}. Uncertainties quoted for the ages are at 1σ confidence level.

Reworking of crystallized alkaline granite complexes. The reworking of previously consolidated hypersolvus granites, resulting in a water-saturated magma, consequently giving subsolvus granites, has been proposed as an explanation of the hypersolvus-subsolvus association (Martin and Bonin 1976). It is possible to test this model in the Cauro–Bastelica ring-complex where subsolvus granites (238+/−4 Ma) cut hypersolvus granites (244+/−5 Ma). Subsolvus granites with initial ratios of 0·7115 can be derived from the near-complete melting of 'Older' hypersolvus granites (Table 3). This remelting has occurred in the magma chamber and not in the ring-complex itself because no sign of *in situ* anatexis has ever been observed. Large influxes of water into the magma chamber where consolidated dry rocks are at temperatures above the wet solidus is responsible for the remelting and development of a water-saturated magma (Martin and Bonin 1976).

Isotopic rehomogenization with hydrothermal fluids. For alkaline complexes with hydrothermal alteration and mineralization, high strontium initial ratios have been ascribed to more or less complete isotopic rehomogenization between consolidated intrusions and their surrounding wall-rocks, the hydrothermal fluids being the isotopic conveyors (Taylor and Forrester 1971; van Breemen *et al.* 1975; Forrester and Taylor 1977; Blaxland *et al.* 1978). Numerical models by Norton and Knight (1977), Norton (1978), have shown that, in an idealized hydrothermal system, the predominant source of fluids flowing through permeable hot plutons is from host rock environments adjacent to the pluton and especially above the pluton. This means that more than 60 per cent of the fluid circulating through plutonic rocks in about 100,000 years has the important consequence that subsolidus hydrothermal alteration and associated mineralization occur essentially at the time of emplacement. This is the case at the Ririwai complex in northern Nigeria (Bonin *et al.* 1979; Kinnaird *et al.* 1985). The whole-rock Rb–Sr isochron therefore indicates the age of mineralization, which is the same as the time of emplacement within the analytical error limits. Initial isotopic ratios, if rehomogenization has occurred between crystallized rocks and fluids, is a function of both the amount of circulating fluids and the isotopic ratio of the crystallized rocks and the fluids (Bonin *et al.* 1979). This feature is clearly shown by isotopic data for the Pastricciola complex (Figure 5). The Pastricciola complex (Bartoli 1979; Bonin 1980) displays a weak mineralization in cassiterite and topaz; the Sr initial ratio is the highest in Corsican alkaline complexes at 0·722, a value close to the strontium isotopic ratio of calc-alkaline granitic country rocks at the time of emplacement 270 Ma ago (see Tables 1b, and Table 2). It is suggested that percolation of hydrothermal fluids, rich in fluorine, has produced a near complete (90–95 per cent) isotopic rehomogenization between the calc-alkaline host rocks (initial strontium isotopic ratio of about 0·725) and the hydrothermal altered alkaline subsolvus granites (pervasive Na-metasomatism in the granite and mineralization in late-stage aplitic dykes).

Life-time of the alkaline granitic magma. Isotopic data gathered on Kerguelen Island (Lameyre *et al.* 1975) lead Vidal *et al.* (1979) to propose that, for magmas rich in Rb and poor in Sr, the Sr isotopic ratio can be enhanced during differentiation. The idea was modelled by McCarthy and Cawthorn (1980) and applied to stratified complexes and batholiths. This attractive model can explain the highly variable isotopic ratios in biotite granites, either hypersolvus or subsolvus, with an original 'magmatic' strontium isotopic ratio of 0·703. In the Sr–Rb plot (Figure 4), hypersolvus and subsolvus biotite granites of the Cauro–Bastellica ring-complexes display the same trend, controlled by sodic plagioclase (oligoclase) and we assume that the mean $^{87}Rb/^{86}Sr$ ratio is approximately 50 and 60, respectively for hypersolvus and subsolvus biotite granites. Thus, it takes 6 Ma time for a hypersolvus magma to reach its final ratio of 0·707 and 9·5 Ma time for subsolvus

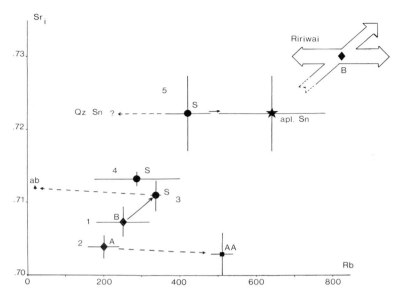

Figure 5. Initial Sr isotopic ratios versus Rb concentration (in ppm) 1. hypersolvus biotite granite (B), Cauro-Bastelica ring-complex, 2. hypersolvus arfvedsonite granite (A) and albitic arfvedsonite-aegirine granite (AA), Evisa complex, 3. subsolvus biotite granite (S), with albitite pocket (ab), Aragnasco ring-dyke, Cauro-Bastelica ring complex. The arrow indicates possible derivation of (S) by remelting of (B), 4. subsolvus biotite granite (S), Tana centre, 5. subsolvus biotite granite (S), Sn-rich aplitic dykes (apl. Sn) and cassiterite-bearing quartz veins (Qz Sn), Pastricciola complex. For comparison hypersolvus biotite granite (B), Ririwai ring-complex, Nigeria (after Bonin et al., 1979).

magma to reach its final ratio of 0·711. Because fractionation processes occur in the magma chamber at depths of about 14 km, the life-time of the magma cannot be the only factor controlling the increase of Sr isotopic ratios. It is likely that during a period of 6–10 Ma postulated for a steady state magma chamber, some contamination by wall-rocks must occur. Moreover, such a long life-time for a magma is unlikely and contradicts with data on present-day magma chambers. A period of time for the magma to reach its solidus is probably more likely to be less than 1 Ma. We conclude that the variation of Sr isotopic ratios is not a product of the life-time of an alkaline granitic magma.

In summary, Corsican alkaline ring-complexes display ages from Middle Permian (Saxonian: 270 Ma) to the Triassic–Liassic boundary (200 Ma). The time interval between the end of orogenic magmatic activity (Autunian, *ca.* 280 Ma) and the beginning of anorogenic magmatism is around 10 Ma. The alkaline magmas are mantle-derived. Their Sr initial ratios are either mantle (0·703 for Evisa peralkaline complex) or higher (from 0·707 up to 0·7222 for metaluminous and peraluminous complexes), evidence for crustal contamination by leaching of the calc-alkaline basement by hydrothermal fluids percolating through alkaline magmas or consolidated rocks.

3. Corsica, part of the alkaline province of the Western Mediterranean

The Corsican alkaline province is about 150 km long from north to south and 60 km wide from west to east. At the same time, a widespread igneous activity of alkaline character is known in many other places, in Europe as well as in Africa.

3a. The geodynamic setting of the Western Mediterranean alkaline province

For the 270–200 Ma period, numerous plutonic complexes and related volcanic lavas have been described with an alkaline affinity. They are chiefly composed of subsolvus biotite granites and K-rich rhyolitic ignimbrites, with minor amounts of monzonites, gabbros, transitional basalts and trachytes. Plutonic alkaline complexes are emplaced in the Variscan fold belt and the Lower Permian calc-alkaline volcanic cover. The volcanic alkaline lava flows lie conformably on calc-alkaline volcanics and they are cut by plutonic complexes of the same composition. The location of alkaline igneous activity is frequently superimposed on the earlier Lower Permian calc-alkaline volcanism (Vellutini 1977).

Permian alkaline provinces. Alkaline magmatism can be recognized in the Upper Permian in the 'European' part of the Alpine fold belt as well as in the 'African' part. For example the external zone of the European part of the Alpine fold belt can be recognized at Carinthia (Austria) (Morauf 1980), Err and Julier-Bernina Decke, Switzerland (Trümpy 1975; Büler 1983; Rageth 1984), Penninic domain, Switzerland (Thélin and Ayrton 1983), Estérel, France (Boucarut 1971), Sardinia, Italy (Vellutini 1977), Badajoz Province, Spain (Dupont and Bonin 1981).

The African part of the internal zone of the Alpine fold belt is illustrated by examples from the Bolzano area, Italy (Mittempergher 1958, 1962), Lugano area, Switzerland and Italy (Buletti 1985), Baveno pink granite, Italy (Gandolfi and Paganelli 1974), and most probably some dyke swarms in Calabria, Italy (Dubois 1976) and granites at Azegour and Western Rehamna, Morocco (Mrini 1985).

Triassic provinces. Triassic ages have been ascribed to alkaline igneous activity at the Brianconnais and Acceglio zone in the French and Italian western Alps (Lefevre 1982). Also at the Monzoni–Predazzo plutonic–volcanic complex, Italy (Emiliani and Balzani 1962; Del Monte *et al.* 1967; Ferrara and Innocenti 1974), Central Alps, Italy (Fiorentini Potenza *et al.* 1975).

Note however that for the same locations Permian alkaline igneous activity is followed in Triassic times by tholeiitic activity for example in Morocco, Spain, and Portugal (Vergely 1984).

Alkaline magmatism, indications of oceanic opening. These igneous massifs are composed of extensive volcanic plateaux (e.g. Esterel, Bolzano area, Lugano area), eroded central volcanoes and ring-complexes, which may cut previous volcanic plateaux. In other places, only subvolcanic ring-complexes are preserved. The main feature of this magmatism lies in the predominance of K-rich rhyolitic ignimbrites and biotite subsolvus granites during the Middle Permian and Upper Triassic times.

The age of igneous activity is essentially around 250–200 Ma with some exceptions at 270 Ma, suggesting a large thermal event, substantiated by overprinting in the mineral ages of the Variscan basement (Ferrara and Innocenti 1974). Thus, we propose to call this collection of alkaline massifs of nearly the same age the Western Mediterranean province.

After a large compilation of radiometric data, Ferrara and Innocenti (1974) have pointed out that a Triassic thermal event is synchronous with alkaline volcanism and plutonism but is distinctly different from the Hercynian metamorphism as well as the older phases of the Alpine episode. They proposed that it corresponds to the period immediately preceeding the breakup and fragmentation of the Pangean continental block with the formation of new oceanic basins named by Aubouin (1977) as the 'Reconquest of Tethys'.

Furthermore, ophiolitic suites, representing the floor of these oceanic basins, are Lower Jurassic (180–135 Ma) in age (e.g. 161+/−3 Ma) for the eastern Corsican Ophiolite (Ohnenstetter et al. 1981) and postdate alkaline magmatic suites by a 20 Ma time interval. However, it is important to note that the layered gabbros and the associated granitic dykes of the Matterhorn and Mont Collon–Dents de Berthol area in Switzerland (Dal Piaz et al. 1977) display transitional characters and yield an age of 250+/−5 Ma suggesting that the switch from alkaline to tholeiitic via transitional magmatism occurred as early as Permo-Triassic times, at the sites of future oceanic opening. Accordingly, Triassic alkaline magmatic activity is widespread far from the line of continental breakup.

A north–south trending alkaline province, linked with the Western Mediterranean province, can be recognized by the occurrence of the following complexes: Combeynot, Rieou Blanc, and Ailefroide (Haut Dauphine, France) (Costarella and Vatin-Pérignon 1985), Kagenfels (Vosges, France) (Rève 1985), alkaline volcanic products recovered in deep-drill holes in the Northern Germany plain and the plutonic–volcanic rocks of the Oslo Graben (Oftedahl 1978). The age range of this alkaline province is from Autunian (*ca.* 280 Ma) to the Permo-Triassic boundary (245 Ma) (Sundvoll 1978a, 1978b; Montigny et al. 1983). The igneous activity remains alkaline, whereas in the Western Mediterranean region, there is a dramatic switch from calc-alkaline to alkaline igneous suites at about 270 Ma. This north–south alkaline province is emplaced either in the Northern European foreland or in the external zone of the Variscan fold belt, suggesting that, under the Variscan external nappes, the basement is made up essentially of formations belonging to the Northern European shield (ECORS deep seismic profile, Cazes et al. 1985).

Together with the Western Mediterranean province, the north–south province constitutes a part of a Y-shaped fault system, where the north–south part is the failed arm and the Western Mediterranean part is the site of Tethys ocean opening.

Tectonic settings of alkaline magmatism in the Western Mediterranean province. The geodynamical model for Permo-Triassic times is based on the palaeogeography (Figure 6) proposed by Vergely (1984) for the 230–200 Ma period and by Dercourt et al. (1985) for the Pliensbachian (190 Ma). These authors insist on the continental character of the crust, where alkaline igneous activity has taken place, but a thinning of the continental crust has probably occurred at the site of the future oceanic basins. According to Vergely (1984), the 'continental bridge' that links Europe to Africa through the Iberian block is subjected to a large sinistral shear movement.

At the beginning, a network of faults is created, by shear movements. Two dislocation zones are situated along the Bay of Biscay and the Straits of Gibraltar together with associated secondary faults. The most important feature of this scheme is the complete reversal of the stress field with large east–west trending Autunian dextral shear zones, and large east–west trending sinistral shear zones in the Triassic. The new regime is controlled by the opening of the Atlantic basin, west of the Iberian block. The 'Tardi-Hercynian' faults (Arthaud and Matte 1975)

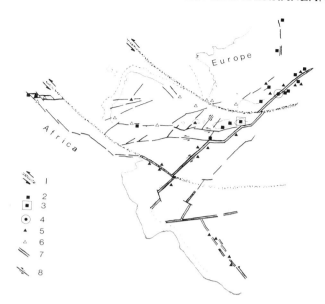

Figure 6. Sketch map of Mediterranean regions during Late Permian and Triassic periods. Symbols: 1. principal shear zones, 2. Permian alkaline province, 3. Corsican alkaline province, 4. Matterhorn transitional province, 5. Triassic alkaline province, 6. Triassic tholeiitic province, 7. line of continental fragmentation and oceanic opening, 8. transcurrent faults with sense of movement.

are reactivated with a reverse sense of movement. This was named the 'harpoon effect' by Black et al. (1985). A reversal in the sense of movement in a pre-existing fault system may trigger or enhance partial melting in the mantle (Lameyre et al. 1985).

The change in the stress field can be substantiated at some places by the elliptic shape of the ring-complexes (Bonin 1982) and by the general trend of dyke swarms associated with alkaline centres. In Corsica, calc-alkaline dyke swarms, restricted to the northwestern part of the island trend N60–70°. Alkaline dyke swarms, widespread in the crystalline basement of the island, trend N60° only in the northwestern area. In the rest of the island the principal direction is N20° with conjugate set at N110° (Arthaud and Matte 1976), some of which are deflected around ring-complexes (Bonin 1980).

Thus, the Western Mediterranean province is located along pre-existing faults, subjected to a reversal in the sense of movement. Along these faults, partial melting of the mantle is enhanced by pressure release and the lithosphere–asthenosphere boundary becomes shallower. The resultant thinning of continental lithosphere is subsequent ocean opening. But not all the alkaline complexes lie on the future zone of continental fragmentation. However it is noteworthy that those complexes located on this zone are frequently older (Middle to Upper Permian) and characterized by biotite subsolvus granites.

3b. The petrological evolution of the Western Mediterranean alkaline province

The alkaline magmatic province displays anorogenic features such as the typical acid-basic bimodality described by Martin and Piwinskii (1972) with basalts, minor trachytes, and rhyolites at the volcanic level; gabbros, minor monzonites, and syenites, and granites at the subvolcanic level.

Although basic liquids were relatively rare, basaltic flows and some acid-basic net-veined complexes are found in the province. In Corsica, basalts have been observed in the Scandola cauldron (Brisset and Cochemé 1976); net-veined complexes are present in the Sorba ring-complex (Bonin 1980), Porto ring-complex and in the Punta Rossa composite dyke in the Porto-Vecchio dyke swarm (Bonin 1980). These rocks are silica-saturated to slightly undersaturated, and plot in the alkaline field of a total alkalies–silica diagram (Platevoet 1983).

On the other hand, many gabbros are often layered, with cumulate textures, and associated with monzonites and syenites (Bavella complex, Bonin 1980; Peloso complex, Platevoet and Bonin 1985). In an idealized cross-section of the Peloso complex (Figure 7), a 2–15 m thick rhythmic layering can be seen in a 500 m thick lower unit. One complete layer is composed of, from bottom to top (Figure 8a): fine-grained gabbro, coarse-grained leucogabbro, and anorthosite. The upper basic unit is made up of layered gabbros and diorites and is capped by a monzosyenite unit.

In the layered basic units, plagioclase and clinopyroxene are clearly cumulus minerals, with minor amounts of olivine. Orthopyroxene, amphibole, and biotite are intercumulus minerals, with increasing amounts of late stage quartz at the top. The mineral compositions imply a crystallization from a parent magma with intermediate composition: An 70–30 plagioclase, $Fe/(Fe+Mg) = 0.3$ calcic clinopy-

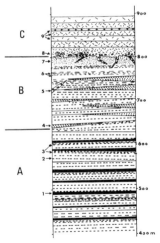

Figure 7. General stratigraphic log in the Peloso complex Corsica. A. lower basic unit: 1. olivine gabbro, 2. leucogabbro-norite, 3. anorthosite, B. upper basic unit: 4. olivine-biotite gabbro, 5. aplite and pegmatite, 6. amphibole-biotite leucogabbro-norite, 7. biotite gabbro and diorite, C. monzosyenitic unit: 8. diorite-pegmatite-monzosyenite net-veined complex, 9. monzosyenite with amphibole-rich layers and granitic layers.

roxene and Fo 60–40 olivine. These compositions (Figure 9) are related to somewhat differentiated liquids (Bonin and Giret 1985), near-silica saturated (modal orthopyroxene), and metaluminous.

The differentiation is governed by plagioclase and calcic clinopyroxene and trapped intercumulus liquid is critically saturated in titanium and phosphorus show by the occurrence of intercumulus ilmenite and apatite. Residual liquids are

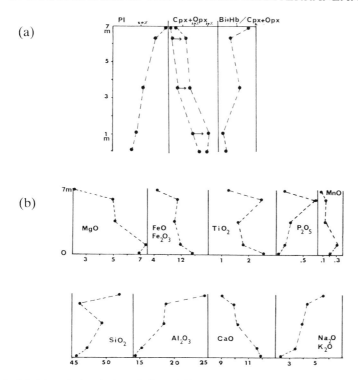

Figure 8. Modal and chemical evolution within one layer of the lower basic unit in the Peloso complex. a) modal data for main rock-forming mineral (Pl, plagioclase; Cpx + Opx, clinopyroxene + orthopyroxene) and anhydrous/hydrous mafic mineral ratio (Bi + Hb, biotite + amphibole). b) major element concentration.

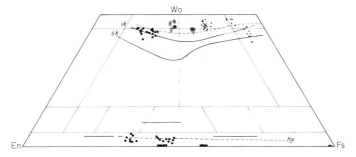

Figure 9. Pyroxenes and olivines from the Peloso complex plotted. Symbols: large closed circles–basic cumulates, light closed circles–monzosyenites and associated cumulates–encircled points-diorite-monzosyenite association. Abbreviations: Ik. Iskou alkaline complex (Niger), Sk. Skaergaard tholeiitic complex (Greenland), Hy. Hyllingen tholeiitic complex (Norway).

represented by the monzosyenite–granite association (Figure 10), where the silica-oversaturation is obtained in Fe- and alkali-enriched liquids by amphibole fractionation (Bonin and Giret 1985).

Thus, for the Corsican example, we propose that the parent magma of ring-complex granites is not basaltic, but intermediate in composition, probably hawai-

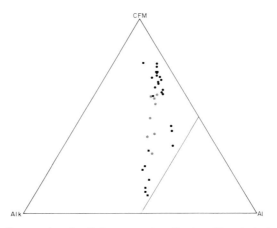

Figure 10. Niggli diagram for the Peloso complex Corsica Closed circle–basic cumulates, closed squares–monzosyenites, associated cumulates and granites, encircled points–intermediate rocks of the diorite–monzosyenite association.

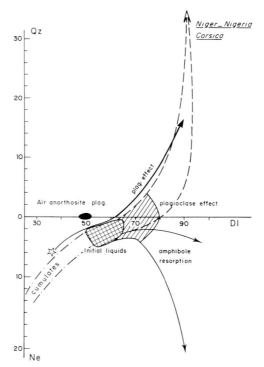

Figure 11. Normative Qz-Ne-DI plot (after Bonin and Giret, 1985). Symbols: DI, differentiation index, Qz, quartz, Ne, nepheline, star: primary alkaline basaltic liquid formed in the mantle source (example taken from the Quaternary volcanic province of Chaîne des Puys, France), cross-hatched area–initial alkaline liquids trapped in the crustal magma chamber; diagonal lines–maximum plagioclase effect in alkaline series. Note that the initial liquids trapped in the magma chamber display a higher differentiation index (DI) than the primary basaltic liquids derived directly from their mantle source, suggesting either differentiation during the ascent from the mantle to the crust, or a derivation from different degrees of partial melting in the mantle.

itic to mugearitic (Bonin and Giret 1985). Even if basic magmas were present, it is likely that partial melting of the mantle results in the production of different primary liquids, the more differentiated ones being trapped in a crustal magma chamber at shallow depths (with plagioclase on the liquidus). The crossing of the critical saturation line (Figure 11) is made easy by amphibole fractionation and is possible only if water is available in significant amounts at the magma chamber level.

The emplacement of shallow magma chambers (depths of about 15 km) in a young continental crust, formed during recent orogenic processes, as recorded in the internal part of the Variscan fold belt, is a favourable factor. Circulation of meteoric and cognate waters at depths through epizonal and mesozonal crustal layers is enhanced by fracturing associated with wrench tectonics. The early availability of water is evidenced by the hypersolvus–subsolvus granite association (Martin and Bonin 1976). In many places, subsolvus granites are the only granitic rocks, without any associated hypersolvus granite, suggesting that water-undersaturated liquids could not be emplaced at the subvolcanic level and have been quickly converted into water-saturated liquids. Subsolvus granites, therefore, are typical of alkaline magmatic suites emplaced in a young continental crust, soon after the end of orogenic processes.

4. Comparison of the Western Mediterranean province with West African alkaline provinces

West Africa is the site of numerous alkaline provinces, of various ages and contrasting compositions (Black 1984; Black et al. 1985). It is not the purpose of this paper to review each alkaline province in West Africa. We shall focus only on features similar to those found in the Western Mediterranean province. First of all, as Black (1984) has pointed out, Phanerozoic oversaturated alkaline provinces occur exclusively in Pan-African domains, whereas mixed or dominantly undersaturated provinces tend to be located on the cratons or in the southern prolongation of the Mozambique belt. That is to say, oversaturated alkaline provinces which are the most similar to the Western Mediterranean province are emplaced in a continental domain stabilized since 550 Ma.

This observation is in disagreement with Walker's statement (1984) that 'practically all known cauldrons and ring-dyke complexes occur in Precambrian cratonic crust' (p. 8412). But it is in agreement with Walker's conclusion that 'most of the world's Quaternary calderas have an epicontinental setting and are developed on relatively young crust'. In this respect, 'relatively young crust' is given a wide sense of meaning to encompass all types of continental crust, with varying ages of cratonization from 500 Ma ago to the present time. Thus, all silica oversaturated alkaline provinces in West Africa are emplaced in a relatively young continental crust, irrespective of the original age of the geological formations constituting the basement, and reworked during the Pan-African orogeny. Alkaline magmatism in Africa began at the close of the Pan-African orogeny during Cambrian times, events which have their parallel in the Western Mediterranean province at the close of the Hercynian orogeny. As in the Variscan fold belt, alkaline complexes are emplaced in the Pan-African fold belt, as well as in other Precambrian regions remobilized during orogenesis.

4a. The Cambrian period and its alkaline magmatism

During this period, alkaline igneous activity occurred in Africa as well as in other parts of the world.

West Africa. In Mali, oceanic closure around 600 Ma led to a collision between the passive margin of the West African craton and an active continental margin to the east (Adrar des Iforas) displaying island arc and cordilleran-type assemblages, bordering a deformed continental mass intruded by a large calc-alkaline batholith (Black *et al.* 1979). Such a geological description for Pan-African events in Mali is identical to that for Hercynian orogeny in Corsica. As in Corsica, post-orogenic alkaline granitoids postdate the uplift and unroofing of a calc-alkaline batholith. A detailed investigation of the relative and absolute chronology of the magmatic events in Mali has determined that collision tectonics ended around 600–590 Ma and that alkaline igneous activity began some 560 to 540 Ma ago with a northward migration of eruptive centres (Liégeois and Black 1984). The rock types are characterized by the presence of subsolvus granites (Ba *et al.* 1985) in more than eight ring-complexes.

The switch from calc-alkaline to alkaline magmatic suites is amplified by spectacular dyke swarms, the earlier striking WNW–ESE with calc-alkaline porphyries and the later striking NNW–SSE with alkaline microgranites and granophyres. The two dyke swarms postdate the unroofing of the batholith and predate the formation of the ring-complexes. Black *et al.* (1985) have pointed out that the appearance of alkaline rocks is related to a change in the stress field at about 560 Ma (Black *et al.* 1985). But in the case of Adrar des Iforas, continental dislocations have not developed into oceanic opening as in the case of the Western Mediterranean province. The reason why is yet unclear and it is suspected that global geodynamics are implicated rather than local tectonic disturbances. Present geochronological data assign the alkaline magmatism in Adrar des Iforas to a time period of only 20–25 Ma which is small compared with the 70 Ma duration time observed in the Western Mediterranean province. The opening of a Palaeozoic Proto-Tethys between the Laurasian block and the Gondwana block (Cogné and Wright 1980) may have induced a reversal of the stress fields in Pan-African domains. The orogenic convergence has an essentially East–West direction (Black *et al.* 1979), and is followed by a dominantly North-South divergence (Cogné and Wright 1980).

Adrar des Iforas is not the only occurrence of Cambrian (post Pan-African) alkaline provinces. A number of other provinces of this age have been reviewed by Black *et al.* (1985), many of them displaying subsolvus granites. It is likely that other provinces will be discovered in the near future, as little attention has been paid until recently concerning subsolvus granites since they are petrographically similar to orogenic leucocratic granites (Bonin 1977). Note in addition that many Cambrian provinces are located in Pan-African domains far from the ancient suture zones and, thus, cannot be used as a guide to delineate palaeosuture zones.

Brazil, France, and Spain. Pan-African orogenesis is not restricted to the African continent but is found in other parts of the world, e.g. Brazil — where the orogenic episode is named Brazilian — and West France — where it is named Cadomian, after the old Celtic name of the city of Caen.

In Brazil, several cratons have been remobilized during the Brazilian orogenesis. The 2 By old Luis Alves craton, in the south of the country, displays about 25 granitic bodies (Serra do Mar intrusive suite, Kaul *et al.* 1984). Rb–Sr whole-rock isochrons yield ages around 500 Ma (Kaul *et al.* 1982). Rock-types are subsolvus biotite granites, peralkaline quartz–syenites, and granites. As in Corsica, peralkaline rocks display a low Sr initial ratio (0·706) while aluminous granites display a higher ratio (0·712). According to Kaul (1984), the Luis Alves craton was subjected around 500 Ma ago to a distensive regime with associated strike-slip movements. Volcanic products are present with rhyolites, trachytes, and basalts.

In the Cadomian domain in France (Brittany and western Normandy), a similar situation is found (Auvray 1979). During the Upper Proterozoic, southeastward subduction processes occurred along a continental margin belonging to the Pentevrian microplate of African affinities (cratonization at 2 By during the Eburnean orogenesis). A calc-alkaline batholith with associated volcanics was emplaced in the Pentevrian formations. Isotopic ages are 640 Ma for volcanics (Vidal 1976) and 610 Ma for plutonites (Graviou 1984). Small alkaline bodies have been described by Auvray (1975): they are made up of ignimbrites, quartz–feldspar porphyries, and granophyric alkaline granites. Alkaline volcanic rocks and plutonic rocks yield the same age (around 550 Ma) but the Sr initial ratio is higher for the plutons (0·7096) than for the rhyolites (0·7064), suggesting a more important crustal contamination for granitic liquids. Auvray (1975) has proposed that a residency period in an intermediate crustal magma chamber can account for the higher ratio in the granitic magma.

In other places where Pan-African orogenesis has occurred in France, but was later obliterated by the Hercynian episodes, it is occasionally possible to recognize alkaline complexes, even when they are strongly deformed. South of the French Massif Central, the Montagne Noire area displays some alkaline orthogneisses (Bard 1979) which have been dated at 530 Ma (Ducrot *et al.* 1979).

Other Cambrian magmatic events have been described elsewhere in the French Massif Central (Boyer 1974; Pin and Lancelot 1978) but, in these cases, their magmatic affinities have not yet been determined. It is likely that antectic, calc-alkaline and alkaline igneous suites have succeeded one after the other, as exemplified in French Brittany (Auvray 1979; Jonin 1981; Graviou 1984).

After the Cambrian period, Ordovician and Silurian bimodal magmatism existed in France and Spain, where peralkaline orthogneisses are well known in the Vigo area (Floor 1966). Leptyno-amphibolites and orthogneisses display volcanic or subvolcanic features (Boyer 1974) and can be typified as tholeiitic or alkaline, according to the different areas surveyed. Bard *et al.* (1980) have proposed that this magmatism, alkaline in Cambrian times, then either alkaline or tholeiitic in Ordovician–Siluran times, is good evidence for distensive regimes developing into oceanic-basin opening.

In summary, the Cambrian period is a time of intense and widespread alkaline magmatism the newly formed Pan-African mobile belt. In cratonic areas, alkaline complexes are comparatively rare. The different provinces are related to reversals in the stress fields and initiation of transcurrent faults systems. Subsolvus biotite granites are typical of this geodynamical setting, forming soon after the consolidation of a young continental crust. But, if some areas are subsequently subjected to distensive, then extensive regimes with oceanic opening as in Europe, the other regions become stabilized like the Luis Alves craton in Brazil or the Trans-Saharan belt and escape the rifting process.

West Africa, with its 2 By old cratons and 600 Ma old fold belts, is one of these places where no subsequent rifting and oceanic opening processes have occurred. However, alkaline igneous activity has been persistent since Cambrian to Recent (Black and Girod 1970).

4b. From Palaeozoic to present: alkaline magmatism in West Africa

Probably the most salient feature of alkaline magmatism in West Africa is the Niger–Nigeria zone with the 'Younger Granite' provinces of Aïr (480–400 Ma), Damagaram (320–290 Ma), and Jos Plateau (Northern Nigeria) (215–140 Ma). The igneous activity is not strictly continuous but age groupings substantiate a southern shift of eruptive centres (Bowden and Turner 1974; Bowden et al. 1976). At the same places, associated with domal uplifts and reactivation of pre-existing fault patterns in Tertiary–Quaternary times, alkaline volcanism with basalt–, trachyte–, rhyolite–phonolite developed in the Hoggar, Aïr, and the Jos Plateau (Black and Girod 1970).

The rock-type distribution varies from north to south, even though similar rocktypes are developed throughout the three subprovinces. In Aïr, peralkaline rocks predominate, associated with anorthosites (Black et al. 1967) and with minor amounts of subsolvus aluminous granites (Pérez 1985). Metaluminous and subaluminous granites prevail in Nigeria, but subsolvus granites are very rare and water influx is active only at subsolidus stages (Martin and Bonin 1976; Martin and Bowden 1981), with successive Na, K, H^+ and Si metasomatic phases (Kinnaird 1985) during the cooling of plutons from 460° to probably less than 70°C (Kinnaird et al. 1985).

An important point is that there is no significant difference between the Adrar des Iforas province and the Younger Granites. There are identical rock-types, magmatic features, tectonic settings, and mode of emplacement. The only significant difference lies in the proportion of the rock-types and in the intensity of postmagmatic hydrothermal alteration and associated mineralization. Therefore, we conclude that it is reasonable to compare the post-orogenic alkaline provinces and the anorogenic (sensu stricto) Younger Granite alkaline provinces. It is also worth noting that in each anorogenic province, the time duration of magmatic activity is of the same order as in post-orogenic alkaline provinces: i.e. only 20 Ma in Adrar des Iforas but 70 Ma in Western Mediterranean province, to be compared with 80 Ma in Aïr, 30 Ma in Damagaram, and 75 Ma in the Jos Plateau. If each anorogenic province is considered individually and not as a part of a larger and hypothetical progressive set of related zone of magmatism on the African continent, it is possible to determine the parameters of one alkaline province. Black et al. (1985) have recalled many observations concerning the synchronism between alkaline magmatic emplacement in within-plate environments and orogenic episodes elsewhere in the world: e.g. Caledonian orogenic episode — Aïr alkaline province; Variscan orogenic episode — Damagaram alkaline province; Jurassic orogenic episodes — Jos plateau province. It seems that a blocking regime in plate relative motions induces intense stress fields at plate margins as well as within-plate environments. Alkaline igneous activity is therefore a within-plate magmatic response when relative plate movements are reduced to zero. In this respect, there is no significant difference between post-orogenic alkaline magmatism and the West African anorogenic provinces.

Southeast of the Niger-Nigeria zone of ring complexes, the Cameroon line displays a number of volcanic and plutonic centres. Some plutonic complexes, chiefly composed of granites, are Cambrian in age (ca. 550 Ma) and emplaced into Pan-African calc-alkaline granitoids (Lasserre 1978). They belong to the same Cambrian event described previously for Mali. Other anorogenic complexes are considerably younger with dates from 80 Ma to the present. Three magmatic episodes can be identified, broadly trending southwestwards, the first period of 75 to 45 Ma, the second from 54 to 27 Ma, and the third from 30 Ma to the present time. The continental magmatic centres are emplaced throughout the three episodes with the volcanic oceanic islands belonging only to the third episode. In order to compare the Cameroon line with other alkaline provinces, we shall only consider the continental magmatic centres. Within each episode, there is no obvious age trend. The magmatic evolution as a function of time is substantiated by a progressive shift from basaltic primary liquids to basanitic and nephelinitic liquids. Accordingly, differentiated rocks are at first silica-oversaturated (granites and rhyolites), then saturated (trachytes and syenites), and lastly undersaturated (phonolites and etindites). The change from silica-(over)saturated to undersaturated residual liquids occurred in continental volcanic centres at around 10–15 Ma, according to a compilation of existing radiometric ages (Dunlop 1983; N'ni 1984; Halliday et al. 1985): the youngest silica-saturated differentiated rock recorded is a 11·8 Ma old quartz–trachyte in Manengouba calderas (Dunlop 1983) and in the Bambouto Mountains, the oldest recorded phonolites belong to an eruptive episode postdating the 16 Ma old quartz–trachyte association (Dunlop 1983). The number of accurate ages however, is insufficient to assign to this period of 10–15 Ma, evidence of a complete reversal of magmatic evolution. It is more likely that some overlap exists between ages of emplacement of the latest saturated volcanic differentiates and the ages of emplacement of the first undersaturated volcanic differentiates. It is also possible that some volcanic areas display the two kinds of volcanic differentiates at the same time and the same places, this situation being frequent at the subvolcanic level (Black et al. 1985).

The critical role of amphibole crystallization on the activity of silica in the residual liquids has already been stressed by Bonin and Giret (1985). For the undersaturated series, it is unlikely that water is available in large amounts. Carbon dioxide is the main fluid phase particularly in associated carbonatites and/or late interstitial carbonates in silicate rocks. Early amphiboles destabilize and dissolve in evolved liquids (Figure 11). In the case of the Cameron line, at each magmatic pulse since 550 Ma, H_2O is decreasing step by step, and after a time interval of around 540 Ma, PH_2O is too low to make amphibole stable until the end of differentiation. It does not imply that late amphiboles cannot crystallize but simply that they do not play a significant role in the differentiation processes. For example, in Mount Cameroun, kaersutite appears only in the most evolved silica-undersaturated tephritic lavas (N'ni, 1984).

If this model is correct it is possible to explain the occurrence of Quaternary phonolitic plugs in Air, where Palaeozoic plutonic differentiates are granitic in composition, and also the development of mixed volcanic provinces during Tertiary–Quaternary times in the Hoggar and on the Jos Plateau (Black and Girod 1970). Thus, it appears that oversaturated complexes predominate during a period of 500–700 Ma after the end of orogenesis, subsequent intrusions being either mixed or strongly undersaturated. Such a model has been emphasized by the alkaline provinces in eastern North America, where a 600 Ma time interval is proposed to account for the rock-type assemblages series of New Hampshire (Black et al. 1985).

5. Conclusion: a general petrological model

A comparison between the Corsican alkaline province and West African provinces has emphasized their similarities and dissimilarities. As previous authors (Barker

1969; Cahen and Snelling 1984) have pointed out, the time of emplacement of alkaline complexes is not random but corresponds to orogenic events and more precisely to the end of these events, preceding continental fragmentation after periods of collision. For these authors, their synchronism between emplacement of alkaline complexes and of orogenic events does not imply that alkaline complexes are directly linked with orogenic processes. This is the reason why they are named 'anorogenic'. Bowden (1985) has proposed a two-fold classification: one association related to a terminal orogenic phase and the other linked with progressive uplift, long-term doming, and subsequent rifting. The first is believed to be a short-time duration suite and the second a long-time duration suite.

In fact, if we consider only one given alkaline province, we find that the time periods are of the same order for the so-called 'short-time duration' province as for the 'long-time duration' province. Therefore, in our opinion, the two types of associations, as proposed by Bowden (1985), are caused by similar processes. However, the magmatic products do not display the same quantitative rock-type assemblage and, in this respect, it is possible to use Bowden's two-fold classification in a somewhat different sense: 'short-time after orogenesis' and 'long-time after orogenesis' suites.

The rapid sequence of a collisional blocking regime and of continental dislocations is responsible for the build up of intense stress fields which provoked reactivation of lithospheric shear zones and rifting far within the continental plates. The 'harpoon effect', proposed by Black et al. (1985), triggered alkaline igneous activity. So, within-plate magmatism location is determined by fracturing from above, i.e. reactivation of older deep crustal faults, and not by lithospheric thinning by the uprising mantle, as postulated by the plume hypothesis. On the other hand, within-plate magmatic products are mantle-derived and have undergone differentiation processes in a crustal magma chamber. At each orogenic event in the world, an alkaline magmatic response occurred when plate relative movements were reduced to zero. Lithospheric faults may subsequently act as preferential lines for continental fragmentation and oceanic basin opening. In this scheme, global tectonics are considered as the framework for alkaline magmatic activity. Considering only one plate, even as large as Africa, is not sufficient to understand the tectonic settings and the geodynamic significance of alkaline provinces.

In one given region, subjected to cratonization processes, the rock-type distribution changes as a function of time elapsed since the end of the last orogenic event. At the beginning, silica-oversaturated complexes predominate with major amounts of subsolvus granites which postdate the last tectonic phases by less than 10 Ma. Then, subsolvus granites become progressively less abundant with hypersolvus granites, suffering subsolidus hydrothermal processes, developed around 500–600 Ma ago. After this period, mixed or silica-undersaturated complexes are emplaced as a rule in rift zones in cratonic areas and become more and more undersaturated, associated with carbonatites.

The differentiation of primary mantle liquids is governed essentially by anhydrous minerals olivine, plagioclase and clinopyroxene, and by hydrous minerals such as amphibole. The stability of amphibole is controlled by water fugacity in the fluid phase. For anorogenic and initially water-undersaturated primary liquids, water fugacity may increase if water is supplied in great amounts from the wallrocks to the magma chamber. The amount of water is critical. If water is so abundant that the evolved granitic liquid becomes oversaturated with a discrete fluid phase, granitic rocks yield subsolvus characters. Conversely, if water is unavailable because host rocks are catazonal, amphibole can no longer crystallize in the evolved liquids. The time interval for a young continental crust to dissipate all its water circulating through deep fractures seems to be about 500–600 Ma. During this period, percolating water is used to contribute to the crystallization of amphibole in each magma chamber emplaced at shallow depths. As amphibole cumulates act as a water source, the host rocks become more and more water depleted. Subsequent magmatic pulses become emplaced in a progressively dehydrated crust causing partial pressure in the magma to collapse.

In this scheme, water is supplied by crustal host-rocks to mantle magmas. The progressive decline of water in the crust over a 500–600 Ma period provokes in the magma chambers a decline in the role of amphibole fractionation. It does not imply that magmas can survive that long without crystallizing, which is thermodynamically impossible; it does imply however that for successive batches of magmas in the same region, water is supplied from outside in less and less amounts.

Thus, we propose for the differentiated rocks of alkaline provinces the following sequence concerning the predominating rock-types: subsolvus granites – hypersolvus granites – (quartz) syenites – nephelinitic syenites $+/-$ carbonatites, with many overlapping sequences after the first 500 Ma period. This sequence is controlled both by the location of partial melting in the mantle (basaltic versus nephelinitic primary liquids), and, by the amounts of water percolating through the magma chamber in the crust. Water from external mesozonal crust sources and carbon dioxide from the mantle and/or lower crust sources, both present in the fluid phase, control the differentiation processes in the magma in opposite ways: thus, the H_2O/CO_2 mole ratio is the critical factor governing the rock-type assemblage in alkaline provinces.

Acknowledgements. We are grateful to Dr. P. Bowden and the organizing committee to present this paper at the 13th Colloquium of African Geology held at St. Andrews (Scotland) in September 1985. One of us (B.B.) is also very grateful to Universities of several countries: U.F.Ba at Salvador, Bahia (Brazil), U.L.B. at Brussels (Belgium), Universitat Bern (Switzerland), and Universite Cadi Ayyah at Marrakech (Morocco) to have given him the opportunity to deliver lectures on the subject of alkaline rocks, and to discuss with many colleagues including Prof. Sabate and Dr. Kaul (Brazil), Prof. Michot, Drs. Demaiffe, Liegeois, and Weis (Belgium), Profs. Hunziker, Jaeger, Niggli, and Streckeisen, Drs. Buhler, Buletti, Oberhaensli, and Stalder (Switzerland), and Drs. Aarab, Gasquet, Lagarde, Rose, and Saquaque (Morocco). We have also benefitted by fruitful discussions with Profs. Ayrton (Lausanne), Boriani (Milano), Lameyre (Paris), Pitcher (Liverpool), and Visona (Padova), Drs. Black (Paris), Boullier (Nancy), Gil Ibarguchi (Bilbao), Kampunzu and Lubala (Lubumbashi), and Van Tellingen (Amsterdam), and many others, especially the graduates from the Universite Paris-Sud at Orsay, France.

References

Arthaud, F. and Matte, Ph. 1975. Les décrochements tardi-hercyniens du Sud-Ouest de l'Europe. Géométrie et essai de reconstitution des conditions de la déformation. *Tectonophysics*, **25**, 139–171.
—— and —— 1976. Arguments géologiques en faveur de l'absence de mouvements relatifs de la Corse par rapport à la Sardaigne depuis l'orogenèse hercynienne. *C. R. Ac. Sc.*, **283**, 1011–1014.
—— and —— 1977. Late Paleozoic strike-slip faulting in southern Europe and northern Africa: result of a right-lateral shear zone between the Appalachians and the Urals. *Geological Society of America Bulletin*, **88**, 1305–1320.
Aubouin, J. 1977. Téthys, Atlantique et Pacifique: Regard tectonique. *C. R. Somm. Soc. Géol.Fr.*, **4**, 170–179.

Auvray, B. 1975. Relations entre plutonisme acide et volcanisme ignimbritique: exemple des manifestations magmatiques cambriennes du Nord de la Bretagne. *Pétrologie*, **1**, 125–138.
—— **1979.** *Genèse et évolution de la croûte continentale dans le Nord du Massif Armoricain.* Thèse Doct. Etat ès-Sci., Univ. Rennes.
Ba, H., Black, R., Benziane, B., Diombana, D., Hascoet-Fender, J., Bonin, B., Fabre, J. and Liégeois, J.P. 1985. La province des complexes annulaires alcalins sursaturés de l'adrar des Iforas (Mali). *Journal of African Earth Science*, **3**, 123–142.
Bard, J.P. 1979. Existence d'une suite granitique alcaline d'âge paléozoïque inférieur dans la zone axiale de la Montagne Noire (Massif Central français) et ses abords immédiats. *C. R. Ac. Sc.*, **288**, 371–374.
——, **Burg, J.P., Matte, Ph., and Ribeiro, A. 1980.** La chaîne hercynienne d'Europe occidentale en termes detectonique des plaques. Colloque C6: Cogné, J. and Slansky, M. (Eds), *Geology of Europe*. 26° I.G.C., Paris, 233–246.
Barker, D.S. 1969. North American feldspathoidal rocks in space and time. *Geological Society of American Bulletin*, **80**, 2369–2372.
Bartoli, P.A. 1979. *Les minéralisations d'étain et de wolfram de la Corse du Sud. Relations avec le magmatisme alcalin permien.* Dipl. Et. Approf., Nice.
Bates, R.L. and Jackson, J.A. 1980. *Glossary of Geology*, 2nd Ed. A.G.I., Falls Church, Va. 751 pp.
Black, R. 1984. The Pan-African event in the geological framework of Africa. *Pangea*, **2**, 8–16.
—— **and Girod, M. 1970.** Late Palaeozoic to Recent igneous activity in the West Africa and its relationship to basement structure. In: Clifford, T.N. and Gass, I.G. (Eds), *African Magmatism and Tectonics*. Oliver and Boyd, Edinburgh. 185–210.
——, **Jaujou, M. and Pellaton, C. 1967.** Notice explicative de la carte géologique de l'Aïr à l'échelle 1/500 000. Dir. Mines Géol. Niger.
——, **Caby, R., Moussine-Pouchkine, A., Bayer, R., Bertrand, J.M.L. Boullier, A.M., Fabre, J. and Lesquer, A. 1969.** Evidence for late Precambrian plate tectonics in West Africa. *Nature*, **278**, 223–227.
——, **Lameyre, J. and Bonin, B. 1985.** The structural setting of alkaline complexes. *Journal of African Earth Science*, **3**, 5–16.
Blaxland, A.B., van Breemen, O., Emeleus, C.H. and Anderson, J.G. 1978. Age and origin of the major syenite centers in the Gardar province of south Greenland: Rb–Sr studies. *Geological Society of America Bulletin*, **89**, 231–244.
Bonin, B.1977. Les complexes granitiques subvolcaniques de Corse: caractéristiques, signification et origine. *Bull. Soc. Géol. Fr.*, **19**, 865–871.
—— **1980.** *Les complexes acides alcalins anorogéniques continentaux: l'exemple de la Corse.* Thèse Doct. Etat ès-Sci., Univ. Pierre et Marie Curie, Paris.
—— **1982.** Les granites des complexes annulaires. *Manuels et Méthodes*, n° 4. Edit. B.R.G.M., Orléans. 183 pp.
——, **Bowden, P. and Vialette, Y. 1979.** Le comportement des éléments Rb et Sr au cours des phases de minéralisation: l'exemple de Ririwai (Liruei), Nigéria. *C. R. Ac. Sc.*, **289**, 707–710.
—— **and Giret, A. 1985.** Contrasting roles of rock-forming minerals in alkaline ring-complexes. *Journal of African Earth Science*, **3**, 41–49.
——, **Grelou-Orsini, C. and Vialette, Y. 1978.** Age, origin, and evolution of the anorogenic complex of Evisa (Corsica): a K – Li – Rb – Sr study. *Contr. Miner. Petrol.*, **65**, 425–432.
——, **Vialette, Y. and Lameyre, J. 1972.** Géochronologie et signification du complexe granitique annulaire de Tolla–Cauro (Corse). *C. R. Somm. Soc. Géol. Fr.*, **2**, 145–150.
Boucarut, M. 1971. *Etude volcanologique et géologique de l'Estèrel (Var, France).* Thèse Doct. Etat ès-Sci., Univ. Nice.
Bowden, P. 1985. The geochemistry and mineralization of alkaline ring-complexes in Africa (a review). *Journal of African Earth Science*, **3**, 17–39.
—— **and Turner, D.C. 1974.** Peralkaline and associated ring-complexes in the Niger-Nigeria province, West Africa. In: Sørensen, H. (Ed.), *The alakaline rocks*, J. Wiley & Sons, Chichester, 330–351.
——, **van Breemen, O., Hutchinson, J. and Turner, D.C. 1976.** Palaeozoic and Mesozoic age trends for some ring complexes in Niger and Nigeria. *Nature*, **259**, 297–299.
Boyer, C. 1974. *Volcanismes acides paléozoïques dans le Massif Armoricain.* Thèse Doct. Etat ès-Sci., Univ. Paris-Sud, Orsay.
Brisset, F. and Cochemé, J. J. 1976. *Etude géologique des presqu'îles de Scandola et du Monte Seninu (Corse).* Thèse Doct. 3° cycle, Univ. Marseille.
Bühler, Ch. 1983. Petrographische und geochemische Untersuchungen im Gebiet La Tscheppa – Lagrev (Julier – Bernina – Decke). *Schweiz. Mineral. Petrogr. Mitt.*, **63**, 457–477.
Buletti, M. 1985. *Petrographisch–geochemische Untersuchungen im Luganer Porphyrgebiet.* Inaugural-dissertation, Univ. Bern. Dokt.
Cahen, L. and Snelling, N.J. 1984. *The Geochronology and Evolution of Africa.* Oxford University Press, Oxford. 550 pp.
Cazes, M., Torreilles, G., Bois, C., Damotte, B., Galdéano, A., Hirn, A., Mascle, A., Matte, Ph., Pham Van Ngoc, and Raoult, J.F. 1985. Structure de la croûte hercynienne du Nord de la France: premiers résultats du profil ECORS. *Bull. Soc. Géol. Fr.*, **27**, 925–941.
Chappell, B.W. and White, A.J.R. 1974. Two contrasting granite types. *Pacific Geology*, **8**, 173–174.
Cogné, J. and Wright, A.E. 1980. L'orogène cadomien. Vers un essai d'interprétation paléogéodynamique unitaire des phénomènes orogéniques fini-précambriens d'Europe moyenne et occidentale,

et leur signification àl'origine de la croûte et du mobilisme varisque, puis alpin. Colloque C6: Cogné, J. and Slansky, M. (Eds), *Geology of Europe*, 26° I.G.C., Paris, 29–55.

Costarella, R. and Vatin-Pérignon, N. 1985. An alkaline complex: the Combeynot massif in the French Alps. *Terra Cognita*, **5**, 318.

Dal Piaz, G.V., De Vecchi, G. and Hunziker, J.C. 1977. The Austro-alpine layered gabbros of the Matterhorn and Mt. Collon — Dents de Bertol. *Schweiz. mineral. petrogr. Mitt.*, **57**, 59–88.

Del Monte, M., Paganelli, L., and Simboli, G. 1967. The Monzoni intrusive rocks. A modal and chemical study. *Miner. petrogr. Acta*, **13**, 75–118.

Dercourt, J., Zonenshain, L.P., Ricou, L.E., Kazmin, V.G., Le Pichon, X., Knipper, A.L., Grandjacquet, C., Sborshchikov, I.M., Boulin, J., Sorokhtin, O., Geyssant, J., Lepvrier, C., Biju-Duval, B., Sibuet, J.C., Savostin, L.A., Westphal, M. and Lauer, J.P. 1985. Présentation de 9 cartes paléogéographiques au 1/20 000 000° s'étendant de l'Atlantique au Pamir pour la période du Lias àl'Actuel. *Bull. Soc., Géol. Fr.*, **1**, 637–652, with an Atlas of 10 maps.

Dubois, R. 1976. *La suture calabro-apenninique crétecé–éocène et l'ouverture tyrrhénienne néogène; étude pétrographique et structurale de la Calabre centrale.* Thèse Doct. Etat ès-Sci., Univ. Pierre et Marie Curie, Paris.

Ducrot, J., Lancelot, J.R. and Reille, J.L. 1979. Datation en Montagne Noire d'un témoin d'une phase majeure d'amincissement crustal caractéristique de l'Europe prévarisque. *Bull. Soc. Géol. Fr.*, **21**, 501–505.

Dunlop, H. 1983. *Strontium isotope geochemistry and potassium – argon studies on volcanic rocks from the Cameroon line, West Africa.* Ph. D. Thesis, Univ. Edinburgh.

Dupont, R. and Bonin, B. 1981. Le massif alcalin de Feria — Sierra Vieja (Sierra Morena occidentale, Province de Badajoz, Espagne): un jalon du magmatisme anorogénique de Méditerranée occidentale. *Bull. Soc. Géol. Fr.*, **23**, 477–485.

Emiliani, F. and Vespignani Balzani, G.C. 1962. Sulla presenza di fayalite e fergusonite nel granito di Monte Mulat presso Predazzo (Valle di Fassa). *Accad. Lincei Rend. Sc. fis. mat. e nat.*, **32**, 111–115.

Ferrara, G. and Innocenti, F. 1974. Radiometric age evidences of a Triassic thermal event in the Southern Alps. *Geol. Rdsch.*, **63**, 572–582.

Fiorentini Potenza, M., Schwander, H., and Stern, W. 1975. Chemical distribution patterns in the Tertiary and Triassic igneous districts of the Central Alps. *Chem. Erde*, **34**, 257–282.

Floor, P. 1966. Petrology of an aegirine–riebeckite gneiss-bearing part of the Hesperian Massif: the Galineiro and surrounding areas, Vigo, Spain. *Leidse Geol. Medel.*, **36**, 1–203.

Forrester, R.W. and Taylor, Jr., H.P. 1977. $^{18}O/^{16}O$, D/H and $^{13}C/^{12}C$ studies of the Teriary igneous complex of Skye, Scotland. *American Journal of Science*, **277**, 136–177.

Gandolfi, G. and Paganelli, L. 1974. Ricerche geologico–petrografiche sulle plutoniti erciniche della zona del Lago Maggiore. *Mem. Soc. Geol. It.*, **13**, 119–144.

Graviou, P. 1984. *Petrogenèse des magmas calco-alcalins: exemple des granitoïdes cadomiens de la région trégorroise (Massif Armoricain).* Thèse Doct. 3° cycle, Univ. Rennes.

Halliday, A.N., Dickin, A.P., Fallick, A.E. and Fitton, J.G. 1985. A Nd, Sr, Pb, and O isotopic study of the Cameroun line volcanics and a comparison of the chemical structure of sub-continental and sub-oceanic mantle. 13th Coll. Afr. Geol., St. Andrews, *C.I.F.E.G. Occasional Publication*, **3**, 271.

Jonin, M. 1981. *Un batholite fini-précambrien: le batholite mancellien (Massif Armoricain, France). Etude pétrographique et géochimique.* Thèse Doct. Etat ès-Sci., Univ. Bretagne Occidentale, Brest.

Kaul, P.F.T. 1984. Significado dos granitos anorogénicos da suite intrusiva Serra do Mar na evoluçao da crosta do sul-sudeste do Brasil, no âmbito das folhas SG.22 — Curitiba e SG.23 — Iguape. *Anais 33° Congr. Brasil. de Geol.*, Rio de Janeiro.

——, **Coitinho, J.B.L. and Issier, R.S. 1982.** O episodio Campo Alegre. *Anais 32° Congr. Brasil. de Geol.*, Salvador, Bahia, **1**, 47–54.

Kinnaird, J.A. 1985. Hydrothermal alteration and mineralization of alkaline anorogenic ring-complexes of Nigeria. *Journal of African Earth Science*, **3**, 229–251.

——, **Bowden, P., Ixer, R.A. and Odling, W.A. 1985.** Mineralogy, geochemistry and mineralization of the Ririwai complex, northern Nigeria. *Journal of African Earth Science*, **3**, 185–222.

Lameyre, J. and Autran, A. 1980. Les granitoïdes de France.Colloque C7: Autran, A. and Dercourt, J. (Eds), *Géologie de la France*. 26° I.G.C., Paris, 51–97.

——, **Black, R., Bonin, B. and Giret, A. 1985.** Dynamique lithosphérique et genèse des magmas mantelliques. *C. R. Ac. Sc.*, **300**, 21–26.

——, **Marot, A., Zimine, S., Cantagrel, J.M., Dosso, L. and Vidal, Ph. 1975.** Chronological evolution of the Kerguelen Islands syenite–granite ring complex. *Nature*, **263**, 306–307.

Lasserre, M. 1978. Mise au point sur les granitoïdes dits 'ultimes' du Cameroun. Gisement, pétrologie et géochronologie. *Bull. B. R. G. M.*, **4**, 143–159.

Lefevre, R. 1982. *Etude géologique de la zone d'Acceglio — col du Longet (Alpes Cottiennes).* Thèse Doct. Etat ès-Sci., Univ. Paris-Sud, Orsay.

Liégeois, J.P. and Black, R. 1984. Pétrographie et géochronologie Rb–Sr de la transition calco-alcaline — alcaline fini-panafricaine dans l'Adrar des Iforas (Mali): accrétion crustale au Précambrien supérieur. *In:* Kleerkx, J. and Michot, J. (Eds), *Géologie Africaine*, Tervuren, 115–145.

Loiselle, M.C. and Wones, D.R. 1979. Characteristics and origin of anorogenic granites. Abstr. *92nd Geology Society of America Annual Meeting*, **11**, 468.

Maluski, H. 1977. *Application de la méthode $^{40}Ar-^{39}Ar$ aux minéraux des roches cristallines perturbées par des événements thermiques et tectoniques en Corse.* Thèse Doct. Etat ès-Sci., Univ. Montpellier.

Marsh, B.D. and Kantha, L.H. 1978. On the heat and mass transfer from an ascending magma. *E. P. S. L.*, **39**, 435–443.
Martin, R.F. and Bonin, B. 1976. Water and magma genesis: the association hypersolvus granite – subsolvus granite. *Canadian Mineralogy*, **14**, 228–237.
—— **and Bowden, P. 1981.** Peraluminous granites produced by rock–fluid interaction in the Ririwai nonorogenic ring-complex, Nigeria: mineralogical evidence. *Canadian Mineralogy*, **19**, 65–82.
—— **and Piwinskii, A.J. 1972.** Magmatism and tectonic settings. *Journal of Geophysical Research*, **77**, 4966, 4975.
McCarthy, T.S. and Cawthorn, R.G. 1980. Changes in initial $^{87}Sr/^{86}Sr$ ratio during protracted fractionation in igneous complexes. *Journal of Petrology*, **21**, 245–264.
Mittempergher, M. 1958. La seria effusiva paleozoica del Trentino – Alto adige. *Com. Naz. Ric. Nucl.*, **1**, 61 pp.
—— **1962.** Rilevamento e studio petrografico delle vulcaniti paleozoiche della Val Gardena. *Atti. Soc. Tosc. Sc. Nat.*, **69**, fasc. 1.
Moïne Vaziri, H. 1985. *Volcanisme tertiaire et quaternaire en Iran.* Thèse Doct. Etat ès-Sci., Univ. Paris-Sud, Orsay.
Montigny, R., Schneider, C., Royer, J.Y. and Thuizat, R. 1983. K–Ar dating of some plutonic rocks of the Vosges. *Terra Cognita*, **3**, 201.
Morauf, W. 1980. Die permische Differentiation und die alpidische Metamorphose des Granitgneises von Wolfsberg, Koralpe, SE-Ostalpen, mit Rb–Sr– und K–Ar-Isotopenbestimmungen. *Tschermaks Min. Petr. Mitt.*, **27**, 169–185.
Mossakovsky, A.A. 1970. Upper Paleozoic volcanic belt of Europe and Asia. *Geotectonics*, **4**, 247–253.
Mrini, A. 1985. *Contribution à l'étude géochronologique des granites hercyniennes de la Meseta marocaine.* Thèse Doct. 3° cycle, Univ. Clermont-Ferrand.
N'ni, J. 1984. *Le volcan actif du Mont Cameroun (ligne du Cameroun): géologie et pétrologie du volcan.* Thèse Doct. 3° cycle, Univ. Paris-Sud, Orsay.
——, **Bonin, B., and Brousse, R. 1985.** Migration de l'activité magmatique de la ligne du Cameroun: réactivation de segments de failles anciennes du socle panafricain. *C. R. Ac. Sc.*, **302**, 453–456.
Norton, D. 1978. Sourcelines, sourceregions, and pathlines for fluids in hydrothermal systems related to cooling plutons. *Economic Geology*, **73**, 21–28.
—— **and Knight, J. 1977.** Transport phenomena in hydrothermal systems: cooling plutons. *American Journal of Science*, **277**, 937–981.
Oftedahl, C. 1978. Cauldrons of the Permian Oslo rift. *J. Volc. Geoth. Res.*, **3**, 343–371.
Ohnenstetter, M., Ohnenstetter, D., Vidal, Ph., Cornichet, J., Hermitte, D., and Mace, J. 1981. Crystallization and age of zircon from Corsican ophiolitic albitites: consequences for oceanic expansion in Jurassic times. *E. P. S. L.*, **54**, 397–408.
Orsini, J. 1980. *Le batholite corso-sarde. Anatomie d'un batholite hercynien. Composition — structure — organisation d'ensemble — sa place dans la chaîne varisque française.* Thèse Doct. Etat ès-Sci., Univ. Aix-Marseille.
Pérez, J.B. 1985. *Nouvelles données sur le complexe granitique anorogénique de Taghouaji (République du Niger). Influence des fluides au cours de la cristallisation.* Thèse Doct. Univ., Nancy.
Pin, C. and Lancelot, J.R. 1978. Un exemple de magmatisme cambrien dans le Massif Central: les métadiorites quartziques intrusives dans la série du Lot. *Bull. Soc. Geol. Fr.*, **20**, 203–208.
Platevoet, B. 1983. Etude pétrologique d'une association acide-basique dans le complexe annulaire anorogénique de Porto (Corse). Univ. Pierre et Marie Curie, *Mém. Sci. Terre*, n° **83**, 21, 200 pp.
—— **and Bonin, B. 1985.** Le massif alcalin du Peloso (Corse): un complexe lité associé à des monzosyénites. *C. R. Ac. Sc.*, **301**, 403–406.
Rageth, R. 1984. Intrusiva und Extrusiva der Bernina-Decke zwischen Morteratsch und Berninapass (Graubünden). *Schweiz. mineral. petr. Mitt.*, **64**, 83–109.
Rève, J.M. 1985. *Répartition de l'uranium et du thorium dans les roches alcalines des Vosges du Nord.* Thèse Doct. 3° cycle, Univ. Paris-Sud, Orsay.
Sørensen, H. 1974. *The Alkaline Rocks.* J. Wiley & Sons, London. 622 pp.
Sundvoll, B. 1978a. Rb/Sr-relationship in the Oslo igneous rocks. In: Neumann, E.R. and Ramberg, I.B. (Eds), *Petrology and Geochemistry of Continental Rifts.* D. Reidel Publ. Comp., Dordrecht, 181–184.
—— **1978b.** Isotope- and trace-element chemistry, geochronology. In: Dons, J.A. and Larsen, B.T. (Eds), *The Oslo Paleorift, a Review and Guide to Excursions. Norges Geol. Unders.*, **337**, 35–40.
Taylor, Jr., H.P. and Forrester, R.W. 1971. Low-^{18}O igneous rocks from the intrusive complexes of Skye, Mull and Ardnamurchan, Western Scotland. *Journal of Petrology*, **12**, 465–497.
Thélin, Ph. and Ayrton, S. 1983. Cadre évolutif des événements magmatico-métamorphiques du socle anté-triasique dans le domaine pennique (Valais). *Schweiz. mineral. petrogr. Mitt.*, **63**, 393–420.
Trümpy, R. 1975. Penninic – Austroalpine boundary in the Swiss Alps: a presumed former continental margin and its problems. *American Journal of Science*, **275A**, 209–238.
Van Breemen, O., Hutchinson, J., and Bowden, P. 1975. Age and origin of the Nigerian Mesozoic granites: a Rb–Sr isotopic study. *Contr. Mineral. Petrol.*, **50**, 157–172.
Van Tellingen, H.W. 1955. *Géologie et pétrologie de la région de Porto (Corse).* Thèse Doct. Univ. Amsterdam, 124 pp.
Vellutini, P.J. 1977. *Le magmatisme permien de la Corse du Nord-Ouest. Son extension en Méditerranée occidentale.* Thèse Doct. Etat ès-Sci., Univ. Aix-Marseille.

Vergely, P. 1984. *Tectonique des ophiolites dans les Hellénides Internes (déformations, métamorphismes et phénomènes sédimentaires). Conséquences sur l'évolution des régions téthysiennes occidentales.* Thèse Doct. Etat ès-Sci., Univ. Paris-Sud, Orsay.

Vidal, Ph. 1976. *L'évolution polyorogénique du Massif Armoricain. Apport de la géochronologie et de la géochimie isotopique du strontium.* Thèse Doct. Etat ès-Sci., Univ. Rennes.

——, **Dosso, L., Bowden, P. and Lameyre, J. 1979.** Strontium isotope geochemistry in syenite–granite complexes. In: Ahrens, L.H. (Ed.), *Origin and Distribution of the Elements* (2nd Symposium), *Phys. Chem. Earth,* **11**, 223–231.

Walker, G.P.L. 1984. Downsag calderas, ring-faults, caldera sizes, and incremental caldera growth. *Journal of Geophysical Research,* **89**, 8407–8416.

Ziegler, P.A. 1983. Caledonian and Hercynian crustal consolidation of the European Alpine foreland. *Terra Cognita,* **4**, 51–58.

Geochemical Characteristics of the Nigerian anorogenic province

Richard A Batchelor

Department of Geology, University of St Andrews, Fife, Scotland

---

The anorogenic granitoids of Nigeria are classic examples of ring complexes associated with migrating mid-plate magmatism. Most complexes represent eroded calderas of once active volcanoes, and range from 1 to 30 km in diameter. Petrology varies from peraluminous biotite granite, through hypersthene and fayalite granites to peralkaline aegirine, riebeckite-arfvedsonite granites rich in perthitic alkali feldspar. The complexes have intruded crystalline basement of Pan-African age, and dating indicates a progressive younging to the south.

The majority of the Nigerian anorogenic ring-complexes represent the roots of volcanoes with dominantly syenitic to granitic compositions. They are A-type granites of long-time duration, petrologically similar to, but chronologically older than, their counterparts to the north in Niger. In a few northern centres the volcanic rocks are preserved indicating an alkaline trend through hawaiite, mugearite, trachyte to rhyolite. Some mixing of contrasting magma compositions has occurred to produce trachyandesitic suites with deisequilibrium mineral reactions. Mixing can be identified from the major and trace chemical variations and petrological instability reactions.

Extensive mineralization by columbite, cassiterite, sphalerite and wolframite suggest fluid interaction on a major scale which has affected much of the original mineral assemblages. In cases of extreme fluid activity, granites have been converted into microclinites, albitites and greisens.

Chemical data for major and trace elements have been obtained for a comprehensive suite of Mesozoic volcanic and plutonic rocks from Nigeria. Analyses of the crystalline basement rocks are presented for comparative purposes. Petrogenetic trends have been highlighted using selected bivariate diagrams in an attempt to reach a consensus view of the chemical evolution of the province, with due regard given to the effects of post-intrusion subsolidus fluid interactions. The series Rb, Th, Nb, La, Ce, Zr, Hf, and Y have been selected to monitor crystal-fluid partitioning, with the more compatible elements showing greater partitioning to the fluid phase.

KEY WORDS   Trace elements   Subsolidus mobilities   Fluids   Geochemical characteristics   Anorogenic granites   Nigeria

1. Introduction

The Nigerian province is one of many Phanerozoic anorogenic provinces on the African continent. The African plate consists of various terranes of different Precambrian ages accreted during major orogenies. The Pan-African orogeny in late Proterozoic, played a major role in re-assembling old cratons and terranes to form the supercontinent Pangaea and its southern extremity, Gondwanaland. It

0072–1050/87/TI0389–14$07.00
© 1987 by John Wiley & Sons, Ltd.

was into this basement that the Phanerozoic complexes were emplaced. The late Precambrian to lower Palaeozoic Pan-African mobile zone (the Benin Sheild) consists of varied rock types from rejuvinated Eburnean crust to Pan-African granitic rocks of orogenic calc-alkaline to subalkaline affinities.

Some of the synorogenic granitoids have a marked metamorphic foliation. Collections have been restricted to specific areas in northern Nigeria for dating purposes (van Breeman et al. 1977) and to assess the degree of Pan-African crustal contribution to the anorogenic granitoids.

The majority of the Nigerian ring-complexes (Figure 1) are Jurassic in age except for the small group of Permo-Carboniferous centres (Daura and Matsena) close to the Nigerian border. They represent the roots of volcanoes which erupted voluminous quantities of rhyolite as lava and ignimbrite, associated with minor but important occurrences of trachytes and basalts. All that remain today in the northern group of structures are examples of dissected calderas and migrating magmatic centres (Turner and Bowden 1979). Further south in the Jos Plateau region and towards the Benue Valley, erosion has removed much of the volcanic evidence leaving a varied array of granitic textures and compositions, primary mineralization, and placer deposits with rich concentrates of cassiterite and other resistant ore minerals. The subvolcanic and volcanic assemblages recorded in the Nigerian Younger province illustrate important geochemical facets of continental magmatism and postmagmatic processes. The dominant chemical changes associated with sodic and potassic metasomatism are readily identified. Part of the metasomatic reactions are desilication processes resulting in the formation of albites and microclinites (Bowden and Kinnaird 1984).

According to the experimental work of Currie (1968) and Mustart (1972) discussed in Taylor et al. (1980) both metaluminous and peralkaline granites exsolve a peralkaline fluid phase from a residual silicate melt at shallow depth. It is the behaviour of this fluid phase in a low pressure environment that dictates the ultimate formation of the ore deposits in Nigeria associated with the Younger granites.

The trace element populations are particularly sensitive indicators of fluid reactions. Although it has been assumed in earlier literature that certain trace elements, especially the rare earths, are immobile under postmagmatic conditions, recent research has shown that many trace elements have considerably different crystal-fluid paritit, coefficients compared with crystal—liquid values. In this paper the behaviour of Rb, Th, Nb, La, Ce, Zr, Hf, and Y show the variable influence of postmagmatic fluid reaction in the Nigerian anorogenic province.

From the economic viewpoint the alkaline granite complexes are important with the abundance of uranium/thorium and rare-earth minerals, columbite, pyrochlore, zircon and complex Ti-Zr silicates. A fluid phase developed during crystallization and the resultant subsolidus reactions, geochemical variations, and mineralization depended on whether this fluid phase was retained or expelled.

Some of the geochemical and petrological characteristics of these A-type granites have comparisons with other A-type provinces not only in Africa but throughout the world. They are all characteristically low in MgO and CaO and high in Na|BOC27,85|G2|BOC27,68|GO and K|BOC27,85|G2|BOC27,68|GO for given silica values. The most significant chemical feature is that slight differences in the ratio of Na+K to Al produced striking changes in the mineralogical composition of the suite of syenites and granites to give peralkaline, metaluminous, and peraluminous variants. When the variation in the suite as a whole is considered,

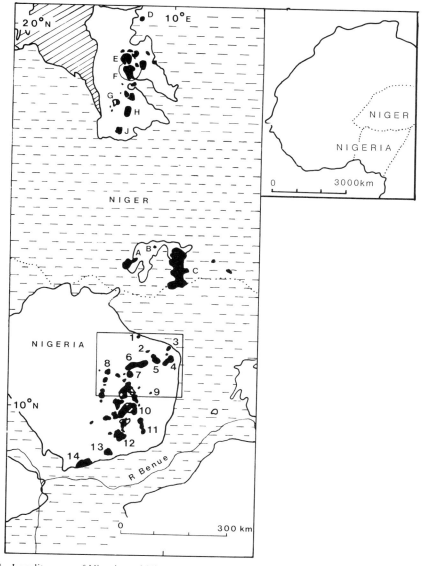

Figure 1. Locality map of Nigeria and Niger showing the distribution of anorogenic intrusions. (taken from Bennett *et al* 1984)

1. Dutse
2. Birnin Kudu
3. Shira
4. Fagam and Dago
5. Kila–Warji

6. Ningi–Burra
7. Tibchi
8. Banke
9. Zaranda
10. Jos–Bukuru

11. Sara–Fier
12. Sha–Kaleri
13. Mada
14. Afu

A. Zinder
B. Zarniski
C. Goure

D. Adrar Bous
E. Tamgak
F. Meugueur

G. Bilete
H. Baguenzane
J. Tagouaji

however, fractional removal of mineral phases only made a very limited contribution to the overall evolution of the magmas, and other factors must have been involved. All the A-type granites are characterized by high concentrations of Sn, Be, Li, U, Zr, Nb, Th, K, Rb, Mo, Zn, F, W and by low K:Rb, Sr:Y and high Rb:Sr, and K:Ba ratios as a result of hydrothermal alteration. Geochemical data used in this paper are available separately (Batchelor and Bowden 1986).

2. Volcanic rocks

2a. De la Roche parameters

Cationic proportions of major oxides have been used for some years by De La Roche and his coworkers (De La Roche 1964; De La Roche and Leterrier 1973; De La Roche et al. 1980). The cationic parameters R1 and R2 are used in this paper to define the magmatic trends in the volcanic rocks, and are also valuable for displaying the major element variations of the Nigerian granites and related rocks. An additional feature of the R1–R2 De La Roche diagram is that mineral fractionation trends can be clearly separated from magma mixing trends (Batchelor and Bowden 1985). This diagram, Figure 2, is well suited for the display of magmatic series. It depicts an alkaline magmatic trend for the volcanic rocks, a

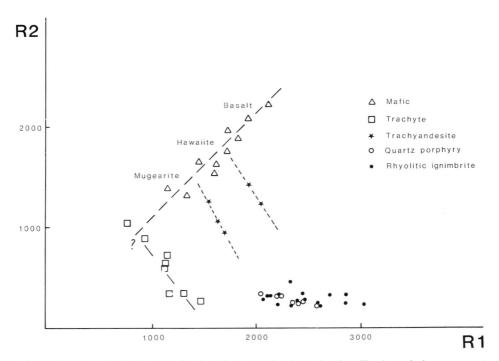

Figure 2. De la Roche diagram for the Nigerian volcanic rock suite. The long dashes represent a fractionation sequence, while the short dashes represent mixing lines between mafic and silicic magmas. $R1 = 4Si - 11(Na+K) - 2(Fe+Ti)$: $R2 = 6Ca + 2Mg + Al$

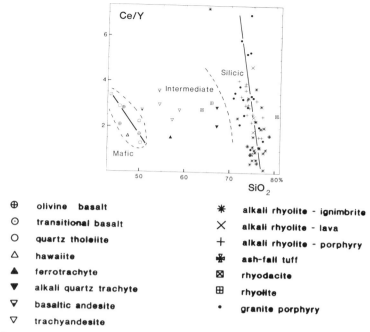

Figure 3. Ce/Y – SiO_2 for volcanic rocks, highlighting three major associations; mafic, intermediate and silicic. Solid lines represent infrared fractional trends

cluster of silicic volcanics, and part of what is inferred to be a mixing trend as represented by the trachyandesite. All three rock types are represented in the Ningi complex.

2b. Cerium/yttrium–silica

The ratio Ce/Y is used in Figure 3 as a fractionation indicator in rocks which may have had less opportunity to be subjected to circulating hydrothermal fluids. The negative correlation seen for the mafic rocks (basalts) implies a fractional crystallization regime involving the removal of pyroxene/amphibole with or without accessory minerals. This trend terminates at $SiO_2 = 55$ per cent. The scatter of intermediate rocks contrasts with the vertical disposition of the silicic volcanics. This trimodal distribution suggests that there is no consanguinous relationship between them.

2c. Chondrite–normalized elements

The inference from Figure 2 is that since three groups of volcanic rocks are not related one to another by fractionation processes, magma mixing should be considered. This diagram strengthens the idea that mafic and silicic magmas mixed to yield intermediate compositions. The consistent mixing proportions of 75–67 per cent (olivine basalt) to 25–33 per cent (rhyolite) for all the elements produced a lava of trachyandesitic composition. A mixed parentage for the Ningi trachyandesite is supported by petrographic evidence of mineral disequilibria (Turner and Bowden 1979).

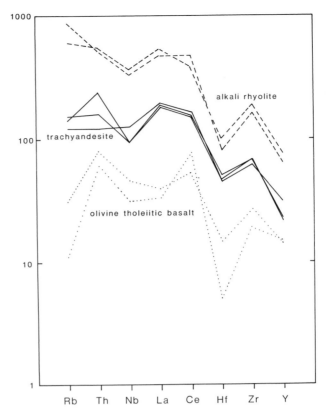

Figure 4. Chondrite–normalized elements for volcanic rocks, showing parallel patterns for alkali rhyolite, trachyandesite and tholeiite. This pattern is consistent with a mixing origin for the trachyandesite.

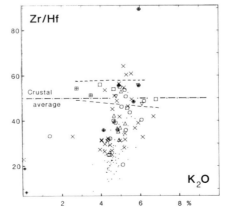

Figure 5. Zr/Hf – K_2O for the Nigerian granitoid suite. The dashed lines ecompass the fractionated suite of intermediate rocks which cluster around Zr/Hf = 50, and show variation in K_2O.

3. Trace element variations in subvolcanic rocks and the surrounding basement

3a. Zirconium-hafnium

The average crustal value for Zr/Hf is approximately 50 (Taylor 1965; Wedepohl 1969). In the Nigerian province, syenites and monzonites, cluster around 50. Values for peralkaline granites show a range 25–65, biotite granites 10–50, fayalite granites 25–50, while the basement granitoids and the basement charnockites range from 20–55. These ranges in Zr/Hf are displayed against K_2O concentrations (Figure 5), which in a fractionating environment would show wide variation.

The deviation in Zr/Hf values from the crustal average is ascribed to differential mobilites for Zr and Hf, in a fluid/vapour medium.

Perez and Rocci (1985) have emphasized the importance of fluorine-rich fluids in determining the subsolidus mineralogy of peralkaline and peraluminous granites. Since it is known that fluorine depolymerizes silicate systems and forms stable complexes with Zr and Hf (Cotton and Wilkinson 1972), particularly in alkaline-rich melts (Collins et al. 1982), an attempt was made to assess the role of F on

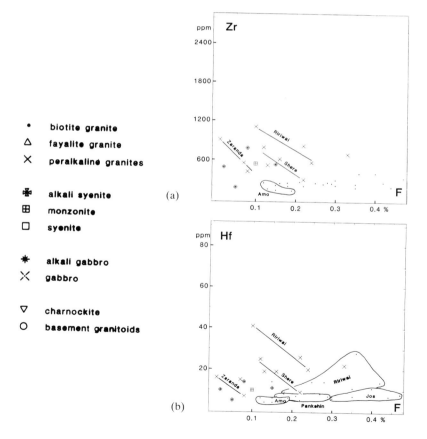

Figure 6. Zr – F(a) and Hf–F(b) plots for various granitoids from different named complexes in Nigeria

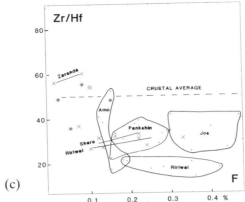

Figure 6 (c) Zr/Hf – F for selected Nigerian granitoids, with names of individual complexes shown where groupings occur.

the behaviour of Zr and Hf (Figures 6a, 6b). Hafnium fluorohydrates are dissimilar to their Zr equivalents, and lose water more readily, implying that the Hf-F bond is stronger (Hf–F 2·02 A, Zr-F 2·10 A in 8 coordination F-OH complexes, Rickards and Waters 1964). Hexachlorohafnates are more soluble in aqueous phases and more resistant to hydrolysis than the Zr equivalents (Bradley and Thornton 1973) and hence implies that Hf will be more stable in a fluid phase, and therefore more mobile than Zr.

In the peralkaline suite, both Zr and Hf correlate negatively with F, while there is no change in the Zr/Hf ratio (Figure 6c). This suggests there was no Zr–Hf fractionation in the peralkaline system. In contrast, biotite granites show little or no change in either element concentration against F, but also show no change in the ratio. (An exception is a suite from the Amo complex, which show a very large vertical scatter.) Since it is well known that these rocks are highly mineralized with columbite, cassiterite, sphalerite,(Kinnaird 1985), the interaction of fluids at various stages in their paragenesis must be assumed. However, with no correlation against F, the implication is that the Zr/Hf ratio has remained undisturbed and may have been inherited, perhaps, by the wholesale incorporation of minerals such as zircon from the basement. Lead isotopic data for rocks from Buji and Pankshin gave a lead age of 500 Ma (Bowden and van Breemen 1970) which approximates to the end of the Pan-African orogeny around 550 Ma ago (van Breeman and Bowden 1973). The wide range in Zr/Hf, with values as low as 10, implies Zr–Hf fractionation at some stage during the formation of the zirconium-bearing phases. The most plausible explanation is that both peralkaline and peraluminous granitoids were derived from a common, probably heterogenous, source which had previously been subjected to disturbance by halogen-rich fluids. Such an environment would tend to fractionate the more stable Hf complexes and give rise to Hf-rich zircons, for which Nigeria is well known (Kogbe and Obialo 1975).

3b. Rubidium-Potassium

In any co-magmatic series, Rb correlates well with K. This is seen for the gabbro–syenite–monzonite rocks found in the Nigerian complexes. Tangential to this trend are the peralkaline and peraluminous granitoids in which Rb shows wide variation for little or no change in K_2O (Figure 7).

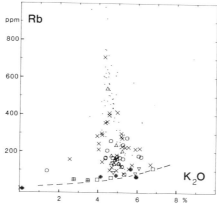

Figure 7. Rb – K_2O for Nigerian granitoids. The dashed line represents the fractionated magmatic suite of intermediate rocks.

This relation is indicative of an independent Rb mobility, most likely within a fluid phase. If this feature is primary, and not secondary, it would imply that the high levels of Rb reflect a concentration of biotite in granitic source rocks which have lost some felsic component (leucosome). If such leucosome-depleted basement granitoids were remobilized, then the resulting rocks would be enriched in Rb. The lack of any coherence between these two elements, K-Rb, in the basement rocks is also indicative that their concentrations have been disturbed, though not as severely as in the anorogenic granitoids themselves.

3c. Rubidium–strontium

The normal negative exponential correlation commonly found in fractionated igneous systems is shown in part by the syenite–monzonite series Figure 8. Enhanced Rb levels with high Sr concentrations in the basement granitoids suggests an accumulation of biotite and plagioclase feldspar. Progressive loss of leucosome, mostly alkali feldspar and quartz, as implied in Figure 7 above, would produce such a 'restite' mineralogy.

3d. Chondrite–normalized elements

The most notable feature of chondrite–normalized trace element diagram (spidergram) is the Nb depletion in the basement granitoids and relatively high levels

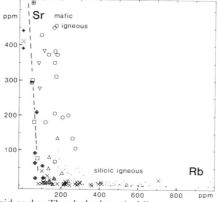

Figure 8. Rb–Sr for granitoid rocks. The dashed vertical line represents the fractional trend displayed by the intermediate rocks.

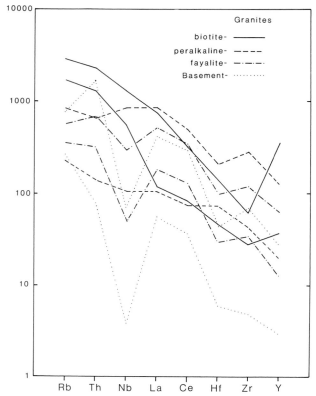

Figure 9. Chondrite–normalized elements for selected Nigerian granitoid suites. Paired lines represent the extreme range in values for each group.

in the anorogenic granitoids (Figure 9).

It can be surmised that Nb was mobilized from the basement rocks and gave rise to the columbite mineralization in the Younger Granite province. Columbite mineralization is an important feature of the Nigerian anorogenic province (Kinnaird 1985). Fluid inclusions studies show a close correlation between CO_2 concentrations and the intensity of Nb mineralization. Nb enrichment is also associated with Cl-rich systems (Kane 1985).

3e. Lanthanum/Yttrium–Niobium

Since lanthanum tends to concentrate in the liquid phase during fractionation of mafic minerals and during partial melting, (Moorhouse et al. 1983) then the ratio will rise in residual liquids after fractionation, and fall in the restite after partial melting. Relative to the syenites, monzonites, and basement charnockites, the basement granitoids display a wide range in the La/Y ratio (Figure 10). The data indicates that this is due to a relative depletion in Y, which could result from mafic and accessory mineral fractionation at the magmatic stage or selective mobilization by fluids at the subsolidus stage. Evidence from other diagrams discussed above, strongly suggests that halogen-rich fluids have been responsible for resetting element abundances, and that mafic mineral fractionation alone cannot explain certain geochemical features such as the high concentrations of Rb

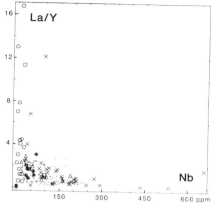

Figure 10. La/Y – Nb for Nigerian granitoids. The intermediate rock suite is enclosed within dashed circle.

and Li with low Nb, and low Zr/Hf ratios. Since Y (and HREE) has a higher charge density than La (and LREE) it tends not to form ionic compounds, but salts of more covalent character which are less soluble than their ionic counterparts. However, covalent HREE compounds are more likely to form complexes (Mineyev 1963) and HREE fluorides are slightly soluble in excess hydrofluoric acid due to the formation of complex molecules (Cotton and Wilkinson 1972).

As both La and Y show a small positive correlation against F, this suggests that F may have been one agent in determining relative REE mobility. The presence of F significantly increases REE mobility (Alderton et al. 1980). During fenitization LREE are enriched in the peralkaline fluid, as Na–REE–F complexes, which, being less stable than the HREE equivalents, break down and precipitate LREE when they migrate into a new environment (Martin et al. 1978). Therefore rocks with high La/Y could represent the precipitation event after REE mobilization at which LREE complexes become unstable. This model has to be considered along with a partial melting episode in the formation of a heterogenous basement.

3f. Beryllium–zinc

The peraluminous granitoids show a distinct bimodal distribution (Figure 11) indicating that the occurrence of beryl and sphalerite, two important minerals in the province (Kinnaird 1985) appears to be mutually exclusive.

4. Discussion

There are two key features of the data set for the peralkaline and peraluminous anorogenic granitoids from Nigeria.
1. A wide range, to high values, for certain trace elements; Rb, Li, Nb, Be, Zn, which are not compatible with simple crystal fractionation.
2. A wide range in Zr/Hf ratios.

In the basement granitoids the key features are variable La/Y and Zr/Hf ratios, and a depletion of Nb and to a lesser extent, Y.

The wide range in certain trace elements is attributed to mobility by fluids. Petrographic evidence for albitization, microclinization and greisenization

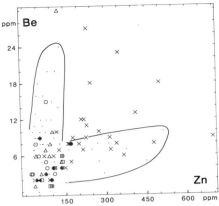

Figure 11. Be – Zn for Nigerian granitoids showing the bimodal distribution within the biotite granite suite (●).

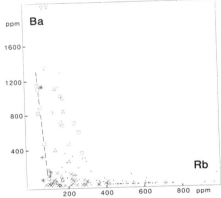

Figure 12. Ba – Rb for Nigerian granitoids. The dashed line represents the fractionation trend for the intermediate rocks.

(Bowden *et al.* 1984; Martin and Bowden 1981) supports this idea. The exponential increase in these elements bears no correlation with normal differentiation trends as reflected by the syenite–monzonite association.

In the peralkaline suite both Zr and Hf show a negative correlation with F, but Zr/Hf does not. Since these two elements are known to fractionate to a small degree as fluoride complexes, the fact that the ratio does not change with increasing F suggests that the ratio is an inherited feature. The crustal average of 50 is well established, and ranges of 25 to 65 imply severe fractionation, probably in halogen dominated environments. The fact that zircons from the peraluminous facies of the Nigerian province are Hf-rich suggests that this is a regional feature. Since the peraluminous granites also show wide ranges in Zr/Hf(10–40) and yet show no correlation with F, in spite of the fact that these rocks have been extensively mineralized by halogen-rich fluids supposes that the Zr/Hf ratio was inherited. Since the basement also shows the same range in Zr/Hf, it may be considered as a potential source of these elements. One must also note that diffusion of CO_2 together with a F-rich peralkaline fluid leads to the removal of Y as a stable Na–Y–F complex, removal of Nb, and fractionation of Zr/Hf.

A notable feature of the province is the low abundance of Ba (Figure 12) in the peraluminous and peralkaline granites (<20 ppm), whereas the levels in the basement rocks are around 363 to 1279 ppm in the calc-alkaline granitoids and from 1104 to 1047 ppm in the charnockites. Yet in spite of an abundance of alkali feldspar in the anorogenic granites Ba is virtually absent. This can be explained by Ba being wholly retained in the restite phase during partial melting, or that Ba was lost to the surface by Cl-rich fluids in volcanic processes.

5. Conclusions

The chemistry of the anorogenic Mesozoic granitoids of Nigeria indicates that the primary mineralogy was influenced in some measure by the surrounding Pan-African basement into which they were emplaced. The basement granitoids themselves show evidence of secondary changes in their element distributions. In particular the Zr/Hf values show deviations from the accepted crustal average of 50, and they have lost Nb and Y. These latter two elements are important in the subsequent mineralization of the peraluminous Mesozoic granitoids. Fluids rich in F with or without CO_2 are suspected of having changed the element signatures in the basement. The chemical heterogeneity in the basement can be ascribed either to the irregular passage of fluids, selectively mobilizing REE, or could result from an earlier partial melting event which generated areas of felsic enrichment (higher Ba, La/Y) and areas of restite rich in incompatible elements.

If the biotite-rich restite phase, with enhanced levels of structural water and F, underwent low temperature fusion, its chemistry, particularly in trace elements, would approximate to that of peralkaline granites. There is petrographic evidence for conversion of biotite to riebeckite. This late event, possibly due to decompression during progressive proto-rifting, could have given rise to columbite mineralization by remobilizing the Nb from destabilized basement. Thus the overall chemistry of the Nigerian Mesozoic granitoids carries the imprint of fluid activity which changed element distributions, altered primary mineralogies, and gave rise to the economic deposits found today.

Acknowledgements.

The samples used in this study were selected from collections made by J. N. Bennett, P. Bowden, E. C. Ike, J. A. Kinnaird, A. B. Moyes, D. C. Turner, D. A. Weir, and O. van Breemen. The editors are thanked for fruitful discussions, amendments to the text, help and encouragement.

References

Alderton, D. H. M., Pearce, J. A., and Potts, P. J. 1980. Rare earth element mobility during granite alteration: evidence from SW England. *Earth and Planetary Science Letters,* **49,** 149–165.
Batchelor, R. A. and Bowden, P. 1985. Petrogenetic interpretation of granitoid rock series using multicationic parameters. *Chemical Geology,* **48,** 43–55.
— and — 1986. Major and trace element analyses of volcanic and sub-volcanic igneous rocks from the Nigeria–Niger anorogenic province. *Supplement B, O.D.A. Research Scheme R2679.* Department of Geology, University of St. Andrews.
Bowden, P. and van Breemen, O. 1970. Isotopic and chemical studies on Younger granites from northern Nigeria. In: *Proceedings of the Conference on African Geology, Ibadan,* 105–120.
— and Kinnaird, J. A. 1984. Geology and mineralisation of the Nigerian anorogenic ring complexes. *Geol. Jahrb.* **B56;** 3–65
Bradley, D. C. and Thornton, P. 1973. Zirconium and hafnium. In: Bailar J.C. (Ed.), *Comprehensive Inorganic Chemistry, vol. 3.* Pergamon Press. 419–490.

Collins, W. J., Beams, S. D., White, A. J. R., and Chappell, B. W. 1982. Nature and origin of A-type granites with particular reference to Southeastern Australia. *Contributions Mineralogical Petrology*, **80**, 189–200.
Cotton, F. A. and Wilkinson, G. 1972. *Advanced Inorganic Chemistry*, Wiley Interscience.
Currie, K. L. 1968. On the stability of albite in supercritical water in the range 400 to 600°C and 3500 bars. *American Journal Science*, **266**, 321–341.
De la Roche, H 1964. Sur l'expression graphique des relations entre la composition chimique at la composition mineralogique quantitative des roches crystallines. Presentation d'un diagramme destine a l'etude chimico-mineralogique des massifs granitiques et granodioritiques — Application aux Vosges crystallines. *Scienc Terre* **9**, 293–337.
—— and Leterrier, J. 1973. Transposition du tetrahedre mineralogique de Yoder et Tilley dans un diagramme chimique de classification des roches basaltiques. *C.R. Acad. Sci. Paris, Serie D*, **276**, 3115–3118.
——, Grand-Claude, P. and Marchal, M. 1980. A classification of volcanic and plutonic rocks using R1–R2 diagrams and major element analyses; its relationship with current nomenclature. *Chemical Geology*, **19**, 183–210.
Kane, J. S. 1985. An inductively-coupled plasma spectrometric method for the determination of six indicator elements for mineralisation processes in granites. *Geostandards Newsletter*, **9**, 181–190.
Kinnaird, J. A. 1985. Mineralisation and hydrothermal alteration of the alkaline anorogenic ring complexes of Nigeria. *Journal of African Earth Science*, **3**, 229–251.
Kogbe, C. A. and Obialo, A. U. 1975. Statistics of mineral production in Nigeria (1946–1974). In: Kogbe, C.A. (Ed.) *Geology of Nigeria* Elizabethan Publishing Co., Nigeria.
Martin, R. F. and Bowden, P. 1981. Peraluminous granites produced by rock–fluid interaction in the Ririwai non-orogenic ring complex, Nigeria: mineralogical evidence. *Canadian Mineralogy*, **19**, 65–82.
—— Whitley, J. E., and Woolley, A. R. 1978. An investigation of rare earth mobility: fenitised quartzites, Borrolan complex, NW Scotland. *Contributory Mineralogical Petrology*, **66**, 69–73.
Mineyev, D. A. 1963. Geochemical differentiation of the rare earth elements. *Geochemistry*, **12**, 1129–1150.
Moorhouse, V. E. and Moorhouse, S. J. 1983. The geology and geochemistry of the Strathy complex of northeast Sutherland, Scotland. *Mineralogical Magazine*, **47**, 123–137.
Mustart, D. A. 1972. Phase relations in the peralkaline portion of the system NA_2O-Al_2O_3-SiO_2-O_2-H_2O. Unpublished Ph.D. thesis, Stanford University, U.S.A.
Perez, J. B. and Rocci, G. 1985. Fluid interaction during crystallisation of granites in the Tagouaji ring complex, Air, Niger. *13th Colloquim African Geology, St. Andrews*, Abstract volume, p. 305.
Rickard, C. E. F. and Waters, T. N. 1964. The Hafnium tetrafluoride — water system. *Journal of Inorganic Nuclear Chemistry*, **26**, 925–930.
Taylor, R. P., Strong, D. F., and Kean, B. F. 1980. The Topsails igneous complex: Silurian–Devonian peralkaline magmatism in western Newfoundland. *Canadian Journal of Earth Science*, **17**, 425–439.
Taylor, S. R. 1965. The application of trace elements to problems in petrology. *Physical Chemistry of Earth*, **6**, 133–213.
Turner, D. C. and Bowden, P. 1979. The Ningi–Burra complex, Nigeria dissected calderas and migrating centres. *Journal of the Geological Society, London*, **136**, 105–119.
van Breemen, O. and Bowden, P. 1973. Sequential age trends for some Nigerian Mesozoic granites. *Nature*, **242**, 9–11.
——, Pidgeon, R. T. and Bowden, P. 1977. Age and isotopic studies of some Pan-African granites and related rocks, north-central Nigeria. *Precambrian Research*, **4**. 307–319.
Wedepohl, K. H. 1969. *Handbook of Geochemistry*. Springer-Verlag.

Accessory mineralogy of the Ririwai biotite granite, Nigeria, and its albitized and greisenized facies

R.A. Ixer, J.R. Ashworth and C.M. Pointer
Department of Geological Sciences, Aston University, Aston Triangle, Birmingham, B4 7ET, U.K.

The Ririwai granitic ring complex has suffered a sequence of post-crystallization alteration and mineralizing processes including albitization and, close to the Ririwai Sn–Zn lode, microclinization, greisenization, and quartz-cassiterite-sulphide veining.

Within these rock types the opaque and accessory minerals of Zr, Hf, U, Th, Nb, Ta, Ti, Sn, and REE have a number of styles of occurrence. They occur in coarse-grained discrete crystals associated with magmatic quartz or as aggregates of small crystals, which are often aligned along the cleavage of micas, or form complex intergrowths outside the micas.

Late stage, possibly magmatic processes within the biotite granite have precipitated Zr, U, Pb, Nb, Ta, Ti as complex intergrowths of uranium-bearing plumbopyrochlore, columbite, ilmenite, TiO_2 minerals, and zircon. Minor to trace amounts of monazite, uranothorite, and cassiterite are also present in these granites.

Albitization has resulted in the crystallization of coarse-grained, early haematite intergrown with slightly later magnetite, ilmenite, and columbite. Both columbite and ilmenite are highly zoned with respect to their iron and manganese, but not their niobium and tantalum contents. Zircon is often strongly zoned with uranium-rich cores and hafnium-rich rims and shows partial replacement and overgrowth by uranothorite and xenotime.

Cassiterite, columbite, zircon, and Fe, Mo, Pb, and Zn sulphides are the main non-silicates developed during greisenization, although thorium too, is mobile, crystallizing mainly within uranothorite. Columbite has a wide range of compositions with tantalum-rich (up to 27 wt% Ta_2O_5) and tungsten-rich (up to 14·5 wt% WO_3) varieties. Cassiterite is commonly colour zoned and this can be related to its iron and niobium content. Zircon shows replacement by, or enclosure within, uranothorite, which itself has later xenotime margins.

The textural and chemical evidence suggests that each alteration process essentially dissolved, or replaced, earlier phases and then reprecipitated them; but with compositions compatible with the new fluids. The dissolution of small accessory phases within, and the alteration of biotite, initially to chlorite, appear to have played important roles in the liberation of rare elements by these fluids.

KEY WORDS   Alkaline granites   Mineralization   Accessory minerals   Uranium-thorium minerals

1. Introduction

The Ririwai complex is amongst the best studied of the Younger granite complexes of Nigeria. Its geology has been described by Jacobson et al. (1963) and Jacobson and MacLeod (1977), and information on petrology, geochemistry, and mineralization are summarized in the comprehensive account of Kinnaird et al. (1985). The complex, which represents the roots of an eroded alkaline volcano, comprises an outer ring-dyke of fayalite granite porphyry, surrounding a peralkaline granite and a central biotite granite emplaced within a collapsed volcanic pile. Both granites have been altered by post-magmatic fluids which have locally converted them to albitites. In addition, the biotite granite has been modified by hydrothermal fluids to produce microclinization, greisenization, and a braided quartz-greisen vein system, which is at its most extensive along the Ririwai lode. This cassiterite–sphalerite lode extends for more than 5 km in an east–west direction, and is entirely enclosed within the biotite granite (Kinnaird et al. 1985). Within the biotite granite an inclined borehole L13 has penetrated to 450 m and encountered albitites in its basal 40 m (Figure 1). These have been interpreted as belonging to an apical part of an underlying granite (Kinnaird et al. 1985).

Regional studies of the Younger Granite province by Bowden and Kinnaird (1978) have shown it to be a zinc-rich tin province, associated with minor uranium, niobium, and tungsten. Kinnaird (1984) has further subdivided the mineralization of the province into (1) disseminated tantalum-bearing columbite, cassiterite, and REE minerals associated with albitization, and later (2) cassiterite–sulphide vein-style mineralization associated with microclinization, greisenization, and quartz-vein infilling.

A preliminary description of the mineralogy and a paragenesis for the mineralization at Ririwai, given in Kinnaird et al. (1985), concentrated upon the lode and its cassiterite–sulphide mineralization. It showed that the albitites, where the biotite has been extensively altered to lithian micas and finally to zinnwaldite, carry accessory columbite, minor cassiterite, thorite, xenotime, thorium-rich monazite and hafnium-rich zircon. The fluids were hot (with homogenization temperatures of 460 – 260°C) and sodium-rich, and introduced sodium and iron together with U, Th, Zr, Nb, and HREE.

Microclinization is a local, minor phenomenon. The greisenization process has altered the feldspars of the biotite granite firstly to chlorite and then to lithian micas, fluorite, topaz, and sericite and the biotites to zinnwaldite. The fluids responsible were hot (with homogenization temperatures of 380 – 360°C) and acidic, and altered the bulk rock geochemistry by lowering the relative Al_2O_3 content, and increasing the contents of Li, Fe, Sn, W, Pb, Zn, and Cu in the form of white mica, cassiterite, wolframite, and sulphides. Th, Ce, and Y contents were also increased as seen by the formation of monazite.

Bowden et al. (1981) analysed bulk samples of the Ririwai biotite granite and albitites from L13, for uranium and thorium. MacKenzie et al. (1984) described the distribution of uranium and thorium within wallrocks and vein material from the Ririwai lode, and suggested that uranium was concentrated during microclinization and was probably held in Hf-rich zircon, and in xenotime in association with Ce and HREE, whereas thorium was concentrated during greisenization and held in thorite, Th-rich monazite, zircon, and cassiterite. Both sets of geochemical data are summarized in Table 1 which gives the bulk rock analyses for uranium and thorium from all the rock types at Ririwai.

This paper presents data on the accessory and opaque minerals found within the unaltered biotite granite and albitite (from borehole L13) and in microclinized and greisenized rocks next to the Ririwai lode, in order to describe more fully the processes of albitization and greisenization and their associated mineralization.

ACCESSORY MINERALOGY OF THE RIRIWAI GRANITE

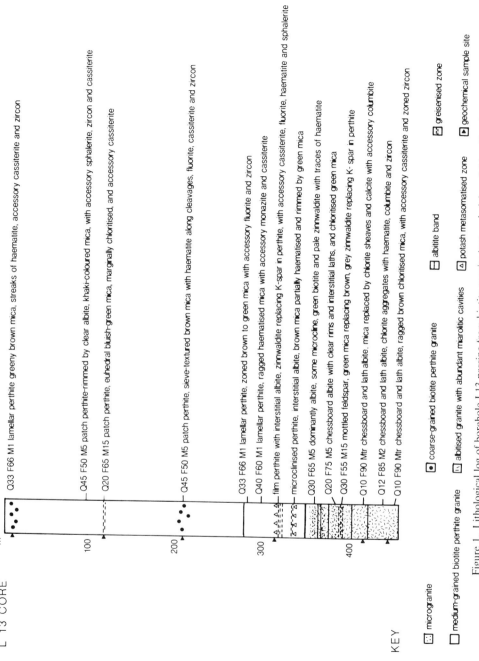

Figure 1. Lithological log of borehole L13 passing from biotite granite into underlying albitite. Depth in metres. Q, quartz; F, total feldspar; M, total mica content. From J. Kinnaird (unpublished data).

Table 1. Bulk rock analyses for U and Th (in ppm) of rocks from Ririwai

		U range	Th range	Th/U	Samples
Biotite granite	L13	30–80	25–72	0·83–0·97	4
Albitite	L13	66–81	69–73	0·90–1·05	2
Microclinite	lode	10–13	39–42	3·25–3·80	2
	(lode)		(17–87)		(13)
Greisen	lode	19	98	5·16	1
	(lode)		(43–191)		(6)
Biotite granite	surface	7–10	41–83	5·5–8·30	4

Data from Kinnaird et al. (1985) and MacKenzie et al. (1984); the latter are given in parenthesis.

ACCESSORY MINERALOGY OF THE RIRIWAI GRANITE

Table 2. Representative analyses of uranium- and thorium-bearing phases in biotite granite.

Sample		ZrO_2	HfO_2	ThO_2	UO_2	Y_2O_3	Ce_2O_3	La_2O_3	Nd_2O_3	CaO	FeO	MnO	SiO_2	P_2O_5	Total
Zircon															
L13–100	core	62·46	2·54	0·96	0·35	0·31	–	–	–	0·23	0·42	0·11	32·43	0·10	99·91
	inner zone	52·20	2·34	4·88	0·91	1·39	–	–	–	1·52	1·76	0·53	26·36	0·22	92·11
L13–100	outer zone	62·79	4·41	0·40	0·21	0·12	–	–	–	0·07	0·37	nd	32·37	0·09	100·83
L13–185	outer zone	59·50	5·79	0·76	1·21	0·37	–	–	–	0·02	nd	nd	32·65	0·21	100·51
L13–315	outer zone	60·80	6·36	0·06	0·29	0·01	–	–	–	0·01	0·23	0·04	30·76	0·08	98·64
Thorite															
L13–315	inclusion	2·30	nd	76·90	5·32	0·66	–	–	–	0·67	1·47	nd	11·27	0·44	99·03
L13–185	inclusion	7·30	0·60	53·32	14·07	2·34	–	–	–	0·47	0·13	nd	12·50	0·43	91·16
L13–305	discrete	11·65	0·74	60·15	3·18	2·83	–	–	–	0·74	0·93	nd	14·74	1·15	96·11
Monazite															
L13–185	discrete	–	–	7·26	0·04	0·07	37·49	17·65	9·66	0·43	–	–	1·45	28·24	102·29
L13–315	crystals	0·09	nd	12·16	0·32	0·07	32·07	26·27	4·90	0·29	nd	–	2·41	26·30	104·88
Coffinite															
L13–100		4·01	0·37	1·94	69·20	0·30	–	–	–	0·81	0·01	–	14·23	0·12	90·99
		4·97	0·44	4·94	64·78	0·33	–	–	–	0·91	0·03	–	14·06	0·11	90·57
	separate grains	11·69	0·43	3·29	57·75	0·54	0·73	–	–	0·97	1·68	–	16·89	0·19	94·16
		2·86	0·43	1·35	75·73	0·26	0·55	–	–	1·07	0·45	–	14·92	0·22	97·84

Sample		Nb_2O_5	Ta_2O_5	ThO_2	UO_2	TiO_2	Na_2O	La_2O_3	PbO	CaO	FeO	MnO	SiO_2	Ce_2O_3	Total
REE phase															
L13–315		nd	nd	2·51	0·19	0·10	40·27	44·52	5·49	0·60	nd	–	0·06	nd	93·74
Pyrochlore															
L13–305 three		19·77	21·85	0·38	15·42	2·21	1·18	nd	26·13	0·89	3·15	0·19	nd	–	91·17
separate grains		19·23	21·16	0·39	14·37	2·52	1·04	nd	28·52	0·42	3·60	0·15	nd	–	91·40
within complex		19·34	20·27	0·41	17·40	2·42	1·44	nd	23·14	0·43	3·44	0·15	nd	–	88·44
intergrowth		19·49	22·70	0·32	19·39	2·00	1·36	nd	18·67	0·49	3·62	0·15	nd	–	88·19
		19·18	22·55	0·47	13·66	2·32	1·32	–	28·57	0·45	3·32	0·14	nd	–	91·98
S81	core	32·04	2·73	0·45	7·19	6·22	–	0·93	31·41	0·54	1·63	–	8·23	5·05	96·42
	rim	34·05	1·64	0·71	3·60	7·08	–	1·82	36·39	0·33	1·49	–	4·03	5·41	96·55

– not determined   nd. not detected   S 81 Arfvedsonite granite.

2. Petrographical studies and mineral analyses

One polished block and polished thin section of several representative examples of unaltered biotite granite, albitite, and greisenized granite were made. Wherever possible, material was chosen which previously has been described in the literature with published bulk rock geochemistry. From L13 borehole material, ten fresh biotite granites, two partially albitized granites (350, 385), three albitites (411, 440, 445), and one incipiently greisenized granite (125) were chosen, as were fourteen samples of microclinized and greisenized granite adjacent to the Ririwai lode. Lexans and autoradiographs were used for finding uranium and thorium sites in the microclinized and greisenized rocks.

Routine transmitted and reflected light petrographical studies were followed by SEM and wavelength-dispersive electron microprobe analysis of the accessory and opaque phases. Quantitative analyses were performed for Fe, Mn, Ti, Nb, Ta, Zn, Sn, W, U, Th, Hf, Zr, REE, P, Ca, Si, and Y on appropriate minerals and using simple standards and full ZAF correction procedures. In common with other studies, poor totals were obtained for uranium- and thorium-bearing phases, due to their metamict nature, poor crystallinity, the presence of unknown quantities of OH and F radicals, and possibly adsorbed H_2O.

3. Results

3a. Accessory and opaque mineralogy of the unaltered biotite granite

The biotite granite contains zircon, columbite, and TiO_2 minerals accompanied by minor amounts of monazite, thorite, ilmenite, marcasite, pyrite, sphalerite, galena, chalcopyrite, and cassiterite. Secondary haematite is ubiquitous.

Zircon is common and is found as interstitial equant grains (up to 500 μm in diameter) between quartz crystals, or as smaller grains, within biotite (up to 50 μm in diameter) associated with columbite, ilmenite, TiO_2 minerals, uranothorite, and fluorite. Many display concentric zoning, which is seen by variations in reflectance. Other lower reflectance areas cross-cut this zoning and extend from the margin of the zircon towards its core. The lower reflectance areas are typically enriched in Th, U, Ca, Y, P, Fe, and Mn compared to areas of the zircon with higher reflectance. Thorite inclusions (up to 10 μm in size) occur in the outer zones of some zircons, while others show an outer inclusion-free zone, 20 μm wide, which is hafnium-rich with up to 6·4 wt% HfO_2 (Table 2).

Thorite is less common than zircon. In addition to forming small inclusions in zircon, it occurs in micas as larger grains (up to 250 × 180 μm) in association with zircon, columbite, and fluorite. Paragenetically it is seen to be later than zircon, upon which it nucleates. Table 2 shows the thorite to be variable in composition: some are uraniferous with up to 14·1 wt% UO_2; some have up to 2·8 wt% Y_2O_3, and some are zirconium-rich with up to 27·1 wt% ZrO_2. Optically, there is no evidence to suggest that the high zirconia values are due to the presence of zircon as an admixed phase.

Thorium-bearing monazite may contain up to 12·2 wt% ThO_2 and 0·3 wt% UO_2. It forms lath-shaped crystals 100 × 30 μm in size, which are often collected into radiating aggregates 300 μm in diameter, but can reach 800 μm in length. Table 2 shows the monazite to have variable ratios of its major rare earth elements, Ce, La, and Nd. Monazite is often intimately intergrown with an unidentified

LREE mineral which also contains thorium (up to 3·5 wt% ThO_2), minor amounts of uranium (up to 0·2 wt% UO_2) and calcium (up to 0·6 wt%) suggesting that it is bastnaesite or fluocerite. Texturally, the LREE mineral appears to be replacing monazite, and fluocerite replacing monazite is recorded from the similar Taghouaji complex in Niger (Perez 1985). All these uranium- and thorium-bearing minerals are stained with haematite and carry small, 2–10 μm diameter, inclusions of haematite, pyrite, and limonite.

Pyrochlore is rare, and was seen as part of a single complex columbite–ilmenite–TiO_2–zircon–pyrochlore–fluorite intergrowth 150 μm in diameter, where it occurs as poorly polished grains, 30 μm in diameter (Plate 1). Table 2 shows that it belongs to the plumbopyrochlore species, and contains up to 0·5 wt% ThO_2 and 19·4 wt% UO_2. The presence of pyrochlore in albitized alkali granite from Ririwai has been known for some time, but it has a different composition from that found in the biotite granite. Beer (1952) in partial analyses for pyrochlore from the albitized alkali granite shows it to have up to 4·3 wt% ThO_2 and up to 3·5 wt% UO_2. Analyses from this study (Table 2) suggest the pyrochlore found in the alkali granite is also a plumbopyrochlore but is more niobium-rich and significantly tantalum- and uranium-poorer than plumbopyrochlores from the Ririwai biotite granite.

Columbite occurs as small grains (up to 30 μm diameter) surrounded by 5–15 μm wide TiO_2 rims, and is seen within biotite, chloritized biotite, or feldspar, usually aligned along cleavages. Similarly rimmed but larger columbite grains, up to 250 μm in diameter, which may be optically unzoned or faintly zoned, are found between quartz crystals; or are associated with the pyrochlore intergrowth. Table 3 shows that the discrete columbite crystals in quartz, and columbite associated with pyrochlore, are iron-rich and niobium-rich. Atomic $Fe/(Fe + Mn)$ ratios are between 0·82 to 0·91 for all but two analyses, $Nb/(Nb + Ta)$ ratios have a range of 0·76 to 0·99. There is no strong correlation between these two ratios, nor any chemical zoning from core to rim other than a slight increase of manganese and decrease of iron at the margins of single crystals. Other elements are present within limited ranges, with up to 2·6 wt% TiO_2, and 1·7 wt% WO_3. Only the tantalum content has a wide range, beyond that given in Table 3, from 1·2 wt% to 23·7 wt% Ta_2O_5.

Ilmenite (which optically is pinker than columbite) is found in minor amounts, as small lath-shaped crystals up to 200 × 20 μm in size, occurring along the cleavage of the micas, or as large aggregates (300 × 200 μm) of typically curved crystals in quartz. It also forms part of the intergrowth with pyrochlore. It is enclosed within columbite or TiO_2 mineral rims and is extensively altered to fine grained haematite and TiO_2 minerals. Table 3 shows the ilmenite to be iron rich but to contain significant amounts of manganese with $Fe/(Fe+Mn)$ in the range 0·45 to 0·99. Although the manganese content is erratic, both the amount of ilmenite and its manganese content increase generally down the drill-hole towards the underlying albitites, with the maximum manganese content of 22·9 wt% MnO occurring at L13–350. Discrete ilmenite carries little niobium (<2 wt% Nb_2O_5) or tantalum (<0·5 wt% Ta_2O_5), but ilmenites which are rimmed by columbite have up to 12·4 wt% Nb_2O_5.

TiO_2 minerals (rutile and ?anatase) occur as discrete acicular crystals up to 200 × 4 μm in length, lying along the cleavages of mica, and as aggregates of equant twinned rutile crystals up to 600 μm in diameter, and as rims around columbite and ilmenite. The TiO_2 minerals have up to 4·1 wt% FeO, 0·1 wt% MnO and 1·0 wt% Ta_2O_5. Table 3 shows that they fall into two groups: a low-niobium

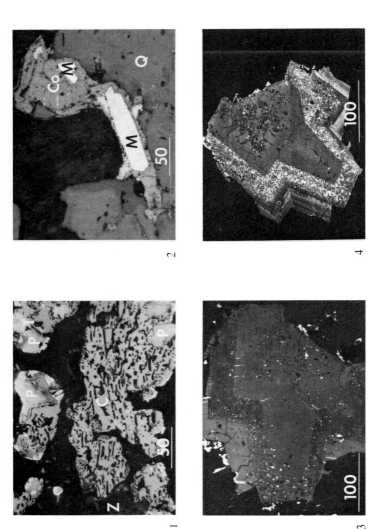

Plate 1. Columbite (C), myrmekitically intergrown with silicates, has narrow TiO$_2$ rims (white, bottom of photograph and top of photograph (T)). It is intergrown with zircon (Z) and plumbopyrochlore (P). Biotite granite, L13–305m. Reflected light, oil immersion. Scale bar 50 microns.

Plate 2. Coffinite (Co) with fine-grained admixed sphalerite (lighter grey patches) encloses molybdenite laths. Gangue minerals are quartz (Q) and mica. Biotite granite, L13-100m. Reflected light, oil immersion. Scale bar 50 microns.

Plate 3. Zoned zircon. The middle zone contains abundant uranothorite, pyrite and haematite inclusions. The outer zone is hafnium-rich (light-grey). Pyrite (white) as patches and veinlets is present around zircon. Albitite, L13–411m. Scale bar 100 microns.

Plate 4. The same zircon. Abundant thorite inclusions (white) are clearly seen. Fine-scale rhythmical zoning of the outer hafnium-rich zone is also well displayed. Back-scattered image. SEM. Scale bar 100 microns.

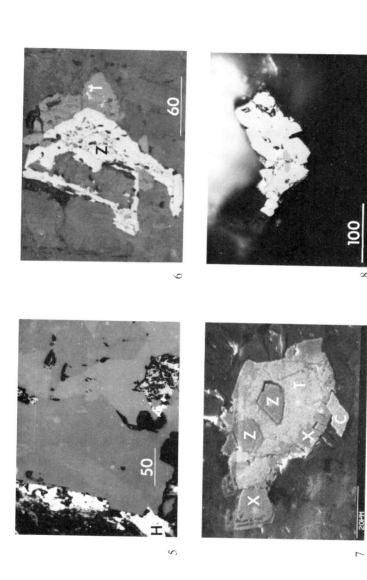

Plate 5. Zoned columbite, darker zones are iron-rich. Haematite (white, H) and TiO$_2$ minerals (grey within columbite) are associated. Albitite L13-440m. Reflected light, oil immersion. Scale bar 50 microns.

Plate 6. Zircon (Z) showing replacement of its inner zones by mica. It is overgrown by uranothorite (T) which contains small inclusions of pyrite (white specks). Gangue is mica. Incipiently greisenized granite L13-125m. Reflected light. Scale bar 60 microns.

Plate 7. Two euhedral zircons (Z) are overgrown by uranothorite (T) which, itself, is replaced by xenotime (X) along its edges. A euhedral grain of cassiterite (C) is present. Lode greisen RS6 (2). SEM image. Scale bar 20 microns.

Plate 8. Columbite core (darker grey in centre) enclosed within subhedral TiO$_2$ (lighter grey). Lode greisen N58. Reflected light, oil immersion. Scale bar 100 microns.

Table 3. Representative columbite, ilmenite and TiO₂ mineral analyses in biotite granite

Sample	FeO*	MnO	TiO$_2$	Nb$_2$O$_5$	Ta$_2$O$_5$	WO$_3$	SnO$_2$	Total	
Columbite									
L13-100 core	18.04	2.03	0.73	72.07	6.95	nd	—	99.82	(Fe$_{0.87}$Mn$_{0.10}$)(Nb$_{1.85}$Ta$_{0.11}$Ti$_{0.03}$)O$_6$
rim	16.76	2.19	1.50	58.32	18.42	1.71	—	98.90	(Fe$_{0.80}$Mn$_{0.11}$)(Nb$_{1.61}$Ta$_{0.31}$Ti$_{0.07}$W$_{0.03}$)O$_6$
L13-155 core	17.64	2.17	0.82	63.68	16.11	—	—	100.42	(Fe$_{0.88}$Mn$_{0.11}$)(Nb$_{1.71}$Ta$_{0.26}$Ti$_{0.04}$)O$_6$
core	17.27	2.28	0.78	59.94	19.72	—	—	99.99	(Fe$_{0.88}$Mn$_{0.12}$)(Nb$_{1.65}$Ta$_{0.33}$Ti$_{0.04}$)O$_6$
rim	17.03	2.53	0.58	68.10	11.25	—	—	99.49	(Fe$_{0.84}$Mn$_{0.13}$)(Nb$_{1.81}$Ta$_{0.18}$Ti$_{0.03}$)O$_6$
L13-155	18.51	2.21	0.56	74.99	3.19	—	—	99.46	(Fe$_{0.88}$Mn$_{0.11}$)(Nb$_{1.94}$Ta$_{0.05}$Ti$_{0.02}$)O$_6$
L13-205 core	17.97	2.33	0.49	65.69	13.53	—	—	100.01	(Fe$_{0.89}$Mn$_{0.12}$)(Nb$_{1.76}$Ta$_{0.22}$Ti$_{0.02}$)O$_6$
rim	17.77	2.40	0.55	67.21	11.83	—	—	99.76	(Fe$_{0.88}$Mn$_{0.12}$)(Nb$_{1.79}$Ta$_{0.19}$Ti$_{0.02}$)O$_6$
L13-256 near ilmenite	18.46	2.22	2.61	74.71	1.69	0.78	0.23	100.70	(Fe$_{0.86}$Mn$_{0.10}$)(Nb$_{1.88}$Ta$_{0.03}$Ti$_{0.11}$W$_{0.01}$)O$_6$
L13-256 myrmekite	18.64	2.19	1.32	75.29	1.78	nd	nd	99.22	(Fe$_{0.89}$Mn$_{0.11}$)(Nb$_{1.93}$Ta$_{0.03}$Ti$_{0.06}$)O$_6$
L13-305 symplectite	17.30	2.50	1.51	64.84	12.83	—	—	98.98	(Fe$_{0.88}$Mn$_{0.10}$)(Nb$_{1.94}$Ta$_{0.03}$Ti$_{0.05}$)O$_6$
L13-350 core	18.37	2.09	1.12	70.74	7.80	0.29	—	100.41	(Fe$_{0.88}$Mn$_{0.10}$)(Nb$_{1.84}$Ta$_{0.12}$Ti$_{0.05}$)O$_6$
rim	17.68	3.11	0.79	67.43	10.69	0.34	—	100.04	(Fe$_{0.87}$Mn$_{0.15}$)(Nb$_{1.79}$Ta$_{0.17}$Ti$_{0.04}$)O$_6$
Ilmenite									
L13-185 lath	35.74	9.77	52.06	1.28	0.14	—	—	98.99	(Fe$_{0.76}$Mn$_{0.21}$)(Ti$_{1.00}$Nb$_{0.01}$)O$_3$
lath	35.22	10.43	52.38	0.55	0.26	—	—	98.84	(Fe$_{0.75}$Mn$_{0.22}$)TiO$_3$
L13-256 core	45.49	1.07	41.41	12.36	0.24	—	—	100.57	(Fe$_{0.99}$Mn$_{0.02}$)(Ti$_{0.81}$Nb$_{0.15}$)O$_3$
rim	46.19	1.28	48.02	2.85	0.14	—	—	98.48	(Fe$_{1.01}$Mn$_{0.03}$)(Ti$_{0.94}$Nb$_{0.03}$)O$_3$
L13-305	39.75	6.60	49.56	1.65	0.49	—	—	98.05	(Fe$_{0.86}$Mn$_{0.15}$)(Ti$_{0.97}$Nb$_{0.02}$)O$_3$
L13-350	19.2	22.9	53.8	4.9	0.2	—	—	101.0	(Fe$_{0.40}$Mn$_{0.48}$)(Ti$_{1.06}$Nb$_{0.05}$)O$_3$
TiO$_2$ minerals									
L13-155 lath	1.62	0.11	94.41	3.42	0.61	—	—	100.17	(Ti$_{0.96}$Nb$_{0.02}$Fe$_{0.02}$)O$_2$
lath	2.68	0.07	91.16	3.94	0.56	—	—	98.41	(Ti$_{0.95}$Nb$_{0.03}$Fe$_{0.03}$)O$_2$
lath	1.73	0.03	89.64	8.00	0.41	—	—	99.81	(Ti$_{0.93}$Nb$_{0.05}$Fe$_{0.02}$)O$_2$
L13-155 nearby smaller grains	1.83	0.05	84.24	13.59	0.61	—	—	100.32	(Ti$_{0.88}$Nb$_{0.09}$Fe$_{0.02}$)O$_2$
L13-205 two adjacent grains in biotite	1.67	0.06	91.94	4.89	0.64	—	—	99.20	(Ti$_{0.95}$Nb$_{0.03}$Fe$_{0.02}$)O$_2$
	3.19	0.01	81.64	15.64	0.58	—	—	101.06	(Ti$_{0.86}$Nb$_{0.09}$Fe$_{0.04}$)O$_2$
L13-256 lath	2.71	0.04	90.38	5.12	0.20	—	—	98.45	(Ti$_{0.94}$Nb$_{0.03}$Fe$_{0.03}$)O$_2$
L13-350 large grain	3.10	0.05	83.44	13.80	0.47	—	—	100.86	(Ti$_{0.87}$Nb$_{0.09}$Fe$_{0.04}$)O$_2$
same	3.48	0.07	82.57	14.49	0.54	—	—	101.15	(Ti$_{0.86}$Nb$_{0.09}$Fe$_{0.04}$)O$_2$

* Total Fe expressed as FeO.

group (≤ 5 wt% Nb_2O_5) and a high-niobium group (13·5 to 15·5 wt% Nb_2O_5). Petrographically these two groups are indistinguishable and, indeed, they are often found next to each other within the same mica crystal. TiO_2 rims about columbite are too narrow for uncontaminated analyses but the data suggest that the TiO_2 belongs to the high-nobium group, whilst TiO_2 rims about low-niobium ilmenites belong to the low-niobium group.

Other phases within the biotite granite are only found in minor amounts and probably have been introduced by later mineralizing fluids. Cassiterite is extremely rare and forms small euhedral crystals up to 60 μm enclosing relict columbite. Galena, which is very close to being pure PbS and in which silver is undetectable, is often intergrown with light-coloured iron-poor (up to 2·5 mol% FeS) sphalerite. Pyrite, marcasite, molybdenite, bismuthinite, and native bismuth are rare. In L13–100 a late-stage complex intergrowth, 200 μm in diameter, of nearly pure end-member molybdenite, minor iron-poor sphalerite, and coffinite is seen within small voids in the granite (Plate 2). The coffinite is variable in composition (Table 2) but has up to 4·9 wt% ThO_2 and 11·7 wt% ZrO_2 with only minor amounts of Y_2O_3 (up to 0·5 wt%).

3b. Accessory and opaque mineralogy of the albitites

The albitites, which have resulted from intense sodium-metasomatism, have a distinctive mineral assemblage which is characteristically coarse-grained. Kinnaird et al. (1985) described the accessory mineral assemblage as early zoned zircon and haematite laths, mixed ilmenite–haematite, columbite–haematite, and magnetite–haematite intergrowths and optically zoned columbite, together with finer-grained uranothorite and TiO_2 minerals. Cassiterite, sphalerite, galena, and pyrite are also present. Haematite, columbite, and zircon are the most abundant of these phases.

Minor amounts of 40–60 μm diameter zircon crystals are found within the biotite and chloritized biotite but most zircon occurs as euhedral, intensely and complexly zoned crystals up to 600 μm in diameter associated with quartz. The zones are between 30 and 100 μm wide, have different reflectances, and some carry abundant, small (<15 μm diameter) inclusions of uranothorite, pyrite, haematite, and marcasite. Individual zircon crystals have inclusion-poor metamict inner zones with up to 6·1 wt% UO_2, surrounded by inclusion-rich near opaque zones (with up to 0·06 wt% UO_2 and 0·4 wt% ThO_2, Table 4) which are themselves enclosed within inclusion-free outer zones of higher reflectance which are hafnium-rich (up to 14·0 wt% HfO_2) with characteristic radial fractures (Plates 4 and 5).

Uranothorite occurs as small inclusions within zircon, and also forms incomplete rims around zircon. It has a variable uranium content (between <5·0 and 28·9 wt% UO_2) and variable amounts of Y_2O_3, usually less than 1 wt% but with a maximum of 5·9 wt% Y_2O_3. Minor amounts of thorium-bearing xenotime (up to 2·2 wt% ThO_2) occur as small (15 μm) crystals within inner zones of zircon and as replacements of zircon. A thorium-bearing (up to 2·4 wt% ThO_2) uniaxial LREE mineral occurs as inclusions (5 – 70 μm in diameter) within yttrofluorite which replaces zircon. Both its composition and optical properties are consistent with those of bastnaesite.

Haematite is the most abundant opaque phase, and forms discrete tabular crystals up to 350×200 μm in size, often in radiating clusters in the micas. Much haematite occurs as complex intergrowths with magnetite, ilmenite, and columbite where it is extensively replaced by them along fractures or grain boundaries. The haematite contains up to 5·3 wt% TiO_2 but only minor amounts of manganese, (normally <0·1 wt% MnO), and niobium (up to 0·8 wt% Nb_2O_5), tantalum

Table 4. Representative analyses of uranium- and thorium-bearing phases from albitites and greisened granite

Sample		ZrO_2	HfO_2	ThO_2	UO_2	Y_2O_3	CaO	FeO	SiO_2	P_2O_5	Total
Albitites											
Zircon											
L13–411	inner	61·34	5·65	0·08	0·08	0·12	nd	nd	32·36	0·14	99·77
	outer	53·94	14·00	0·42	0·06	0·24	nd	nd	32·13	0·31	101·10
	inner	55·53	3·07	0·51	6·10	–	–	–	28·11	–	93·32
	outer	61·50	2·94	0·58	0·98	0·38	–	–	31·99	–	98·37
Thorite											
L13–411	inclusions	2·30	1·40	72·07	1·47	1·07	1·06	2·04	15·33	0·59	97·33
	in zircon	–	–	53·30	28·92	0·78	1·12	–	16·85	–	100·97
		–	–	80·08	4·96	–	0·39	–	16·32	–	101·75
		1·89	–	53·00	19·97	5·07	0·94	0·73	16·47	1·15	99·22
		0·20	–	58·82	23·72	0·81	1·42	0·77	15·68	0·38	101·80
Xenotime											
L13–411		–	–	2·19	0·04	42·74	0·24	–	4·05	23·88	(73·14)
L13–411 REE phase in zircon		0·57	–	1·65	0·19	1·07	0·11	0·61	0·48	nd	(75·11)

plus 38·02 Ce_2O_3, 20·31 La_2O_3 and 12·10 Nd_2O_3

ACCESSORY MINERALOGY OF THE RIRIWAI GRANITE

Greisens										
Zircon										
L13–125 dark core	59.33	4.99	2.58	0.51	0.34	0.10	2.21	30.26	0.21	100.53
High R%	59.32	3.34	2.32	0.63	1.28	0.17	0.67	30.31	0.33	98.37
Low R%	44.62	2.69	7.35	0.96	5.07	0.61	1.93	24.59	0.50	88.32
RS14(1)	62.32	4.97	0.27	1.17	0.19	nd	1.12	30.84	–	100.88
RS6(4)	63.96	2.94	0.17	0.41	0.18	–	–	33.66	–	101.32
Thorite										
L13–125 inclusions	24.74	1.77	40.08	3.09	7.09	0.77	0.62	19.10	1.07	98.33
	29.20	2.10	35.17	7.33	0.88	0.84	0.97	23.78	0.51	100.78
RS6(2) overgrowth of thorite	2.38	–	47.43	23.45	6.82	1.05	–	13.84	1.23	96.20
on zircon	1.04	–	51.12	18.57	5.06	1.18	–	12.91	1.19	91.07
	0.55	–	63.57	4.53	7.51	0.69	0.10	11.63	1.96	90.54
Xenotime										
RS6(2) overgrowth on thorite			0.89	0.10	44.93	0.08		3.86	27.00	(76.86)
			3.91	0.43	48.30	0.33		5.34	23.68	(81.99)
Coffinite										
RS6(4) coffinite	0.07	–	2.68	59.94	11.61	1.33	–	13.05	1.94	90.62
	0.04	–	2.90	64.54	9.01	1.66	–	14.20	0.72	93.07

R%, reflectance

(normally up to 0.25 wt% Ta_2O_5) and trace amounts of tin (Table 5).

Ilmenite is present in minor amounts as small laths 60 × 10 μm in size along cleavages in mica, and more commonly in intergrowths with haematite as tabular crystals up to 100 μm in length. The iron–manganese composition of ilmenite is variable, with two ranges of Fe/(Fe + Mn) ratios (0.24 to 0.25 and 0.86 to 0.88) suggesting that 'single' grains are a mixture of ilmenite and pyrophanite. Although optical zoning is seen in these ilmenite grains, neither colour nor reflectance variations correlate well with composition. Ilmenite contains up to 2.3 wt% Nb_2O_5, 0.9 wt% Ta_2O_5, and trace amounts of tin. Locally, ilmenite is extensively altered to haematite and TiO_2 minerals. Magnetite, too, is intimately intergrown with haematite and replaces haematite, to give mixed grains up to 300 × 150 μm in size. Magnetite is manganese-poor, up to 0.4 wt% MnO, but has up to 3.5 wt% TiO_2 and trace amounts of niobium, up to 0.2 wt% Nb_2O_5 and tantalum, up to 0.2 wt% Ta_2O_5, but no detectable tin or tungsten (Table 5).

Small columbite crystals up to 40 μm in diameter are found in micas and feldspar, but most columbite is coarse-grained (up to 500 μm in diameter), euhedral, and optically highly zoned with brown cores and bluer, lower reflectance margins (Plate 6). Relict haematite is common within columbite. Unlike the columbite analyses obtained from the surface albitite samples at Ririwai (Kinnaird et al. 1985), which showed a very restricted range of compositions, the present analyses show that columbite from the drill-hole albitites is extremely variable in composition. This compositional variation can be related to the mineral's optical properties, for the inner, browner iron-rich cores have Fe/(Fe + Mn) ratios between 0.80 and 0.96 and the outer, lower reflectance manganese-rich rims between 0.16 to 0.40. The boundaries between iron-rich and manganese-rich columbite are sharp and distinct. The variation in niobium and tantalum contents is less extreme; the ratios Nb/(Nb + Ta) varies between 0.86 and 0.98 showing little correlation with the iron and manganese content of the columbites.

Small acicular crystals of TiO_2 minerals up to 20 μm in length lie along cleavages of biotite and chloritized biotite, or form equant crystals up to 60 μm in diameter surrounding columbite and ilmenite. Table 5 shows the TiO_2 minerals to be niobium-poor with up to 4.7 wt% Nb_2O_5 together with up to 2.7 wt% Ta_2O_5, 0.2 wt% SnO_2 and 0.04 wt% WO_3.

Minor amounts of pyrite and marcasite are associated with zircon and uranothorite frequently as rims to, or inclusions within them. Iron-poor sphalerite (up to 1.4 mol% FeS) is intergrown with galena, both of which enclose haematite laths.

3c. Accessory and opaque mineralogy of the microclinized and greisenized granites.

The mineralization, especially the sulphide phases, accompanying both processes has been extensively described in Kinnaird et al. (1985). Both processes have produced the same mineralogy which is essentially cassiterite as the main oxide phase accompanied by sphalerite, galena, molybdenite, chalcopyrite, pyrite, and marcasite. Zircon, columbite, wolframite, thorite, monazite, and TiO_2 minerals are also present, as are trace amounts of ilmenite. Fourteen samples of microclinized and greisened rocks adjacent to the Ririwai lode, and L13–125, an incipiently greisened granite from the drill-hole, were studied.

Zircons from L13–125 occur as 200 μm diameter crystals which have been extensively corroded, and typically show total replacement of their inner zones by silicates and fluorite to leave 30 μm wide outer zones with 30 μm uranothorite overgrowths (Plate 7). In material from the Ririwai lode the zircons are equant crystals up to 120 μm in diameter within quartz crystals and display fine-scale

Table 5. Representative analyses of columbite, haematite, ilmenite, magnetite and TiO_2 minerals of the albitites

Sample	FeO	Fe_2O_3	MnO	TiO_2	Nb_2O_5	Ta_2O_5	WO_3	SnO_2	Total	
Haematite										
L13–440 lath within columbite	2.10	95.58	0.06	2.27	0.02	0.10	—	—	100.13	$(Fe_{1.95}Ti_{0.05})O_3$
L13–445 associated with magnetite	4.34	89.24	0.24*	4.15	0.47*	0.56*	—	—	99.00	$(Fe_{1.90}Ti_{0.08})O_3$
	4.47	90.80	0.04	4.03	0.69	0.22	—	—	100.25	$(Fe_{1.91}Ti_{0.08}Nb_{0.01})O_3$
L13–440 close to ilmenite	5.17	89.39	0.05	5.26	0.29	0.24	nd	0.09	100.40	$(Fe_{1.89}Ti_{0.10})O_3$
	3.28	93.74	0.06	2.39	0.84	0.22	nd	0.09	100.62	$(Fe_{1.94}Ti_{0.05}Nb_{0.01})O_3$
	2.25	94.29	0.85	2.84	0.30	0.21		0.07	100.81	$(Fe_{1.92}Ti_{0.06}Mn_{0.02})O_3$
Ilmenite										
L13–440 intergrown with haematite	41.15	1.36	6.11	49.30	2.26	0.88	nd	nd	101.06	$(Fe_{0.90}Mn_{0.13})(Ti_{0.94}Nb_{0.03}Ta_{0.01})O_3$
	40.09	1.68	6.67	50.62	1.07	0.20	nd	nd	100.33	$(Fe_{0.86}Mn_{0.14})(Ti_{0.96}Nb_{0.01})O_3$
	11.27	1.04	34.76	50.97	0.44	0.20	nd	nd	98.68	$(Fe_{0.25}Mn_{0.75})(Ti_{0.98}Nb_{0.01})O_3$
	10.39	1.67	35.58	50.47	0.63	0.39	nd	nd	99.13	$(Fe_{0.24}Mn_{0.79})(Ti_{0.97}Nb_{0.01})O_3$
Magnetite										
L13–445	32.97	62.58	0.41	2.81	0.19	0.11	nd	nd	99.07	$(Fe_{1.08}^{2+}Fe_{1.82}^{3+}Ti_{0.09}Mn_{0.01})O_4$
Columbite										
L13–440 single zoned crystal	16.31		3.15	0.79	63.34	16.21	—	—	99.80	$(Fe_{0.82}Mn_{0.16})(Nb_{1.72}Ta_{0.26}Ti_{0.04})O_6$
	16.40		3.51	0.50	70.66	9.08	—	—	100.15	$(Fe_{0.80}Mn_{0.17})(Nb_{1.85}Ta_{0.14}Ti_{0.02})O_6$
	16.49		3.95	0.18	76.05	3.73	—	—	100.40	$(Fe_{0.78}Mn_{0.19})(Nb_{1.95}Ta_{0.06}Ti_{0.01})O_6$
dark	6.94		12.49	0.46	69.38	10.51	—	—	99.78	$(Fe_{0.34}Mn_{0.62})(Nb_{1.83}Ta_{0.17}Ti_{0.03})O_6$
dark	3.18		16.13	0.58	71.48	7.37	—	—	98.74	$(Fe_{0.16}Mn_{0.80})(Nb_{1.88}Ta_{0.12}Ti_{0.03})O_6$
L13–445 single zoned crystal	18.39		2.65	0.56	74.18	4.87	—	—	100.65	$(Fe_{0.87}Mn_{0.13})(Nb_{1.91}Ta_{0.08}Ti_{0.02})O_6$
	16.91		3.60	0.16	73.85	5.78	—	—	100.30	$(Fe_{0.81}Mn_{0.17})(Nb_{1.91}Ta_{0.09}Ti_{0.01})O_6$
dark	7.90		11.90	0.20	70.24	9.73	—	—	99.97	$(Fe_{0.39}Mn_{0.59})(Nb_{1.85}Ta_{0.15}Ti_{0.01})O_6$
dark	5.71		13.95	0.67	71.36	7.53	—	—	99.22	$(Fe_{0.28}Mn_{0.69})(Nb_{1.87}Ta_{0.12}Ti_{0.03})O_6$
dark	3.05		16.39	0.88	75.21	3.13	—	—	98.66	$(Fe_{0.15}Mn_{0.79})(Nb_{1.95}Ta_{0.05}Ti_{0.04})O_6$
TiO_2 minerals										
L13–440 four separate grains in chlorite	2.26		0.35	88.56	4.74	2.71	nd	0.22	98.84	$(Ti_{0.93}Fe_{0.03}Nb_{0.03}Ta_{0.01})O_2$
	4.02		0.13	88.06	4.68	1.79	0.04	0.12	98.84	$(Ti_{0.93}Fe_{0.05}Nb_{0.03}Ta_{0.01})O_2$
	4.32		0.12	89.10	4.25	1.24	0.10	0.01	99.14	$(Ti_{0.95}Fe_{0.05}Nb_{0.03})O_2$
	4.60		0.12	91.77	2.43	0.65	nd	0.05	99.62	$(Ti_{0.95}Fe_{0.05}Nb_{0.02})O_2$

* 'contamination', by secondary X-ray generation in columbite is suspected. In haematite, ilmenite, and magnetite, the proportion of FeO to Fe_2O_3 is calculated on the basis of the stoichiometric ratio of total cations to oxygen (2:3 in haematite, and ilmenite, 3:4 in magnetite). In columbite and TiO_2 minerals, total Fe is expressed as FeO.

zoning. In the white micas and chlorite they occur as aggregates ('welded clumps') up to 400 μm in diameter in association with columbite, cassiterite, and TiO_2 minerals. Compositionally, the zircons contain little uranium (up to 1·2 wt% UO_2) or thorium (up to 0·3 wt% ThO_2) but have up to 5·0 wt% HfO_2 (Table 4).

Uranothorite in L13–125 overgrows zircon or infills dissolution voids within it. It is zirconium-rich with up to 29·2 wt% ZrO_2. In the greisened wallrocks of the lode uranothorite is widespread and forms elongated crystals up to 300 μm long in quartz. In the white micas it forms part of the loose aggregates of zircon, monazite, xenotime, columbite, and cassiterite which are accompanied by fluorite; here uranothorite overgrows zircon and is itself replaced by xenotime (Plate 7).

Locally, cross-cutting veinlets carry abundant 60 μm long thorite crystals. Large uranothorites are optically zoned, with cloudy cores, but show little or no chemical zoning. Table 4 shows that the uranothorite in the greisens of the Ririwai lode has variable uranium contents with up to 23·5 wt% UO_2, and contains significant amounts of other elements, notably yttrium (between 0·6 to 10·2 wt% Y_2O_3) and phosphorus (up to 3·0 wt% P_2O_5), but relatively little zirconium (up to 2·4 wt% ZrO_2). Monazite is common as laths up to 600 × 160 μm in size between quartz and mica crystals. Xenotime is rare: it replaces uranothorite and contains up to 3·9 wt% ThO_2 and 0·4 wt% UO_2. Trace amounts of LREE minerals are associated with zircon, monazite, and yttrofluorite, and may belong to the bastnaesite–fluocerite groups of minerals.

All the uranium- and thorium-bearing phases are haematitically stained and are surrounded by pleochroic halos when in mica.

Columbite in L13–125 is seen as small crystals (up to 50 μm) surrounded by 5–10 μm wide TiO_2 mineral rims within mica, or typically, as relict crystals up to 200 μm, within very coarse-grained cassiterite. These relict columbites show a wide range of compositions with Fe/(Fe + Mn) ratios between 0·60 and 0·90 and Nb/(Nb + Ta) between 0·76 and 0·98; they also contain up to 1·0 wt% WO_3 and up to 0·5 wt% SnO_2 (Table 6). Within the greisenized lode the columbites are small, up to 150 μm in diameter, enclosed within TiO_2 mineral rims, and have variable compositions with Fe/(Fe + Mn) between 0·64 and 0·90 and Nb(Nb + Ta) of between 0·79 and 0·99.

TiO_2 minerals typically form overgrowths, up to 40 μm in width, on columbite (Plate 8) and contain significant amounts of niobium, up to 13·6 wt% Nb_2O_5 (L13–125) and 8·5 wt% Nb_2O_5 (Ririwai lode); of tantalum, up to 1·2 wt% Ta_2O_5 (L13–125) and 1·3 wt% (Ririwai lode), with up to 8·6 wt% WO_3 and 0·5 wt% SnO_2 next to the lode. Discrete laths of TiO_2 up to 360 μm in length, and twinned rutile crystals up to 200 μm in diameter occur in the wallrocks of the lode. They have variable concentrations of niobium (2·4 to 10·7 wt% Nb_2O_5), tantalum (0·1 to 0·8 wt% Ta_2O_5), tin (0·1 to 0·7 wt% SnO_2), and tungsten (0·9 to 5·5 wt% WO_3) but little manganese (up to 0·04 wt% MnO). In general, TiO_2 mineral overgrowths on columbite have higher minor element concentrations than do discrete TiO_2 crystals.

Cassiterite is characteristic of the greisens and forms abundant small, 60 μm crystals, whereas larger crystals, up to centimetres in diameter, are complexly zoned and twinned, and carry relict columbite. Zoning is seen optically by lighter and darker internal reflections. Table 6 shows there to be a correlation between intensity of the body colour and the increase in iron and to a lesser extent, tantalum concentrations.

Rare laths, up to 300 × 60 μm in size, within quartz have been optically identified as wolframite. Trace amounts of ilmenite too, are present in quartz.

ACCESSORY MINERALOGY OF THE RIRIWAI GRANITE

Table 6. Representative analyses of columbite, TiO_2 minerals and cassiterite from greisenized granite.

Sample		FeO*	MnO	TiO_2	Nb_2O_5	Ta_2O_5	WO_3	SnO_2	Total	
Columbite										
L13–125 columbite		13.49	5.52	0.35	66.08	11.63	1.03	0.43	99.53	$(Fe_{0.68}Mn_{0.28})(Nb_{1.79}Ta_{0.19}Ti_{0.02}W_{0.02}Sn_{0.01})O_6$
replaced		18.54	2.10	0.59	75.88	2.60	0.08	nd	99.79	$(Fe_{0.88}Mn_{0.10})(Nb_{1.95}Ta_{0.04}Ti_{0.03})O_6$
by cassiterite		12.09	8.11	0.57	71.38	7.05	0.24	0.04	99.48	$(Fe_{0.59}Mn_{0.40})(Nb_{1.87}Ta_{0.11}Ti_{0.03})O_6$
		16.16	1.98	0.95	52.69	27.07	0.99	0.46	100.30	$(Fe_{0.85}Mn_{0.11})(Nb_{1.49}Ta_{0.46}Ti_{0.05}W_{0.02}Sn_{0.01})O_6$
RS14 vein: in quartz		18.92	2.05	0.80	76.22	2.17	nd	nd	100.16	$(Fe_{0.89}Mn_{0.10})(Nb_{1.94}Ta_{0.03}Ti_{0.03})O_6$
associated TiO_2		16.15	5.70	2.74	59.27	1.25	14.44	0.35	99.90	$(Fe_{0.80}Mn_{0.29})(Nb_{1.63}Ta_{0.02}Ti_{0.12}W_{0.22})O_6$
in quartz		17.13	2.19	1.92	54.64	24.16	0.59	0.28	100.91	$(Fe_{0.88}Mn_{0.11})(Nb_{1.52}Ta_{0.40}Ti_{0.09}W_{0.01})O_6$
RS6(3) assoc. TiO_2		13.75	7.04	1.29	73.02	3.82	–	–	98.92	$(Fe_{0.66}Mn_{0.34})(Nb_{1.90}Ta_{0.06}Ti_{0.06})O_6$
TiO_2 minerals										
L13–125		2.55	0.07	84.75	12.17	1.22	nd	tr	100.76	$(Ti_{0.88}Fe_{0.09}Nb_{0.08}Ta_{0.03})O_2$
RS14(2) single rutile		2.49	0.04	81.87	13.61	0.96	nd	–	98.97	$(Ti_{0.87}Nb_{0.09}Fe_{0.03})O_2$
grain in mica		5.45	0.03	88.93	3.81	0.35	1.31	0.29	100.17	$(Ti_{0.93}Fe_{0.08}Nb_{0.02})O_2$
RS14(2) single		7.08	0.04	76.74	10.72	0.78	4.26	0.60	100.22	$(Ti_{0.84}Nb_{0.09}Fe_{0.07}W_{0.01})O_2$
grain in mica		5.07	0.01	91.28	2.35	0.17	0.88	0.11	99.87	$(Ti_{0.95}Fe_{0.06}Nb_{0.02})O_2$
RS14 vein. overgrowths		6.89	0.04	80.52	6.98	0.41	5.53	0.66	101.03	$(Ti_{0.87}Fe_{0.08}Nb_{0.05}W_{0.02})O_2$
on columbite		6.09	0.05	77.71	7.41	0.70	8.62	0.45	101.03	$(Ti_{0.85}Fe_{0.07}Nb_{0.05}W_{0.03})O_2$
		6.14	0.13	76.86	8.53	1.32	6.57	0.32	99.87	$(Ti_{0.85}Fe_{0.08}Nb_{0.06}W_{0.03})O_2$
Cassiterite										
L13–125 replacing	dark	2.05	nd	0.18	3.30	5.04	nd	89.74	100.31	$(Sn_{0.89}Fe_{0.04}Nb_{0.04}Ta_{0.03})O_2$
columbite	paler	0.55	nd	0.10	0.57	2.16	nd	97.35	100.73	$(Sn_{0.97}Fe_{0.01}Nb_{0.01}Ta_{0.01})O_2$
	palest	0.17	nd	0.05	0.12	0.52	nd	99.29	100.15	$Sn_{0.99}O_2$
RS6 (3)	dark	0.64	0.04	0.15	2.39	0.40	–	96.79	100.41	$(Sn_{0.95}Fe_{0.01}Nb_{0.03})O_2$
	paler	0.29	0.04	0.16	0.33	0.25	–	98.02	99.09	$(Sn_{0.98}Fe_{0.01})O_2$
	palest	0.10	nd	0.04	0.13	0.32	–	99.68	100.27	$Sn_{0.99}O_2$
RS6 14 vein	dark	0.85	nd	0.05	2.16	0.50	–	95.93	99.49	$(Sn_{0.96}Fe_{0.02}Nb_{0.02})O_2$
	paler	0.35	nd	0.12	2.26	1.33	–	98.07	102.13	$(Sn_{0.95}Fe_{0.01}Nb_{0.03}Ta_{0.01})O_2$

*Total Fe expressed as FeO
tr. trace

Molybenite, as curved laths within mica, is associated with the oxide phases. Sphalerite up to two centimetres in diameter, is iron-poor and encloses trace amounts of pyrrhotine and chalcopyrite. Galena, pyrite, and marcasite occur as intergrowths about cassiterite.

Late stage coffinite is rare and was found in microclinized samples where it forms 30 to 150 μm diameter, poorly crystalline grains which are interstitial to quartz. It has a variable composition (Table 4) with significant thorium, up to 3·5 wt% ThO_2, yttrium up to 16·0 wt% Y_2O_3 and phosphorus up to 3·2 wt% P_2O_5.

4. Discussion

4a. Biotite granites

This study confirms the observations of earlier workers that the biotite granite contains accessory zircon and columbite with minor amounts of thorite and monazite (Kinnaird et al. 1985), but shows that these are accompanied by ilmenite and pyrochlore. Phases associated with groundmass quartz and which are relatively coarse-grained, are interpreted as being magmatic, possibly late-stage magmatic rather than the result of hydrothermal alteration. However, the presence of chloritized biotite, of columbite (with TiO_2 rims) and ilmenite aligned along these altered mica cleavages, of altered zircon, trace amounts of cassiterite and sulphides, coffinite with sulphides, and fluid inclusion data, all suggest that the granite has suffered pervasive hydrothermal alteration and mineralization.

Table 1 shows that the biotite granites within L13 are rich in uranium, with a Th/U ratio of 0·8 to 1·0. Primary pyrochlore and late stage coffinite are both found within the biotite granite, and although neither is common they are the most significant uranium carriers. Minor amounts of uranium are found within uranothorite and, to a lesser extent, within zircon and monazite. Thorium, too, is found within these phases but is mainly concentrated within uranothorite, but also in significant amounts in monazite (Kinnaird et al. 1985) and trace amounts in other REE minerals. The presence of coffinite with minor amounts of thorium is evidence that uranium and perhaps thorium have been remobilized within the granite. The lower uranium but similar thorium contents of the biotite granites from the surface, compared to those from drill-hole L13, show that uranium has been remobilized from the surface biotite granites and presumably lost by weathering.

Niobium, tantalum plus iron, and manganese are found principally within columbite, within pyrochlore, and in minor amounts in ilmenite and TiO_2 minerals. Figure 2, which shows the plots of Nb/(Nb + Ta) against Fe/(Fe + Mn) for columbite, ilmenite, and TiO_2 minerals, suggests that there is a limited range of iron:manganese substitution but a wide range of niobium:tantalum substitution within columbite. Ilmenite shows a similar range of niobium:tantalum but a wider range of iron:manganese substitution. There is no obvious relationship between the two ratios.

Titanium is hosted in coarse-grained ilmenite, and in minor amounts within columbite and pyrochlore, but much occurs within the small TiO_2 grains often enclosing columbite, associated with the alteration of biotite. Semiquantitative analyses of biotite show them to have 0·2 wt% Nb_2O_5 and 0·1 wt% Ta_2O_5 and 0·6 to 3·2 wt% TiO_2. It is suggested that these metals are released during alteration, especially chloritization, and are reprecipitated as columbite, followed

ACCESSORY MINERALOGY OF THE RIRIWAI GRANITE

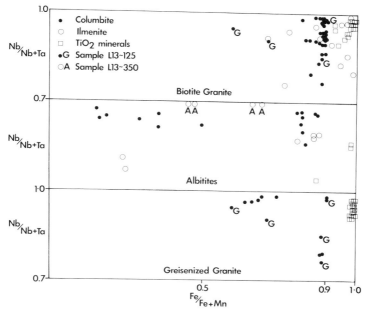

Figure 2. Plot of Fe/(Fe + Mn) ratio against Nb/(Nb + Ta) ratio for columbite, ilmenite and TiO$_2$ minerals from unaltered biotite granite, albitites, and greisened granites.

by niobium-and tantalum-bearing TiO$_2$ minerals and finally more stoichiometric TiO$_2$.

Figure 3 shows that the TiO$_2$ minerals in the biotite granite plot in two fields: most plot in a field which has the greatest niobium and tantalum contents of any TiO$_2$ from Ririwai and are interpreted as being the result of incipient alteration, but approximately 25 per cent plot towards the Ti corner of the diagram and may represent more intense alteration and differentiation of niobium and tantalum from titanium.

In the biotite granites, magmatic fluids have precipitated Nb, Ta, Ti, Pb, U, and Zr as columbite, pyrochlore, ilmenite, and zircon which are intergrown as complex symplectites. Zr, U, Th, Hf, and REE elements have precipitated in zircon associations. Later hydrothermal fluids have altered and redistributed the elements from biotite and also precipitated minor amounts of cassiterite, sulphides, and coffinite.

4b. Albitites

This study shows that haematite, columbite, and zircon were formed extensively during albitization, accompanied by minor amounts of ilmenite, magnetite, cassiterite, uranothorite, and xenotime. Chloritization of biotite was accompanied by crystallization of fine-grained TiO$_2$ minerals along the cleavages of the mica.

Table 1 suggests that both the bulk analyses and Th/U ratio for the albitites are similar to those of the biotite granite from L13. No discrete uranium minerals have been found in the albitites to explain the ratio, but the maximum uranium contents, both in zircon and uranothorite, are higher than those found in the biotite granites. Most of the thorium content of the albitites is within uranothorite,

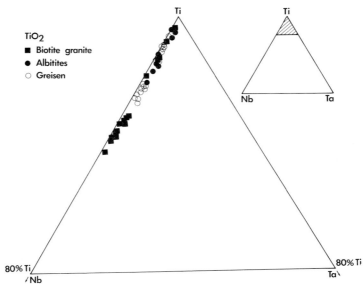

Figure 3. Triangular plot of Ti, Nb, and Ta in TiO$_2$ minerals from unaltered, albititized and greisened granite.

and the high values within inner zones of zircon are due to small uranothorite inclusions. Many zircons have inner uranium-bearing core and hafnium-rich outer margin. Numerous zircons show dissolution accompanied by precipitation of fluorite, xenotime, and other REE minerals which carry minor amounts of uranium and thorium (mainly thorium).

The addition of iron by the fluids is evidenced by the abundant, early, coarse-grained haematite which is very close to being stoichiometric, followed by crystallization of magnetite, ilmenite, and columbite. Figure 2 shows that in both columbite and ilmenite there is a wide range in the iron to manganese ratio and this is optically seen by the zoning displayed by single crystals. Niobium to tantalum ratios are less variable, but the data show that the manganese-rich phase (pyrophanite) is more tantalum-rich than ilmenite. No correlation of iron to manganese and tantalum to niobium is evident, however, in columbite. Comparisons between the columbite found in albitite and those of the biotite granite show they plot in different fields, suggesting that none of the albitite columbites are inherited/relict from the biotite granite.

Figure 3 shows that the TiO$_2$ minerals in the albitite carry less niobium and tantalum than most of those from the unaltered granites (and greisenized granite), which is consistent with the degree of alteration of the micas. Analyses of the TiO$_2$ show them to have significantly higher average manganese contents (0.18 wt% Mn) than those in either biotite granite or greisens (<0.05 wt% MnO). The mobility of the manganese in the albitites is well documented.

During albitization the fluids precipitated Zr with minor U, followed by Zr plus Hf; and later fluids crystallized Th and REE. Alongside this, the fluids precipitated Fe followed by Fe, Mn, Nb, Ti, Ta.

4c. Greisens and microcline-rich rocks

The present study shows that both cassiterite and sulphides were introduced during greisenization, and that the greisens also carry minor amounts of zircon, monazite, uranothorite, xenotime, and columbite. The extensive alteration of the micas is accompanied by abundant fine-grained columbite and TiO_2.

Table 1 suggests the bulk uranium contents of the greisens to be less than those of the biotite granites, but that the thorium content is enhanced, and hence the Th/U ratio is increased for the greisens. Bowden *et al.* (1981) suggested that this enhanced Th/U ratio showed that uranium was mobile during greisenization, and that thorium was essentially fixed. The presence of paragenetically late yttrium-rich coffinite does suggest that uranium was relatively mobile during greisenization (or in post-greisenization times), but the presence of enhanced bulk rock thorium values together with the occurrence of abundant uranothorite (intergrown with cassiterite and within cross-cutting veinlets) suggests that it too, was mobile and crystallized during greisenization.

Coffinite is the only discrete uranium mineral which has been identified, although much of the uranium appears to be within uranothorite. Both zircon and xenotime carry only minor amounts of uranium and an early uranium-rich zircon phase is absent. Uranothorite is the main thorium carrier; lesser amounts of thorium are found in coffinite, monazite, xenotime, and zircon. The uranothorite is characteristically zirconium-poor but has a wide range of yttrium contents, and in Figure 4 is seen to plot in a distinct field away from those found in unaltered biotite granites. Uranothorites in the incipiently greisened granites (L13–125) plot away from both those fields, and are both zirconium- and yttrium-rich. These distinctly different compositions suggest that the greisen uranothorites are not

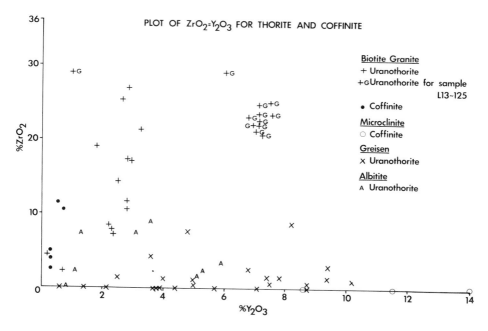

Figure 4. Plot of wt% ZrO_2 against wt% Y_2O_3 for uranothorite and coffinite in biotite granite, albitites, and greisenized granite.

inherited from the biotite granite, but result from greisening fluids. Monazite contributes to the whole-rock LREE content and xenotime, uranothorite, and yttrofluorite to the whole-rock yttrium content.

Niobium, tantalum, iron, and manganese are found together in columbite and TiO_2 minerals (which are almost exclusively found within the altered micas), and presumably also in ilmenite although this was too small for analysis. Figure 2 shows the iron to manganese ratios of columbite in greisens to be more variable than those in the biotite granite but less variable than those in the albitites. Niobium to tantalum ratios are, however, similar to those from the biotite granite columbites. Iron is also present within pyrite, marcasite, chalcopyrite, and, in minor amounts, in sphalerite, cassiterite, and TiO_2 minerals.

Titanium is concentrated within the TiO_2 phases, which have variable amounts of niobium and tantalum. These are accompanied by significant amounts of tin and especially tungsten, when compared to their concentrations in the TiO_2 minerals from the albitites. From the present study it is not possible to show that the tin and tungsten were progressively released during the alteration of the micas, although this is likely.

Tin is found mainly within the abundant cassiterite, which characterizes the greisenization. It is zoned with respect to iron and niobium but also carries tantalum and titanium. The relationship seen at Ririwai, namely that the darkest zones are iron- and niobium-rich, has been found for cassiterite elsewhere (Greaves et al. 1971).

The incipiently greisenized granite, L13–125 clearly shows the dissolution of zircons, especially of their inner (U-rich?) cores, and the dissolution/replacement of columbite by cassiterite and that fluids precipitating uranothorite were both zirconium- and yttrium- rich. Zircon, uranothorite, and xenotime within material from the greisens close to the Ririwai lode show that the fluids precipitated Zr, followed by Th plus minor U and Y, followed by Y with minor Th. The same fluids precipitated Nb and Ta followed by Sn, and by Zn, Pb, Mo, and W.

The accessory mineralogy and styles of mineralization of the Ririwai ring complex are shared by many other alkali granite complexes world wide, for example those of the Arabian Shield (Drysdall et al. 1984; Jackson 1986), as well as others in Nigeria (Kinnaird et al. 1985) and Niger (Kinnaird 1985).

In particular the mineralogy of the Ririwai biotite granite, its altered facies and later quartz–tin–sulphide mineralization is closely similar to that from the Taghouaji ring complex in Niger, as described by Perez (1985), and in both complexes the role of post-magmatic fluids has been responsible, in a large part, for the variety of rock types and associated mineralization.

5. Conclusions

Textural and paragenetic studies of the accessory and opaque minerals of the biotite granite and its altered varieties shows that they can be divided into two, according to the relationship between the ore phases and silicates.

1. Fine-grained columbite, ilmenite, and TiO_2 minerals are associated with the breakdown of biotite and liberation of Ti, Fe, Nb, Ta, Mn, and probably W and Sn, and their local reprecipitation along the new mica cleavage. An essentially similar association is seen in all the rocks. Analyses of the TiO_2 phases suggest that, with an increase in the alteration of the biotite, there is an increased differentiation of niobium and tantalum from titanium (this is represented in Figure 3).

2. Coarse-grained phases are typically associated with quartz, but here the association differs between the rock types. The biotite granite carries zircon, with columbite, minor ilmenite, and pyrochlore and late zirconium-rich coffinite; the albitites carry haematite, minor magnetite, iron-manganese zoned columbites and ilmenite; the greisens carry cassiterite plus iron, zinc, lead, and molybdenum sulphides, uranothorite, and late yttrium-rich coffinite. The textural and chemical evidence suggests that both alteration processes (albitization and greisenization) resulted in the destruction of earlier phases followed by their reprecipitation; but with different compositions compatible with the new fluids.

Table 7 summarizes the relative abundancies of the primary ore phases found in the biotite granite and its altered equivalents. Although the different fluids produced essentially very similar assemblages the proportions of the minerals vary markedly.

Overall, the following sequence of metal precipitation has occurred, as summarized in Figure 5.

Late granitic fluids, which may be magmatic, precipitated Zr, U, Nb, Ta, Ti, and Fe, albitizing fluids precipitated Fe, Mn, Nb, Ta, Zr, with minor U and Th, and greisenizing fluids precipitated Sn, Th, Zn, Pb, Mo, and Y.

Each process shows a fixed and similar paragenetic relationship between Zr, Hf, U, Th, and Y which is: Zr $\pm$U, Th and Hf; followed by Th $\pm$ U and Y; followed by Y$\pm$ Th. This is seen as zircon crystals, overgrown by thorite/uranothorite, overgrown by xenotime. The paragenetic position of monazite is unclear.

Uranium is concentrated early, and has its maximum concentration in the biotite granites where it forms discrete minerals. Some additional uranium may have been introduced during albitization, but it is lost during greisenization (and microclinization), and leached during surface weathering. Thorium too, may be further introduced during albitization, but its concentrations are enhanced during greisenization, and it is not lost by low temperature surface processes.

Table 7. Relative abundances of the accessory and opaque phases from Ririwai

	Biotite granite	Albitite	Greisen
Zircon	major	major	major
Uranothorite	minor	minor	minor
Xenotime	not seen	rare	rare
Pyrochlore	rare	not seen	not seen
Monazite	minor	rare	minor
'Bastnaesite'	rare	rare	rare
Coffinite	rare	not seen	rare
Haematite	rare	major	rare
Ilmenite	minor	minor	minor
Magnetite	not seen	minor	not seen
Columbite	minor	major	minor
TiO$_2$ minerals	minor	rare	minor
Cassiterite	minor	minor	major
Wolframite	rare	rare	minor
Sulphides	minor	minor	major

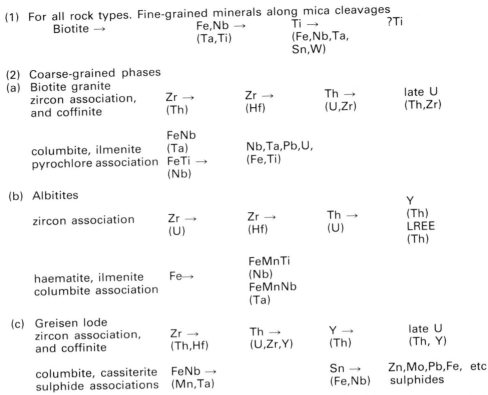

Figure 5. Sequence of precipitation of elements from ore fluids as deduced from paragenetic studies. Minor elements are in parentheses.

Acknowledgements. Peter Bowden and Judith Kinnaird are thanked for their encouragement and for supplying all the material. The staff at the SURRC are thanked for their help, in particular Dr Gus MacKenzie for the preparation of fission-track plates. Drs N. Jackson and P. Webb are thanked for many useful suggestions. C.M.P. acknowledges a NERC research studentship. Chris Gee typed the manuscript, and Sue Knox prepared the text-figures.

References

Beer, K.E. 1952. The petrology of some of the riebeckite-granites of Nigeria. *Report Geological Survey UK Atomic Energy Division*, **116**, HMSO London, 38 pp.

Bowden, P., Bennett, J.N., Kinnaird, J.A., Whitley, J.E., Abaa, S.I., and Hadzigeorgiou-Stavrakis, P.K. 1981. Uranium in the Niger–Nigeria Younger granite province. *Mineralogical Magazine*, **44**, 379–389.

— and Kinnaird, J.A. 1978. Younger granites of Nigeria — a zinc-rich tin province. *Trans. Inst. Min. Metal. Sect. B*, B66–B69.

Drysdall, A.R., Jackson, N.J., Douch, C.J., Ramsay, C.R., and Hackett, D. 1984. Rare metal mineralization related to Precambrian alkali granites in the northwest Arabian Shield. *Economic Geology*, **79**, 1366–1377.

Greaves, G., Stevenson, B.G., and Taylor, R.G. 1971. Magnetic cassiterite from Herberton, North

Queensland, Australia. *Economic Geology*, **66**, 480–487.
Jackson N.J. 1986. Mineralization associated with felsic plutonic rocks in the Arabian Shield. *Journal of African Earth Science*, **4**, 213–227.
Jacobson, R.R.E. and MacLeod, W.N. 1977. Geology of the Liruei, Banke, and adjacent younger granite ring-complexes. *Bulletin of the Nigerian Geological Survey*, **33**, pp. 117
—, **Snelling, N.J., and Truswell, J.F. 1963.** Age determinations in the geology of Nigeria with special reference to the Older and Younger granites. *Overseas Geol. Mineralogical Research*, **9**, 168–182.
Kinnaird, J.A. 1984. Contrasting styles of Sn–Nb–Ta–Zn mineralization in Nigeria. *Journal of African Earth Science*, **2**, 81–90.
— **1985.** Hydrothermal alteration and mineralization of the alkaline anorogenic ring complexes of Nigeria. *Journal of African Earth Science*, **3**, 229–251.
— **Bowden, P., Ixer, R.A., and Odling, N.W.A. 1985.** Mineralogy, geochemistry, and mineralization of the Ririwai complex, northern Nigeria. *Journal of African Earth Science*, **3**, 185–222.
MacKenzie, A.B., Bowden, P., and Kinnaird, J.A. 1984. Combined neutron activation and particle track analysis of element distribution in a rock slice of mineralized granite. *Journal of Radioanalytical and Nuclear Chemistry Articles*, **82**, 341–352.
Perez, J.B. 1985. *Nouvelles donnés sur le complexe granitique anorogénique de Taghouji (République du Niger) influence des fluides au cours de la cristallisation.* Unpublished PhD thesis, University of Nancy. 317 pp.

Section 5
Continental rifting and continental margins: implications from structural and sedimentological studies

Coincident with the fragmentation of Gondwanaland there was continued exploitation of lineaments leading to continental separation and the development of continental margins from the Cretaceous onwards. As well as the development of marginal basins ancient lineaments are exploited leading to continental rifting. Such features are an important facet of African geology which is still evolving to the present day. This section therefore covers all aspects pertinent to the study of lineaments, sedimentary basins, magmatism, and the evolution of continental margins on the African plate.

During the evolution of the Alpine fold belt in North Africa, deformation occurred within the African plate, and Precambrian/Palaeozoic fracture systems were reactivated as major lineaments. One set of lineaments follow east-west trends in North Africa, and from Guinea to the Nubian Province in west, central and north-eastern Africa. The other set constitute major megashears orientated approximately north-south. This study which is presented as the first paper in Section 5 is a welcome review by Guiraud and his co-workers on post-Hercynian tectonics, particularly in North Africa and West Africa. The authors suggest that magmatic activity with alkaline affinities, and folding, are considered to be related to major fault systems especially in northern Cameroon and the Mamfe basins. In the Benue trough an intracontinental chain with abundant igneous intrusions can be correlated with the opening of the Atlantic near the Gulf of Guinea. The authors suggest that the whole of the African continental domain from the Alpine chain to Cameroon and southern Chad, was subjected to post-Hercynian deformation. If the Alpine continental margin has structures which demonstrate compression, then in contrast the Atlantic margin off the western coast of Africa can be modelled as a contrasting extensional regime, linked to shearing along transform fault systems. Intraplate deformation can be correlated with six tectonic episodes in various regions of the North African continental domain identified as occurring

© 1987 by John Wiley & Sons, Ltd.

during the Lias, Aptian, Santonian, late Cretaceous, upper Eocene, and Miocene times. Giuraud *et al* conclude that intraplate deformation in Africa is a consequence of both plate collision between Africa and Europe combined with the extensional effects of the opening of the Atlantic.

The Benue trough is a rift system developed along major lineaments in West Africa on which detailed structural studies have been completed. A microtectonic analysis in the Abakaliki anticlinorium by Benkhelil has revealed various types of deformations of different ages. Furthermore, the thermal history of the lower Benue trough is characterized by two major episodes of metamorphism. A dominantly contact thermal metamorphism in Albian times is related to the emplacement of volcanic and subvolcanic centres mainly with alkaline affinities, but with some characteristics of continental tholeiites. A later regional metamorphism up to greenschist facies occurred during folding in the Santonian 81 Ma ago.

Such deformation is consistent with the development of shear zones penetrating the continental crust. It is clear that major lineaments have played an essential role during the opening of the Benue trough in Albian times associated with the formation of the Gulf of Guinea. These lineaments also acted as pathways for the emplacement of igneous rocks. Benkhelil concludes that transcurrent movement was the critical motion related to the development of the Benue trough and not a classical rifting mechanism as assigned to East Africa.

Turning now to the East African rift system, a tectonic survey has been completed by Chorowicz and his coworkers in Eastern Africa, using remote sensing as well as structural analysis in the field. The geomorphology of the region suggests that deformation occurred in the western and eastern branches of the East African Rift system, but compilations of earthquake data show that recent activity is widespread over a large part of Eastern Africa. Many ancient structures have been reactivated during Neogene times, especially large Precambrian lineaments which are evident on satellite imagery. Field observations, earthquake recordings, together with data concerning the geometry of the deformation, are used by Chorowicz *et al.* to suggest that there are four distinct stages in the evolution of the East African Rift system.

Related to the evolution of rifting are distinct periods of magmatism. Some aspects of the variations in mineralogy and chemistry of the lavas can be correlated with the early stages in the formation of the western branch of the East African Rift system. Lubala and his colleagues have studied some of the youngest volcanic flows in the Kahuzi–Biega region, which developed within the central zone of the Western Rift, Kivu province, in Zaire. They are composed of ankaratrites, alkaline basalts, basanites and hawaiites, belonging to a Na-rich alkaline suite. These products are closely associated with the marginal faults of the rift and show a space–time evolution expressing their emplacement during the early stages of opening. Lubala *et al.*, consider that the age of the alkaline lavas (6 Ma) is a chronological constraint on the early events in the evolution of the Western branch of the East African Rift system.

Recent research has shown that many of the ancient lineaments exploited by continental rifting also control the evolution of continental margins. Marine studies of selected sheared-type margins have been undertaken by Mascle and his coworkers along both the West and East African coastal regions in order to determine the geological structure and the evolution of a specific type of passive margin. Geological and geophysical data have been obtained in the coastal waters off Western Equatorial Africa (Guinea, Ivory Coast, and Ghana) and off Eastern Africa (Mozambique). Mascle and his coworkers conclude that when shearing commenced between two continental margins, it generated either tensional features as in the Ivory Coast, or transgressive features seen off Mozambique. Later, basement-involved structures and sedimentary wedges developed at the boundary between a thinned continental crust and a thick continental crust before connecting laterally with a typical oceanic transform fracture zone.

To conclude this Section a sedimentological study of Quaternary deposits in a deep-sea channel off the South African continental margin is reported by Frances Westall. Her research reveals that terrigenous and shelf components from the sediment–sand fraction are concentrated on the upper continental slope and the channel floor. In comparison, sediments from the lower flanks of the channel contain relatively low amounts of terrigenous and shelf components.

Post-Hercynian tectonics in Northern and Western Africa

R. Guiraud
Laboratoire de Géologie dynamique et appliquée, Faculté des Sciences, 33 rue Louis Pasteur, 84000 Avignon, France

and

Y. Bellion, J. Benkhelil and C. Moreau
Département de Géologie, Faculté des Sciences, Dakar-Fann, Senegal

---

During the evolution of the Alpine fold belt in North Africa, important deformations occurred within the African plate. Precambrian/Palaeozoic fracture systems were reactivated as four major lineaments with trends close to East–West (from Guinea in North Africa to the Nubik Province, in Central Africa) and also north–south as megashears between 2° and 6° E.

Magmatic activity, in part attributed to folding, and expressed as chains or roots of volcanoes with alkaline affinities, are considered to be related to major fault systems especially in the North Cameroon and Mamfe basins. In the Benue Trough an intracontinental chain with abundant igneous intrusions can be correlated with the opening of the Atlantic near the Gulf of Guinea.

Intraplate deformation is linked to six major periods of tectonic activity identified as Lias, Aptian, Santonian, End Cretaceous, Upper Eocene, and Miocene. From the Aptian onwards, the stress regime remained relatively constant. Intraplate deformation is a consequence of both the plate collision between Africa and Europe and the opening of the Atlantic.

KEY WORDS   Lineaments   Africa   Atlantic opening   Plate collision   Deformation   Magmatism

1. Introduction

In this paper we present a review of post-Hercynian tectonics, activation of major lineaments, and associated magmatism in continental Africa. The following geographical regions are considered, and indicated in Figure 1.
  1. The Alpine margin, limited to the South by the Sahara.
  2. The Atlantic margin, from Morocco to the Gulf of Guinea with coastal basins and an overview of the Canaries and Cape Verde islands
  3. The intraplate domain from Libya to the Gulf of Guinea and Cameroon (Figure 1).

0072–1050/87/TI0433–34$17.00
© 1987 by John Wiley & Sons, Ltd.

These events not only confirm the existence and part played by the major lineaments found on the northern boundary of the African plate and in the Gulf of Guinea, but also the intraplate activity from Guinea and Senegal to the Nubian Province, and from the El Biod ridge to the borders of Benin and Togo.

These major tectonic episodes linked to the deformation of the north African active margin and the so-called passive Atlantic margin, as well as the continental domain, are considered to be important major features comparable to elsewhere on the African plate.

2. Structure of the Alpine margin: the Alpine chain of North Africa

In North Africa the Alpine chain can be divided from Morocco to Tunisia into three domains (Figure 2):
 1. The Tellian domain is characterized by thrust sheets, mainly slide sheets but also with some units showing movements of several hundred kilometres; the

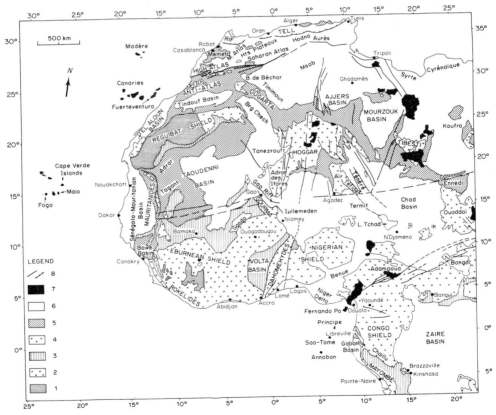

Figure 1. Schematic map of the major geological features of Western and North–Central Africa. 1— Lower Precambrian; 2— middle Precambrian; 3— Upper Precambrian; 4— Pan-African mobile belt; 5— Palaeozoic; 6— Mesozoic, Tertiary, and Quaternary; 7— Tertiary to Recent volcanism; 8— major faults and thrusts.

maximum displacements are to be found in the Constantine region, where they measure at least 200 km and where the Tellian domain nearly reaches the Atlas domain, in the Aures region, and in the prolongation of the Saharan Atlas.

2. The Pre-Atlas domain from the Moroccan tablelands to the South Constantine region, around Hodna where very rigid carbonate rocks can be found, and where rocks are mainly fractured with some folding. The intensity of the folds increases nearer to the Constantine ridge with its large mass overthrusting the foreland.

3. The Atlas domain with the High Atlas and the Middle Atlas in Morocco, the Aures–Nementcha in Algeria, and the Tunisian Atlas to the east.

2a. Main structural characteristics of the North African domains

The Tellian domain is composed of thrust sheets, usually of flysch and sometimes also carbonate rocks (Figure 3). These complex structures, with their polyphase deformation have been well described in the literature.

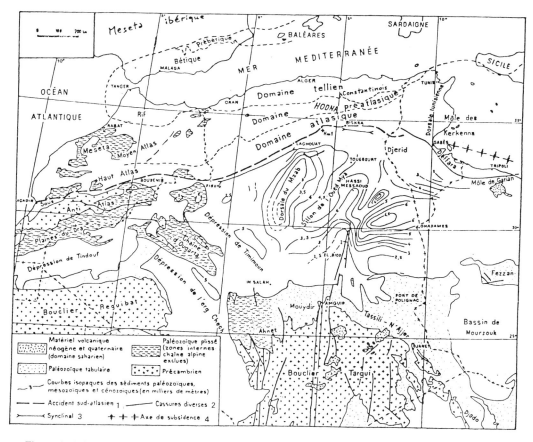

Figure 2. Principal structural units in Northwest Africa. 1— South Atlas fault system; 2— synclinal structures; 3— various fractures; 4— axis of subsidence.

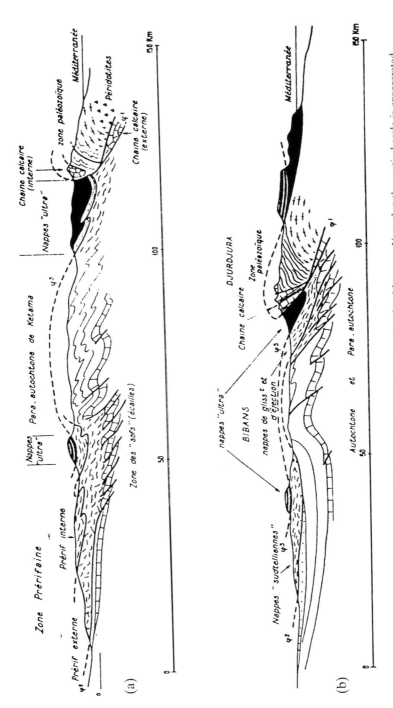

Figure 3. Schematic cross-section of the Rif(a) and Tellian(b) domains after Mattauer. Note that the vertical scale is exaggerated.

Schistose blocks within the thrust sheets are important because they date the major deformation episodes identified as:
— Upper Aptian to Albian (in the Babors and Cheliff regions)
— Mid-Santonian in the massifs of Alger, Cheliff, Babors, and probably in West Oran and Blida
— Upper Lutetian in Oran and at Blida
— Lower Miocene in the Rif region.

The Atlas chain is mainly composed of groups of buckle folds (Figure 4) cut across by strike-slip faults contemporaneous with the folding. These are typical medium level structures. Their southern limit corresponds to the Sahara 'flexure' or south-Atlas terrane which comprises of a succession of flexures, strike-slip faults and more rarely reverse faults.

Pseudo–diapirs with Triassic gypsiferous shales often pierce their cover irrespective of composition. These diapirs are always localized on fault zones with their emplacement facilitated by compressive tectonic episodes.

During all these periods the N80° to N90° oriented faults played a major role. They represent the fractures which delimit the northern edge of the African plate. This North African feature is marked by a major slide which has not been previously well documented.

An important structural characteristic of North Africa can be interpreted from the neotectonics and seismicity (see Orleansville, Agadir, El Asnam, etc.). Most of the uplift is relatively recent since it is linked to normal faulting, or reverse faulting, or to Upper Miocene, Pliocene, or Quaternary strike-slip faults.

2b. Structural evolution of North Africa

The first stage in the structural evolution of the region consists of the creation of basins during the Triassic–Liassic extension, to which the opening of the central Atlantic is linked. This phenomenon is mainly shown by interplay of east–west faults and oblique faults. The great east–west slides linked to the north end of the central Atlantic are sinistral slides (Figure 5). Since the Atlantic did not open again (near Spain), West Africa moved rapidly east. This relative movement at the junction between the two plates appears as a sinistral strike-slip displacement.

Simultaneous strike-slip movements and northeasterly orientated fractures (rejuvenated Hercynian fractures) lead to the formation of a certain number of pull-apart basins. Mattauer *et al.* (1977) were the first to describe these basins in Morocco. Laville (1981, 1985) has clearly shown the existence of this phenomenon for the Triassic and Liassic High Atlas. One can explain in the same way the formation of the sedimentary basin from which the Saharan Atlas series originated (Duée and Kazi Tani 1982). Gradually this model has been adopted to explain nearly all the domains of the Alpine chain in North Africa. A well-developed basic volcanism emphasizes the extensional characteristics.

The Jurassic–Cretaceous boundary is sedimentologically marked by discontinuities, proof of an important tectonic event (Donze *et al.* 1974). However, to this day, the only structures attributed to this event in North Africa are the folds that have been found in the Babors (Obert 1981) as well as in the north eastern limit of the Saharan Atlas (Kazi-Tani 1984). Submarine volcanic events have also been described in the Berriasian series of Grande Kabylie, Petite Kabylie, and the Rif (Bouillin 1983).

The first important compressional event corresponds to the so-called Albian phase responsible for the appearance of schistosity in the Babors and in some

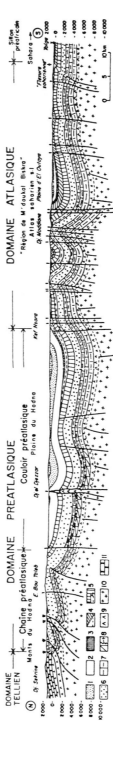

Figure 4. Geological section across the Alpine chain foreland in Algeria around eastern Hodna. 1— Pliocene; 2— Miocene; 3— Eocene; 4— Senonian; 5— Cenomanian-Turonian; 6— Lower Cretaceous; 7— Malm and Berriasian; 8— Lias and Dogger; 9— Trias; 10— Palaeozoic; 11— Jurassic and Lower Cretaceous carbonate horizon of the Constantine horst block.

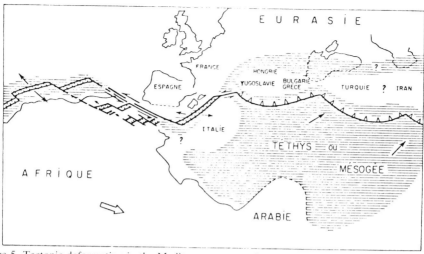

Figure 5. Tectonic deformation in the Mediterranean area due to the opening of the Atlantic towards the end of the Lias (approximately 180 Ma). Slightly modified after Tapponnier (1977).

massifs of the Cheliff. The compression direction was approximately north–south, an episode corresponding to the Austrian phase, described in the European Alps.

The following mid-Satonian compressional phase was responsible for the schistosity in various massifs in Kabylie, at Alger, in part of the Cheliff, etc., and to a well-dated metamorphic event (85 +/−2 Ma) in part of the High Atlas, corresponding to the beginning of the erosion of the Mesozoic cover (Laville et al. 1977). More locally this phase was also responsible for the formation of folds (Nador and Oulad Nail regions in Algeria) and for some strike-slip faults (Hodna Mountains) (Guiraud 1973).

The major shortening direction was approximately north–south. There then followed the Pyrennean–Atlas phase, very important in the Atlas and Aures, which by then achieved their main structural identity. This phase also occurs in the Tellian chain where one finds large overthrusts with some schistosity developed in Oran and in the Blida Massif, in Algeria. The Pyrennean–Atlas phase is widespread on the whole of the North African margin, whether in the allochthonous Tellian domain or the autochthonous domain. The most important dextral sliding episode for the E–W uplift belongs to this phase, forming the Tunisian ridge at the junction with the Kerkenna horst (Guiraud 1970). Chronologically these deformations date from the end of the Middle Eocene, confirmed by recent palaeontological discoveries (Coiffait et al. 1985). The average direction of the movement was between NNW–SSE and N–S.

More locally one finds traces of a compression phase at the lower Burdigalian. This Lower Miocene phase shows thrusting and schistosity in the Rif and the Tell, as well as deformation in the High Atlas and in Tunisia. Immediately after this Lower Miocene phase, around 25 Ma, the calc-alkaline volcanism found on the northern border of the African plate began.

During the Miocene, the Tortonian compressional phase was the last and the most important one, with nappe formation, traces of which are found in the autochthon.

Then the neotectonic events occurred.

Firstly, there was an important extensional period shown by the formation of the well-known Messinian basins, with some large normal faults as seen at the limits of the Cheriff basin. This extension, emphasized by alkaline volcanism on the Algerian coast and in Morocco, continued during the Pliocene until the beginning of the Quaternary.

Then during the Lower Quaternary about 1·5 Ma ago, there occurred a final compressional episode with modest consequences but still noticeable nowadays by earthquakes. The study of the earthquake epicentres shows that σ^1 is between N25°W and N–S.

To conclude, the major horizontal shortening direction for North Africa has remained constant since Albian times and is, apart from a few exceptions, between NNW and N–S.

3. Structure of the Atlantic passive continental margin

The development of the passive continental margin began with the opening of the Atlantic Ocean (Figure 6) in several stages. This can be shown by the various

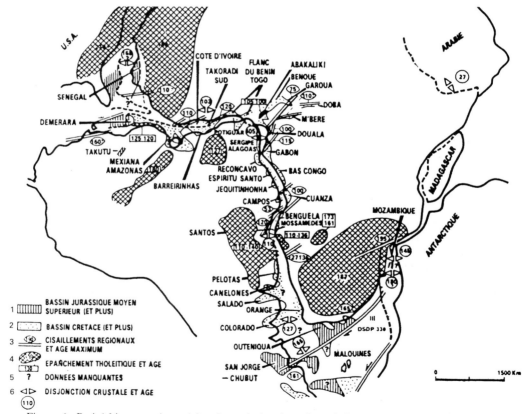

Figure 6. Peri-African continental break-up during Jurassic and Cretaceous times. 1— Middle to Upper Jurassic (and later) sedimentary basins; 2— Cretaceous (and later) sedimentary basins; 3— Regional shearing and oldest age; 4— tholeiitic lava flows and age; 5— lack of data; 6— age of continental separation.

ages of the earliest formation of Mesozoic sedimentary rocks, and by tholeiitic magmatism emplaced during the rupture of the Gondwanan continental domain.

The structure of the West African coastal basins, from Morocco to Nigeria are described briefly below with some examples.

3a. The Moroccan coastal basins (Ruellan 1985)

The history of the Alpine basins north of Agadir was strongly influenced by the active continental margin. In contrast, the more southerly basins south of Agadir show only a weak tectonic deformation, similar to the Senegal basin.

3b. The Mauritania–Senegal–Guinea Basin (Bellion and Guiraud 1984)

This basin is quite well known from petroleum exploration studies. In general the strata are horizontal (Figure 7) with normal faults of variable directions. Faulting was more frequent near the continental slope and is mainly orientated between N–S and N20°E. In this domain two horsts (Dakar and Ndiass) are developed between Gambia and the Mauritanian Sahara. The origin of the horsts could be linked with blocktilting related to the opening of the central Atlantic.

Diapirs with Upper Triassic or Lower Liassic saliferous material have been identified on the continental shelf through seismic studies and drilling off shore at Casamance and Guinea-Bissau, and by seismic work offshore near Mauritania.

One should also remember that several fracture zones oriented E–W affect the continental margin between Nouadhibou and Conakry (Figure 8). These are the continuation of oceanic transform faults towards the continent. However though these fractures are clear on gravimetric and magnetic maps they have little effect on the Mesozoic and Cenozoic cover. Their movements seem to have taken place mainly during the opening of the Central Atlantic, the southern limit of which corresponds exactly with the Guinea fault. Magnetic anomalies (Reyre 1984) suggest that basic magmatism occurred at the same time. The study of post-Hercynian magmatism is very important in this basin. In the bordering regions, Guinea, Sierra Leone, Mauritania, and Mali the intrusives are mainly Triassic to Jurassic gabbros and tholeiitic microgabbros. However May (1971) suggests that some of the doleritic dykes may have been emplaced between the Upper Permian and the Lower Cretaceous. There are also some rare alkaline magmatism occurrences, in the Campano–Maastrichtian syenitic dome south of Saint Louis and the 104–80 Ma old Los Islands complex in Guinea (Lazarenkov and Sherif 1975). Undifferentiated basic alkali volcanism, Upper Eocene to Quaternary in age (Crevole 1980) can be found in the Cape Verde region (Dakar–Thies). This magmatism constitutes one of the stages in the development of the Guinea–Nubian lineament (Guiraud et al. 1985).

To conclude, the tectonic style and history of the Mauritania–Senegal–Guinea basin are characterized by a predominance of extensional structures (horst blocks, listric faults, diapirs) and the interplay of polyphase faulting.

One can distinguish several major tectonic episodes during which the various structures originated or were developed. These episodes were often followed by erosion, or unconformities. They show the effects of tectonic phases well defined in the Alpine chain at the base of the Upper Aptian; during the Lower Senonian; at the Cretaceous–Palaeocene boundary; at the Palaeocene–Eocene boundary, at the base of the Upper Eocene, but not so clearly defined in the Neogene. The same episodes can be found in the El Aioun basin (Ranke et al. 1982).

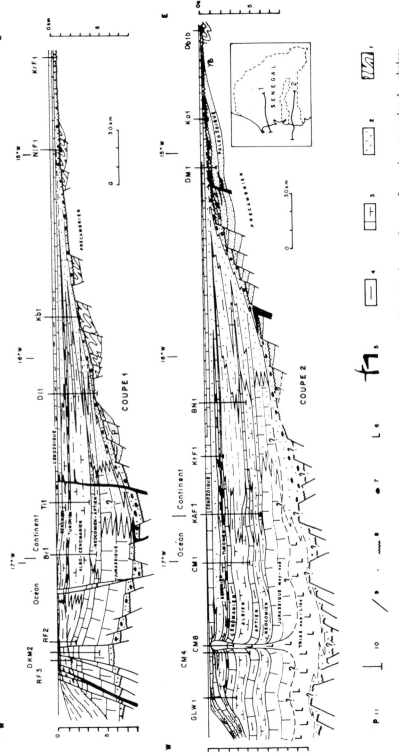

Figure 7. Interpretative geological cross-sections of the Senegal basin. 1— metamorphic rocks; 2— sandstones; 3— carbonate rocks; 4— shales; 5— magmatic rocks; 7— conglomerate; 8— unconformity; 9— fault; 10— petroleum exploration drill-hole; 11— Palaeozoic succession.

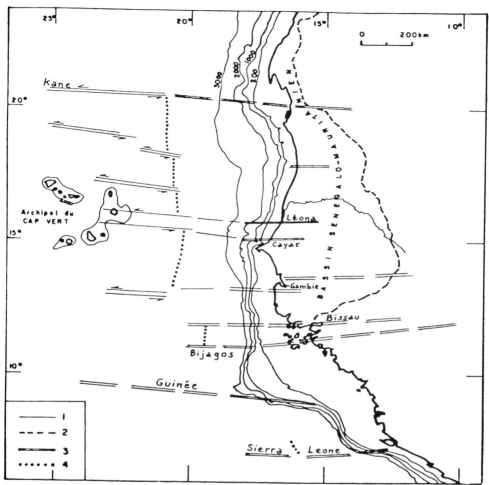

Figure 8. Transform faults at the continental margin of Mauritania and Sierra Leone. 1— depth curves; 2— eastern limit of the sedimentary basins; 3— transform fault and its continuation inshore; 4— western limit of the quiet magnetic zone.

3c. Basin development at the margin of the Gulf of Guinea

The basinal history at the margin of the Gulf of Guinea is linked to tectonic sliding during the formation of transform faults, to vertical movements linked to subsidence, and to continental accretion.

The Ivory Coast basin illustrates this feature. It is limited in the south by a basement ridge overlying the Romanche fracture zone (Figure 9) and appears polygonal which might indicate a pull-apart feature. Several discontinuities affect the sedimentary series such as the marine Albian discordantly overlying the moderately folded and faulted Lower Cretaceous (Blarez 1986); and the Senonian sometimes overlying the Lower Cenomanian. A general subsidence would have taken place at the end of the Maastrichtian (de Klasz 1978). The Oligocene is unconformable on the Middle Eocene (Simon and Amakou 1984) and the Lower Miocene is unconformable on the Upper Cretaceous or the Eocene (Simon and Amakou 1984). Furthermore the Pliocene and the Quaternary were also transgressive fossilizing the topography of the region.

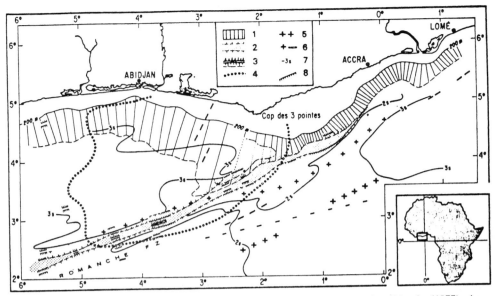

Figure 9. Principal features of the eastern margin of the Ivory Coast. After Mascle (1977). 1— continental slope; 2— buried ridge; 3— outcropping ridge; 4— limit of the quiet magnetic zone; 5— positive magnetic anomaly; 6— positive and negative gravimetric anomalies; 7— 2 and 3 second contours; 8— probable faults.

3d. The Canary and Cap Verde Archipelagoes

The Canary and Cape Verde Archipelagoes are mainly composed of Neogene and Quaternary volcanic rocks. However in the eastern region, there are rather thick Mesozoic sedimentary sequences. The facies are strongly tectonized and similar to those of the West African coastal basins. These original formations are to be found in the Islands of Fuerteventura (Canary) and Maio (Cape Verde), the structural style of which have recently been studied by the authors.

The structure of Fuerteventura Island consists of a dominant Upper Jurassic to Senonian sedimentary series, mainly detrital, outcropping in the western part of the Island (Figure 10). It is conformably overlain by Upper Senonian to Palaeogene volcanosedimentary rocks. This assemblage, called the basal complex, is strongly cut by a dense network of Miocene basaltic dykes but remains nevertheless relatively homogeneous. It constitutes a large reversed limb formed before the Lower Miocene. The emplacement age of the structure and its mechanism of formation are not very well known. The most likely hypothesis seems to be that of an overturning of the sedimentary series caused by the emplacement of a volcanic massif during the Oligocene.

After these events a Lower Miocene carbonate reef-type formation is linked to volcanosedimentary deposits. The island was then partially submerged and continued to develop in a volcanic environment. The sedimentary series was then slightly tilted (dips of 25°) before being cut by a basaltic dyke network with directions N–S to N20°E, that are to be found in the whole western part of Fuerteventura. These dykes are 19 to 24 Ma old (Féraud 1981). They are contemporary with other similar formations and with other isolated massifs that were then emplaced in the centre of the island. Such volcanic events indicate a strong

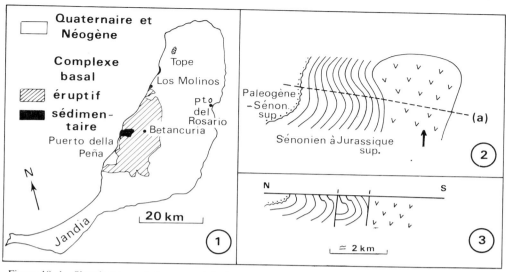

Figure 10. 1—Sketch map and structure of the basal complex of Fuerteventura. 2—uprise of a volcanic massif during the Oligocene: (a) present land surface. 3—cross-section to the east of Puerto de la Pena.

extensional regime oriented E–W to WNW–ESE. Later the whole of the sedimentary and volcanic formations were cut by minor faults. This microfracturation, clearly seen in various sectors of the western coast, presents small strike-slips and reverse faults, sometimes concentrated in shear zones which belong to two compression episodes oriented N–S to NNE–SSW. These episodes date from the Tortonian, i.e. around 10 Ma, if one considers similar events in the west of the Alpine chain linked with the collision of the African and Iberian plates.

The Pliocene–Quaternary period is marked by important basaltic flows with related dyke and sill intrusions indicating the continued existence of an extension regime during this period.

3e. The structure of Maio Island

The Island of Maio looks like a large anticline, well defined by the Mesozoic sedimentary series, uplifted by a central volcanic complex (Figure 11). In several sectors the marly Jurassic–Cretaceous formations, and more locally Miocene volcanic sills, define dense folds of metric to decametric amplitude with a more or less tilted axis. These small structures show relatively similar directions to the main anticline, within a 20° margin (Bellion et al. 1988).

Numerous faults affect the various assemblages between the Upper Jurassic and the Middle Miocene. The major faults cut the Mesozoic, mainly in the north of Monte Branco where we have identified sinistral strike-slips oriented E–W. Some rare strike-slips also affect the Middle Miocene of Pedro Vaz with some weak throws. Reverse faults create repeated sequences on the slopes of Monte Branco (Stillman et al. 1982).

According to Serralheiro (1970), the tectonics deformation on Maio Island results from the emplacement of a central volcanic complex. However the genesis of some structural elements seems to imply the action of other mechanisms. This

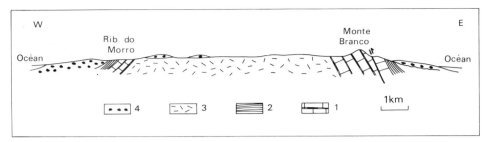

Figure 11. Schematic cross-section of Maio Island. 1— Upper Jurassic to Lower Cretaceous carbonate rocks; 2— Cretaceous shales; 3— Miocene central volcanic complex; 4— Neogene volcano-sedimentary conglomerates.

is the case for the strike-slips oriented E–W because these faults can be found on both sides of the complex, and because some affect already subvertical horizons of the Pedro Vaz formation which on the whole post-dates it. Besides, the strike-slips show a sinistral interplay and one is then tempted to link them with the fracture zones of similar direction found at the periphery of the Cape Verde Archipelago. Some authors attribute the recent volcanism of this region to movements along these faults. They are therefore strike-slips indicating late fault movement linked with the opening of the Atlantic (Cape Verde fracture zone, Le Pichon *et al.* 1977).

The existence at Maio of tholeiites (pillow-lavas) and of numerous basaltic sills injected layer by layer into the Jurassic–Cretaceous deep water carbonates give this island its original character. It corresponds in part to a sample of the oceanic bottom as was shown by Celestino *et al.* (1979) and developed in the schematic reconstruction of Maio island by Robertson (1984).

3f. Major structural characteristics of the Atlantic margin of Africa

The Western African margin is composed of subhorizontal strata cut by normal faults or more locally by diapirs. Major faults are parallel to the coast and represent the rather rapid passage from the continental crustal domain to an oceanic crustal domain. In some cases these faults define coastal horsts in Senegal and Nigeria.

Some important fractures can be oblique or even perpendicular to the shore lines; they represent the marginal prolongation of transform faults that continue into the continental domain (South Atlas and Benue fault systems) or in contrast weaken near the Guinea shore. This fracture type initiated horizontal slides, at least during the initial opening stages of the basin, often related to magmatism and uplift (Reyre 1984).

Detailed studies of this series enable us to identify several major tectonic events: during the Upper Aptian or at the end of the Aptian; during the Senonian; at the Cretaceous–Tertiary boundary; and at the base of the Upper Eocene.

However deformation mechanisms are quite different along the Atlantic margin, between Morocco and Cameroon. One can define several original domains. In the Moroccan margin, north of the South Atlas fault system, compressive structures (folds and transverse faults) coexist with structures linked to the oceanic extension. The Mazagan ridge faults are a good example. This original domain corresponds to a mixed margin, both active and passive in which the simultaneous influences of the Alpine deformation and the Atlantic opening can be observed.

The Central Atlantic margin, between the South Atlas and Guinea zones is a simple stable margin of the divergent type (Schlee 1981). It is a particular type of domain, between Sierra Leone and Nigeria, in which the tectonic slides linked to the prolongations of transform faults led mainly to the formation of diamond shaped basins which one can identify as a transform margin.

The volcanic archipelagoes of Canary and Cape Verde and in particular the islands of Fuerteventura and Maio, owe the main part of their structure to Tertiary volcanic activity. However while the Fuerteventura lavas were subject during the Middle to Upper Miocene to a compression phase of Alpine origin, the Maio lavas at the same period were subjected to deformation linked to mid-oceanic transform movements.

4. The structure of the continental domain

One can distinguish in northwest Africa and west Africa various sedimentary basins which separate the cratons, shields, ridges, and volcanic chains (Figure 1). Locally volcanic and subvolcanic massifs often aligned along major lineaments zones transect these assemblages.

Four units can be distinguished in the post-Hercynian domain, all presenting some structural homogeneity. They are the northern Saharan or Algerian–Libyan basins; the Taoudenni basin; the Nigerian–Chadian basin; the Benue and the central and northern Cameroon basins.

The Precambrian shields and mobile belts were variably affected by the orogenic events linked to the Alpine cycle, for example in the Hoggar, the Nigerian shield, and the northern fringe of the Congo shield.

4a. The north Saharan basins

From southern Morocco to the Egyptian–Libyan border one can observe a sequence of basins, troughs or grabens, ridges, horsts, blocks, or uplifted areas with various orientations (Figures 1, 2, and 12). Such examples are the basins of Tindouf, Timimoun, Oued Mya, Ghadames, Mursuk, Hun, Dor el Goussa, Syrte, and Kufra. To these should be added the Gulf of Syrte and the Pelagian Sea; and ridges at the Ougarta, Mzab, Gassi Touil, Garian–Gargaf–Tibesti, Cyrenaic, and Djebel Uweinat.

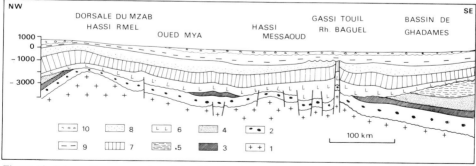

Figure 12. Schematic cross-section of the Algerian Sahara. 1— Precambrian; 2— Cambro–Ordovician; 3— Silurian; 4— Devonian; 5— Carboniferous; 6— Permo-Trias; 7— Jurassic; 8— Lower Cretaceous; 9— Upper Aptian to Senonian; 10— Tertiary.

The ridges show considerable tectonic instability during the Mesozoic and the Cenozoic, especially the El Biod-Gassi Touil-Rhourd el Baguel ridge (Busson 1970), the remains of which often correspond to dragfolds linked with a submeridian sinistral strike-slip fault. This is true also of the Garian horst and Cyrenaic ridge with a structure similar to the Atlas folds (Rohlich 1978). This is logical considering the proximity of the Aegean Arc. Basins can sometimes be deformed, especially near the Alpine chain, for example between the Mzab and Hodna. Deformation took place near the Bedoulian–Gargasian boundary (Austrian phase) in a very large domain; also at the Cretaceous–Palaeocene boundary in Cyrenaic and Tademait and Tinrhert. After the Lower to Middle Eocene and before the Miocene, very general events of the Pyrenean–Atlas phase were of weak intensity except for the extreme north of the Algerian Sahara. During Miocene to Quaternary times, the South Aures ridges and troughs formed where the Neogene series can be up to 2000 m thick.

Basaltic magmatism is known in the Trias–Lias in the Algerian Sahara to the north of Hassi Messaoud (Busson 1970), at different times (Trias to Neogene) in the Pelagian Sea and the Gulf of Syrte, and during the Neogene and Quaternary in Libya. Some syenitic intrusive complexes, of Cretaceous to Eocene in age have been found in the far south of Libya, in the Djebel Uweinat region. These complexes of deep origin are structurally quite important.

4b. The Taoudenni basin

The post-Hercynian formations are situated mainly in the eastern half of the basin and constitute a very flat basin with fault systems on the east and south sides where the series thickens rapidly into tectonic troughs. Geophysical, geological, and photogeological studies have shown the presence on the south border of the Taoudenni basin of an important major lineament oriented N80° E (Bellion et al. 1984; Guiraud et al. 1985; and Figure 13). Such faulting can affect the Precambrian, Palaeozoic, and Mesozoic formations. Their interplay at the base of the Mesozoic has been linked to the emplacement of dolerite sills, dykes, and flows, mainly during the Lias (180 to 190 Ma). In the Nara region this volcanic assemblage is overlain by sandstone formations from the Middle Jurassic ('Continental intercalaire') which thickens in a trough reminiscent of a diamond shaped

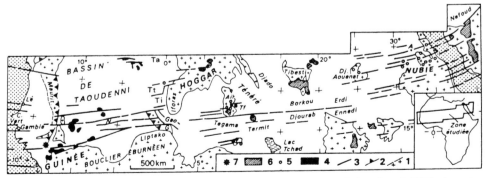

Figure 13. Schematic map of the Guinea–Nubian lineament zone (after Guiraud et al. 1985). 1— edge of Precambrian metamorphic rocks; 2— anormal contact zone; 3— tectonic lineament; 4— Liassic dolerites; 5— intrusive massifs; 6— Neogene and Quaternary volcanic rocks; 7— Guinea seismic epicentre for 22 December 1983. A— Aswan; G— Gourma; L— Los; Le— Leona; N— Nara; Ta— Tanezrouft; Tf— Tefidet; Ti— Tilemsi; Tt— Timetrine.

basin. Between the Upper Cretaceous and the present time, several periods of ferruginization have affected the whole of the series. The ferruginized continental formations show the effects of two tectonic compressive episodes. The first tectonic episode shows reverse faults orientated N40 to N80° E, which are often linked to dragfolds and less often strike-slip movements. Structural analysis shows that the major shortening direction is between N120E and N165°E, an average of N140°E.

The second tectonic episode is characterized by strike-slips linked to folding, and some reverse faults: the shortening direction varies between N05 and N60°E, and averages N30°E.

The style of deformation varies between tectonic episodes, a logical occurrence considering the respective orientation of the existing major fault systems (N80°E) and the shortening direction. As for the age of the tectonic episodes, one can consider the data from the east on the Taoudenni basin in the Timetrine region and from the west in the Senegal basin. We consider that the first tectonic episode corresponds to the Mid-Eocene phase, the second tectonic episode immediately following the first, or it may be even more recent.

To the west of Hoggar one has known for a long time that the secondary and Tertiary formations are folded and/or faulted in a strip 100 to 200 km wide from Tanezrouft to the Gao Trough (Figure 13). If the outcrops are not well exposed near Tanezrouft, they are excellent in the Timetrine–Tilemsi zone. The folds are weakly indicated, but the strata often upturn at the contact of the fault plane which corresponds nearly always with reverse faults. These faults limit in particular the massifs or the basement wedges which overlie the Mesozoic cover. In fact the passage from the Maastrichtian to the Palaeocene is continuous within the well dated marine series, and the Lower to Middle Eocene sometimes is affected by these fault zones. Thus it is clear that the major tectonic phase is Mid-Eocene. The microtectonic analysis shows that shortening indicated by these structures followed a direction of N140°E, similar to the one shown for the oldest phase in the Nara–Nema region (Bellion *et al.* 1984). Also west of Timetrine, in Azaouad, faults have been found in the 'Continental terminal' and the Quaternary (Caby and Riser personal communication) but the structural study remains to be undertaken.

If one also considers the existence of numerous Liassic doleritic dykes in the northern part of the Taoudenni basin one concludes that this basin has seen important post-Hercynian deformation.

4c. The Niger–Chad basin

The post-Hercynian Niger–Chad basin can be subdivided into two smaller basins divided by the Tegama limit into the Iullemeden basin of Western Niger and the Chad basin in the east (Figure 1). These two basins are structurally quite complex because of the interplay at different periods of major basement fractures, which created tectonic grabens or megashear zones. One of these zones cuts across the two basins and extends both to the west and the east (Figure 13).

The Western Niger basin. The Western Niger basin is in fact half a synclinal basin, truncated in the southwest by fault scarps oriented NW–SE to N–S (Figure 14). Various geophysical studies in the western part of the basin have discovered sediment-filled troughs. One trough shows a submeridian orientation and is linked to the reactivation of faults along the Kandi lineament zone. These fractures continue northwards towards In Guezzam. Their last appearance corresponds to the development of sinistral strike-slips dating from the end of the Cretaceous. A second sediment-filled trough is located between Adrar des Iforas and Gourma.

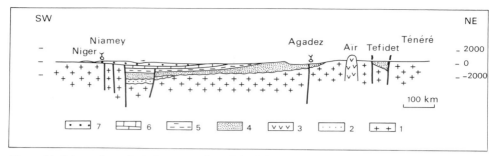

Figure 14. Schematic geological cross-section of the Iullemeden basin in the Tenere. 1— rejuvenated Birrimian; 2— Palaeozoic sedimentary rocks; 3— Palaeozoic ring complexes; 4— Trias to Lower Cretaceous; 5— Middle and Upper Cretaceous; 6— Palaeocene and marine Eocene; 7— 'Continental terminal'.

This is the Gao trough (Radier 1959) shaped like a half-graben tilting southwest with a Cretaceous infilling 3500 to 4000 m thick.

West and south of the Air massif, wrench faults are oriented N80 to N90°E. They are more continuous than is shown on the maps (Joulia 1963; Faure 1966). Also truncated and plunging folds are systematically linked with faulting. A microtectonic study has shown that these were dextral shear zones (Guiraud et al. 1981b). These strike-slips are associated with similar type faults, oriented N10°W to N30°E, with sinistral slips (mainly the Arlit fault). The major shortening direction can be determined rather precisely in western Aïr where it is between N120 and N130°E.

These directions are known in the Precambrian basement of the Aïr massif and correspond to reactivation of faults from the Palaeozoic up to the present time (Black et al. 1967). The megashear zones are more important. They define the limits of the Air massif and correspond to the major schistosity of the metamorphic deformation, and to the alignment of the Palaeozoic subvolcanic complexes (Moreau 1982). During this era the N–S, NW, and NE directions are active and facilitate the magmatic intrusions. They also are vital to the structural evolution of the Western Niger basin. The N80°E faulting was developed later and regularly presents a dextral strike-slip movement posterior to the emplacement of the ring structures in central and southern Aïr (Moreau 1982). During the Quaternary, various volcanic massifs were emplaced in southern Aïr (Cantagrel and Karche 1983).

The dextral shear zones seen immediately south of Air and indicated by Precambrian basement wedges found amid the Cretaceous, continues in the east towards the Tenere. Towards the west one can observe a succession of minor faults affecting the 'Continental terminal' succession, and the east–west branch of the Gao trough. It is therefore a major fracture zone. Earlier expressions of this zone are responsible for the uplift of the Aïr Precambrian basement and perhaps in the west for the formation of the major Palaeozoic syncline of Tin Serine.

The Chad basin. The Chad basin is larger than the Western Niger basin, but outcrops are much rarer. Nevertheless following the geophysical work of ORSTOM condensed by Louis (1970), the study of Faure (1966) and more localized research by several petroleum companies—and especially deep drillings— the major structural features of the basin have been defined. A network of rather complex grabens have been defined (Figure 15) by Louis (1970). The Eastern Niger graben is the best known. It is characterized by the large Termit–Ténéré trough, and by three smaller troughs at Kafra, Tefidet, and N'Geledji. These troughs or grabens, are separated by horst blocks where the Precambrian and Palaeozoic basement outcrops, for example the Grein–Achegour– Fachi ridge or the Takolokouzet ridge.

The Termit–Ténéré trough was formed by the polyphase interplay of a multitude of normal faults oriented NW–SE to N–S. The sedimentary infilling can reach a depth 7000 m (Figure 16). The series started very probably during the Lower Cretaceous, and more precisely during the Upper Aptian. This theory is shared by the geologists of CONOCO (Mathieu 1983) and is based on analogies with the Benue trough and the North Cameroon graben. In the Termit basin, one notices the presence of an angular unconformity between the upper Cretaceous, and the Lower Eocene (Faure 1966 and Bellion et al. 1983). Also the oolithic formation is slightly folded and cut by faults. A major structural feature separates the Ténéré graben in the north from the Termit graben in the south. This high zone is clearly defined on gravimetric maps. At this location one can find Neogene basaltic

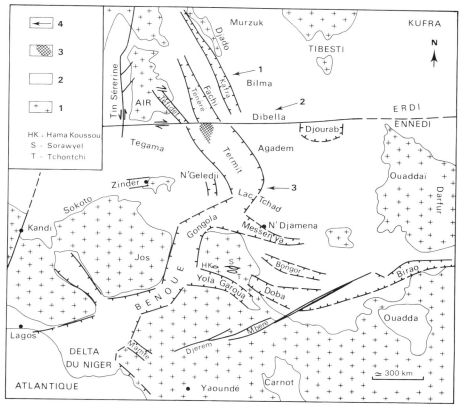

Figure 15. Structural evolution of Lake Chad basin and the neighbouring regions. 1— Precambrian basement; 2— sedimentary basins; 3— Recent volcanism in the Termit region; 4— locations of cross-sections in Figure 16.

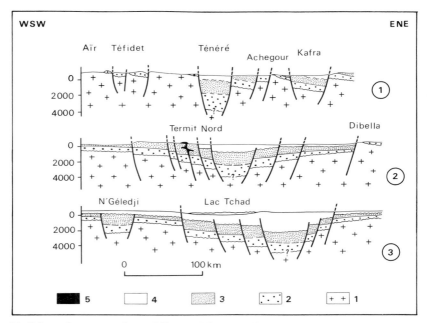

Figure 16. Schematic cross-sections of Eastern Niger. 1— Precambrian; 2— Lower Cretaceous; 3— Upper Cretaceous; 4— Cenozoic; 5— Recent volcanism. Numbers in circles refer to the locations of the cross-sections in Figure 15.

volcanoes linked to sills found by drilling and dated as 8·6 +/− 0·5 Ma). The gravimetric and seismic data also underline the presence of structural complications, which are not surprising considering that we are exactly in the eastern prolongation of the shear zone described in the south of Aïr. Further to the east, along the same direction, there is the small Dibella basement horst and the northern limit of the Djourab graben (Figure 15).

The Kafra, Tefidet, and N'Geledji grabens have structural directions similar to the Termit–Ténéré graben. The very narrow and deep Kafra graben, containing approximately 3000 m of post-Aptian sandstone, extends far to the north into the Touareg shield.

The Tefidet graben is more accentuated. A recent field study showed that its western border was not faulted but simply flexed. It is therefore more a basin limited in the east by normal faults, the throw of which does not exceed a few hundred metres (Figure 16). The northern end of this basin is interesting because it has two main characteristics. The basin is limited by a dextral shear zone, expressed by a network of faults oriented N90° to N130°E to which are often linked some dragfolds. One finds numerous basaltic volcanic assemblages, emplaced from the Miocene to the Quaternary on north-south structural alignments, corresponding to deep fractures with little or no movement at the Cretaceous sedimentary cover.

This Tefidet basin extends and deepens towards the southeast, before intersecting the South Aïr–North Termit fracture zone. Thus the graben of the Tefidet basin present structurally the appearance of a large tension crack limited in the north and south by dextral strike slips. This assemblage belongs to a constrained environment similar to that in West Aïr in which the major horizontal shortening direction is between N120 and N130°E, a direction which is similar to the orientation of Tefidet graben (N130 to N140°E).

What can be said about the structural history of eastern Niger? The grabens, generally orientated N140°E, and which began to develop during the Lower Cretaceous, seem to have an origin linked to the Austrian, intra-Aptian phases. These same tectonic episodes are recorded in North Hoggar and occurred during the opening of the Benue trough. There then follows an extension episode, during the Santonian, emphasized by the interplay in the Termit graben of numerous faults, and terminated by the end of the Upper Santonian.

The fact that a major tectonic phase occurred at the Cretaceous–Tertiary boundary can be shown by the enlargement of the basins due to the interplay of normal faults and numerous wrench faults oriented N140°E. A slight tilting of the layers at the edges of the basin, at Termit and Agadem can also be noticed. Furthermore the dextral strike-slip motion of the North Termit shear zone and the faults limit the development of the northern part of the Tefidet graben.

In the Termit basin certain deformations provide a record of the Pyrennean–Atlas phase. And finally, from the Upper Eocene to the Quaternary the extension regime continued with periods of increased intensity, as shown by the volcanic eruptions found in North Termit, North Tefidet, and southern and central Air.

Data on other regions. In the southern and eastern regions of the Chad basin, other grabens have been found but their orientations differ from those in Eastern Niger (Figure 15). For example the Maiduguri graben or basin which links the Termit graben and the Benue trough, and the Gongola branch which connects with the South Chad grabens. There are also the southern Chad and northern Central African Republic grabens, as well as the Djourab graben, isolated in northern Chad.

Structurally, there are few data on these grabens. The important Doba–Birao graben, orientated N80°E exhibits a positive gravimetric anomaly orientated N70°E. This narrow and very straight gravimetric zone constitutes a major feature (Louis 1970). This anomaly might correspond to a shear zone, with basement wedges and which moved several times during the Cretaceous. According to Louis (1970) the Doba–Birao graben is composed of a succession of diamond shaped small basins, slightly offset from one another. These would be pull-apart basins, likely to have originated during the Upper Aptian, and similar in age to the Bongor basin (Mathieu 1983). These basins were slightly folded at the end of the Cretaceous.

The Djourab graben is also interesting. It is infilled by Cretaceous and Tertiary sediments 5000 to 6000 m thick, limited in the north by a fracture zone oriented E–W. These fractures probably moved during the Mio-Pliocene (Wacrenier *et al.* 1958) indicating the presence of faults in the 'Continental terminal' WSW of Faya–Largeau. The same authors suggest that the 'Continental intercalaire' sandstones are strongly influenced by strike-slip movement with dragfolds, south of Tibesti (Wacrenier personal communication).

In conclusion one should not be surprised by the existence of structural deformation in northern Chad. The northern limit of the Djourab graben is very precisely located in the western prolongation of southern Air–North Termit–Dibella fracture zone which extends in the east as far as the Sudan–Libia border.

4d. The Benue Trough and the basins of Central and Northern Cameroon

The Benue Trough is really a deformed intracontinental sedimentary basin (Benkhelil 1982). It is in reality a graben with structural characteristics belonging to the upper to middle structural domains. From the Niger delta to the Lake Chad basin, the trough is nearly 800 km long and divided into three major structural assemblages: the Abakaliki anticlinorium; the Middle Benue; the Upper Benue.

The Abakaliki anticlinorium, sometimes called the Lower Benue, lies immediately to the northeast of the Niger delta province (Figure 17). It corresponds to a network of folds, 200 km long and 60 km wide, influencing Albian to Coniacian terranes.

A median strip, 20 km wide, is clearly identified. At this level folds are dense, inclined towards the northwest and linked to a schistosity, with a strongly tilted axial plan (80°) towards the southeast (Figure 18). The zone of deformation is the result of a noticeable flattening with mineral crystallization (chlorite, quartz, calcite) indicating anchizonal to epizonal metamorphism. Shear zones oriented N30°E to N50°E cut this strongly deformed zone presenting a sinistral strike-slip component. The anticlinorium results, on the whole, from a shortening oriented N140°E. Also very weakly angled monoclinal layers are found on the southeast border where they are only deformed by a network of conjugated strike-slip faults. The deformation and metamorphism is Santonian in age (Benkhelil this volume).

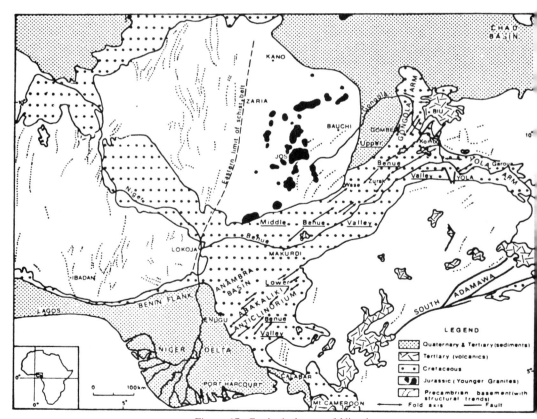

Figure 17. Geological map of Nigeria.

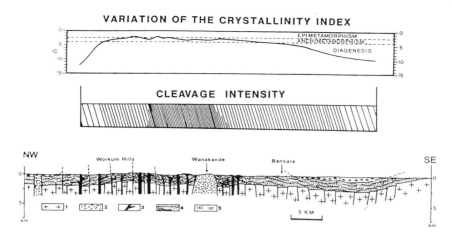

Figure 18. Interpretative section of the deep structure of the Abakaliki anticlinorium, Nigeria, showing the variations in the illite crystallinity index with the intensity of cleavage (after Benkhelil et al. 1986). 1— Precambrian basement; 2— syenite or diorite; 3— intrusive rocks; 3— shale and siltstone of the Asu River group; 5— shale (Eze Aku formation)

Northwest of the Abakaliki anticlinorium lies the Anambra syncline, a large structure with little deformation. During the sedimentary infilling of the Abakaliki Trough, from Albian to Senonian, the Anambra region was a shallow shelf. Its evolution into a subsiding basin began after the Santonian tectonic event which uplifted the Abakaliki anticlinorium. Thus one can consider its role as a flexural basin.

The large quantity of volcanic material in the Abakaliki region, and its relative diversity have led some authors to suggest the existence of an oceanic crust at shallow depth. But geochemical results do not support this hypothesis. Recent geophysical interpretations have nevertheless shown the existence of a thinned crust under the trough (Benkhelil et al. in press).

The Middle and Upper Benue. The passage from the Abakaliki domain to the Middle Benue does not happen in a continuous way; both domains seem to be offset in space with the Middle Benue situated in the prolongation of the Anambra basin. However some narrow basins, enclosed in the basement, represent the trace of the Abakaliki domain towards the northeast along the southern border of the basin in the Middle Benue and possibly the Upper Benue.

The Keana anticline is the Middle Benue main structure. This concentric type folding shows by a weak schistosity in a thick argillaceous cover.

In the Upper Benue, the chain is divided into two orthogonal arms, Gongola and Yola. The Gongola arm is composed of a thick and little deformed Cretaceous series overlain unconformably in the west and north by the Tertiary Kerri–Kerri formation and the Tertiary deposits of the Lake Chad basin.

As in the Middle Benue, the cover is marked by folding but with uplifted and sometimes vertical margins (Figure 19). The Yola branch is continued into the Cameroon by the Garoua basin which is weakly tectonized. The passage from the folded domain to the tabular domain is related to the formation of a very important N–S fracture zone, on which was developed the Faro river at the border between Nigeria and Cameroon. The folds are of various lengths and sometimes as long

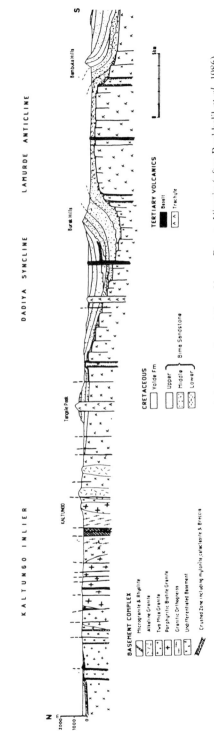

Figure 19. Schematic cross-section of the Kaltungo inlier and the Lamurde anticline, Upper Benue, Nigeria (after Benkhelil et al. 1986).

as 60 km, but more often folds are relatively short with parallel axes, inclined and truncated by strike-slip events. The axial directions are approximately N50°E on average, except in the Yola and Gongola branches, where the axis tends to be east–west.

Besides the folding, the brittle fractures are typical of the Upper Benue structures. Numerous faults played an important role at all stages of basin evolution. They can be classified into two main groups. The first group includes those faults orientated N50°E controlling the whole of the Trough. Their importance in the formation of the sedimentary Albo-Aptian basins has been mentioned by Benkhelil (1982), Benkhelil and Robineau (1983), and Maurin et al. (1985). They follow the late Pan-African phases during which they had a dextral shear motion generating the mylonites. During the opening phases of the Benue, these same faults were reactivated creating a sinistral motion. The motion is mainly strike-slip and was responsible for the formation of subsiding subbasins, either in the echelon extension zones or in the extensional curvature of the fault systems. In the sediments that were deposited at the same time, small synsedimentary to pre-diagenetic tectonic movements have been noted. Faults orientated between N10°W and N20°E show normal motions and are sometimes mineralized.

The second group of faults includes a group of dextral strike-slip systems oriented N110 to 120°E and sinistral ones oriented N160°E to N–S. Their genesis is linked to a late-Cretaceous compressive phase, also responsible for the folding and formation of fault planes oriented N50°E. Also some folds situated along N50°E faults began in early Albian times during the infilling phase of sub–basins.

The Middle and Upper Benue constitute an assemblage corresponding to a 'middle structural level chain' (Benkhelil and Guiraud 1980; Benkhelil 1982) formed during the late-Cretaceous. The Maastrichtian is an important component in the assemblage, while the Palaeocene detrital formations are horizontal and lie unconformably on the earlier structures in the northern regions. Important fault systems during more recent periods have been defined. The very straight contact between the Palaeocene and the basement of the Jos Plateau corresponds to a fracture where the Palaeocene is strongly upturned suggesting that the Kerri-Kerri Tertiary basin is a half-graben. A north-south fracture zone has oriented the Gongola lineament and continues on the Biu Plateau where it is expressed as an alignment of Plio-Quaternary volcanic cones (Benkhelil 1982).

Volcanic activity in the Benue was particularly important during the Cretaceous and took place during several phases. Prior to the opening of the Benue Trough (Popoff et al. 1982) basaltic and occasionally rhyolitic flows were erupted around 147 +/− 7 Maago. These are contemporary with the end of the emplacement of the anorogenic alkaline granitic ring complexes in the Jos Plateau region (186 to 141 Ma) (Van Breemen et al. 1975; Bowden et al. 1976; Turner et al. 1975).

Umeji and Caen-Vachette (1983) gave an Aptian age (113 +/− 3 Ma) to the rhyolites of Yandev (Middle Benue). Ages of 100 to 105 Ma (Albian) were obtained from basalts of the Gwol region (Upper Benue) by Popoff et al. (1982). They could be transitional olivine tholeiites. This episode was also identified in the Abakaliki anticlinorium where basaltic products were dated as Albian (Uzuakpunwa 1974). These events probably continued during the Cenomanian. A similar age was found for the intrusive syenitic complex at Wanakande in the Workum Hills, Abakaliki (Snelling 1965; Benkhelil this volume).

Towards the end of the Santonian or at the beginning of the Campanian (81 Ma), the Albian volcanic activity in the Abakaliki area could have been rejuvenated by epimetamorphic transformation of the Pre-Santonian sediments (Benkhelil 1985).

From the Palaeocene to the end of the Eocene, between 63 and 35 Ma, granites, were emplaced along the Cameroon Line (Tempier and Lasserre 1980; Regnoult et al. in press). They are anorogenic ring complexes, similar to those described from Nigeria. Some of these complexes are in the Yola Arm, southwest of Garoua, others are to be found near the Mamfe basin, east of the Abakalikis.

In the Upper Benue, trachytes and phonolites are dated between 11 and 22 Ma, while the Biu Plateau basalts were emplaced between 23 and 7 Ma, and from 3 Ma to the present time. The occurrences are very numerous, several hundred necks have been found in this region. This volcanism is found not only in the neighbouring regions (Cameroon, Jos Plateau) but also in the whole of Africa.

The history of the Benue can therefore be divided into several periods related to the formation and infilling of the Trough. Early traces of instability from the Jurassic onwards are shown by the existence of some contemporary volcanic occurrences at the end of anorogenic magmatism in the Jos Plateau region. The first basins in the Upper Benue were structures oriented E–W, following those known in northern Cameroon. The mechanism responsible for this episode to which are linked important magmatic events, and thermal anomalies (Allix and Popoff 1983) appears to be a north-south extensional lineament system (Popoff et al. 1982).

But the important thermal anomaly under the Abakaliki anticlinorium between Albian and Santonian times, appears to be controlled by a major crustal fault system probably related to the Charcot fracture zone hidden beneath more than 12 km of sediments in the Niger delta. Most of the basins originated during the Albian. Indeed the Albian basins of the Benue show generally a diamond shape and a structural control linked to two types of fault systems (Benkhelil 1982; Benkhelil and Robineau 1983). They are NE–SW faults corresponding to some strike-slip movement, and N–S to N15°E faults often mineralized.

This system seems to result in a general sinistral slip motion, causing the reactivation of ancient lineaments and the opening of pull-apart basins. These major lineaments belonging to the continental domain are situated in the prolongation of transform faults in the Gulf of Guinea, which appeared at the same time during the opening of the South Atlantic. Strike-slip movement in the Gulf of Guinea and the Benue can be explained by the development of the central and south Atlantic on both sides of the Romanche, Chain, and Charcot equatorial fractures. The Albian volcanic occurrences found in these various regions, from the Abakaliki area to the Upper Benue, support the continued opening of the Atlantic. The extensional regime continued until Turonian times in the Abakaliki area and the Maastrichtian in the Middle and Upper Benue. Two rather short compressive episodes explain the formation of the Abakaliki anticlinorium, and the Middle and Upper Benue. Folding associated with schistosity took place during the Santonian in the Abakaliki area. The shortening/flattening followed a N140°E direction and affected mainly the central part of the basin. The existence of an anchizone metamorphism in a rather large domain illustrates the presence at shallow depth of igneous intrusions. Shortening was linked to a slight shear motion along the axis of the Abakaliki anticlinorium during the Santonian times.

In the Middle and Upper Benue folding occurred at the end of the Maastrichtian. The average direction of shortening was approximately N150°E, but there are local variations because of the influence of older structures (mainly E–W grabens). The compressional diachronism between the Middle and Upper Benue, as well as the small variation in the direction of shortening may have resulted from a change in the position of the pole of rotation for the Central Atlantic, an event which took place about 80 Ma ago (Le Pichon and Fox 1971).

The development, or the rejuvenation, of major fault systems took place during the Tertiary. In particular, fractures at the contact between the Gongola branch and the Jos Plateau; fractures linked with the Palaeocene and Eocene alkaline granite intrusions in Cameroon; numerous faults which controlled the emplacement of Neogene volcanic occurrences in Upper Benue, all indicate a dominant E–W extension. It seems therefore that the Benue domain had been subjected to a dominantly extensive regime from the beginning of the Tertiary.

The Central and Northern Cameroon basins. Some mainly Cretaceous basins are preserved by faulting within Precambrian country rocks (Figure 15). One of the most characteristic is the Mbere and Djerem trough, partly covered by the Neogene volcanic formations of the Adamawa Plateau. This trough lies on a fracture zone that can be followed laterally westwards until the Cameroon Line and eastwards towards Sudan. The Cretaceous sandstones and conglomerates are folded and cut by many faults, in particular conjugated strike-slips, the analysis of which enabled Ngangom (1983) to prove the effects of a compressive phase characterized by a shortening direction of N145°E. The Cretaceous was also in part subjected to thermal metamorphism of the Pyrennean type (Vincent 1968).

In the north of the country, the Garoua basin represents an eastern continuation of the Yola arm of the Upper Benue. However, the strata are not folded but cut by lineaments which may have acted as strike-slip and/or normal faults. Tertiary to Quaternary magmatism is associated with these lineaments as doleritic sills, Palaeocene (65 to 60 Ma) alkaline complexes, and trachy–phonolitic necks and dykes (Neogene). In the Cameroon arm of the Upper Benue, one finds numerous small syenite outcrops, alkaline granites, basalts, and trachyphonolites.

Three small isolated basins are preserved further north near Hama Koussou, Sorawyel, and Tchontchi. They should be mentioned because they correspond to half-grabens, oriented E–W, limited by strike-slip fracture zones. The Tchonti basin is folded and slightly metamorphosed during end-Cretaceous times. Doleritic and trachytic dykes extending for 20 to 30 km in an E–W direction with occasional small anorogenic granite outcrops localized along strike-slips near the basins, show the effects of Miocene extension.

To conclude, these basins, and others not mentioned here, have been subjected to late post-Cretaceous deformation, similar in style and chronology to the Benue trough.

The Cameroon line. The Cameroon line is a major structure on the African plate orientated N50°E in the south, with a curved zone in its central part and orientated N25°E to the north. It is interpreted as an oblique megastress fault zone similar to the Adamawa lineament (Regnoult *et al.* in press).

Magmatism has been reactivated along the Cameroon line. For instance between 73 and 60 Ma anorogenic granites were emplaced in Northern and Central Cameroon. Similar rocks were developed between 45 and 35 Ma in Central and South Cameroon. Then, around 30 Ma ago, alkali volcanism began and has continued intermittently until the present day. The last eruption of Mount Cameroon in 1982 (Déruelle *et al.* 1983) occurred along a fault zone orientated N50°E. All these features point to sequential reactivation of the megastress fault systems controlling the location of magmatism.

4e. The main structural characteristics of post-Hercynian tectonics

In the domain between the Alpine chain and the Gulf of Guinea are post-Hercynian formations that are generally subhorizontal and cut by numerous fractures. These fractures have acted as normal faults, then strike-slip faults, and less frequently as reverse faults. One can delineate networks of fracture zones where slip motions were followed by dragfolds.

Some troughs or half-grabens look like tensional megafaults (Gao, Tefidet, Termit–Tenere, Kafra, Hun). But the Benue Trough constitutes an original feature as the only really folded domain on the African plate. In contrast the folds found in the Northern Algerian Sahara and in Cyrenia are related to the Alpine chain.

Magmatic volcanic and subvolcanic occurrences on the African plate are important tectonically, and chronologically linked to movement along major lineaments. For example, the Triassic–Liassic (220 to 180 Ma) tholeiitic dolerites known from the Northern Sahara to the Ivory Coast, and localized mainly to the west of the West African craton, are related to the opening of the Western Tethys and of the Central Atlantic. Furthermore, the emplacement of the Triassic–Jurassic (186 to 141 Ma) anorogenic alkaline granites of Nigeria preceded the opening of the Gulf of Guinea and the Benue Trough. Some kimberlites in Guinea and Liberia (140 to 90 Ma) are linked to the rupture of the Liberian–Guyanese continental isthmus. The Cretaceous volcanism of the Benue in Aptian–Santonian times is mainly alkaline in the south but also includes tholeiitic basalts of transitional type in the northeast. The anorogenic alkaline granites in Cameroon were emplaced during Palaeocene to Upper Eocene times. While the Oligocene to Quaternary alkali basalts are found east of the West African craton in the Pan-African mobile domain.

One should note that tholeiitic magmatic occurrences are mainly localized along the passive Atlantic margin, while those of the continental domain are nearly all alkaline.

The tectonic instability of the northwest African continental domain occurred during episodes that can be considered stratigraphically rather short and synchronous, with the exception of Triassic–Liassic and Miocene–Quaternary extensional periods.

Between Jurassic and Neogene times, the main tectonic events took place, for example, during the Aptian, around 110 Ma, with important strike-slip motions at the North Saharan ridges, in the Benue, and probably in Southern Chad. The creation of the Niger troughs is also linked to this episode. During the Santonian the Abakaliki antoclinorium developed in the Lower Benue. While at the end of the Cretaceous tectonic movement occurred in many sectors particularly in the Middle and Upper Benue, and finally during Eocene times in most of the basins.

The major shortening direction linked to these phases is similar, and apart from a few local deviations, is near N140°E.

5. Major lineament zones

The African plate is cut by several major lineament zones along the limits of the great structural domains (Figure 20). Four groups of lineaments can be identified in western and northern Africa where they played a major role from post-Hercynian times.

5a. The North African lineaments

In North Africa and South Iberia (Iberian Cordilleras) a group of N80 to N90°E orientated lineaments defined the principal structural domains. In the internal

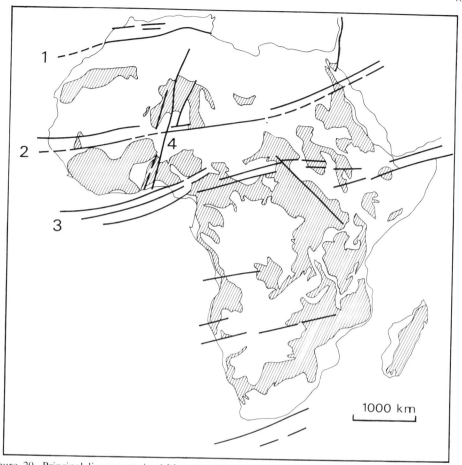

Figure 20. Principal lineaments in Africa. 1— North African lineament zone; 2— Guinea–Nubian lineament zone; 3— Central African lineament zone; 4— N–S megashear zones. Diagonal lines—Precambrian metamorphic rocks.

zones of the Alpine chain, some of them, which mark palaeogeographical boundaries, can only be individualized on palinspatic maps. One can recognize from north to south, as depicted in Figure 21, the Guadalquivir lineament forming the southern limit of the Iberian mesetas, the Crevillente lineament, the boundary faults of the Moroccan and Algiers–Oran tablelands; the North Atlas faults; the Sahara flexure or South Atlas lineament which corresponds in fact to a rather complex succession of varied structures (Guiraud 1973). During Alpine history, these lineaments acted alternatively as sinistral strike-slips until the Aptian, and then as dextral ones from Santonian times and as normal faults or more rarely reverse ones. The strong seismicity now registered in this part of the Alpine chain is concentrated near these major lineaments or sometimes near other faults orientated in the same direction but apparently of less importance.

Magmatism which occurred sequentially since the Triassic is, for the main part, linked to the motion of the lineaments. This is quite clear for the Miocene calc-alkali volcanism found concentrated along a narrow strip on the North African margin.

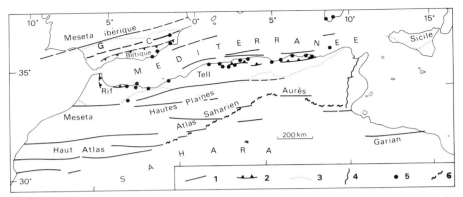

Figure 21. East–west lineaments in North Africa and Southern Spain. 1— major lineament; 2— thrust front of the internal zone; 3— allochthonous front; 4— Tunisian ridge; 5— calc-alkaline volcanics; 6— 'Saharan flexure'. C— Crevillente; G— Guadalquivir.

5b. The Guinea–Nubian lineaments.

This is a very important lineament zone generally orientated N70 to N80°E, that can be followed from the Senegal–Guinea margin to the Red Sea (Guiraud et al. 1985) (Figure 13). There are many aligned fractures in the continental and oceanic domains. They are old pre-Mesozoic fractures which will rejuvenate episodically from the Upper Triassic. The main episodes took place during the Liassic, in Northern Guinea and south of the Taoudenni basin where many doleritic tholeiitic dykes, sills, and flows were emplaced, linked with the opening of the Central Atlantic, also at the end of the Cretaceous where in the north of the Niger–Chad basin the 17th parallel lineament and related fractures act as a dextral strike-slip fault linked to dragfolds. Movement also occurred at the base of the Upper Eocene, in most areas. Even today there are strong seismic tremors off Guinea (22 December 1983) and in the Aswan region. The nature of the magmatism linked with these lineaments—especially in the syenitic complexes of Leona, Djebel Uweinat, and in the Nubian province— shows that these are surface traces of a major fracture zone affecting the whole of the lithosphere.

5c. The Central African lineaments

Many authors have emphasized the importance of the Cameroon and Southern Chad lineaments, and have considered their links with the Benue Trough and the transform faults in the Gulf of Guinea. The best description has been written by Cornacchia and Dars (1983). The authors hypothesize on the existence of a lineament network over Africa from the Gulf of Guinea to the Gulf of Aden. As confirmatory evidence, the principal stages of the geodynamic history of this lineament network can be identified. From the Late Pan-African to Aptian times these lineaments have operated mainly as dextral strike-slip faults. From the post-Hercynian period the movement along the lineaments can be associated with the opening of the Central Atlantic. From the Upper Aptian to the Turonian, the movements are reversed, linked to the opening of the Gulf of Guinea. And then a sinistral shear motion occurred to which can be linked the formation of the Benue and Southern Chad basins. One also notes, during this period, the tholeiitic and alkali volcanic activity in the Benue. During the Santonian, the strike-slip motions slowed down considerably because of the change in trajectory of the African plate (Olivet et al. 1980).

The southernmost continental basins were then strongly folded (Abakaliki, Djerem–Mbere) with shortening orientated NW–SE. At the end of the Cretaceous, similar deformations are registered in the Middle and Upper Benue as well as in Northern Cameroon. Then vertical movements dominated, in a general extensional context to which was linked the emplacement of anorogenic granites in Cameroon during the Eocene, and finally the Miocene to Quaternary alkali volcanism in Cameroon and Nigeria.

5d. Northsouth lineaments of the Central West African mobile zone

Very important and continuous submeridian lineaments have been found in the West African Pan-African mobile belt by several authors. The main fractures constitute a network that can be followed from the Gassi Touil ridge in the north to Togo in the south (Guiraud and Alidou 1981a). These fractures have acted in a more or less continous way from the end of the Precambrian to the end of the Cretaceous, and then, more locally during the Tertiary and even in the Quaternary (Amguid fault in the Hoggar, and Akwapim fault in Ghana showing a strong seismicity). The most frequent motions have been as normal faults, of synsedimentary age in the basins. But strike-slip movements with some dragfolds also took place during certain phases such as the sinistral strike-slip movement at the end of the Cretaceous.

The Miocene to Quaternary volcanism in the Hoggar illustrates the control these lineaments have on the location of magmatism.

6. Conclusions

The whole of the continental domain situated between the Alpine chain and the Central African lineaments was subjected to post-Hercynian deformation. If the Alpine margin has structures which illustrate an important shortening, then in contrast the Atlantic margin can be modelled as a permanent extensional regime. The exception is in its transform zone, where shearing played a major role during the opening of the Gulf of Guinea.

In the continental domain the main structural features are linked to the density of fractures and to the fact that some important fractures will show an episodic instability during long periods of geological time up to the present. The zones of weakness in the African plate have thus materialized from a complex tectonic history. Two groups of structures are particularly common. Firstly the megashear zones linked to dragfolds, and secondly the troughs. The structures associated with the evolution of the Benue Trough constitute an original assemblage which illustrates the importance of intraplate tectonics in Africa.

Several tectonic episodes have been noted in numerous regions of the North African continental domain. This data contributes to the understanding of the dominant influence of the Africa–Europe collision on intraplate tectonics. Nevertheless links exist between the structural discontinuities associated with the ocean opening in the North Atlantic and the preceding tectonic episodes (Schwan 1980). The study of relationship between the opening of the Atlantic and the collision of the African and European plates are being studied in several groups and new and more precise informations should be made available in the next few years. But to obtain satisfactory models one will have to take into account data on intraplate tectonics in East Central Africa and South Africa, regions for which a synthesis such as the one given here remains to be written.

Acknowledgements. Many field trips preceded the elaboration of this synthesis. They were made possible through financial assistance from SNEA (P) and the CNRS. We also wish to thank both reviewers for their constructive comments, Mme Daniel for typing, and the editors for the translation of this paper into English.

References

Allix, P. and Popoff, M. 1983. Le Crétacé inférieur de la partie nord-orientale du fossé de la Bénoué (Nigéria): un exemple de relation étroite entre tectonique et sédimentation. *Bull. Centres Rech. Explor-Prod. Elf-Aquitaine,* **7**, 1, 349–359.
Andrews Jones, D. A. 1971. Structural history of Sierra Leone. *In: Tectonics of Africa. Earth Sciences,* **6**, UNESCO, 203–207.
Aubouin, J. 1977. Méditerranée orientale et Méditerranée occidentale: esquisse d'une comparaison du cadre alpin. *Bull. Soc. Géol. France,* **3**, 421–435.
Bellion, Y., Benkhelil, J., Faure, H., Guiraud, R., Le Theoff, B., and Ousmane, B. 1983. Le bassin du Niger oriental: observations structurales et évolution géodynamique. *In: Bassins sédimentaires en Afrique, Soc. Géol. France,* Marseille, résumé, **57**.2, 57.
——, **Benkhelil, J. and Guiraud, R. 1984.** Mise en évidence de déformations d'origine compressive dans le Continental intercalaire de la partie méridionale du bassin de Taoudenni (Hodh oriental, confins mauritano-maliens). *Bull. Soc. géol. France,* **6**, 1137–1147.
—— **and Guiraud, R. 1984.** Le bassin sédimentaire du Sénégal: Synthèse des connaissances actuelles. *In: Plan Minéral de la République du Sénégal.* BRGM and DMG Dakar éd., **1**, 4–63.
—— **and Guiraud, R. 1982.** Sur la tectonique de l'île de Maio (Archipel du Cap-Vert, Atlantique Central). *C.R.Acad. Sc., Paris,* **294**, 1017–1020.
—— **and Guiraud, R. 1985.** Tectonique intraplaque—Déformations d'origine compressive d'âge intra-éocène à l'Ouest de l'Adrar des Iforras (Mali). *13th Congr. African geol.,* St. Andrews, abstract and this volume.
Benkhelil, J. 1982. Benue Trough and Benue Chain. *Geological Magazine,* **119**, 155–168.
—— **1985.** Metamorphism and deformation in the Cretaceous of the Southern Benue Trough (Nigeria). *13th Congr. African geol.,* St. Andrews, abstract and this volume.
——, **Dainelli, P., Ponsard, J. F., Popoff, M., and Saugy, L. 1986.** The Benue Trough: wrench fault related basin on the border of the Equatorial Atlantic. *In: 'AAPG Rift Basin Memoir'.* In press.
—— **and Guiraud, R. 1980.** La Bénoué (Nigéria): une chaîne intracontinentale de style atlasique. *C.R.Acad. Sc., Paris,* **290**, 1517–1520.
—— **and Robineau, B. 1983.** Le fossé de la Bénoué est-il un rift? *Bull. Centres Rech. Explor.–Prod. Elf-Aquitaine,* **7**, 1, 315–321.
Biju-Duval, B., Letouzey, J., and Montadert, L. 1979. Variety of Margins and Deep Basins in the Mediterranean. *American Association Petroleum Geological Memoir,* **29**, 293–318.
Black, R. and Girod, M. 1970. Late Palaeozoic to Recent igneous activity in West Africa and its relationship to basement structure. In: Clifford and Gass (Eds), *African Magmatism and Tectonics,* 185–210.
——, **Jaujon, M., and Pellaton, C. 1967.** *Notice explicative de la carte géologique de l'Aïr.* B.R.G.M. éd. Direction des Mines et de la Géologie du Niger. 57 pp.
Blarez, E. 1986. *La marge continentale de Côte d'Ivoire–Ghana. Structure et évolution d'une marge continentale transformante.* Thèse, Univ. Paris 6, 188 pp.
Bouillin, J. P. 1983. Nouvelles hypothèses sur la structure des Maghrébides. *C.R. Acad. Sc., Paris,* **296**, 1329–1332.
Bowden, P., Van Breemen, O., Hutchinson, J., and Turner, D. C. 1976. Palaeozoic and Mesozoic age trends for some ring complexes in Nigeria and Niger. *Nature, London,* **259**, 297–299.
Busson, G. 1970. Le Mesozoïque saharien. Deuxième partie. Essai de synthèse des données des sondages algéro-tunisiens. *Publ. Centre Rech. Zones arides.* CNRS, Paris, Sér. Géol. n° 11, 2 vol. 811 pp.
Cantagrel, J. M. and Karche, J. P. 1983. Le volcanisme quaternaire du massif des Todgha (Aïr–Niger): étude géologique et géochronologique. *Bull. Soc. géol. France,* (7), **25**, 557–562.
Celestino, S. L., Serralheiro, A., Macedo, J. R., Matos Alves, C. A. and, Peixoto Faria, A. F. 1979. L'île de Santiago, Cap Vert, dans le cadre pétrologique de l'archipel et d'autres îles de l'Atlantique. *Boll. Mus. Lab. Mineral Geol. Fact. Ciencias, Lisbonne,* **16**, 81–100.
Coiffait, P. E., Coiffait, B., Jaeger, J. J., and Mahboubi, M. 1985. Un nouveau gisement à Mammifères fossiles d'âge éocène supérieur sur le versant sud des Nementcha (Algérie orientale): découverte des plus anciens rongeurs d'Afrique. *C.R. Acad. Sc., Paris,* **299**, 893–898.
Cornacchia, M. and Dars, R. 1983. Un trait structural majeur du continent africain. Les linéaments centrafricains du Cameroun au golfe d'Aden. *Bull. Soc. géol. France,* **7**, 101–109.
Crévola, G. 1980. Principaux caractères du volcanisme de la presqu'île du Cap-Vert (Sénégal). *8e Réunion Ann. Sc. Terre,* Marseille, Soc. Géol. France édit., 114 pp.

Déruelle, B., Moreau, C., and Nsifa, E. N. 1983. Sur le récente éruption du mont Cameroun (16 octobre 1982–12 novembre 1982. *C.R. Acad. Sc., Paris*, **296**, 2, 807–812.

Donze, P., Guiraud, R., and Le Hégarat, G. 1974. A propos du passage Jurassique–Crétacé en domaine mésogéen: révision des principales coupes du Sud-Ouest constantinois. *C.R. Acad. Sc., Paris*, **278**, 1607–1700.

Dorbath, C., Dorbath, L., Gaulon, R., George, T., Mourgue, P., Ramdani, M., Robineau, B., and Tadili, B. 1984. Seismotectonics of the Guinean Earthquake on December 22, 1983. *Geophysical Research Letters*, **11**, 971.

Duée and Kazi-Tani, N. 1982. Cinématique des plaques et tectonique intraplaque. Exemple des chaînes intracontinentales d'Afrique du Nord. 9e *Réunion Ann. Sc. Terre*, Paris, résumé, 215.

Faure, H. 1966. Reconnaissance géologique des formations sédimentaires post-paléozoïques du Niger oriental. *Publ. Dir. Mines Géol.* (Rép. Niger), n° 1, éd. BRGM, 630 pp.

Féraud, G. 1981. *Datation de réseaux de dykes et de roches volcaniques sous-marines par les méthodes K–Ar et ^{40}Ar–^{39}Ar. Utilisation des dykes comme marqueurs de paléocontraintes.* Thèse Sciences, Université de Nice.

Girardin, N., Hatzfeld, D., and Guiraud, R. 1977. La séismicité du Nord de l'Algérie. *C.R. Somm. S.G.F.*, **2**, 95–100.

Guiraud, R. 1970. Sur la présence de décrochements dextres dans l'Atlas saharien. Interprétation mégamétrique. *C.R. Somm. S.G.F.*, **8**, 316–318.

—— 1973. *Evolution post-triasique de l'avant-pays de la chaîne alpine en Algérie, d'après l'étude du bassin du Hodna et des régions voisines.* Thèse Sciences, Université de Nice, 270 pp.

—— 1974. A la recherche du rhegmatisme de l'Afrique du Nord et des régions voisines. *Ann. Sc. Univ. Besançon, Géol.*, **22**, 135–153.

—— and Alidou, S. 1981a. La faille de Kandi (Bénin), témoin du rejeu fini-crétacé d'un accident majeur à l'échelle de la plaque africaine. *C.R. Acad. Sc., Paris*, **293**, 779–782.

——, Ousmane, B., and Robert, J. P. 1981b. Mise en évidence de déformations traduisant un raccourcissement dans le Mésozoïque de la périphérie de l'Aïr (Niger). *C.R. Acad. Sc., Paris*, **292**, 753–756.

——, Issawi, B., and Bellion, Y. 1985. Les linéaments guinéo-nubiens: un trait structural majeur à l'échelle de la plaque africaine. *C.R. Acad. Sc., Paris*, **300**, 17–20.

Joulia, F. 1963. Carte géologique du reconnaissance de la bordure sédimentaire occidentale de l'Aïr au 1/500.000. Ed. B.R.G.M.

Kazi-Tani, N. 1984. *Evolution géodynamique du segment alpin d'Algérie.* 10e Réunion Ann. Sc. Terre, Paris, résumé, 315 pp.

Klasz de, I. 1978. The West African sedimentary basins. *In*: Moullade, A. M. and Nairn, A. E. M. (Eds), *Phanerozoic geology of the World. The Mesozoic*. Elsevier. 371–399.

Laville, E. 1981. Rôle des décrochements dans le mécanisme de formation des bassins d'effondrement du Haut Atlas marocain au cours des temps triasique et liasique. *Bull. Soc. Géol. France*, **7**, 303–312.

—— 1985. *Evolution sédimentaire, tectonique et magmatique du bassin jurassique du Haut Atlas (Maroc): modèle en relais multiples de décrochements.* Thèse Sciences, U.S.T.L. Montpellier.

——, Lesage, J. L., and Seguret, M. 1977. Géométrie, cinématique (dynamique) de la tectonique atlasique sur le versant sud du Haut Atlas marocain. Aperçu sur les tectoniques hercyniennes et tardi-hercyniennes. *Bull. Soc. Géol. France*, **7**, 527–539.

Lazarenkov, Y. G. and Sherif, M. 1978. Chemical composition of the Los Pluton, Republic of Guinea. *Dokl-Akad. Nauk., SSSR*, **224**, 922–924.

Le Pichon, X. and Fox, J. P. 1971. Marginal offsets, fracture zones and the early opening of the North Atlantic. *Journal of Geophysical Research*, **76**, 6294–6308.

——, Sibuet, J. C., and Francheteau, J. 1977. The fit of the continents around the North Atlantic ocean. *Tectonophysics*, **38**, 169–209.

Louis, P. 1970. Contribution géophysique à la connaissance géologique du bassin du lac Tchad. *Mém. ORSTOM, Paris*, **42**, 311 p.

Mascle, J. 1977. Le golfe de Guinée (Atlantique Sud): un exemple d'évolution de marges atlantiques en cisaillement. *Mém. Soc. Géol. France*, **128**, 104 pp.

Mathieu, P. 1983. Le Post-Paléozoïque du Tchad. In: *Afrique de l'Ouest. Introduction géologique et termes stratigraphiques*. Pergamon. 143–146.

Mattauer, M. 1963. Le style tectonique des chaînes tellienne et rifaine. *Geol. Rundsch*, **53**, 296–313.

——, Tapponnier, P., and Proust, F. 1977. Sur le mécanisme de formation des chaînes intracontinentales. L'exemple des chaînes atlasiques du Maroc. *Bull. Soc. Géol. France*, **7**, 521–526.

Maurin, J. C., Benkhelil, J., and Robineau, B. 1985. Fault rocks of the Kaltungo lineament, NE Nigeria, and their relationship with Benue Trough tectonics. *Journal of the Geological Society, London*, **143**, 587–599.

Moreau, C. 1982. *Les complexes annulaires anorogéniques à suites anorthositiques de l'Aïr Central et Septentrional (Niger).* Thèse Sciences, Nancy I, 356 pp.

May, P. R. 1971. Pattern of Triassic–Jurassic diabase dikes around the North Atlantic in the context of prerift position of continents. *Geological Society of America Bulletin*, **82**, 1288–1292.

Ngangom, E. 1983. Etude tectonique du fossé crétacé de la Mbéré et du Djérem, Sud-Adamaoua, Cameroun. *Bull. Centres Rech. Explor.–Prod. Elf-Aquitaine*, **7**, 339–347.

Obert, D. 1981. *Etude géologique des Babors orientaux (Domaine tellien, Algérie).* Thèse Sciences, Paris VI. 635 pp.

Olivet, J. C., Bonnin, J., Beuzart, P., and Auzende, J. M. 1984. Cinématique de l'Atlantique Nord

et Central. Cartes et notice explicative. *Ed. CNEXO COB, Paris*, **54**, 108 pp.
Popoff, M., Kampunzu, A. B., Coulon, C., and Esquevin, J. 1982. Découverte d'un volcanisme mésozoïque dans le nord-est du Nigéria: datations absolues, caractères magmatiques et signification géodynamique dans l'évolution du rift de la Bénoué. *In: Rifts et Fossés Anciens, Résumé. Trav. Lab. Sci. Terre, Marseille St-Jérôme* (B), **19**, 47–49.
Pouclet, A. and Durand, A. 1983. Structures cassantes cénozoïques d'après les phénomènes volcaniques et néotectoniques au nord-ouest du Lac Tchad (Niger oriental). *Ann. Soc. Géol. Nord*, **111**, 143–154.
Radier, H. 1959. Contribution à l'étude géologique du Soudan oriental (A.D.F.). Le bassin crétacé et tertiaire de Gao. Le détroit soudanais. *Bull. Serv. géol. pros. min., Dakar*, **26**.
Ranke, U., Von Rad, U., and Wissman, G. 1982. Stratigraphy, facies and tectonic development of the on-and offshore Aaiun–Tarfaya Basin—a review. *In:* Von Rad, U. *et al.* (Eds), *Geology of the Northwest African Continental Margin*. 86–105.
Regnoult, S. N., Moreau, C., and Deruelle, B. in press. A new tectonic model for the Cameroon line. *Tectonophysics.*
Reyre, D. 1984. Remarques sur l'origine et l'évolution des bassins sédimentaires africains de la côte atlantique. *Bull. Soc. géol. France*, **6**, 1041–1059.
Robertson, A. H. F. 1984. Mesozoic deep-water and Tertiary volcanoclastic deposition of Maio, Cape Verde Islands: implications for Atlantic paleoenvironments and ocean island volcanism. *Geological Society of America Bulletin*, **95**, 433–453.
—— **and Bernouilli, D. 1982.** Stratigraphy, facies and significance of Late Mesozoic and Early Tertiary sedimentary rocks of Fuerteventura (Canary Islands) and Maio (Cape Verde Islands). *In:* Von Rad, U. *et al.* (Éds), *Geology of the Northwest African Continental Margin*. 408–525.
—— **and Stillman, C. J. 1979.** Late mesozoic sedimentary rocks of Fuerteventura, Canary Islands: implications for West African continental margin evolution. *Journal of the Geological Society, London*, **136**, 47–60.
Rohlich, P. 1978. Geological development of Jabal al Akhdar, Libya. *Geol. Rundsch.*, **67**, 401–412.
Ruellan, E. 1985. *Evolution de la marge atlantique du Maroc (Mazagan): étude par submersible, seabeam, et sismique réflexion*. Thèse, Univ. Brest.
Schlee, J. S. 1978. A comparison of two Atlantic-type continental margins. *U.S. Geological Survey Professional Paper*, **1167**, 21 pp.
Schwan, W. 1980. Geodynamic Peaks in Alpino-type Orogenies and Changes in Ocean Floor-Spreading During Late Jurassic – Late Tertiary Time. *American Association Petroleum Geological Bulletin*, **64**, 359–373.
Serralheiro, A. 1970. Geologie da Ilha de Maio. *Junta Investig. Ultramar*, Lisbonne, 103 pp.
Simon, P. and Amakou, B. 1984. la discordance oligocène et les dépôts postérieurs à la discordance dans le bassin sédimentaire ivoirien. *Bull. Soc. géol. France*, **7**, 1117–1125.
Snelling, N. J. 1965. *Overseas geol. Surv. Ann. Rept. for 1964*, 101 pp.
—— **1966.** Age determination unit. *Institute of Geological Science (London) Report*, 48–51.
Stillman, C. J., Furnes, H., Lebas, M. J., Robertson, A. H. F., and Zielonka, J. 1982. The geological history of Maio, Cape Verde Islands. *Journal of the Geological Society London*, **139**, 347–361.
Tapponnier, P. 1977. Evolution tectonique du système alpin en Méditerranée: poinçonnement et écrasement rigide-plastique. *Bull. Soc. géol. France*, **3**, 437–460.
Tempier, P. and Lasserre, M. 1980. Géochimie des massifs 'ultimes' du Cameroun: rapports entre l'évolution magamtique, l'âge et la position géographique. Comparaisons avec les 'Younger granites' du Nigéria. *Bull. Soc. géol. France*, **7**, 203–211.
Turner, D. C. and Bowden, P. 1975. •
Umeji, A. C. and Caen-Vachette, M. 1983. Rb–Sr isochron from Gboko and Ikyen rhyolites and its implications for the age and evolution of the Benue Trough, Nigeria. *Geological Magazine*, **120**, 529–650.
Uzuakpunwa, A. B. 1974. The Abakaliki pyroclastics—Eastern Nigeria: new age and tectonic implications. *Geological Magazine*, **111**, 65–70.
Van Breemen, O., Hutchinson, J., and Bowden, P. 1975. Age and origin of the Nigerian Mesozoic granites: a Rb–Sr isotopic study. *Contrib. Mineral. Petrol.*, **50**, 157–172.
Vincent, P. M. 1968. Attribution au Crétacé de conglomérats métamorphiques de l'Adamaoua (Cameroun). *Ann. Fac. Sc. Cameroun*, **1**, 69–76.
Wacrenier, P., Hudeley, H., and Vincent, P. 1958. Notice explicative de la carte géologique provisoire du Borkou–Ennedi–Tibesti au 1/1.000.000. Publ. DMG–AEF.
Wildi, W. 1983. La chaîne tello-rifaine (Algérie, Maroc, Tunisie): structure, stratigraphie et évolution du Trias au Miocène. *Rev. géol. dynam. Géogr. phys.*, **24**, 201–297.

Cretaceous deformation, magmatism, and metamorphism in the Lower Benue Trough, Nigeria

J. Benkhelil

Departement de Geologie, Universite de Dakar, Dakar, Senegal

A microtectonic analysis carried out in the Abakaliki Anticlinorium has revealed various types of deformation of different ages. Early soft-sediment deformation (normal faulting, slumping) occurred during the Albian–Pre-Santonian period and is related to active N50°E trending shear zones. A compressional tectonic phase of Santonian age was responsible for the shortening of the sedimentary cover and resulted in tight folding and axial plane cleavage. The thermal history of the Lower Benue Trough is characterized by two major episodes. A first phase (Albian) is related to the emplacement of volcanic and subvolcanic rocks. These rocks have alkaline affinities, with some characteristics of continental tholeiites. Contact metamorphism has occurred around the intrusions with the formation of andalusite in the shales. This thermal phase has been dated by the K-Ar method, giving an age of 104 Ma corresponding to the emplacement of the Wanakande syenite and associated basic sills. A second thermal phase occurred during the Santonian tectonic event resulting in low-grade metamorphism characterized by a higher illite crystallinity values, and by a paragenesis including quartz, chlorite, calcite, white mica, and epidote. A K-Ar age of 81 Ma obtained on igneous and sedimentary rocks in the Lower Benue Trough corresponds to the regional metamorphic phase contemporaneous with the Santonian tectonic episode.

KEY WORDS Benue Trough Tectonics Abakaliki anticlinorium Igneous intrusions K-Ar dating

1. Introduction

The southern part of the Benue Trough is separated from the Atlantic ocean by the important Niger delta which covers the relationship between the sedimentary basin, lying on a continental crust, and the Atlantic oceanic basin. Knowledge of the geological history of the southern part of the Benue Trough helps in reconstructing the first stages of its formation in relation to the opening of the Gulf of Guinea. The particular feature of this sedimentary basin is its final evolution into a small folded chain. The geology of the Lower Benue Trough is relatively well known from surveys of economic interest (lead/zinc, coal, petroleum) carried out since the beginning of the 20th century. The biostratigraphy of the area was established by Bain 1924; Wilson and Bain 1928; McConnell 1949; and Farrington 1952. From 1960 onward, the first syntheses were published (Reyment 1965; Cratchley and Jones 1965; Murat 1972; Whiteman 1982) and models based on the global tectonics concept were proposed Stoneley, 1966; Wright 1968; Burke et al. 1971). Despite the fact that the stratigraphy of the area is well established, no detailed descriptions of the deformation are available for the Cretaceous and Tertiary terranes. This paper is a contribution to the knowledge of the study of the deformation, the metamorphism and magmatic events in the Lower Benue Trough, during Cretaceous times.

2. Main geological and structural features

The basement underlying the Lower Benue Trough is dominated by granitic and migmatitic rocks including rare relics of metasedimentary rocks. All these basement rocks have been affected by the Pan-African orogeny around 600 +/−70 Ma (McCurry 1971). Post-Pan-African geological units lying near the Lower Benue include the Jurassic Younger Granites of Jos to the North and the Tertiary Cameroon Volcanic line to the East.

The southwestern end of the Benue Trough is concealed by the Niger Delta (Figure 1). The Mamfe basin connected to the Lower Benue is a Lower Cretaceous structure stretching in a southeasterly direction. On the other edge of the Trough, the Niger Basin (Bida Basin) is a northwest trending sedimentary basin of Maastrichtian age. In the northeast direction, the Lower Benue Trough connects with the Middle Benue and further north with the Upper Benue Trough.

The structural framework of the Lower Benue includes two main units: the Anambra Syncline and the Abakaliki Anticlinorium. The Anambra Syncline is a vast sedimentary basin trending N30°E and mainly filled by Upper Cretaceous

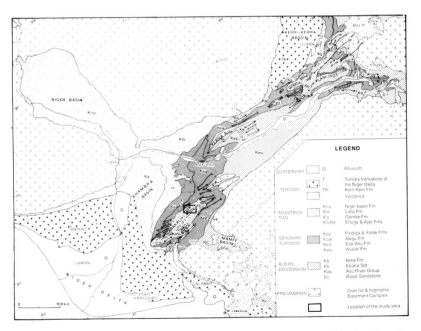

Figure 1. Main geological and structural units of the Benue Trough. X Y is the line of section of figure 2.

and Tertiary sediments. Tectonic deformations, rare in this basin, are restricted to normal faults most of them being of synsedimentary origin. Located to the southeastern edge of the Anambra Syncline, the Abakaliki Anticlinorium is a 50 km wide structure trending northeast-southwest for more than 200 km, from the Niger Delta hinge to the Gboko area. The core of the anticlinorium is occupied by the oldest sediments known in this part of the Benue Trough, the shales of the Asu River Group of Albian Age.

The Abakaliki Anticlinorium may be divided into three structural domains according to the degree and the nature of deformation (Figure 2). A narrow band, including the southeastern edge of the basin, displays monoclinal strata gently dipping toward the northwest. This domain includes sets of strike-slip faults, rare normal faults, and joints. Northwestward, the Cretaceous strata increase to form a 2500 m thick basin (Benkhelil et al. 1986). The sedimentary sequence is gently folded, but faulting is well indicated due to the dominance of shales. The northwestern edge of this domain displays an incipient fracture cleavage. Bounded to the northwest by the Anambra Syncline, the third domain occupies a central position in the anticlinorium. The sediments are tightly folded, cleaved, and intruded by igneous rocks. They are also slightly metamorphosed.

The present study has been carried out in the Workum Hills which are located 50 km northeast of Abakaliki (Figure 1). The area is hilly, contrasting with the surrounding plains which characterize the topography of the whole Lower Benue. The main relief is an elongate ridge 300 m high trending northeast-southwest incurving toward the east at its northern end. The bulk of the Asu River Group is formed here by a monotonous sequence of alternating shales and siltstones. Occasional sandstones and calcareous siltstones are interbedded in the series on the northern hillside.

3. Structural analysis

The most striking feature of this part of the Lower Benue Trough is the cleavage which locally turns clays and siltstones into slates. Folding is also a major structural element and its intensity is the most important not only within the Benue Trough but in all West African Post-Pan-African sedimentary basins of the same age. Other deformations, such as faults and shear zones, are not so prominent but are combined with folding and cleavage to give a relatively complex structural setting which is not usual in this intracontinental framework.

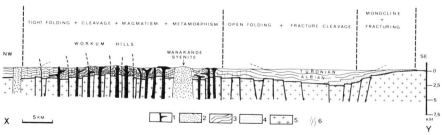

Figure 2. Cross section showing the various structural domains across the northeastern part of the Abakaliki Anticlinorium. Line of section X Y is shown on Figure 1. 1: Basic sill; 2: Subvolcanic intrusion; 3: Shale siltstone and carbonate of the Asu River Group and Eze Aku Formation; 4: Basal sandstone; 5: Precambrian Basement; 6: Cleavage.

Early deformations have been detected in the whole of the study area but are best concentrated along the northern hillside (Figure 3). These deformations include small scale faulting Figure 4A which are well displayed within thin laminated carbonate siltstones. The faults which generally affect only part of the strata are emphasized by calcite fillings. Their arrangement and geometry indicates that they have formed in a partially consolidated sediment. Slumping occurs at all

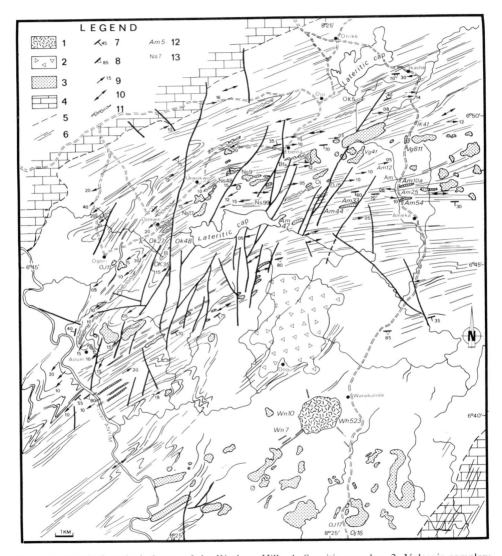

Figure 3. Sketched geological map of the Workum Hills. 1: Syenitic complex; 2: Volcanic complex; 3: Undifferentiated igneous rocks; 4: Eze Aku Formation; 5: Geological boundary between the Asu River Group and the Eze Aku Formation; 6: Trace of bedding; 7: Strike and dip of bedding; 8: Strike and dip of cleavage; 9: Direction and plunge of the intersection lineation; 10: Horizontal lineation; 11: Synsedimentary deformation; 12: Location of the samples collected for K/Ar dating and geochemical analyses; 13: Location of the deformed ammonites; AB: Line of section of the figure 6.

Figure 4. A: Normal microfaults in alternating siltstone-carbonate series, B: Microfold (probably of synsedimentary origin) with an early cleavage S', C: Cleavage/bedding relationships in a laminated siltstone, D: Cleavage and shear planes in a laminated siltstone.

levels and in all lithologies. It is much more common in siltstones (Figure 5) and calcareous pelites. In some cases an early axial plane cleavage S' may be associated with folds affecting the calcareous pelites (Figure 4B). Evidence for this early cleavage S' is given by the orientation of the pressure fringes which develop parallel to regional cleavage S1 of Santonian age. In the adjacent silty layers, the pressure fringes are found to grow within the S1 cleavage indicating the X strain direction. In the calcareous material, the S1 cleavage is only marked by spaced fractures, but pressure fringes develop parallel to the direction of S1 which is here oblique to the S' cleavage (Figure 6). On a larger scale, these deformations may be important and affect the whole strata several metres thick. The rock type is a monomictic conglomerate with a disorganized internal structure with slumping, and stretched and folded pebbles.

The distribution of the early deformations clearly indicates their relation with tectonic movements. They are all located along narrow strips near the major shear zone of the northern Workum Hills. The evidence for early tectonic movements is found along the Ukwokwu shear zone where sandstones are affected by small scale pitching drag folds. The tight folding at this small scale was most probably formed in an uncompletely lithified material rather than in brittle rocks. The early deformations are not restricted to the depositional period but may have been produced on complete lithification of the sediments. In fact, at the time of the major tectonic episode i.e. Santonian, it is possible that the lithification was not completely achieved, at least for the clay-rich rocks.

3a. Folding and cleavage

The folding of the strata and in the Workum Hills and more generally in the Abakaliki Anticlinorium is attributed to a compressional episode of Santonian age (Simpson 1954; Cratchley and Jones 1965). The folding appears progressively from

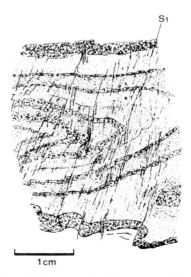

Figure 5. Synsedimentary microfold cross cut by the cleavage S_1 in a laminated siltstone (Ukwokwu area).

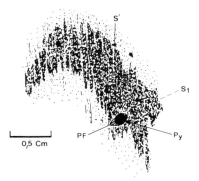

Figure 6. Synsedimentary microfold with an early axial plane cleavage S'. The pressure fringe (PF) around a pyrite grain (Py) stretching parallel to the S1 cleavage direction is posterior to the microfold.

the southeastern edge of the basin (Figure 2), first with broad open structures then passing into tight folds in the core of Workum Hills.

Folds are concentric but in shaly sequences where the cleavage is well expressed a slight tendency towards similarity is observed. Very tight folds are located along the northern hillside which is the most deformed area. The wavelength of folding may be reduced to some metres and microfolding is commonly found in the inner part of the fold hinges one metre or less in size. In the mica-rich layers, the microfolding develops a fine lineation that differs from the S_1/S_0 intersection lineation. Field cross-sections are only available along river beds but are always discontinuous. One of these sections (Figure 7) clearly shows the general asymmetry of the folding. Axial plans of folds trend northward and are in agreement with the attitude of the S_1 cleavage which has a mean dip of 80° southward.

The S_1 cleavage, related to the Santonian compressional phase, is widespread in the shales of the Asu River Group and partially in the shales and siltstones of the Eze Aku Formation. The entire Abakaliki Anticlinorium displays a cleavage which is generally a fracture cleavage marked in the shales by pencil structures. When approaching the Workum Hills, the cleavage passes to a slaty cleavage which affects the shales and the siltstones but does not show in sandstones and carbonates. Under the microscope, the cleavage appears as close-spaced planes emphasized by thin films of iron oxides. Depending on the amount of phyllosilicates, the fabric may vary from regular parallel discontinuities to a braided net of cleavage planes. In some cases, offsets of layering along the cleavage planes are observed (Figure 4C). The displacements which are sometimes important (Figure 4D and 8A) may result from a volume reduction along the cleavage planes during a process of dissolution (Gray 1979). In the northern hillside, offsets along

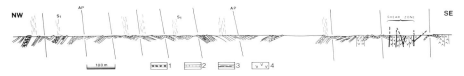

Figure 7. Field cross section along a river bed near Ojekwe (location of line of section in Figure 3). 1: Monomictic conglomerate in calcareous siltstone or pelite; 2: Sandstone; 3: Siltstone and shale; 4: Gabbro; S_1: Attitude of cleavage trace; AP: attitude of the axial plane trace.

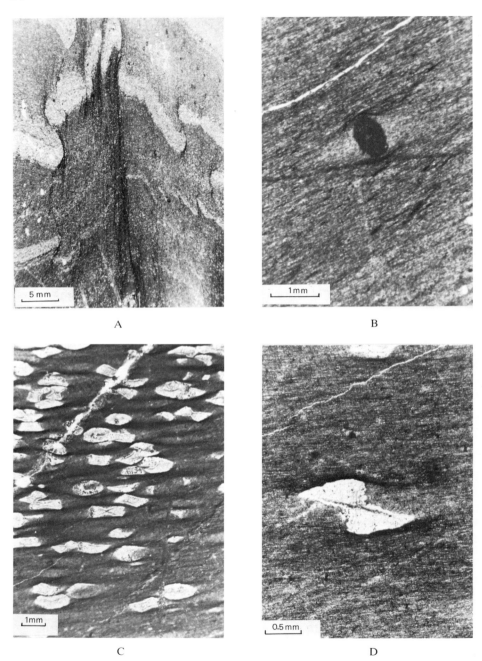

Figure 8. A: Detail of shear planes in the siltstones of the Ukwokwu area. B: Pressure shadow around a clast in a silty matrix, C: Andalusites (chiastolite variety) in a contact metamorphic aureole (Ameka area), D: Detail of a chiastolite showing the characteristic black cross and the pressure shadows around the mineral.

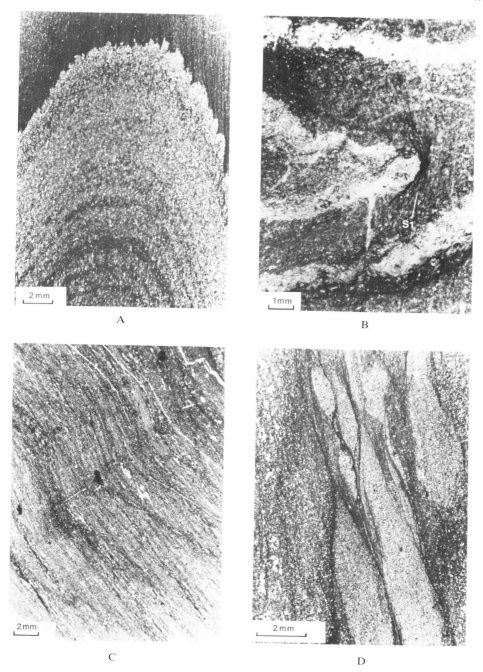

Figure 9. A: Microfold and axial plane cleavage. Note the injection of the coarse-grained material along the cleavage planes, B: Microslump hinge cross cut by the cleavage S_1 (detail of Figure 4), C: Kink band affecting the bedding and the cleavage which are parallel, D: Transposition of the sandstone layers in the shear planes (shear zone near Ukwokwu).

cleavage planes are accentuated by the proximity of shear zones. Volcanic clasts and foraminifera are enclosed by the cleavage and pressure shadows commonly develop (Figure 8B). Aggregates of quartz and phyllosilicate are the commonest minerals which recrystallize in the shadows. The most striking pressure shadows are observed in the aureoles of contact metamorphism produced by the magmatic intrusions. Contact metamorphic minerals, such as andalusite (Figure 8C) were formed prior to the major tectonism and were subsequently deformed and emphasized by the cleavage S_1 (Figure 8D).

Numerous examples of the relationship between the cleavage and the bedding are found in the fold hinges and flanks at the outcrops (Figures 4B and 9A). The cleavage is clearly related to the folding and is aligned parallel to the axial surface. The intersection between the cleavage and the bedding is often at a high angle except along the southern hillside where the strata is highly dipping. The same relationships can be observed on a smaller scale especially in the shales which have been subjected to a marked microfolding (Figure 9A). However, at this scale, confusion may arise due to the presence of abundant microfolds of soft-sediment deformation. In some cases, the bedding/cleavage relationships clearly indicate a pretectonic origin for the microfolding as shown in Figure 5 and Figure 9B. The relatively constant southward dip is associated with the slight northward asymmetry of the folding. The cleavage trend and the attitude of the intersecting lineation show variations from N50°E in most of the area to an east-west direction in the northwestern end of the Workum hills.

3b. Other structures

Structures common in the deformed areas of low-grade metamorphism are found in the Workum Hills. Kink bands, for instance, are present at the various scales (Figure 9C) and occur in all lithologies except in the carbonates and the volcanic rocks. They are late deformations affecting the cleavage. South dipping reverse faults are located north of the hills where they are generally best displayed in the shaly sequences. They confirm the general structural asymmetry and are related to the late stage of deformation as they cut across and drag the cleavage.

Shear zones trending N30°E to N50°E are visible in some rivers of the northern hillside. They may affect either the sedimentary strata or gabbroic intrusions and doleritic sills. Within the sediments, a shear zone is formed by a highly sheared central band 1 to 2 m wide flanked by two wide strips where the deformation intensity decreases away (Figure 10). The most intense deformation affects a shale–siltstone–sandstone alternating sequence and is marked by A shearing in which the bedding is almost completely vanished and transposed along the shear planes (Figure 9D). The shearing affects a sequence which bears soft sediment deformations and the thin sandstones layers are truncated (Figure 11) and re-aligned in the shear planes which display a bedding like aspect. In the flanking zones, the shearing disappears, and pitching drag folds develop in sandstones. In some instances, the shape of minor drag folds indicates a sinistral sense of movement (Figure 12).

Southeast of Ojekwe a shear zone affects a gabbroic intrusion. The progressive increase in the deformation may be observed from the edge toward the central part of the intrusion where the maximum deformation is attained. The rock is completely laminated and consequently too weathered for thin section study. Besides, the deformation is less intense and the internal structure, which displays an anastomosing network of shear planes, can be easily studied. The texture is mylonitic with elongated bands formed by chlorite and muscovite moulding sigmoi-

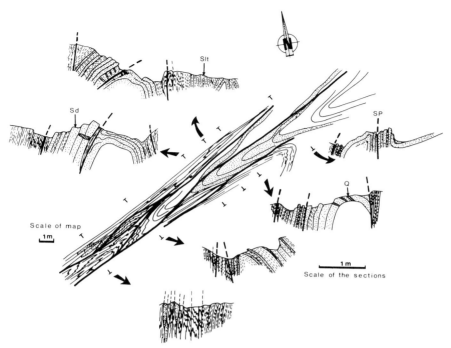

Figure 10. A shear zone near Ukwokwu. Plane view and cross sections. Slt: siltstone; Sd: sandstone; Q: Quartz vein; SP: Shear plane.

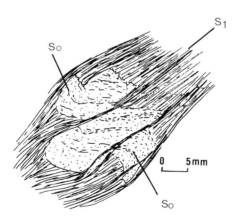

Figure 11. Truncated sandstone layer in alternating siltstone–sandstone–shale series. The layers undergo a beginning of reorientation along the shear planes.

dal-shaped amphibole porphyroblasts. The 'S–C' surfaces described by Berthe *et al.* (1979) were used as criteria for sense of movement. The resulting movement observed is sinistral and confirmed by another criterion of displaced broken grains (Simpson and Schmid 1983). In this case, the grains are amphibole porphyroblasts which break along microfractures oblique to the foliation plane. Since the mapping

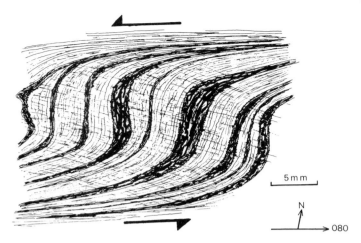

Figure 12. Minor drag fold in a laminated siltstone indicating a sinistral sense of movement.

of the shear zones was not possible due to the bad field conditions, the extent of the shear zone and the amount of displacement could not be evaluated.

3c. Determination of the strain ellipse from deformed ammonites

The cleavage is particularly well expressed in an elliptical-shaped zone, slightly sigmoidal, 60 km by 30 km, stretching in a mean N50°E direction. Such a large area affected by a slaty cleavage must have undergone a noticeable flattening. The observation indicates that some markers included in an incompetent matrix have been flattened. Fossils were found to be good markers, especially ammonites which were abundant and easy to use for the evaluation of the strain ratio. A statistical study of this ratio has been performed on several populations collected in various areas of the anticlinorium using the method proposed by Quiblier (1980). The results shown in Figure 13 indicate a relatively constant mean axial ratio of the strain ellipse for the fossils of the Workum Hills with, however, important variations within a same group of measurements. The reason is that fossils recorded in one outcrop of finely laminated siltstones several square metres in size may be made of material which is different in composition and grain size. Therefore, the mechanical behaviour during the deformation was different according to the matrix composition. Other examples of differential mechanical behaviour are seen in alternating shale and siltstone sequences where cleavage refraction or cleavage fans are common on all scales. All the fossils used in this study have the same matrix composition and texture. The calculated strain ellipse is therefore fairly representative of the total deformation undergone by the rocks. The average axial ratio of the strain ellipse is 1·8 in the Workum Hills. In a first approximation, the finite strain ellipse may be considered as the result of a pure shear. The method used here for the determination of the strain ellipse does not allow one to distinguish the possible influence of a rotational component.

Other evidence indicates that the finite strain is rotational. Syntectonic pressure fringes which have grown around pyrite grains have a slight sigmoidal shape indicating a rotation during the process of crystallization (Figure 14A). The strain ratio values obtained on ammonites collected 20 km northeast of the Workum Hills are much lower and are in agreement with the weaker deformation observed in this area. In the southwestern end of the anticlinorium the values obtained are

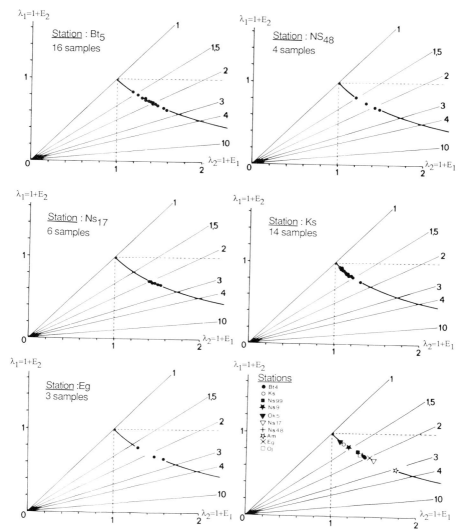

Figure 13. Strain ellipse axes ratio for deformed ammonites from the Workum Hills (Bt5, NS17, NS48), Northeast (KS) and Southwest (Eg) of the Abakaliki Anticlinorium (Graph from Ramsay, 1967).

relatively high but are still compatible with the tight folding in the dominantly shaly series of the Asu River Group.

4. Magmatism

Igneous rocks occupy an important place in the geological setting of the Workum Hills. Numerous intrusive bodies form scattered outcrops south of the Workum Hills. The rock types ranging from basic to intermediate in composition consist of gabbro, monzonite, diorite, and syenite. Their volcanic equivalents are mainly

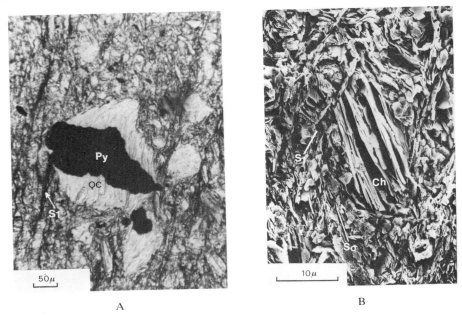

Figure 14. A: Sigmoidal pressure fringes formed by quartz and chlorite (QC) around a pyrite grain (Py) stretching parallel to the cleavage S1. B: Pretectonic chlorite stack (Ch) parallel to the bedding (So). The cleavage S_1, secant on the stack, is emphasized by syntectonic phyllosilicates (SEM).

basalt and trachyte. The igneous bodies may occur commonly as discordant intrusive stocks, or as swarms of small-sized sills or dykes.

Within the hills, the dominant rock type is gabbroic in composition occuring as massive concordant lenses interbedded in the shales. A small circular pluton, 2 km in diameter located near Wanakande, is mainly composed of syenite. Mineralogically, the syenite is a grey, nepheline–alkali feldspar rock with mafic minerals such as aegirine–augite, kaersutite, and some biotite. Sphene and apatite may be abundant. In the same massif, titanomagnetite and a brown amphibole with a green rim can be observed. The texture is medium-grained with locally a discrete vertical layering and cumulus fabric. The diorites occur as small intrusive bodies. The rock is generally medium-grained and greenish, consisting of weakly zoned oligoclase and brown and green amphibole. The accessories include biotite, opaques, and titanomagnetite. Gabbros are dark-green massive rocks occurring as large stocks. In one occurrence, magmatic layering was observed near Ojekwe consisting of an accumulation of amphibole. Plagioclase, titaniferous augite, and brown amphibole are the essential minerals. Accessories include ilmenite, apatite, and in one instance, a few sodalite grains were observed.

The igneous activity also includes a volcanic phase characterized by tuffs, agglomerates, breccias, and lava flows. Near Dogu, an important massif consists of pyroclastic material with interbedded lava flows. Volcanic activity is contemporaneous with the deposition of Albian sediments. Volcanic ashes are found in all rock types forming thin interlayers or generally appear scattered within the groundmass of the sedimentary rocks.

Chemical analyses were performed on 46 samples of igneous rocks collected along the Abakaliki Anticlinorium in the vicinity of the Workum Hills. The analyses indicate that most of the rocks have alkaline affinities (Benkhelil 1986), with many of the analysed rocks containing normative nepheline. All the basic rocks have normative olivine and some show normative hypersthene. When plotted in the alkali/silica diagram (Figure 15) most of the rocks are located in the alkaline domain defined by Miyashiro (1978) and Irvine and Baragar (1971). For comparison, analyses of the igneous rocks from the Ugep area (Hossain 1981) were plotted on the same diagram. However the tholeiitic character of part of the magmatism of Ugep–Afikpo area cannot be extended to the whole anticlinorium. In the AFM diagram (Figure 16), the rock analyses define a continuous trend parallel to the boundary between the calc-alkaline and tholeiitic domains as defined by Irvine and Baragar (1971). Monzosyenitic rocks plot in this zone. The Fe-enrichment of the alkaline series with MgO relatively constant, is generally interpreted as a tholeiitic trend but this statement is contested (Bowden 1985). Other geochemical characteristics (Benkhelil 1986) indicate that the igneous rocks are sodic-rich and have a relatively high content of Al_2O_3. A slight bimodal tendency with alkaline and aluminous trends characterizes the syenites of Wanakande. The rocks of the southwestern anticlinorium fall partly into the field of continental tholeiite and become more alkaline northeastward.

The magmatism of the Lower Benue indicates pronounced alkaline undersaturated affinities. However, the analysis of EG30 situated in the southwestern end of the Anticlinorium confirms the tholeiitic tendency found by Hossain (1981) for some of the Ugep–Afikpo magmatic rocks. The magmatism in the Lower Benue therefore belongs to same evolution trend resulting from fractional crystallization of a transitional or mildly alkaline basaltic magma.

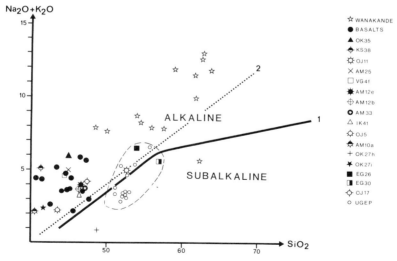

Figure 15. $Na_2O–K_2O/SiO_2$ diagram for the igneous rocks of the Workum Hills, Eziator-Eziagu (EG26–Eg30) and Konshisha river (KS38, located in figure 1); Wanakande includes the analyses of the syenitic massif; Basalts include the lava flows and sills from the Workum Hills — Dogu area (location in figure 3); UGEP corresponds to the analyses of the Ugep area (HOSSAIN, 1981).

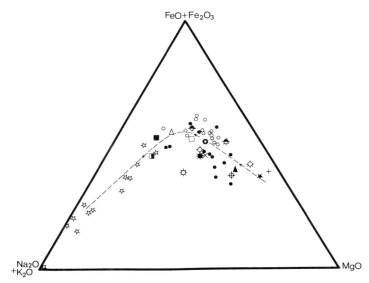

Figure 16. AFM diagram. The dashed line separates the tholeiitic (above) from the calc-alkaline (below) domains after Irvine and Baragar (1971).

5. Metamorphism

The induration of the Albian sediments results from the combined action of tectonic movement and low-grade metamorphism. The latter was intepreted as resulting from contact metamorphism due to the intrusion of igneous rocks (Shell B.P. 1957). In fact, the thermal history of the area is more complex and may be divided into two distinct episodes: an early phase of Albian age characterized by a very low-grade regional metamorphism overprinted by extensive contact metamorphism; and a further period of low-grade regional metamorphism contemporaneous with the Santonian tectonic event.

Regional metamorphism is best displayed in the shaly facies. It is characterized by the crystallization of fine mica flakes parallel to the S_1 cleavage planes. Measurements of illite crystallinity across the hills have been plotted in Figure 17. The crystallinity index varies from 1·5 to 8. Most of the samples collected in the Workum Hills therefore lie in the anchizonal domain defined by Esquevin (1969) belonging to the epizone. Index values greater than 4 are found in samples located either north or south of the Workum Hills. A composite section (Figure 18) shows the variation of the crystallinity index across the Workum Hills and the evolution of the cleavage intensity. There is a good correlation between the cleavage distribution and the crystallinity indices of which highest values coincide with the most deformed part of the hills. The increase in crystallinity is progessive from south to north following the cleavage evolution. To the north, the passage from metamorphosed sediments to unmetamorphosed sediments is sharp and the curve shows the same asymmetry observed for the structures. In the northern hillside which is the most affected area, syntectonic crystallization is common in the shales, siltstones, and carbonates. The mineral assemblage includes quartz, chlorite, white mica, and calcite. Quartz and chlorite are the commonest minerals and occur

CRETACEOUS DEFORMATION, MAGMATISM AND METAMORPHISM

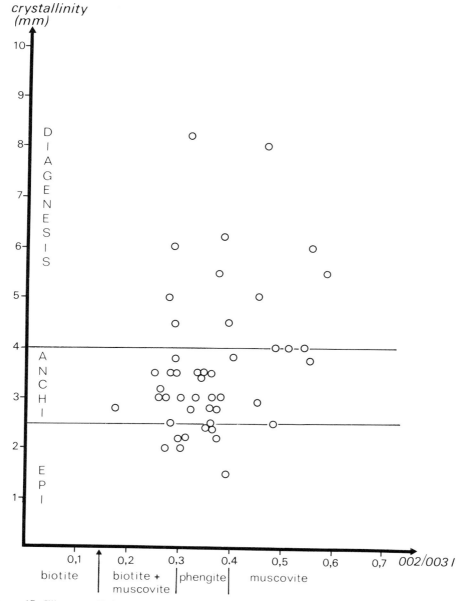

Figure 17. Illite crystallinity for samples of the Workum Hill area (Diagram after Esquevin, 1969).

mainly as pressure fringes around pyrite or iron oxides (Figure 19 and Figure 14A).

Contact metamorphic aureoles are particularly widespread especially along the southern hillside near Ameka. The aureoles are related to the igneous intrusions emplaced during the Albian magmatic phase. At Ameka, a 30 m wide aureole, displays spotted slates. The mineral forming the spots is andalusite (Figure 8C)

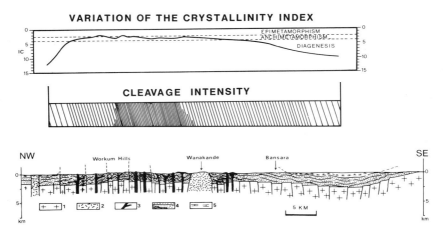

Figure 18. Composite section across the northeastern Abakaliki Anticlinorium showing the variation of the crystallinity index and the distribution of the cleavage intensity. Note the general asymmetry toward the Northwest. Legend of the geological cross section; 1: Basement Complex; 2: Subvolcanic intrusion; 3: Basic sill; 4: Shales, siltstone and sandstone of the Asu River Group; 5: Shale and carbonate of the Eze Aku Formation.

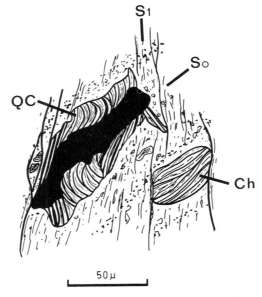

Figure 19. Sigmoidal quartz-chlorite (QC) pressure fringes around a pyrite grain. The S1 cleavage is deflected around a pretectonic chlorite stack (Ch). So: bedding.

with black crosses of organic inclusions characteristic of the chiastolite (Figure 8D). The minerals are only recognizable by their crystallographic shape as they have undergone retrograde metamorphism during the Santonian–Campanian. Outside of the contact metamorphic aureoles, the shales and siltstones commonly bear stacks of chlorite which are mainly pretectonic (Figure 19 and Figure 14B). The stacks are generally intergrowths of chlorite and white mica. There are many theories proposed to explain their formation (Craig et al. 1982). In the present case, their origin is most probably related to diagenesis and low-grade metamorphism before the onset of deformation as proposed by Craig et al. (1982), through minetic growth on a primary bedding fabric. The restricted distribution of these stacks to the Workum Hills area is evidence for their metamorphic origin. This very low-grade metamorphic phase is probably contemporaneous with the Albian magmatic event and both are interrelated.

5a. Geochronology

K-Ar dating has been completed on several igneous rocks from Workum Hills representing part of the Abakaliki Anticlinorium. The <2 μ fraction of the Cretaceous sediments were also analysed by the same method. From the analyses on four samples of basic and intermediate rocks (Table 1) and 12 samples of sediments (Table 2) the following conclusions can be made. During Albian times (104 Ma on amphiboles), the Wanakande syenitic complex and associated dykes and sills intrude the sediments producing contact metamorphic aureoles. The whole rock ages obtained on the basic rocks (Figure 20) are aligned along an 80 Ma isochron (Lower Campanian) which is the age of the low-grade metamorphic phase contemporaneous with the Late Santonian tectonic phase. Retrograde metamorphism of 'greenschist' facies (albite, green amphibole, epidote) or characterized by phyllosilicates and calcite, has affected the igneous rocks, but no age for this event is currently available. Field evidence and microtectonic analysis indicate that the retrograde metamorphism is pretectonic and certainly related to the Albian phase.

An age of 94 Ma obtained on three samples from the Workum Hills, may reflect the effects of contact metamorphism related to basic intrusions. But one must not exclude a possible rejuvenation due to the Santonian–Campanian thermal phase. Samples located in the northern and southern hills give an age of $83 +/- 3$ Ma (Figure 21) corresponding to the Santonian–Campanian metamorphism. The two samples collected outside the Workum Hills indicate an age of 81 Ma which is also related to the metamorphic episode.

Furthermore the age of 41 Ma (Figure 21) obtained on sediments located near a basaltic intrusion at Ohana is in agreement with the Tertiary age commonly attributed to the Ohana massif, part of the Cameroon Volcanic line.

6. Interpretation and conclusions

The Cretaceous history of the Abakaliki Anticlinorium shows tectonic, sedimentary, and magmatic events. The tectonic activity was more or less continuous since the formation of the basin in Aptian–Albian times until the Santonian paroximal phase which has definitively sutured this part of the Lower Benue Trough. The tectonic regime seems to have varied between the Albian and the Santonian. During the deposition of the Asu River Group, a general extensional regime prevailed in the Abakaliki Trough where synsedimentary deformations occur along N50°E trending fault zones, one being located along the northern Workum Hills. The transcurrent nature of the movement along the fault zone is obvious but the sinistral sense of movement during the sedimentary phase can only be inferred from microtectonics. At the same time, an important magmatism occurred concen-

Table 1. K/Ar analytical data for the Wanakande syenite and an associated basic dyke.

Samples and fractions	K%	$^{40}Ar^*$	$^{40}Ar/^{36}Ar$	$^{40}K/^{36}Ar$ 10^{-5}	Air%	Apparent age
Wanakande syenite						
RT (100–180μ)	2.20	7.407	2968	5.412	10	84.6
0.4A (100–180μ)	1.90	8.099	2037	2.785	15	106.5
0.5A (60–100μ)	1.62	6.768	6462	10.064	5	104.4
2A (100–180μ)	2.76	9.833	2488	4.195	12	89.4
2A (60–180μ)	2.54	9.164	1746	2.740	17	90.6
RT (100–180μ)	2.10	7.808	1693	2.562	18	93.2
RT (100–180μ)	3.94	11.817	1806	3.435	16	75.6
0.3A (60–100μ)	4.66	17.796	2117	3.252	14	95.7
2A (100–180μ)	4.00	12.512	2596	5.015	11	78.8
2A (60–100μ)	3.74	11.933	1590	2.767	19	80.3
Basic dyke						
RT (100–180μ)	0.34	1.234	500	0.385	59	91.1
RT (100–180μ)	0.40	1.675	989	1.139	30	104.7

RT: whole rock A: mineral.

Table 2. K/Ar analytical data for sediments of the Workum Hills (Am33, Am44, Am47, Am54, Oj16, Ok27e, Ok48, VG211), Bansara (BAN22) and Ohana (Oj23, Oj27) (^{40}Ar*: radiogenic argon)

Samples	Fraction (μ)	K%	^{40}Ar* 10^{-6}	^{40}Ar/^{36}Ar	^{40}K/^{36}Ar 10^{-5}	Air%	Apparent age
OJ23	<0.3	5.6	18.486	2187	3.907	14	81.5
OJ27	<0.3	4.2	6.927	1823	6.210	16	41.2
BAN22	<0.3	3.4	10.77	1209	1.934	24	78.3
OJ16	<0.3	3.5	11.873	781	0.920	38	90.3
VG811	<0.3	4.9	18.261	2934	4.746	10	91.7
OK27e	<0.3	2.8	10.05	791	0.925	37	88.4
Am33	<0.2	4.37	0.433	7602	13.26	4.3	92.4
Am33	0.2–0.5	4.84	0.522	11160	18.12	2.65	100.7
Am44	<0.2	2.63	0.253	2560	4.236	11.5	89.7
Am44	0.2–0.5	3.67	0.375	4457	7.333	6.60	95.1
Am47	<0.2	3.43	0.330	3384	5.776	8.73	89.7
Am47	0.2–0.5	3.58	0.397	5099	7.787	5.49	103.1
Am54	<0.2	3.90	0.353	10190	19.67	2.9	84.5
Am54	0.2–0.5	4.78	0.472	12890	22.96	2.37	92.0
OK46	<0.2	4.15	0.378	2461	4.279	12	85.0
OK46	0.2–0.5	4.96	0.454	3024	5.367	9.37	85.4
OK48	<0.2	1.09	0.196	402	0.108	73.50	90.5
OK48	0.2–0.5	1.10	0.120	415	0.197	71.1	101.6

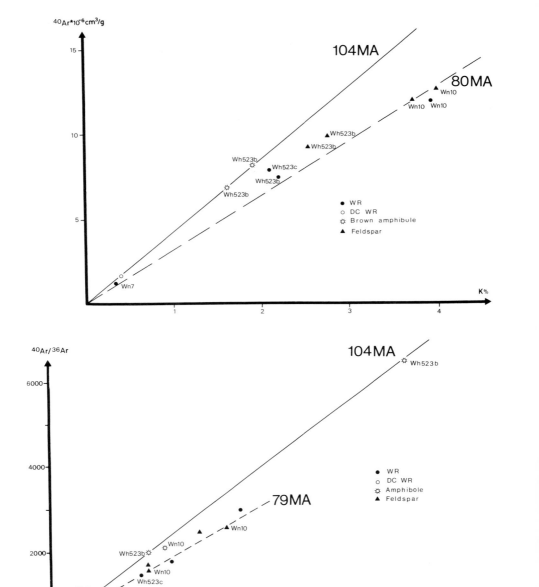

Figure 20. K/Ar diagrams for the Wanakande syenites (Wh523b, Wh523C, Wn10) and an associated basic dyke, Wn7).

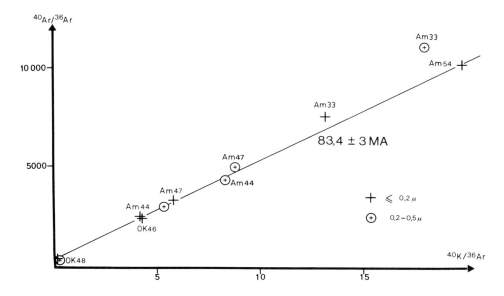

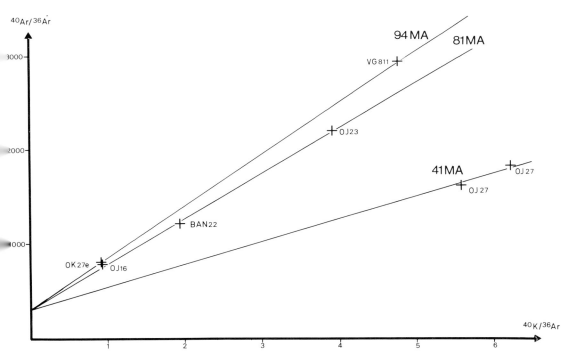

Figure 21. $^{40}Ar/^{36}Ar$ versus $^{40}K/^{36}Ar$ diagrams for the fine fraction of sediments from the Workum Hills (Am33, Am44, Am47, Am54, Ok27e, Oj16, Ok46, Ok48, VG811), Bansara (BAN 22), and Ohana (Oj23, Oj27). Sample locations in Figures 1 and 3.

trated in the Workum Hills. The main structural trend N50°E controlling the emplacement of intrusions, visible at surface, reflects the deep-seated lineaments in the Pre-Albian basement. However, the overall distribution of most of the igneous intrusions does not show a clear alignment. The geographic distribution of the intrusions was analysed using an autocorrelation method (Leistel 1984) based on 104 intrusive bodies in an area of 700 square kilometres centred on the Workum Hills region. The resulting image displays an S-shape with an elongate central area (Figure 22A). The arrangement of the igneous bodies defines a sigmoidal pattern (Figure 22B) which can be integrated with the general wrenching model already proposed for the opening of the Benue Trough (Benkhelil 1982; Benkhelil and Robineau 1983). Considering the Workum Hills as a model for basins formed along strike slip faults, this area would correspond, during the Albian times, to a depression that formed at a 'releasing bend'. Stretching and sagging of the crust were responsible for the intensive igneous activity, high heat flow, and thick sedimentation observed in this area.

The distribution and the relationships between the various structural elements of the Workum Hills show a relative homogeneity resulting from a single compressive

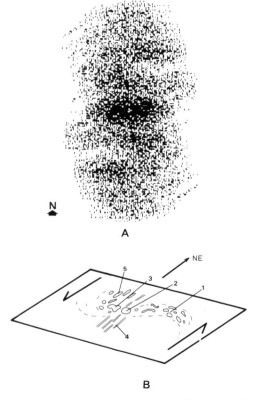

Figure 22. Autocorrelation analysis of the magmatic bodies distribution in the Workum Hills. A: Image obtained by computing the geographical location of 104 igneous bodies; B: Interpretation of the image. The general sigmoidal shape corresponds to a large scale tension gash within a sinistral wrench zone. 1: South Wanakande area; 2: Wanakande complex; 3: Dogu complex; 4: Basic sills of the River Anyim; 5: Basic intrusions of the hills.

phase. This phase of Santonian age, has acted in a NNE–SSW direction. However, in the Workum Hills area, the S_1 cleavage and the lineation with a mean N50°E trend show a clockwise rotation toward the east near Ameka. The resulting geometry demonstrates a general sigmoidal shape centred on the Workum Hills (Benkhelil *et al.* 1986). This arrangement is interpreted as a large scale shear lens with the most severe deformation and the highest crystallinity indices located inside. The existence of a sinistral shear component during compression is probable especially along the shear zones of the Workum Hills. The shortening calculated from folds is evaluated at 10 km for the anticlinorium. The thermal history of the area is marked by the existence almost permanently from Albian to Campanian times of a high gradient with a maximum at the end of the Santonian (Figure 23). An early Campanian phase forms part of the main tectonic phase. If the effects of the Albian metamorphism are relatively restricted, those of the Santonian–Campanian phase are much more widespread and still present 20 km away on the anticlinorium flanks.

In the general framework of the Lower Benue Trough, such important deformations associated with the N50°E trend and located in a narrow band, confirm the existence of deep-seated basement discontinuities. They have played an essential role during the opening and the closure of the basin and were pathways for the intrusion of magmatic rocks. The period of opening of the Benue Trough was in Albian times, when the Gulf of Guinea was also opening and the major equatorial fault zones were initiated. The relationships between these discontinuities of the continental basement and the incipient oceanic fracture zones have been discussed (Mascle 1977; Wilson and Williams 1979; Benkhelil 1982). It appears that transcurrent movement was the basic role of these major faults rather than a pure rifting mechanism for the Benue Trough. Recently, it has been shown that the sub-basins of the Upper Benue Trough have formed along transcurrent fault zones during Aptian–Albian times (Maurin *et al.* 1986).

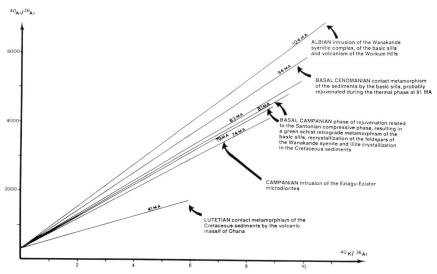

Figure 23. Schematic K/Ar diagram showing the age of the various magmatic and metamorphic events of the Lower Benue Trough.

Acknowledgements. I thank Mr F. Walgenwitz and SNEA (P) for the K/Ar dating analyses and for permission to publish the results.

References

Bain, A.D.N. 1924. The Nigerian Coalfield, Section 1, Enugu area, *Bull. geol. Surv. Nigeria*, **6**, 81 p.
Benkhelil, J. 1982. Benue Trough and Benue Chain, *Geol. Mag.*, **119**, 155–168.
——, 1986. Caractéristiques structurales et évolution géodynamique du bassin intracontinental de la Bénoué (Nigéria), Thèse d'état, Nice.
——, Dainelli, P., Ponsard, J.F. Popoff, M. and Sougy, L. 1986. The Benue Trough: Wrench fault related basin on the border of the Equatorial Atlantic, in Triassic-Jurassic Rifting, AAPG Memoir (in press).
——, and Robineau, B. 1983. Le fossé de la Bénoué est-il un rift?, *Bull. Centres Rech. Explor. -Prod. Elf-Aquitaine*, **7**, 315–321.
Berthé, D., Choukroune, P. and Jegouzo, P. 1979. Orthogneiss, mylonite and non coaxial deformation of granites: the example of the South Armorican Shear Zone, *J. Struct. Geol.*, **1**, 31–42.
Bowden, P. 1985. The geochemistry and mineralization of alkaline ring complexes in Africa (a review), *J. Afr. Earth Sci.*, **3**, 17–39.
Burke, K.C.A., Dessauvagie, T.F.W. and Whiteman, A.J. 1971. Opening of the Gulf of Guinea and geological history of the Benue Depression and Niger Delta, *Nature*, **233**, 51–55.
Cratchley, C.R. and Jones, J.P. 1965. An interpretation of the geology and gravity anomalies of the Benue Valley, Nigeria, *Overseas geol. Surv. Geophys. Paper* **1**, 1–26.
Craig, J., Fitches, W.R. and Maltman, A.J. 1982. Chlorite-mica stacks in low-strain rocks from central Wales, *Geol. Mag.*, **119**, 243–256.
Esquevin, J. 1969. Influence de la composition chimique des illites sur leur cristallinité, *Bull. Centre Rech. Pau – SNPA*, **3**, 147–153.
Farrington, J.L. 1952. A preliminary description of the Nigerian lead-zinc field, *Econ. Geol.*, **47**, 583–608.
Gray, D.R. 1979. Microstructure of crenulation cleavages: an indicator of cleavage origin, *Am. J. Sci.*, **279**, 97–128.
Hossain, M.T. 1981. Geochemistry and petrology of the minor intrusives between Efut Eso and Nko in the Ugep area of Cross River state, Nigeria, *J. Min. Geol.*, **18**, 42–51.
Irvine, T.N. and Baragar, W.R.A. 1971. A guide to the chemical classification of the common volcanic rocks, *Canad. J. Earth Sci.*, **8**, 523–548.
Leistel, J.M. 1984. Evolution d'un segment de la chaîne hercynienne dans le nord-est du Massif Central français — le fossé volcano-tectonique de la Loire —, Thèse de Doctorat, Nancy.
Mascle, J. 1977. Le Golfe de Guinée (Atlantique Sud): un exemple d'évolution de marges atlantiques en cisaillement, *Mém. Soc. Geol. Fr.*, **128**, 104 p.
Maurin, J.C., Benkhelil, J. and Robineau, B. 1986. Fault rocks of the Kaltungo lineament, NE Nigeria and their relationship with Benue Trough tectonics, *J. Geol. Soc. London*, **143**, (in press).
McConnell, R.B. 1949. Notes on lead-zinc deposits of Nigeria and Cretaceous stratigraphy of Benue and Cross River valleys, *Geol. Surv. Nig. Report*, **752**.
McCurry, P. 1971. Pan-african orogeny in Northern Nigeria, *Geol. Soc. Amer. Bull.*, **82**, 3251–3261.
Miyashiro, A. 1978. Nature of alkalic rock series, *Contr. mineral. Petrol.* **66**, 91–104.
Murat, R.C. 1972. Stratigraphy and Palaeogeography of the Cretaceous and Lower Tertiary in Southern Nigeria. In: Dessauvagie, T.F.J. and Whiteman, A.J. (Eds), *African Geology*, pp. 251–276, Ibadan University Press, Ibadan.
Quiblier, J. 1980. Quels renseignements peut-on tirer d'une ammonite déformée?, *Rev. Inst. franc. Pétrole*, **35**, 737–738.
Ramsay, J.G. 1967. *Folding and Fracturing of Rocks*, McGraw-Hill, New York.
Reyment, R.A. 1965. *Aspects of the Geology of Nigeria*, Ibadan University Press, Ibadan.
Shell, B.P. 1957. Geological maps, 1:250 000 Sheets; Makurdi (64); Ankpa (63); Enugu (72); Ogoja (73); Umuahia (79); Oban Hills (80); Calabar (85); *Geological Survey of Nigeria*.
Simpson, A. 1954. The Nigerian Coalfield: the geology of parts of Owerri and Benue Provinces, *Bull. geol. Surv. Nigeria*, **24**, 85 p.
Simpson, C. and Schmid, S.M. 1983. An evaluation of criteria to deduce the sense of movement in sheared rocks, *Geol. Soc. Amer. Bull*, **94**, 1281–1288.
Snelling, N.J. 1965. *Overseas geol. Surv. Ann. Rept* (for 1964), p. 101.
Stoneley, R. 1966. The Niger delta region in the light of the theory of the continental drift, *Geol. Mag*, **103**, 385–397.
Whiteman, A. 1982. *Nigeria: Its Petroleum Geology, Resources and Potential*, Graham and Trotman, London.
Wilson, R.C. and Bain, A.D. 1928. The Nigerian coalfield. Section II. Parts of Onitsha and Owerri Provinces. With an appendix by L.F. Spath on the Albian Ammonoidea of Nigeria, *Bull. geol. Surv. Nigeria*, **12**, 54 p.

Wilson, R.L. and Williams, C.A. 1979. Oceanic transform structures and the development of Atlantic continental sedimentary basins — a review. *J. geol. Soc. Lond.*, **136**, 311–320.
Wright, J.B. 1968. South Atlantic continental drift and the Benue Trough. *Tectonophysics*, **6**, 301–310.

A model for rift development in Eastern Africa

Jean Chorowicz
Département de Géotectonique, Université Pierre et Marie Curie, 4 Place Jussieu, 75252 Paris Cedex 05, France

Jacques Le Fournier
Laboratoire de Geologie, SNEA (P) Boussens, 31360 St Martory, France

and

Gerard Vidal
Departement de Geotectonique, Universite Pierre et Marie Curie, 4 Place Jussieur, 75252 Paris Cedex 05, France

---

A tectonic survey has been completed in Eastern Africa, using remote sensing as well as structural analysis in the field. New data have been collected in the eastern and western branches of the East African Rift system.

The geomorphology suggests that the main deformation occurred in the two principal branches of the East African Rift system, but compilations of earthquake data show that recent movements are widespread over a large area in Eastern Africa. Many ancient structures have been reactivated during Neogene times, especially large Precambrian lineaments which are evident on satellite imagery. The Tanganyika–Rukwa–Malawi lineament, along its segment between Lakes Tanganyika and Malawi, now serves as an intracontinental transform zone, with folds and transcurrent faults. The Aswa lineament is, at the present time, only partly reactivated between Lake Mobutu and the Gregory Rift, and also along its southeastern continuation to the Indian Ocean.

In many areas, tension gashes or striations on the slickensides have given the local palaeostress orientation. It appears to be changing with time, but the most typical orientation indicates horizontal extension striking NW–SE or WNW–ESE. At the earliest stage of rifting compression was horizontal, then it later became vertical. Compilation of earthquake foci from publications, confirms these results.

All these observations together with data concerning the geometry of the deformation, as well as the mechanisms of the Cenozoic intracontinental deformation, implies a near NW–SE movement of the Somalian block, relative to the African continent. A dynamic tectonic model for the rift opening and propagation of the main fracturing across Africa is proposed. It suggests that opening is more important in the north than in the south. Four stages can be identified in the evolution of the East African Rift. The pre-rift stage corresponds to a dense fracturing, characterized by strike-slip faults, while a shallow but wide depression formed, and open tension gashes gave way to tholeiitic volcanism. The initial rifting stage is recognized by oblique-slip faults, bounding tilted blocks, but the uplift of the rift shoulders is not very important. The following 'typical' rift formation stage is associated with normal faults bordering the main tilted blocks, while subsidence of the rift floor and uplift of the shoulders are important. The advanced rifting stage, shown by important magmatic intrusions along the rift axis, corresponds to the initial formation of oceanic crust. These stages correspond to different present day aspects of various rift segments in Eastern Africa.

KEY WORDS   Rifting   East Africa   Western Rift   Remote sensing   Palaeostress

0072–1050/87/TI0495–19$09.50
© 1987 by John Wiley & Sons, Ltd.

1. Introduction

The East African Rift has been the subject during the last ten years of tectonic studies by satellite remote sensing analysis (Mohr 1974; Chorowicz and Mukonki 1979; Mukonki 1980; Chorowicz and Mukonki 1980; Chorowicz 1984), seismic profiling (Rosendhal and Livingstone 1983; Ebinger et al. 1984; Le Fournier et al. 1985), and structural analysis in the field (Chorowicz et al. 1983). The present survey has been focussed, firstly along the north–south segments of the Western branch and in the Kenyan part of the Eastern branch, and secondly along the NW–SE striking transverse structures such as the Aswa lineament, the Tanganyika–Rukwa–Malawi lineament, and the Zambezi lineament (Figure 1). Following interpretation of aerial photographs (Hepworth 1967) and using Landsat MSS and RBV satellite images, remote sensing analysis has produced new structural maps covering large areas in eastern Africa. Due to the synoptic view offered by these images, ancient lineaments can be observed. Despite the presence of deeply weathered soil and a thick vegetation cover, geological interpretation of these images is possible thanks to a good geomorphological expression of the structures. Seismic profiling on the East African lakes has given data concerning the geometry of faults and folds, evolution of deformation with time, basin structures, and sedimentary layering. Structural analysis in the field has been carried out at a large number of outcrops which were easier to locate after examination of satellite imagery. Precise study of tension gashes and striations on the slickensides allows one to understand the mechanisms of faulting, to determine the palaeostress orientation, and to know the evolution of stress with time. The 'right dihedral' method (Angelier and Mechler 1977) has been used.

This new information concerning the rift geometry and the trends of the successive palaeostress fields, together with the support of new data from the literature concerning seismic activity of the rift and the focal mechanisms of earthquakes, allows us to propose a new model for the opening of the East African Rift system.

2. Rift geometry

Initial ideas about the evolution of the East African Rift (Gregory 1896) were first of all due to geomorphological observations, which outlined the two obvious eastern and western branches. Each branch is made of a series of successive grabens which can be considered as the 'main' grabens. It is necessary to add the transverse structures shown by satellite images and also the less evident 'subsidiary' structures.

2a. The main grabens

The main grabens are the most evident from the morphotectonic point of view. They result from a long structural evolution which began in Miocene or Late Oligocene times, and continued through to the present time. In addition to Afar and the Ethiopian Rift (Figure 1), the main grabens are the Gregory Rift and, in the western branch, Lakes Mobutu, Edward Edouard, Kivu, Tanganyika, and Malawi. It is necessary to add to this list the Kerimbas and Lacerda basins (Figure 1), discovered offshore from the Tanzanian coastline by Mougenot et al. (1986). Most of the grabens have a north–south trend.

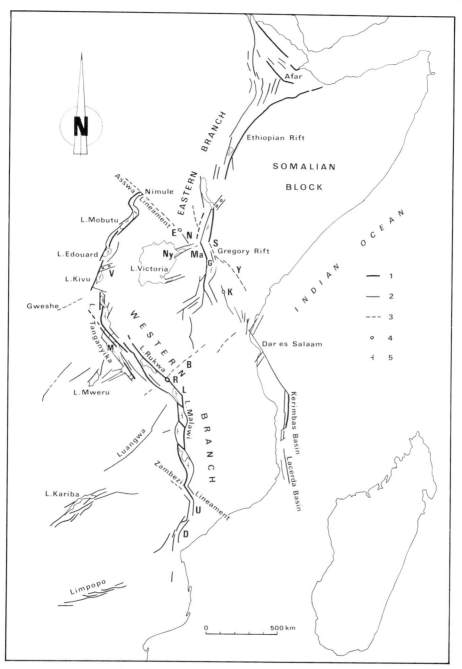

Figure 1. Structural map of the East African Rift. (1) major fault zone, presently active; (2) other active fault zone; (3) fault, inactive at present time; (4) major volcano; (5) major dip. B: Buhoro Flats; D: Dombe trough; E: Elgon volcano; G: Gilgil fault; K: Kilimanjaro; L: Livingstone fault; M: Mahali Mounts; Ma: Mau scarp; N: Nandi fault; Ny: Nyanza rift; R: Rungwe volcano; S: Sattima scarp; U: Urema trough; V: Virunga; Y: Yatta Plateau. Edouard ≡ L. Edward; Asswa ≡ Aswa

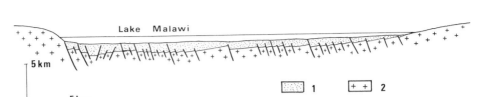

Figure 2. Cross-section of Lake Malawi, from a seismic profile located in the northern part of the lake. (1) recent sediments; (2) basement.

Each of these main grabens is asymmetrical (Figure 2). One side (W) is bordered by a zone of large unconformable normal faults whose throw is, sometimes, more than 3000 m. The other rift margin (E) is wider and is composed of tilted blocks, whose surface dips towards the rift axis, and which is offset by smaller normal faults. The other side (E) then has the general pattern of a faulted flexure, or when unfaulted, of a simple monocline. All the tilted blocks are generally dipping in the same direction and emphasize the asymmetry.

Along each graben, the main structure is therefore a zone made up of large normal faults showing a throw of many kilometres on one side. It corresponds to a major crustal fracture along which the African continent is breaking apart. From maps showing the distribution of earthquake epicentres, it appears that seismic activity lies on the major fault, as shown in the north Tanganyika basin (Wohlenberg 1975). This major fault zone has been drawn on Figure 1.

Field observations as well as seismic profiling on the Lakes show that the older the Cenozoic sediments, the denser is the synsedimentary fault spacing. The Lake Malawi trough shows this pattern (Figure 2). At the beginning of the rift evolution, fracturing is dense while vertical throws along faults are small. During later stages, tectonic activity is focussed along the major fault zone and along a few faults belonging to the fault-flexured side.

2b. The NW–SE transverse structures

The East African Rift segments corresponding to the main troughs are linked or cut by transverse structures, generally striking NW–SE.

There are two types of NW–SE transverse structures (1) large lineaments linking the main segments of the Rift; and (2) fault zones dividing the main rift segments into elementary parallelogram-shaped basins (Figure 1).

The large transverse lineaments. Large lineaments represent Precambrian shear zones which have been reactivated and serve as intracontinental transform structures during the opening of the Cenozoic Rift (Chorowicz and Mukonki 1979; Kazmin 1980). Although not yet true oceanic type transform faults, the geometry and kinematics are similar.

The Aswa lineament forms the northern limit of the Western branch of the rift, near Nimule (Figure 1). It crosses the Gregory Rift at the point where the rift changes direction. The lineament continues southeastwards to the Indian Ocean where it may form the northern limit of the Kerimbas trough.

The observed Neogene structures, located along this lineament, are discontinuous. North of Lake Mobutu, accentuated by the Nile, lies the Aswa lineament fault zone (Almond 1969) which seems to continue towards the volcanic cone of

Mt Elgon. This fault zone shows evidence of movement since the Early Miocene (Baker *et al.* 1971), allowing magma to migrate up the fault. The Nandi fault system lies southeast of Elgon volcano. These fault systems have had a rather important Neogene tectonic activity as shown by vigorous scarps and the 'fresh' aspect of the slickensides. The striations are subhorizontal and indicate left-lateral motions with a slight normal component. In the NNW–SSE transverse part of the Gregory Rift, the lineament corresponds to the Mau active fault and to the less mobile Sattima scarp. On these two faults, the observed motions are mainly normal, with a dextral or sinistral component, depending on the place and the structural stage. 'En echelon' faults are located southeastward and correspond to volcanic alignments, including Kilimanjaro. On the Tanzanian continental shelf, near Dar-es-Salaam, the lineament zone continues southeastwards and links several offshore troughs, including the Kerimbas basin. A little to the north, the Yatta Plateau lineament is accentuated by a phonolitic lava flow forming inverse relief, preserved in a nearly rectilinear valley, superimposed upon a fault. This area shows evidence of Cenozoic dextral strike-slip activity. On satellite imagery, large open gashes can be observed. These fractures were used by the lavas to reach the surface. The field evidence for dextral movements is small folds and microfaults. The Yatta Plateau fault is presently inactive. The Aswa lineament is thus discontinuous, partly active along some faults and volcanoes. All the faults end longitudinally and they dissect the lineament zone into strips bordered by parallel faults.

The Tanganyika–Rukwa–Malawi lineament is also composed of successive faults, all ending longitudinally and proved to be right-lateral strike-slip faults (Chorowicz *et al.* 1983). Among the most important faults are, from southwest to northeast: (1) the Mahali fault, also forming the central basement high in Lake Tanganyika; (2) the southwest and northeast border faults of Lake Rukwa. Most of the faults in this area are folded strips which form large anticlines and synclines corresponding to hills and topographic depressions. This pattern is different from the ramp structure described by Griffiths (1980) in the Gregory Rift. These folds are Neogene in age, affect the basement (Figure 3), and allow this tectonic zone to have an overall dextral transcurrent motion. It thus serves as an intracontinental transform zone linking the northern and southern segments of the western branch of the East African Rift.

The Zambezi lineament is an active fault zone in the area where it intersects the Lake Malawi Rift. The lineament continues to the southeast. It is also a dextral strike-slip fault acting as a link between southern Lake Malawi and the small Urema and Dombe grabens.

Transverse faults inside the grabens. Transverse faults situated within the grabens are smaller than those previously described. Generally striking NW–SE, they dissect the troughs into primary basins and give these basins a rhomb shape (Figure 1). They are a link between the major border faults, and thus act as transfer faults (Gibbs 1984).

In the Lake Malawi trough, for instance, the transverse fault zone may link two major faults situated along opposite sides of two succeeding basins. In this case, the dip directions of the tilted blocks are opposite from one basin to the other. Sometimes, the transverse fault links two offset major border faults located along the same side, as it can be seen in the northern Tanganyika basin.

Generally, a basement high separating the basins corresponds to these transverse faults along which a little compression occurs, expressed by the folding of lacustrine

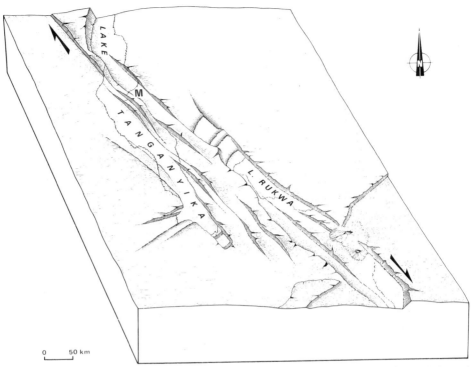

Figure 3. Oblique view of the Tanganyika–Rukwa–Malawi area, showing the folding of the strips situated between the dextral strike-slip faults. M: Mahali Mounts.

deposits. Reverse faults have been observed in the uplifted area between Lake Kivu and northern Lake Tanganyika (Chorowicz and Mukonki 1980).

2c. The subsidiary structures

The major fault zones bordering the main grabens, linked by the transverse faults inside these grabens, along with the Tanganyika–Rukwa–Malawi lineament and, to a lesser extent, with the Aswa and Zambezi lineaments, form the principal directions of fracturing in East Africa. This fracture zone is complex and discontinuous but it is clearly expressed geomorphologically, resulting from a long morphotectonic evolution. It is still seismically active along most of its segments. Nevertheless, many faults and troughs are not situated along this main zone of fracturing and therefore the features should be considered as subsidiary structures. This also concerns NE–SW striking faults, generaly related to early Cenozoic volcanic activity. We interpret these fractures as open tension gashes lying perpendicular to the NW–SE extension, formed during the earliest stages of the opening of the rift. Other faults, striking NW–SE, or in different directions, correspond to an important seismic activity, but have only a faint geomorphological expression. This feature gives the structures a juvenile aspect.

Early fractures equivalent to open tension gashes. The Virunga volcanic chain (Figure 1) represents a typical example of these structures. Covering a small area and proved to have begun its volcanic activity in Eocene–Oligocene times

(Bellon and Pouclet 1980), the volcanic chain lies perpendicular to the near NW–SE regional tensional stress (Chorowicz 1983). The Nyanza Rift, also striking NE–SW, was associated with volcanic activity dating from Late Oligocene or Early Miocene to Middle Miocene times (Pickford 1982). The Buhoro Flats graben, void of sediments, has influenced the development of the Rungwe volcanic massif at its intersection with the dextral strike-slip Livingstone fault.

Juvenile structures. Maps defining the locations of seismicity (Anderson 1984; Bath 1975; Chapman and Pollack 1977; De Bremaecker 1959; Fairhead and Girdler 1972; Khan and Swain 1981; Molnar and Aggarwal 1971; Rykouvov *et al.* 1972; Sutton and Berg 1958; Sykes 1967; Sykes and Landisman 1964; Wolhenberg 1969, 1975; Zana and Hamagushi 1978) show that the present tectonic movements affect, not only the East African Rift system, but also very large areas in Eastern and Southern Africa. Apart from the seismic zones along the main troughs, four other zones of seismic activity are recognized, all of them striking NE–SW (Figure 4). One is located in Eastern Zaire and seems to coincide with small faults in the Gweshe area. A second one passes by Lake Mweru, southwest of Lake Tanganyika, in a region of small faults, and continues southwestwards. A third one seems to reactivate the Luangwa Karoo trough and extends southwestwards in the Lake Kariba area where many small faults are exposed. The fourth one can be found in Southern Africa, following NE–SW striking faults, especially in the High Limpopo valley.

These faults, active but associated with shallow and wide depressions that are geomorphologically poorly expressed, represent areas of the most recent rift activity, and they therefore may be considered as structures having a juvenile character. They may be interpreted as troughs in a very early stage of development, at a pre-rift stage, characterized by a dense spacing of faulting with each throw being small.

South of Lake Malawi, the seismically active area, striking approximately N–S, continues far southwards. But beyond the Zambezi transverse structure, the faults are far less important than along Lake Malawi (Figure 1). The small Urema and Dombe troughs seem also to be structures with a juvenile character, through which the East African Rift is propagating southwards.

3. The Stress Fields

3a. Superimposed motions along the faults

Some slickensides expose two or sometimes three families of striations and tension gashes (Figure 5). The study of their intersections gives the succession order. On each slickenside the oldest striations generally are the most horizontal.

From the many areas studied along the rift from Kenya to Malawi (Figure 6), three palaeostress orientations have been distinguished. Their dating is only relative and we could not classify them into precise tectonic phases. They most likely correspond to different stages whose age differs from one segment to the other along the rift.

In some locations the oldest orientation shows horizontal extension striking approximately NW–SE or rarely E–W, compression being horizontal. After that, a new orientation corresponds also to near NW–SE horizontal extension, but compression turns to vertical. This second stage is widely observed everywhere

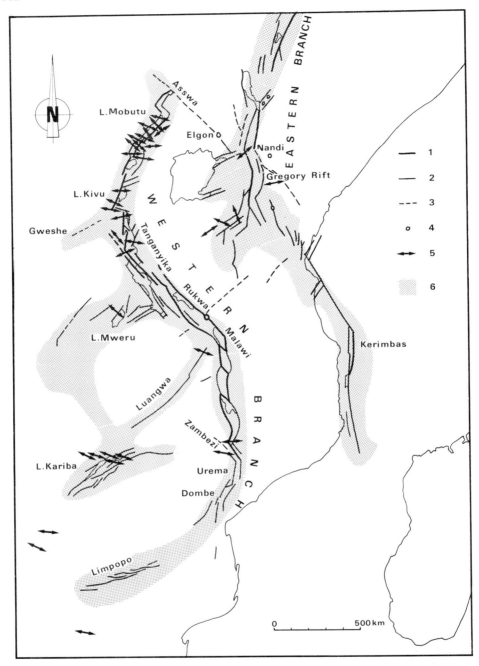

Figure 4. The present activity in the rift area. (1) major fault zone bordering the main troughs; (2) active fault zone; (3) inactive fault; (4) the three major volcanoes in Kenya (Elgon, Kenya, and Kilimanjaro) and Mt Rungwe in Tanzania; (5) focal mechanism of earthquake extension trend; (6) seismically active area.

RIFT DEVELOPMENT IN EASTERN AFRICA

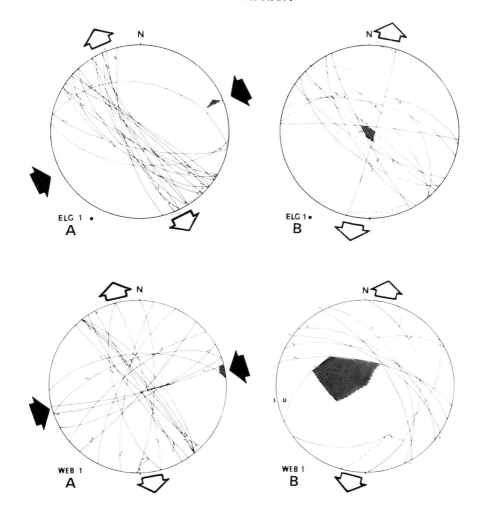

Figure 5. An example of fault analysis in Kenya (along the Elgeyo Escarpment — ELG 1, and along the Nandi Fault near the Webuye Falls — WEB 1). Stereograms of faults and striations, Schmidt project, lower hemisphere. Dark grey area: compression; dashed white area: extension (right dihedral method). Small arrows: striations; large black arrows: compression; large white arrows: extension.
In each of the stations (ELG 1 and WEB 1) two populations of faults and striations give two palaeostress settings. A: first setting corresponding to an early stage; B: second setting of a later stage.

and is followed in several areas by the most recent orientation, corresponding to horizontal extension, nearly perpendicular to the local scarp, compression being vertical. The first orientation is well exposed when the graben is relatively young, as for instance in the southern Tanganyika Rift, and along NW–SE faults. The last type of orientation is always found along the borders of the more developed grabens, characterized by important scarps. This leads us to interpret the last stress field as resulting from local motions due to gravity effects in the scarps, extension being perpendicular to the main slope (Figure 5). After a certain amount

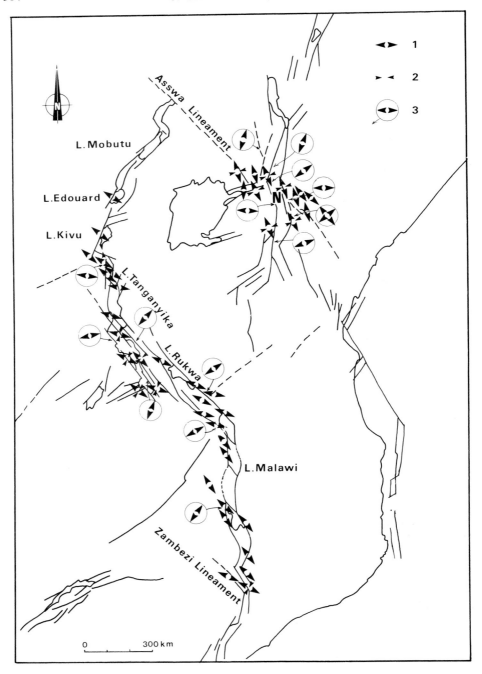

Figure 6. Palaeostress orientation in the East African Rift. 1 and 2: earliest stages; 1: extension; 2: compression; 3: latest stage: extension (compression is vertical). N: Nakuru trough.

of movement along oblique-slip faults, the system becomes 'open' enough to allow gravity tectonics to act. A certain displacement is required before compression rotates to vertical.

Considering their near ubiquitous NW–SE trend, the first two orientations may be interpreted as closely related with regional NW–SE movements.

3b. The earliest palaeostress orientations

In transverse zones. In the three NW–SE tectonic transverse zones forming the Aswa, the Tanganyika–Rukwa–Malawi, and the Zambezi lineaments, the earliest extensions are generally striking close to the NW–SE direction (Figure 6). In the Tanganyika–Rukwa–Malawi and Zambezi zones, the associated compression is more often vertical: these two zones are both dextral oblique-slip structures. Along the Aswa lineament, there is a more complex setting, some areas showing, for the first movements, a vertical compression, while others show a horizontal one. This zone is a transcurrent structure which can be tensile or compressive depending on the segment and the time.

The palaeostress setting found in these transversal zones is very significant in so far as extension is generally striking very close to the major faults' NW–SE direction. When the NW–SE major faults are dextral, extensional trends are WNW–ESE, and when the faults are sinistral, extensional trends are NNW–SSE. This situation is a general feature of an intracontinental transcurrent tensile fault zone, acting to some extent as a transform zone.

Along the main rift segments. The main rift segments mostly strike N–S, or more rarely NE–SW (Lake Mobutu for instance). For the two first stages, the extensional trend is generally near NW–SE. It is clear that the first extensions are typically non-perpendicular to the main troughs.

When the very first palaeostress orientation is observed along N–S faults, it corresponds to strike-slip motions having a small vertical component, exactly as in the NW–SE zones. This is based on the observation of striations. These facts show that the very first rift opening movements are everywhere horizontal and not vertical.

3c. The present deformation

Many papers have been devoted to the focal mechanism of earthquakes (Bath 1975; Chapman and Pollack 1977; Fairhead and Girdler 1972; Fairhead and Stuart 1982; Maasha 1975; Maasha and Molnar 1972; Scholz *et al.* 1976; Shudofsky 1985; Sykes 1967; Tanaka *et al.* 1980). Most of the mechanisms are compatible with a NW–SE to WNW–ESE extension, some with an E–W one or, very infrequently, with a NE–SW extension (Figure 4). In the juvenile structures (*cf.* section 2c), such as in the Lake Kariba or Mweru areas, the compatible extension trends are exclusively NW–SE to WNW–ESE. On the other hand, in more developed rift zones such as the Northern Tanganyika Rift, or the Kivu and Mobutu Rifts, the extensional trends may strike E–W to NE–SW or NW–SE. By analogy with the interpretation of the palaeostress orientation, two types of deformations may be considered:

    (a) Those having a regional significance, related to the overall movements in East Africa, typically striking NW–SE to WNW–ESE, uniquely exposed as juvenile structures.

(b) Those having a local significance, whose NW–SE, E–W or NE–SW trend is perpendicular to the major faults, resulting from the combination of the previous deformations with simultaneous gravity tectonics along the steep scarps.

The homogeneous NW–SE to WNW–ESE orientation of most of the ancient or present extension trends, can be explained by an approximately NW–SE relative movement of large crustal blocks in Eastern Africa. This conclusion is in agreement with the transform character inferred for the NW–SE transcurrent lineaments that are parallel to this movement. The parallelism is not perfect, but this can be considered as a normal feature in an intracontinental transform zone.

4. Model for the opening of the rift system in Eastern Africa

A better knowledge of the structures, palaeostress setting, and recent deformations in the East African Rift, allows us to propose a new model for the evolution of the rift system in Eastern Africa.

4a. Summary of the previous models of rifting

The first model devised by Gregory (1921) presented a compressive crustal doming accompanied by a surficial extrados extension. Willis (1936) proposed a thermal origin linked with the intrusion of a subcrustal asthenolith for the doming. The mantle intrusion may result from convection cells (Holmes 1944; Freund 1966; Matsuzawa 1969), or from hot spots (Morgan 1971; Burke and Dewey 1973; Burke and Whiteman 1973; Schilling 1973; Burke 1978). Theories linked with plate tectonics appeared more recently. For example, McKenzie *et al.* (1970) have modelled the rift formation in terms of a simple NW–SE lithospheric stretching. Baker (1970) interpreted the East African Rift as corresponding to a shear zone between two plates, while Oxburgh and Turcotte (1974) took into account the distension that appears within plates moving on a non-spherical ellipsoid. Anderson (1984), noting an important geoid gradient along Eastern Africa, assumed that this gradient induced a tendency of continent fragmentation toward the Indian Ocean. Chorowicz and Mukonki (1979) and Kazmin (1980), have improved the model of 'stretching blocks' by the proposition that the major rift segments are linked by large intracontinental transform zones and by new evidence of a NW–SE extension.

The following model, also considers the basins or grabens linked with the East African Rift system but located in the Mozambique Channel (Mougenot *et al.* 1986), and linked with an approximately NW–SE relative movement of individual continental blocks.

4b. The different stages of rift evolution

The geometry of the structures as well as the mechanism allows us to distinguish four successive stages in the formation of each given rift segment (Figure 7).

1st stage: pre-rift. It seems that the very first motions are horizontal. Since Late Oligocene–Early Miocene times, a horizontal slip movement with a divergent component occurred along fractures and lineaments inherited from older tectonic phases. Generally, the direction of this movement, striking approximately NW–SE, is not perpendicular to the palaeostructures. Nevertheless, the NE–SW faults, as

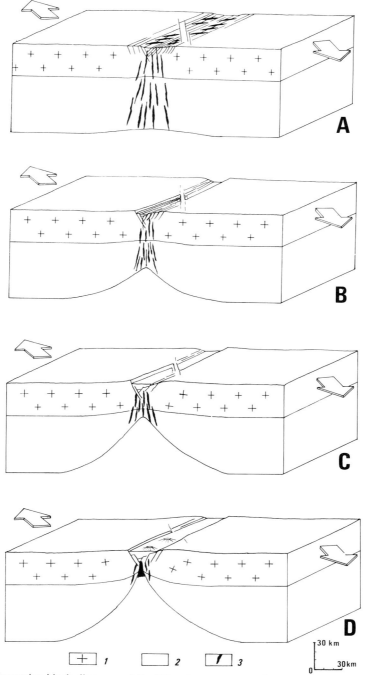

Figure 7. Successive block diagrams of the lithosphere corresponding to the different stages of rift evolution. Arrows show the inferred motion. Faults and tension gashes control the location of volcanism. 1: crust; 2: upper mantle; 3: tension gashes. A: pre-rift stage; B: initial rift stage; C: typical rift stage; D: advanced rift stage.

tension gashes, are the first to open, resulting in initial tholeiitic volcanism (Kampunzu *et al.* 1983) as, for instance, in the Virunga volcanic chain. Similarly, 'en echelon' tension gashes may open (Lameyre *et al.* 1984) perpendicular to the extensional trend, along all the ancient discontinuities that are reactivated, whatever their direction, allowing the uprise of tholeiitic magmas (Zanettin *et al.* 1974). This happened for instance in the Eastern branch of the East African Rift where (Miocene?) dyke swarms have been found striking approximately N45°.

Subsequently, many alignments of tension gashes evolved into transcurrent faults having a small vertical throw, with no appreciable uplift of the rift shoulders. Fracturing was dense and covered a large area, compression and extension being horizontal. The result may be similar to the present topographic depressions that are poorly delineated, characterized by swamps and shallow lakes, similar to those situated in Lakes Mweru and Kariba. There is not yet a true graben but rather a shallow and wide basin. In the deep lithosphere, a possible ductile extension may have permitted the intrusion of light asthenosperic material, giving a slight but widespread negative gravity anomaly (Wohlenberg 1975; Fairhead 1975; Fairhead and Reeves 1977).

2nd stage: initial rifting. At this stage, some faults become more normal, but still with an appreciable strike-slip component. Compression rotates to vertical, with extension remaining horizontal striking NW–SE. Subsidence increases while deformation becomes limited to a smaller number of faults bordering tilted blocks. The graben is well defined but the shoulders are not yet sufficiently uplifted. The main rift segments are formed of successive elementary rhomb-shaped basins, each of them bordered along one side by a major fault, and along the other side by a faulted flexure. These basins are separated by basement highs corresponding to transverse faults striking near NW–SE and linking the major border faults.

This stage corresponds to the situation in Lake Malawi and in northern Lake Tanganyika. If the tension gashes are open, an alkaline to peralkaline magma reached the surface, as is the case in the eastern branch of the East African Rift.

3rd stage: typical rift formation. The faults become mainly normal, with a slight strike-slip component, extension being horizontal and often striking perpendicular to the main scarps, while compression remains vertical. The tension gashes change trend and lie parallel to the rift axis, as observed in the Eastern Rift. Deformation concentrates into a smaller number of faults, mainly along the major border fault. Subsidence is rapid while important uplift affects the shoulders. The main troughs are still composed of successive elementary basins separated by ridges. If some ridges tend to disappear, others still form important ecological barriers (Denys *et al.* 1985). The negative gravity anomaly is steeper and narrower, mainly due to the asthenospheric intrusions. The whole of the northern part of the Western Rift including the northern Tanganyika basin, is representative of this stage.

4th stage: advanced rifting. Similarly to the preceding stage, motions along the faults seem to result from a combination of local and regional stresses, acting simultaneously or alternately. Subsidence is fast and alkaline volcanism important. Along the axis of the large negative regional anomaly, there is a narrow positive gravity anomaly, interpreted as indicating the intrusion of dense material, and considered as the first appearance of oceanic crust (Baker and Wohlenberg 1971; Darracot *et al.* 1972; Fairhead 1976; Green 1976; Khan and Swain 1978). Most of the ridges situated between the basins have disappeared. This situation essentially corresponds to the present eastern branch of the East African Rift system.

The succession of these stages is not synchronous all along the East African Rift system, it differs from one segment to the other. The most advanced development of rifting occurs in the eastern branch, except for the Afar graben which already corresponds to an oceanic stage (Barberi and Varet, 1975). The stages outlined above coincide with the successive tectonic and volcanic phases noted by McCall (1967); Baker et al. (1972); King (1978); Shackleton (1978), and Mohr (1983).

The pre-rift stage (A), recognizable by the fissural eruption of basaltic flood lavas from tension gashes, started during Late Oligocene times in Eithopia (Baker et al. 1971; Bellieni et al. 1981; Zanettin and Justin-Visentin 1974, 1975; Zanettin et al. 1980) and during Late Oligocene times (Fitch et al. 1985) or Early Miocene times (Baker et al. 1971) in the Gregory Rift. The initial rifting stage (B) began during Middle Miocene times 15–16 Ma ago (Baker et al. 1971; Black et al. 1975; Roeser 1975; Ross and Schlee 1977). The typical rift formation stage (C), characterized by major uplift of the shoulders, began around 9–10 Ma, during Late Miocene times. The advanced rifting stage (D) is emphasized by an important volcanic and tectonic activity around 4 Ma (Pliocene) and 2 Ma ago (Figure 7)

The ages of the successive stages of rift development in the Western branch of the East African Rift are less well known. However, since the advanced rifting stage has not yet been reached, the Western Rift is obviously younger.

4c. Propagation of rifting across the African continent

In the Oligocene, fissuring, mainly comprising of tension gashes, gave way to volcanism in Afar and Ethiopia. In the Early Miocene, while faults were forming in the north, fissuring accompanied by volcanism appeared in separate regions widely dispersed from each other in the Nyanza Rift, the Virunga Chain and the Gregory Rift. Later in the Miocene, the pre-rift stage spread along the northern part of the Western branch while the Eastern branch reached the initial rifting stage. The southern part of the Western branch developed later in the Miocene (Crossley and Crow 1980).

As oceanic crust is presently forming in Afar, the Eastern branch of the East African Rift system has arrived at the advanced rifting stage (D); the northern part of the Western branch has come to the typical rift formation stage (C), and its southern part to the initial rifting stage (B). Structures with a juvenile character, corresponding to the pre-rift stage (A), may be interpreted as indicating present day rift propagation. As they are located in the far south, arranged into several branches outlined by seismicity, one can assume that rifting is distributed along several paths of propagation (Figure 1).

The classical eastern and western branches of the East African Rift system, together with the Kerimbas trough correspond to several directions of fracturing. The opening movement is transferred from one segment to the other by intracontinental transform zones. For example, the dextral Tanganyika–Rukwa–Malawi lineament zone links the northern segment of the western branch with its southern part.

The influence of the Aswa lineament on the East African Rift system is not so evident. There is important seismic and tectonic activity along the region linking the Gregory Rift to the Kerimbas trough (Fairhead 1980). But along the segment, located between the eastern and the western branches, seismicity and geomorphological expression are relatively weak. Nevertheless, the fresh scarps situated along the Aswa lineament and the Nandi fault, form with the Elgon volcano an 'en echelon' alignment, proving that activity occurred in the Neogene. Striations have recorded a sinistral motion for this episode.

5. Conclusions

Fracture zones on the East African continent, comprise not only of the main rift segments, but also the transverse structures. Transverse zones are complex, multiple, and still active.

The principal relative motion is a NW–SE separation of developing African blocks with uplift as a later consequence of opening. Some secondary movements correspond with local gravity tectonics along the main scarps.

Faulting is widespread in the eastern and southern parts of the African continent. Fracturing reactivated ancient lineaments situated within the Pan-African mobile belts using various structural elements inherited from a polycyclic thermotectonic history.

Rifting avoids the large cratons. This is the case for the Eastern branch of the East African Rift system, which, after crossing the Aswa lineament, ends southwards in the Tanzanian shield, as a series of small half-grabens forming the North Tanzanian divergence. This latter area is one of the most seismically active along the Rift and it seems to be a major 'zone of resistance' (Courtillot 1982).

The large transverse NW–SE structures serve as intracontinental transform zones, linking the main segments of the Rift system. The Aswa lineament zone is particularly complex. In the Neogene, it linked the western and eastern branches of the East African Rift, by left-lateral faults. It presently links the eastern branch with the Kerimbas trough.

The various rift segments correspond to different stages of opening. The younger structures, poorly geomorphologically exposed but still active, are located in the southern part of Africa.

Acknowledgements. We are very grateful to Elf-Aquitaine for supporting this research and allowing the publication of this paper. Anonymous reviewers made useful comments and Christopher Sorlien helped to improve text presentation. The English text has been further scrutinized by the editors.

References

Almond, D.C. 1969. Structure and metamorphism of the basement complex of North-east Uganda. *Overseas Geol. Min. resources*, **10**, 146–163.
Anderson, D.L. 1984. The earth as a planet: paradigms and paradoxes'. *Science*, **223**, 4634, 347–354.
Angelier, J. and Mechler, P. 1977. Sur une méthode graphique de recherche des contraintes principales également utilisable en tectonique et en séismologie: la méthode des dièdres droits. *Bull. Soc. Géol. France* (7), **XIX**, 1309–1318.
Baker, B.H. 1970. The structural pattern of the Afro-Arabian rift system in relation to plate tectonics. *Royal Society, London, Philosophical Transactions Series A*, **267**, 383–391.
——, **Mohr, P.A. and Williams, L.A.J. 1972.** Geology of the Eastern Rift System of Africa. *Geological Society of America, Special Paper*, **136**, 67 pp.
——, **Williams, L.A.J., Miller, J.A. and Fitch, F.J. 1971.** Sequence and geochronology of the Kenya rift volcanics. *Tectonophysics*, **11**, 191–215.
——, **Wohlenberg, •. 1971.** Structure and evolution of the Kenya Rift Valley. *Nature*, **229**, 538–542.
Barberi, F. and Varet, J. 1975. The nature of the Afar crust: a discussion. In: *Afar Between Continental and Oceanic Rifting. Proc. Int. Symp. Afar reg. related rift problems*, Bad Bergzabern, **1**, 375–378.
Bath, M. 1975. Seimicity of the Tanzania region. *Tectonophysics*, **27**, 4, 353.
Bellieni, G., Justin-Visentin, E., Zanettin, B., Piccirillo, E.M., Radicati, Dl., Brozolo, F., and Rita, F. 1981. Oligocene Transitional Tholeiitic Magmatism in Northern Turkana (Kenya): Comparison with the Coeval Ethiopian volcanism. *Bulletin of Volcanology*, **44–3**, 411–427.
Bellon, H. and Pouclet, A. 1980. Datations K-Ar de quelques laves du Rift-Ouest de l'Afrique Centrale; implications sur l'évolution magmatique et structurale. *Geol. Rundschau.*, **69**, 1, 49–62.
Black, R., Morton, W.H. and Rex, D.C. 1975. Block tilting and vocanism within the Afar in the light of recent K/Ar age data. In: Pilger, A. and Rosler, A. (Eds), *Afar Depression of Ethiopia*, Schweizerbart, Stuttgart. 296–300.

Burke, K. 1978. Evolution of continental rift systems in the light of plate tectonics. *In*: Ramberg, I.B. and Neumann, E.R. (Eds), *Tectonics and Geophysics of Continental Rifts*. 1–9.
—— **and Dewey, J.F. 1973.** Plume-generated triple junctions: key indicators in applying plate tectonics to old rocks. *Journal of Geology*, **81**, 406–433.
—— **and Whiteman, A.J. 1973.** Uplift, rifting, and break-up of Africa. *In*: Tarling, D.H. and Runcorn, S.K. (Eds), *Implications of Continental Drift to the Earth Sciences*. Academic Press, London. Vol. 2, 735–753.
Chapman, D.S. and Pollack, H.N. 1977. Heat flow and heat production in Zambia: evidence for lithospheric thinning in Central Africa. *Tectonophysics*, **41**, 1/3, 79–100.
Chorowicz, J. 1983. Le rift est-africain: début d'ouverture d'un océan? *Bull. Centres Rech. Explor-Prod. PAU–SNPA*, **(7)**, 155–162.
—— **1984.** Cartographie géologique à partir d'images spatiales de la zone transformante Tanganyika–Rukwa–Malawi (rift est-africain). 27 ème Congrès Géologique International, Moscou. *Géochronique*, supp. no. **10**, Paris, 48–49.
——, **Le Fournier, J., Le Mut, C., Richert, J.P., Spy-Anderson, F.-L., and Tiercelin, J.J. 1983.** Observation par télédétection et au sol de mouvements décrochants NW–SE dextres dans le secteur transformant Tanganyika–Rukwa–Malawi du rift est-africain. *C.R. Acad. Sc. Paris*, **296**, sér. II, 997–1002.
—— **and Mukonki, M.B. 1979.** Linéaments anciens, zones transformantes récentes et géotectonique des fossés de l'Est africain, d'après la télédétection et la microtectonique. *Mus. roy. Afr. centr., Tervuren (Belg.), Dépt. Géol. Min.*, Rapp. ann. 1979, 143–167.
—— **and Mukonki, M.B. 1980.** Apport géologique des images MSS Landsat du secteur autour du lac Kivu (Burundi, Rwanda, Zaïre). *C. R. Acd. Sc. Paris*, **290**, 1245–1247.
Courtillot, V. 1982. Propagating rifts and continental breakup. *Tectonics*, **1**, 239–250.
Crossley, R. and Crow, M.J. 1980. The Malawi Rift. *In: Geodynamic Evolution of the Afro Arabic Rift System*. Roma: Accademia Nazionale dei Lincei (Atti dei Convegni Lincei, 47), 77–87.
Darracot, B.W., Fairhead, J.D. and Girdler, R.W. 1972. Gravity and magnetic surveys in northern Tanzania and southern Kenya. *Tectonophysics*, **15**, 1/2, 131–141.
De Bremaecker, J.C. 1959. Seismicity of the West African Rift Valley. *Journal of Geophysical Research*, **64**, 1961–1966.
Denys, C., Chorowicz, J. and Jaeger, J.J. 1985. Tectogenèse et évolution des faunes de rongeurs et autres mammifères du rift est-africain au Mio-Pliocène. *Bull. Soc. Géol. France*, **(8)**, I, no. 3, 381–389.
Ebinger, C.J., Crow, J.J., Rosendahl, B.R., Livingstone, D.A., and Le Fournier, J. 1984. Structural evolution of Lake Malawi, Africa. *Nature*, **308**, 627–629.
Fairhead, J.D. 1975. The regional gravity field of the Eastern Rift, East Africa. *In: Afar Between Continental and Oceanic Rifting, Proc. Int. Symp. Afar reg. related rift problems, Bad Bergzarbern*, **1**, 113–120.
—— **1976.** The structure of the lithosphere beneath the eastern rift, East Africa, deduced from gravity studies. *Tectonophysics*, **30**, 3/4, 268–298.
—— **1980.** The structure of the cross-cutting volcanic chain of northern Tanzania and its relation to the East African Rift System. *Tectonophysics*, **65**, 3/4, 193–208.
—— **and Girdler, R.W. 1972.** The seismicity of the East African rift system. *Tectonophysics*, **15**, 1/2, 115–122.
—— **and Reeves, C.V. 1977.** Teleseismic delay times, Bouguer anomalies and inferred thickness of the African lithosphere. *Earth and Planetary Science Letters*, **36**, 63–76.
—— **and Stuart, G.W. 1982.** The seismicity of the East Africa rift system and comparison with other continental rifts. *In*: Palmason, G. (Ed.), *Continental and Oceanic Rifts. Geodynamic series*, **8**, 41–61.
Fitch, F.J., Hooker, P.J., Miller, J.A., Mitchell, J.G. and Watkins, R.T. 1985. Reconnaissance potassium-argon geochronology of the Suregei–Asille district, northern Kenya. *Geological Magazine*, **122**, 609–622.
Freund, R. 1966. Rift Valleys. *In: The World Rift System*. Rept Symp. Ottawa 1965. *Geological Survey of Canada, paper*, **66–14**, 300–344.
Gibbs, A.D. 1984. Structural evolution of extensional basin margins. *Journal of the Geological Society, London*, **141**, 609–620.
Green, A.G. 1976. Raytracing through models of the East African Rift. *Earth and Planetary Science Letters*, **31**, 403–412.
Gregory, J.W. 1896. *The Great Rift Valley*. London, John Murrary. 422 pp.
—— **1921.** *The Rift Valleys and Geology of East Africa*. Seeley Service, London. 479 pp.
Griffiths, P.S. 1980. Box fault systems and ramps: atypical associations of structures from the eastern shoulder of the Kenya rift. *Geological Magazine*, **117** (6), 579–586.
Hepworth, J.V. 1967. The photogeological recognition of ancient orogenic belts in Africa. *Quarterly Journal of the Geological Society, London*, **123**, 253–292.
Holmes, A. 1944. *Principles of Physical Geology* (1st Ed.). London, Thomas Nelson and Sons Ltd. 532 pp.
Kampunzu, A.B. Vellutini, P.J., Caron, J.P., Lubala, R.T., Kanika, M. and Rumvegeri, B.T. 1983. Le volcanisme et l'évolution structurale du Sud-Kivu (Zaïre). *Bull. Centres Rech. Explor-Prod. Elf–Aquitaine*, **7**, 1, 257–271.

Kazmin, V. 1980. Transform faults in the East African Rift System. *In: Geodynamic Evolution of the Afro-Arabic Rift System.* Roma: Accademia Nazionale dei Lincei (Atti dei Convegni Lincei, 47). 65–73.

Khan, M.A. and Swain, C.J. 1978. Geophysical investigations and the rift valley geology of Kenya. *In:* Bishop, W.W. (Ed.), *Geological Background to Fossil Man.* Geological Society of London, Scottish Academic Press, 71–83.

—— **and Swain, C.J. 1981.** The structure of the rift valley in central Kenya from seismic and gravity data. E.O.S. Transactions American Geophysical Union, **62** (45), 1033.

King, B.C. 1978. Structural and volcanic evolution of the Gregory Rift Valley. *In:* Bishop, W.W. (Ed.), *Geological Background to Fossil Man.* Geological Society of London, Scottish Academic Press. 29–54.

Lameyre, J., Black, R., Bonin, B. and Giret, A. 1984. Les provinces magmatiques de l'Est américain, de l'Ouest africain et des Kerguelen. Indications d'un contrôle tectonique et d'une initiation superficielle du magmatisme intraplaque et des processus associés. *Ann. Soc. Géol. Nord,* **CIII**, 101–114.

Le Fournier, J., Chorowicz, J., Thouin, C., Balzer, F., Chenet, P.Y., Henriet, J.-P., Masson, D., Mondeguer, A., Rosendahl, B., Spy-Anderson, F.-L., Tiercelin, J.-J. 1985. Le bassin due lac Tanganyika: évolution tectonique et sédmentaire. *C.R. Acad. Sc., Paris,* **301**, II, 14, 1053–1058.

Maasha, N. 1975. The seismicity of the Ruwenzori region in Uganda. *Journal of Geophysical Research,* **80**, 1485–1496.

—— **and Molnar, P. 1972.** Earthquake fault parameters and tectonics in Africa. *Journal of Geophysical Research,* **77**, 5731–5743.

McCall, G.J.H. 1967. Geology of the Nakuru-Thomson's Falls — Lake Hannington area. *Geological Survey of Kenya,* 122 pp.

McKenzie, D.O., Davies, D., and Molnar, P. 1970. Plate tectonics of the Red Sea and East Africa. *Nature,* **226**, 243–248.

Matsuzawa, I. 1969. Formation of the African Great Rift System. *Journal of Earth Sciences, Nagoya University,* **17**, 11–70.

Mohr, P.A. 1974. ENE trending lineaments of the African Rift System. *In: Proc. 1st Int. conf. New Basement Tectonics.* Salt Lake City, 1974; Salt Lake City, Utah. Geol. Assoc., 1976.

—— **1983.** Volcanotectonics aspects of Ethiopian rift Evolution. *Bull. Centres Rech. Explor. Prod. Elf–Aquitaine,* **7**, 1, 175–189.

Molnar, P. and Aggarwal, Y.P. 1971. A microearthquake survey in Kenya. *Bulletin of the Seismological Society of America,* **61**, 195–201.

Morgan, W.J. 1971. Convection plumes in the lower mantle. *Nature,* **230**, 42–43.

Mougenot, D., Recq. M., Virlogeux, P. and Lepvrier, C. 1986. Une ramifiaction sous-marine du rift est-africain: les grabens Kerimbas et Lacerda (marge continentale nord-Mozambique). *Bull. Soc. Géol. France* (in press).

Mukonki, M. na B. 1980. Application de la télédétection à l'étude structurale du système de fossés de l'Est-africain. *Thèse 3e cycle,* Univ. P.M. Curie, 168 pp.

Oxburgh, E.R. and Turcotte, D.L. 1974. Membrane tectonics and the East Africa Rift. *Earth and Planetary Science Letters,* **22**, 133–140.

Pickford, M. 1982. The tectonics volcanics and sediments of the Nyanza rift valley. *Z. Geomorph. N.F.,* suppl. Bd, **42**, 1–33, Berlin, Stuttgart.

Roeser, H.A. 1975. A detailed magnetic survey of the Southern Red Sea. *Geol. Jahrb.,* **13**, 131–153.

Rosendahl, B.R. and Livingstone, D.A. 1983. Rift lakes of East Africa: New seismic data and implications for future research. *Episodes,* **83–1**, 14–19.

Ross, D.A. and Schlee, J. 1977. Shallow structure and geologic development of the southern Red Sea. *In: Red Sea Research 1970–1975. Bulletin,* **22**, 18 pp.

Rykounov, L.N., Sedov, V.V., Savrina, L.A. and Bourmin, V.J. 1972. Study of microearthquakes in the rift zones of East Africa. *Tectonophysics,* **15**, 1/2, 123–130.

Schilling, J.G. 1973. Afar mantle plume: rare earth evidence. *Nature,* **242**, 2–5.

Scholz, C.H., Koczynski, T.A. and Hutchins, D.G. 1976. Evidence for incipient rifting in southern Africa. *Geophysical J.R. astr. Society,* **44**, 135–144.

Shackleton, R.M. 1978. Structural development of the East African Rift System. *In:* Bishop, W.W. (Ed.), *Geological Background to Fossil Man.* Geological Society of London, Scottish Academic Press. 19–28.

Shudofsky, G.N. 1985. Source mechanisms and focal depths of East African earthquakes using Rayleigh wave inversion and body wave modelling. *Reprint* ••

Sutton, G.H. and Berg. E. 1958. Seismological studies of the Western Rift Valley of Africa. *Transactions of the American Geophysical Union,* **39**, 474–481.

Sykes, L.R. 1967. Mechanism of earthquakes and nature of faulting on the mid-ocean ridges. *Journal of Geophysical Research,* **72**, 2131–2153.

—— **and Landisman, M. 1964.** The seismicity of East Africa, the Gulf of Aden and the Arabian and Red Seas. *Bulletin of the Seismological Society of America,* **54**, 1927–1940.

Tanaka, K., Horiuchi, S., Sato, T., and Zana, N. 1980. The earthquake generating stresses in the western rift valley of Africa. *Journal of Phys. Earth (JPN),* **28**, no. 1, 45–57.

Willis, B. 1936. East African plateaus and rift valleys. *Carnegie inst. Washington, publication,* **470**, 358 pp.

Wohlenberg, J. 1969. Remarks on the seismicity of East Africa between 4°N – 12°S and 23°E – 40°E. *Tectonophysics*, **8**, 4/6, 567–577.

—— **1975.** Geophysikalische Aspekte der ostafrikanischen Grabenzonen. *Geol. Jb, E, Dtsch.* no. **4**, 1–82.

Zana, N. and Hamaguchi, H. 1978. Some characteristics of aftershock sequences in the western Rift Valley of Africa. *Science Report Tohoku University*, **5**, JPN, 124, part 2, 38 pp.

Zanettin, B. and Justin-Visentin, E. 1974. The volcanic succesion in central Ethiopia. *Mem. Geol. Miner. University of Padova*, **XXXI, 3–19.**

—— **and Justin-Visentin, E. 1975.** Afar Depression of Ethiopia. *Inter Union Commission on Geodynamics* Scientific. Report no. 14. Pilger, A. and Rosler A. (Eds), Geologisches Institut der Technischen Universitat Clausthal Germany. 299–310.

——, **Justin-Visentin, E., Nicoletti, M. and Piccirillo, E.M. 1980.** Correlations among Ethiopian volcanic formations with special references to the chronological and stratigraphical problems of the 'trap series'. *Geodynamic evolution of the Afro Arabic Rift System.* Roma: Accademia. Nazionale dei Lincei. 231–252.

——, **Gregnanin, A., Justin-Visentin, E., Mezzacasa, G. and Piccirillo, E.M. 1974.** Petrochemistry of the volcanic series of the central Eastern Ethiopian plateau and relationships between tectonics and magmatology. *Mem. Geol. Miner. University of Padova.*, **XXXI**, 1–34.

Petrology and geodynamic significance of the Tertiary alkaline lavas from the Kahuzi–Biega region, Western Rift, Kivu, Zaire

R. T. Lubala and A. B. Kampunzu
Laboratoire de Pétrologie, Universite d'Aix-Marseille III, 13397 Marseille Cedex 13, France and Laboratoire de Pétrologie, Université de Lubumbashi, B. P. 1825, Lubumbashi, Zaire

J. P-H. Caron
Mission Universitaire de Géologie, B. P. 756, Lubumbashi, Zaire and Université d'Aix-Marseille III, 13397 Marseille Cedex 13, France

and

P. J. Vellutini
I.S.E.R.S.T., Djibouti, République de Djibouti

The youngest volcanic flows in the Kahuzi–Biega region of the Western Rift, are made up of ankaratrites, alkaline basalts, basanites, and hawaiites, belonging to a Na-rich alkaline suite. These lavas are closely associated with marginal faults and their development can be related to the early stages of formation of the rift. Thus, the geochronological age of *ca.* 6 Ma for this alkaline lava suite provides a constraint on the early events in the evolution of the western branch of the East African Rift system.

KEY WORDS   Western Rift   Alkaline lavas   Zaire   Bukavu rift zone

1. Introduction

Between 2° and 3°30'S the western branch of the East African Rift is composed of two main volcanic provinces covering 5000 km² (Figure 1). The larger province is found in the Bukavu area south and southwest of Lake Kivu, with the

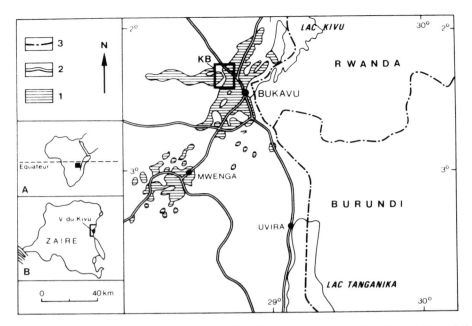

Figure 1. Distribution of volcanism associated with the western branch of the East African Rift, southern Kivu, Zaire. 1. volcanic rocks; 2. major roads; 3. International boundaries. The square KB denotes the Kahuzi–Biega region discussed in the text.

Mwenga–Kamituga area 60 km south–southwest of the Bukavu province.

In both areas volcanism was discovered at the turn of the century but has been little studied apart from an inventory of the various outcrops of lavas. However during these mapping surveys some petrographical and geochemical studies were undertaken for the major elements (Denaeyer et al. 1965) which contributed to the first attempt at defining the magmatic characteristics of the volcanic provinces.

Work on the Bukavu volcanic province was published a few years ago (Kampunzu and Vellutini 1981; Kampunzu 1981; Lubala 1981; Kampunzu et al. 1981, 1983, 1984; Kanika et al. 1981; Lubala et al. 1982, 1984; Caron et al. 1983) Three volcanic cycles were identified as characteristic of the various evolutionary stages of the western branch of the East African Rift.

The first cycle extends from Cretaceous times to Miocene (Kampunzu 1981; Lubala 1981). It is represented exclusively by tholeiitic to transitional basalts (Kampunzu 1981; Lubala et al. 1982, 1984). The second cycle is divided into two stages. The first stage is Miocene to Pliocene in age (Bagdasaryan et al. 1973; Lubala 1981). In the Haute–Ruzizi (southern end of the Bukavu province) it is represented by an alkaline series differentiated from ankaratrites to phonolites and trachytes (Kanika et al. 1981; Caron et al. 1983). The second stage of the second cycle is Pleistocene to Recent in age (Bellon and Pouclet 1980) with little-evolved basalts (Kampunzu et al. 1981). This alkaline cycle is seen as an important feature of the evolution of the Western Rift in the Lake Kivu region.

The third cycle began during the Pleistocene (Degens et al. 1972). It is composed of transitional lavas representing increasing extension and thinning of the continental crust under the Western Rift (Kampunzu et al. 1984).

The Tertiary alkaline lavas of the Kahuzi–Biega region belong to the second Bukavu volcanic cycle. They were mentioned for the first time by Boutakoff (1939) who linked them with the 'Lugulu flow'. Chemical analysis of major elements and microscopic studies later defined them as alkaline basalts (Thoreau and Chen 1943; Meyer 1954). This paper presents the first detailed mineralogical and petrological characteristics of these alkaline lavas.

2. Field studies

The lavas outcrop near the Upper Proterozoic alkaline complexes of Kahuzi–Biega. In these two regions they overlie tholeiites which constitute the major part of the Lugulu lavas emplaced prior to the formation of the Western Rift (Kampunzu 1981; Lubala 1981; Lubala et al. 1982, 1984).

The alkaline lavas can be divided into two spatially defined structural units (Figure 2). The first is composed of the Ciniondju and Burale flows which outcrop in a basin bordered in the west by the metamorphic Gaka ridge. They show a flow direction similar to the earlier tholeiitic and transitional lavas (Kampunzu and Vellutini 1981; Lubala et al. 1982).

The second unit which lies more to the East is composed of the Musisi and Lushandja flows which outcrop on the Mpuse and Mulume–Munene hills. These lavas overflowed onto the shoulders of the rift. The absence of any pyroclastic deposits associated with these flows implies that they are fissure eruptions, similar to the earlier tholeiitic and transitional lavas. The change in the flow direction of these lavas at the same time as the trend of active volcanism from outside the rift fractures to within the rift zone demonstrates the link between the structural evolution of rifting and the distribution of volcanic activity.

The Lushandja flow gives a K/Ar age of 5·9 Ma (Lubala 1981), and a relative timing of the beginning of rifting in the Kahuzi–Biega region. Miocene and earlier ages were proposed for the opening of the Lake Tanganyika rift 100 km to the south (Degens et al. 1971), and for the development of the Amin (Lake Edward) trough and the Mobutu (Lake Albert) trough 200 km to the north (Hopwood and Lepersonne 1953; Pouclet 1975, 1976, 1979). This variation in age shows the relative independent formation of troughs within the western branch of the East African Rift system.

3. Petrology

3a. General petrographic characteristics

The Tertiary alkaline lavas of Kahuzi–Biega are petrographically more diversified and thus contrast with the underlying hy-normative basalts represented only by quartz or olivine tholeiites (Kampunzu 1981; Lubala 1981; Lubala et al. 1982, 1984).

The Kahuzi–Biega volcanic assemblage can be divided into three main facies based on their mineralogical compositions (Table 1). The Ciniondju facies is characterized by microporphyritic texture composed of zoned olivine phenocrysts (18 to 22 per cent) and clinopyroxene (6 to 8 per cent). The groundmass consists of plagioclase, clinopyroxene, and iron–titanium oxides in a glassy matrix. The Burale–Musisi facies also has a microporphyritic texture with flow textures near

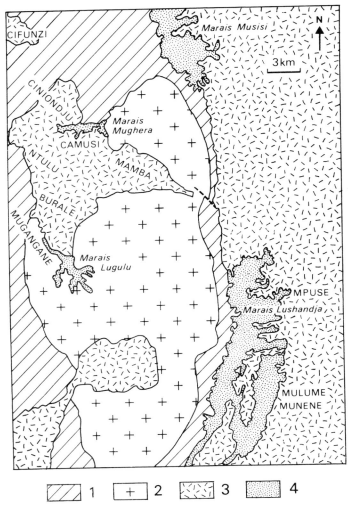

Figure 2. Sketch geological map of the Kahuzi–Biega region, Zaire. 1. Precambrian basement; 2. Upper Proterozoic anorogenic magmatism; 3. Tertiary volcanism; 4. swamp.

the top. In the first stages of crystallization the facies contains olivine (9 to 13 per cent) and clinopyroxene (7 to 10 per cent) with groundmass plagioclase, olivine, and iron–titanium oxides. The Lushandja–Mulume–Munene facies with its microporphyritic texture consists of olivine microphenocrysts (3 to 4 per cent) and clinopyroxene (1 to 2 per cent) with plagioclase, iron–titanium oxides, and glass in the groundmass.

On the whole there is a progressive diminution of the proportion of phenocrysts from the Ciniondju lavas to the Lushandja and Mulume–Munene facies.

Olivine is always stable and particularly abundant in the first facies but its modal abundance diminishes in the other lavas. Plagioclase, as well as iron–titanium oxides crystallized late indicating equilibrium crystallization conditions at moderate water vapour pressures.

Table 1. Chemical and modal parameters for the alkaline lavas of Tertiary age from Kahuzi–Biega Western Rift, Zaire.

			Ciniondju facies	Burale – Musisi facies	Lushandja – Mulume–Munene facies
PHENOCRYSTS	SiO_2 (rock) %		42·90–44·44	43·06–45·95	45·57–45·85
	D.I. "		19·55–24·89	25·42–33·38	35·46–41·81
	mg "		59·58–65·11	55·88–63·67	53·20–57·04
	Olivine Fo %		84·73–85·69	72·94–82·18	77·34
	modal vol % ol		18–22%	9–13%	3–4%
		Ca %	48·50–49·97	45·46–47·16	48·77–50·60
	Clinopyroxene	Mg %	40·97–37·66	42·47–37·31	38·67–32·90
		Fe + Mn %	12·07–15·35	12·07–15·53	12·56–16·43
	modal vol % CPX		6–8%	7–10%	1–2%
GROUNDMASS	Olivine Fo %		79·10–79·63	69·25	71·20
		Ca %	48·35–52·91	46·93–50·17	50·86
	Clinopyroxene	Mg %	37·70–32·46	41·74–34·31	36·44
		Fe + Mn %	15·95–14·63	11·65–15·52	12·70
	Plagioclase An	%	66·63–69·79	51·82–63·58	43·15–51·13
	Ti–magnetite	Usp %	72·40–76·90	85·50–90·30	86·10–91·70
		TiO_2 %	3·40–4·52	1·31–1·45	1·28–1·32

3b. Geochemistry

The chemical analyses (major elements) are shown on Table 2. All the lavas described above are alkaline and undersaturated with sodic affinities (Figures 3 and 4). This is reflected in their differentiation index (Figure 5) which varies from 20 in the ankaratrites (Ciniondju facies), alkali basalts and basanites (Burale–Musisi facies) to 42 in hawaiites from the Lushandja–Mulume–Munene region.

On an AFM diagram (Figure 6) the lavas define a trend characterized by a moderate iron enrichment, compatible with their alkali nature and with the beginning of magmatic differentiation.

Eleven samples of basic lavas (ankaratrites, alkali basalts and basanites) from Kahuzi–Biega were selected for trace element studies. Representative and average analyses are given in Table 3 which also includes, for comparison, the average compositions of transitional (MT2) and tholeiitic (MT1) basalts of Kahuzi–Biega which are slightly older. In the alkaline lavas the decrease of Ni and Cr is directly proportional to Mg (Figure 7) and confirms the part played by fractionational crystallization of olivine and clinopyroxene.

The ratio La/Th lies in the usual range for basic alkaline rocks (La/Th: 6–10; Wood et al. 1979) and Ba/La is near 10. A comparison can be made with the values obtained by Lloyd and Bailey (1975) on the strongly alkaline lavas of Toro–Ankole region to the south-west of Uganda and by Hertogen et al. (1985) on the olivine nephelinites from Nyiragongo in the Virunga–Zaire (western branch of the East African Rift). The main difference is due to the enrichment of light rare earths compared to heavy rare earths which increase from South Kivu (La/YbN=10–20) to the Virunga (La/YbN=30–45; Hertogen et al. 1985) then to Toro–Ankole (La/Yb_N=80–200; Mitchell and Bell 1976). The comparison between

Table 2. Chemical analyses and CIPW norms for Tertiary lavas: Kahuzi–Biega, Zaire.

	Ankaratrites			Basanites												Hawaiites		
	RTL 134	RTL 299	RTL 280	AK 256	RTL 138	AK 900	RTL 133	RTL 189	RTL 105	RTL 63	RTL 225	RTL 185	RTL 209A	RTL 291	RTL 184	RTL 3	RTL 2	RTL 8
SiO_2	42·90	43·74	44·04	43·06	43·85	43·86	43·88	43·96	44·44	44·52	44·93	44·97	45·10	45·28	45·70	45·83	45·95	45·57
TiO_2	2·09	2·29	1·84	3·19	1·92	2·80	2·62	2·40	1·98	1·93	2·29	2·49	2·39	1·61	2·13	1·76	1·77	2·48
Al_2O_3	13·91	14·54	13·79	15·30	13·83	14·45	14·31	14·54	13·58	14·05	14·08	14·71	15·14	12·97	14·66	14·77	15·00	13·40
Fe_2O_3	3·82	3·71	5·10	2·77	2·93	2·98	2·24	3·38	4·20	4·98	3·63	3·68	2·78	3·34	2·47	1·92	1·62	3·78
FeO	7·82	7·52	7·01	9·60	8·27	8·73	9·55	8·91	8·59	7·74	8·64	8·27	9·63	8·35	8·71	8·34	8·55	7·16
MnO	0·21	0·23	0·24	0·21	0·18	0·20	0·17	0·20	0·22	0·22	0·20	0·27	0·22	0·29	0·20	0·18	0·17	0·20
MgO	10·47	11·26	9·50	10·25	10·62	10·20	10·68	9·96	11·21	10·53	8·64	8·15	9·48	10·10	9·02	6·45	6·32	7·79
CaO	11·06	12·46	13·20	8·32	10·62	10·90	10·06	9·47	10·53	10·47	10·22	10·00	8·37	11·40	10·38	10·57	9·83	9·45
Na_2O	2·74	2·67	3·34	2·92	3·29	2·97	2·94	2·96	2·78	2·81	3·47	3·92	3·58	3·14	3·55	3·80	3·96	4·94
K_2O	1·02	0·81	0·79	0·98	1·14	1·09	1·05	0·77	1·15	1·04	1·29	1·31	1·23	1·10	1·12	1·27	1·32	2·10
P_2O_5	0·86	0·43	0·43	1·16	1·10	0·73	1·01	1·00	0·78	0·97	0·98	1·14	0·91	0·47	1·11	1·04	1·01	1·25
H_2O^+	1·75	0·33	0·13	0·91	1·14	0·05	0·15	1·50	0·72	0·14	0·21	0·52	0·00	1·15	0·00	3·58	3·94	1·32
H_2O^-	0·54	0·30	0·12	0·66	0·28	0·12	0·91	0·23	0·44	0·36	0·39	0·16	0·20	0·43	0·03	0·24	0·28	0·31
Total	99·19	100·29	99·53	99·33	99·17	99·08	100·57	99·28	100·62	99·76	98·97	99·59	99·03	99·63	99·08	99·75	99·72	99·75

CIPW norms

Or	6·21	4·80	4·70	5·92	6·88	6·50	6·29	7·05	6·82	6·18	7·74	7·82	7·35	6·62	6·67	7·81	8·16	12·63
Ab	10·16	5·56	4·13	21·05	12·00	11·40	15·14	15·51	11·83	14·66	16·20	16·77	20·72	11·24	18·35	20·78	22·56	13·43
An	23·35	25·36	20·43	26·32	20·04	23·11	23·08	20·49	21·28	22·80	19·33	18·86	21·85	18·39	20·94	20·31	10·15	8·34
Ne	7·45	9·26	13·18	2·28	8·92	7·58	5·47	5·95	6·39	5·03	7·38	9·07	5·37	8·58	6·48	6·89	6·78	15·79
Aeg	0·00	0·00	0·00	0·00	0·00	0·00	0·00	0·00	0·00	0·00	0·00	0·00	0·00	0·00	0·00	0·00	0·00	0·00
Di	22·62	27·40	34·83	7·18	21·94	21·91	17·34	17·68	21·51	19·23	21·39	19·90	11·94	29·51	19·77	22·83	20·19	25·88
Hyp	0·00	0·00	0·00	0·00	0·00	0·00	0·00	0·00	0·00	0·00	0·00	0·00	0·00	0·00	0·00	0·00	0·00	0·00
Ol	20·61	18·94	14·58	24·83	20·68	19·06	21·95	23·00	22·86	22·42	17·70	16·73	22·56	17·99	17·99	12·52	13·77	13·03
Mt	3·04	2·85	3·06	3·23	2·92	3·02	3·30	3·12	3·25	3·22	3·17	3·06	3·21	3·03	2·89	2·91	2·46	2·81
Ilm	4·10	4·37	3·53	6·21	3·74	5·39	5·06	4·39	3·79	3·70	4·43	4·79	4·60	3·12	4·09	3·49	3·53	4·81
Ap	2·10	1·02	1·03	2·81	2·66	1·75	2·43	2·46	1·86	2·31	2·36	2·73	2·18	1·13	2·65	2·57	2·50	3·02
DI	23·82	19·62	22·00	29·25	27·80	25·48	26·90	28·51	25·05	25·87	31·32	33·66	33·43	26·44	31·50	35·48	37·49	41·85
mg	66·40	68·71	63·45	64·33	67·47	65·54	66·19	66·42	65·84	64·69	60·71	59·79	62·42	65·21	63·68	57·43	56·36	61·04

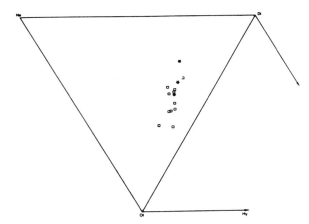

Figure 3. Tertiary alkaline lavas from the Kahuzi–Biega region plotted in the normative diagram Ne–Di–Ol. Black stars–ankaratrites; open squares–basanites.

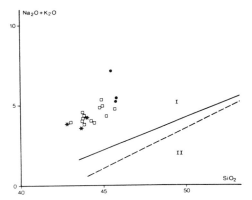

Figure 4. Kahuzi–Biega alkaline lavas plotted in an alkali–silica diagram. Continuous inclined line separates the tholeiitic field (I) from the alkaline field (II) for hawaiian basalts (after MacDonald and Katsura 1964); the dashed lined defines the limits of the subalkaline field (I) and the alkaline field (II) after Miyashiro (1978). Black stars–ankaratrites; open squares–basanites; closed circles-hawaiites.

basanites and pre-rift saturated basalts which constitute the first volcanic eruptions in Kahuzi–Biega shows that the ratios Ba/La, Ba/Th, and Zr/Nb decrease progressively from tholeiites to transitional basalts to alkaline lavas. The reverse is true for Zr/Hf and La/Yb. The rare earth spectrum (Figure 8) show that all these lavas are enriched in light rare earths and that, obviously the enrichment level increases from the tholeiites to the transitional basalts to the basanites, i.e. from the base to the top of the volcanic series of the region.

The fractionation of the heavy rare earths together with variations in the concentrations of incompatible elements can be attributed to changes in the level of partial fusion in a relatively homogeneous mantle to produce lavas ranging in composition from tholeiitic to alkaline (Auchapt 1985). The constant ratios La/Th (7·2–7·4) and Nb/La (0·9–1·0) in the various basalt types agrees with this hypothesis. In contrast, the Ba/Rb ratio shows a significant increase from tholeiitic basalts (approximately 18) to transitional basalts (approximately 23) to alkali lavas

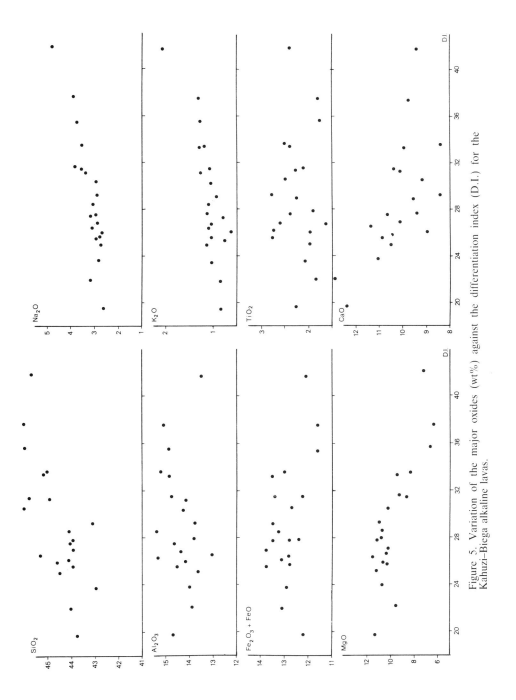

Figure 5. Variation of the major oxides (wt%) against the differentiation index (D.I.) for the Kahuzi–Biega alkaline lavas.

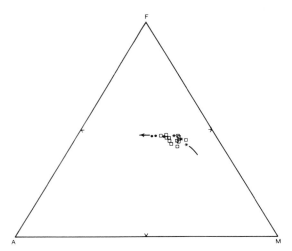

Figure 6. AFM diagram for the Tertiary alkaline lavas, Kahuzi–Biega, Zaire. Same symbols as in Figure 4.

(approximately 26). Such a continuous evolution cannot be ascribed to a crustal contamination. It more likely indicates a weak vertical mantle heterogeneity. In all cases the trace element geochemistry (Figures 8 and 9) shows that the mantle processes include a degree of enrichment possibly due to metasomatism (Auchapt 1985).

It is worth noting that in the Virunga Pliocene–Quaternary volcanic province, less than 100 km NNE of Kahuzi–Biega, an isotopic study of the potassic alkali lavas established their origin as resulting from a metasomatized mantle which took place approximately 500 to 600 Ma ago (Vollmer and Norry 1983). This age corresponds to the terminal stages of the Pan-African orogeny and correlates with K–Ar 'uplift' ages recorded in the Upper Proterozoic alkaline and carbonatitic complexes emplaced along the same lineaments which were reactivated in Tertiary times. It is likely that some of the geochemical features identified from the study of the Kahuzi–Biega lavas can be ascribed to this thermal and metasomatic end-Pan-African event.

The chondrites normalized multielement spectra (Figure 9a) show some strong similarities with those of the Kenya basalts (Figure 9b); the Cameroon Line (continental and oceanic domains – Figure 9c), and even those of the oceanic islands (Figure 9d). One must stress these similarities because they indicate that the initial material and even the processes in the asthenosphere are probably the same in the oceanic islands as well as in the first stages in the formation of intracontinental rifts (Kampunzu et al. 1986).

3c. Mineralogy

Three samples (R.T.L. 8, 209A and 299) were selected for microprobe analyses of olivine, clinopyroxene, iron–titanium oxides, and plagioclase. Chemical analyses of the host lavas are given in Table 2, with mineral analyses recorded in Tables 4, 5, and 6.

Compositions of the olivine phenocrysts (Table 4) vary from Fo 86–80 for the ankaratrites, to Fo 82–69 in the basanites to Fo 77–71 in the hawaiites.

Table 3. Representative analyses of trace elements in the Tertiary alkaline lavas from Kahuzi–Biega, compared with mean compositions of tholeiitic and transitional basalts in the same area.

Sample	AK 900	RTL 63	RTL 184	MB 3	Mt 2	MT 1
SiO_2	43·86	44·93	45.70	44·50	47·80	50·60
mg	65·50	60	64	65	60·80	58
Li	7	7	6	6·80	6·20	5·70
Rb	28	38	29	34	23·8	15·30
Sr	780	808	934	821	468	285
Ba	705	nd	1030	880	557	281
Sc	29	27	25	27	27	25
V	250	237	201	239	230	203
Cr	230	280	278	262	323	206
Co	52	50	47	50	55	69
Ni	198	204	173	192	172	165
Cu	68	72	66	71	51	67
Zn	109	156	143	106	112	144
La	75·69	69·50	94·80	80	55	27·20
Ce	139·89	131·50	162·80	145	96	51·4
Sm	9·39	10·19	9·19	9·60	6·80	5·25
Eu	2·81	3·10	2·68	2·94	2·77	1·78
Tb	1·19	1·28	1·27	1·25	1·09	0·88
Yb	3	2·92	2·67	2·86	2·70	2·66
Lu	0·44	0·45	0·39	0·43	0·44	0·41
Y	35	nd	33	38	35	30·00
Hf	4·50	5·46	4·92	5·00	3·70	3·20
Zr	227	nd	254	246	158	123
Nb	89	nd	114	94	50	27
Th	8·75	8·14	15·84	10·90	7·50	3·70

AK900: RTL 63: RTL 184 – basanites (see Table 2 for major element data)
MB 3 average composition of 10 analyses of alkaline lavas
Mt 2 average composition of 5 analyses of transitional lavas
Mt 1 average composition of 3 analyses of tholeiitic lavas
All lavas collected from the same region of the Western Rift.

Because of their high magnesian characteristics (Fo>80) some of the olivine phenocrysts might be considered as mantle xenocrysts. However their CaO contents between 0·22 and 0·32 per cent are much higher than for high pressure crystallized olivines (CaO<0·1 per cent) and thus indicate their volcanic derivation (Simkin and Smith 1970; Stormer 1973). Most magnesian phenocrysts from the various samples are in equilibrium with their respective liquids (Roeder and Emslie 1970). With these equilibrium conditions, temperatures determined by the Leeman and Scheidegger (1977) method are approximately 1250 ±50°C for ankaratrites and basanites, and 1100 ±30°C for hawaiites.

All the analysed clinopyroxenes (Table 5 and Figures 10, 11, and 12) are salites (Figure 10). According to Leterrier et al. (1982) they belong to the alkali rock domain (Figure 11). Generally in this magma type, compositional variations are quite limited and often within the margin of analytical error. The lack of plagioclase

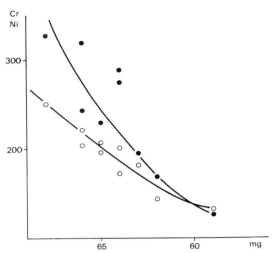

Figure 7. Distribution of Cr and Ni in relation to the magnesium–iron ratio (Mg) for the Tertiary alkaline lavas from the Kahuzi–Biega region. Cr variation – closed circles; Ni variation – open circles. Lines represent Olivine and Clinopyroxene fractionation.

phenocrysts is reflected in the evolution of the pyroxene compositions during the various crystallization stages and particularly when the crystals show an excess of wollastonite. The clinopyroxene richness in calcium and titanium reflects the weak silica activity in the crystallization environment. The magma undersaturation conditions are compatible with the relative richness in aluminium of the pyroxenes. This is reflected in Figure 12.

The opaque minerals (Table 6) are mainly titanomagnetites, with compositional variations ranging from Usp 79–81 in the ankaratrites; Usp 84–90 in the basanites; Usp 86–92 in the hawaiites. However there is a significant evolution of these iron–titanium oxide compositions across the various types of the lavas, characterized in particular by the increase in Mn levels and the decrease in Mg and Al.

The plagioclase microlites (Table 7 and Figure 13) show a decrease in their basicity from ankaratrites (An 70–67) to basanites (An 64–57) then to hawaiites in which labradorites coexist with andesines (An 51–43). The K_2O content of the microlites, although low, increases proportionally with differentiation index of the host rock.

The geochemical and mineralogical data suggest that the Kahuzi–Biega Tertiary alkaline lavas belong to the same magmatic series produced by fractionated crystallization under average pH_2O conditions. A geochemical comparison of the more basic end members (Mg> 65) of the alkaline series with the transitional and tholeiitic basalts erupted earlier in the Kahuzi–Biega region, suggests that all these volcanic products could have originated from the same metasomatized mantle.

4. Structural interpretation and conclusions

The final Tertiary volcanic events in the Kahuzi–Biega region of the Western branch of the East African Rift system are represented by ankaratrites, alkaline basalts, basanites, and hawaiites belonging to a sodic alkaline series. These lavas postdate the tholeiitic flows linked to the first volcanic cycle of South Kivu (Kampunzu 1981; Lubala et al. 1982, 1984; Kampunzu et al. 1983) and are closely associated with the boundary faults of the rift.

TERTIARY ALKALINE LAVAS

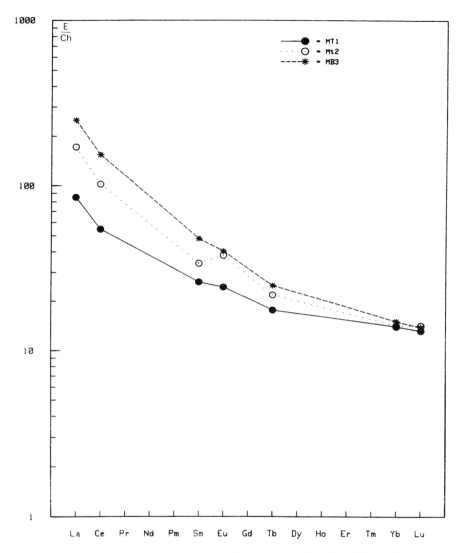

Figure 8. Average concentrations of rare earths in compositionally different lavas from the Kahuzi–Biega region, Zaire. Ree normalized to chondrites (Frey et al., 1968). MT1–tholeiites; MT2–transitional basalts; MB3–basanites.

The initial flows from external fractures show a centrifugal pattern in relation to the actual rift axis indicating an emplacement during the first extensional stages. Later eruptions overflowed onto the shoulders of the rift valley and represent the initial period of opening of the Western Rift in the Kivu region. An age of 6 Ma allows us to date this rifting stage. Later evolution of the rift in this region is characterized by the eruption of Quaternary transitional basalts (Kampunzu et al. 1984).

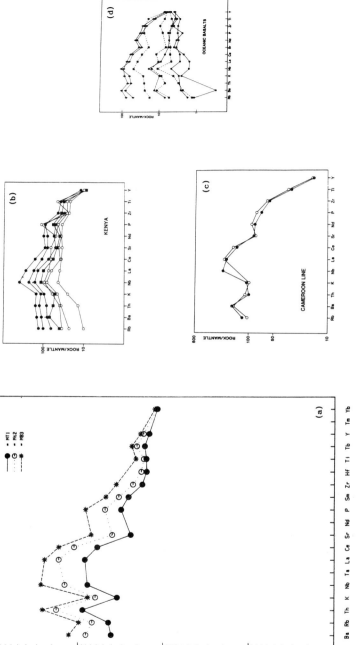

Figure 9. Spidergrams for compositionally different lavas from the Kahuzi-Biega region, Zaire. MT1–tholeiites; MT2–transitional basalts; MB3–basanites. Comparative spidergrams are shown for (b) Kenya volcanic rocks; (c) continental domain volcanics (closed circles), and oceanic domain volcanics (open circles); and oceanic basalts (d) after Norry and Fitton (1983).

Table 4. Representative analyses of olivines from alkaline lavas· Zaire.

Analyses	1(c)	2(b)	3(c)	4(b)	5(c)	6(b)	7(m)	8(m)	9(m)
SiO_2	38·43	34·56	40·75	38·67	39·26	38·06	37·29	39·26	35·25
TiO_2	0·02	0·02	—	—	—	—	—	—	0·04
Al_2O_3	0·14	—	—	—	0·16	0·18	0·13	0·02	0·05
FeO*	14·91	20·78	13·36	19·03	16·69	24·14	27·05	12·47	27·16
MnO	0·25	0·28	0·18	0·28	0·27	0·46	0·63	0·66	0·54
MgO	45·98	43·71	44·45	41·35	42·75	36·15	33·85	47·34	36·28
CaO	0·32	0·36	0·22	0·27	0·31	0·29	0·29	0·26	0·42
Na_2O	0·06	0·06	0·06	0·13	0·14	—	0·17	0·04	0·09
K_2O	—	—	—	—	—	—	—	—	0·01
Total	100·11	99·77	99·02	99·73	99·58	99·28	99·41	99·55	99·84
Fo %	84·73	79·10	85·69	79·63	82·18	72·94	69·25	71·20	77·34

1–4 ankaratrites; 5–7 basanites; 8–9 hawaiites *FeO as total iron
Analyses by CAMECA electron microprobe Science Faculty University Aix–Marseille III.
c–phenocryst core b–border rim of phenocryst: m–microlite in the groundmass.

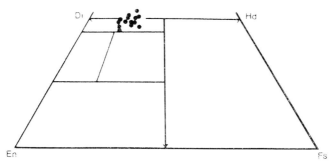

Figure 10. Compositional variations in clinopyroxenes from the Tertiary alkaline lavas from the Kahuzi–Biega region, Zaire.

The genesis of alkaline magmas after the tholeiitic and/or transitional basalts in South Kivu and before the Quaternary transitional basalts present some problems concerning the source regions of magma generation during the continental pre-rift stage to the initiation and evolution of intracontinental rifting. Indeed the succession of tholeiitic and/or transitional basalts, alkaline basalts, then transitional basalts is characteristic of several volcanic provinces within the Western Rift (Kampunzu et al. 1985) and is difficult to conceive in the classical model of asthenospheric diapirs as proposed by Green and Ringwood (1967). Neither is it satisfactorily explained by the lithospheric model suggested by Giret and Lameyre (1985) and Lameyre et al. (1985). A better model linked with the available data has been suggested by Kampunzu et al (1986) in which there was a rapid rise of the lower velocity zone up to the mantle–crust interface at the beginning of the extensional regime. The tholeiitic lavas/alkaline lavas succession which is characteristic of initial intraplate magmatism in continental and oceanic settings is, according to this model, an expression of the necessary time period for these magmas to reach the surface.

Table 5. Representative analyses of pyroxenes from alkaline lavas, Zaire.

Analyses	1(c)	2(b)	3(c)	4(b)	5(c)	6(b)	7(b)	8(c)	9(b)	10(c)	11(b)	12(m)	13(m)	14(m)	15(m)
SiO_2	46.25	45.49	50.42	44.70	49.27	48.61	47.01	49.05	46.87	47.91	46.19	48.17	49.89	45.89	49.95
TiO_2	3.10	3.30	1.64	3.94	1.58	2.91	2.69	1.43	2.55	1.79	2.80	2.98	1.86	3.69	2.05
Al_2O_3	9.38	8.82	5.16	9.55	6.24	5.67	8.07	5.58	6.78	5.77	7.00	5.26	2.90	6.44	4.21
FeO^*	7.15	8.45	5.96	7.97	6.31	8.32	8.27	6.64	10.14	7.07	8.91	7.38	7.46	9.19	7.54
MnO	—	—	—	—	—	—	0.21	0.22	0.28	0.17	0.20	0.11	0.15	0.17	0.13
MgO	11.36	10.97	12.87	9.82	12.90	11.19	10.42	13.90	10.96	14.16	12.16	11.76	12.76	10.24	11.91
CaO	20.78	20.80	21.33	21.47	22.19	20.83	21.33	21.82	20.74	21.22	21.52	22.98	22.53	22.01	22.09
Na_2O	0.56	0.52	0.48	0.47	0.45	0.61	1.10	1.10	0.97	0.94	0.86	0.72	0.75	0.97	0.75
K_2O	0.02	0.01	—	—	0.01	—	0.01	0.03	—	—	—	0.05	—	0.02	0.01
Total	98.60	98.36	97.86	97.92	98.95	98.14	99.11	99.77	99.29	99.03	99.64	99.41	98.30	98.62	98.64
Ca %	49.97	48.35	48.50	52.91	49.14	48.48	50.17	46.81	46.93	45.46	47.16	50.86	48.77	50.60	49.50
Mg %	37.66	35.70	40.97	32.46	39.99	36.46	34.31	41.74	34.72	42.47	37.31	36.44	38.67	32.96	37.36
Fe + Mn %	13.17	15.95	10.54	14.63	10.87	15.06	15.52	11.65	18.35	12.07	15.53	12.70	12.56	16.43	13.14

1–6 ankaratrites; 7–11 basanites; 12–15 hawaiites *FeO as total iron
c, b, m – same as in Table 4.

TERTIARY ALKALINE LAVAS

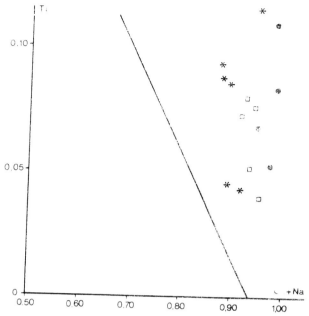

Figure 11. Clinopyroxene variations in terms of Ca+Na vs Ti for the alkaline lavas from Kahuzi–Biega. Black stars–ankaratrites; open squares–basanites; closed circles–hawaiites. The inclined line separates the alkaline domain to the right of the diagram from the tholeiitic domain C after Leterrier et al. 1982)

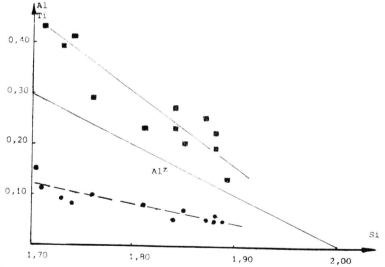

Figure 12. Correlation between Al (closed squares), Ti (closed circles), and Si in clinopyroxenes from the Kahuzi–Biega region. The continuous line marked by Alz represents $Si+Al_{IV}=2$.

Table 6. Representative analyses of groundmass titanomagnetities, alkaline lavas, Zaire.

Analyses	1	2	3	4	5	6
SiO_2	0·78	0·41	0·15	0·32	0·55	0·37
TiO_2	16·22	18·06	24·97	25·98	26·30	24·52
Al_2O_3	4·52	3·40	1·31	1·45	1·32	1·28
Fe_2O_3	37·01	32·99	19·13	16·75	16·53	17·65
FeO	38·48	41·55	49·56	50·66	49·25	51·12
MnO	0·22	0·90	2·81	2·65	3·80	2·92
MgO	2·74	2·05	1·15	1·10	0·84	0·93
CaO	0·68	0·69	0·06	0·13	0·10	0·15
Na_2O	—	—	0·45	0·88	0·08	0·06
K_2O	—	—	0·02	0·02	—	—
Total	100·65	100·05	99·60	99·93	98·87	99·00
Usp	78·90	81·26	83·50	90·30	91·70	86·10

1–2 ankaratrites; 3–4 basanites; 5–6 hawaiites

Table 7. Representative analyses of groundmass plasioclase crystals, alkaline lavas, Zaire.

Analyses	1	2	3	4	5	6	7	8	9
SiO_2	49·64	50·51	50·88	51·08	51·94	54·44	58·39	58·16	57·33
Al_2O_3	31·14	31·24	30·48	30·20	24·73	28·95	25·42	25·67	26·18
FeO*	0·34	0·40	0·59	1·05	0·69	0·59	1·10	1·16	0·95
CaO	14·27	13·64	13·64	12·82	12·05	10·28	8·63	8·46	10·15
Na_2O	3·62	3·11	3·65	3·85	4·94	5·06	5·45	5·80	5·00
K_2O	0·20	0·24	0·20	0·22	0·27	0·35	0·46	0·56	0·53
Total	99·21	99·14	99·44	99·22	99·62	99·67	99·45	99·81	100·14
Ca %	67·39	69·79	66·63	63·98	56·58	51·82	45·37	43·15	51·13
Na %	31·48	28·75	32·21	34·71	41·90	46·08	51·75	53·44	45·50
K %	1·13	1·46	1·16	1·31	1·50	2·10	2·88	3·40	3·36

1–3 ankaratrites; 4–6 basanites; 7-9 hawaiites
*FeO as total iron

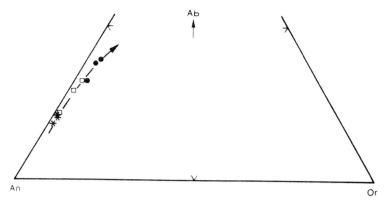

Figure 13. Variation in plagioclase groundmass compositions from the alkaline lavas of the Kahuzi–Biega region. Black stars–ankaratrites; open square–basanites; closed circles–hawaiites.

Acknowledgements. The authors wish to thank MM. Aguirre (Université d'Aix–Marseille III) and Bernard Bonin (Université de Paris XI) for reviewing the manuscript, Mme Trensz for the chemical analysis (major elements), MM. Auchapt and Dupuy (Université de Montpellier) and Dostal (St Mary's University, Halifax) for the trace and minor elements analysis. Mlle Issa (Marseille) typed the manuscript.

This paper is a contribution to IGCP project 227 and was translated from French into English by the editors of this volume.

References

Auchapt, A. 1985. Les éléments traces dans les basaltes des rifts continentaux: exemple de la province du Sud-Kivu (Zaïre) dans le rift Est-Africain. *Thèse 3ème cycle, Univ. Sciences Techn. Languedoc* (Montpellier). 81 pp.
Bagdasaryan, G. P., Gerasimovskiy, V. I., Gerasimovij, V. I., Polyakov, A. I. and Gukasyan, R. K. H. 1973. Age of volcanic rocks in the rift zones of East Africa. *Geochem. International. U.S.A.*, **10**, (1), 66–71 trad. de Geokhimija S.S.S.R. 1973 no. 1.
Bellon, H. and Pouclet, A. 1980. Datations K-Ar de quelques laves du Rift-Ouest de l'Afrique Centrale; implications sur l'évolution magmatique et structurale. *Geol. Rdsch.*, **69**, 49–62.
Boutakoff, N. 1939. Géologie des terrains situés à l'Ouest et au Nord-Ouest du fossé tectonique du Kivu. *Mém. Inst. géol. Univ. Louvain*, Belg., t. IX, fasc. **1**, 7–207.
Buddington, A. F. and Lindsley 1964. Iron–titanium oxide minerals and synthetic equivalents. *Journal of Petrology*, **5**, 310–357.
Caron, J.P.—H, Kampunzu, A. B., Kanika, M., Makutu, M. N., and Vellutini, P. J. 1983. Les éléments en traces dans les laves de la Haute-Ruzizi (Rift Valley du Kivu, Zaïre). *C.R. Ac. Sc. Paris, série II*, **296**, 269–274.
Degens, E. T., von Herzen, R. P., and Wong, H. K. 1971. Lake Tanganyika: water chemistry, sediments, geological structure. *Naturwissenschaften*, **58**, 229–241.
——, E. T., von Herzen, R. P., Wong, H. K., Deuser, W. G., Jannasch, H. W., and Kanwisher, J. W. 1972. *Lake Kivu, Anatomy of rift lake.* Final report of Woods Hole Oceanographic Institution, mission lac Kivu 1971. (Unpublished manuscript).
Denaeyer, M. E. 1972. Les laves du fossé tectonique de l'Afrique Centrale. *Ann. Mus. Roy. Afr. Centrale, Tervuren*, Sc. Géol., n° 72.
——, Schellinck, F., and Coppez, A. 1965. Recueil d'analyses des laves du fossé tectonique de l'Afrique Centrale (Kivu, Rwanda, Toro-Ankole). *Ann. Mus. Roy. Afr. Centrale, Tervuren*, série in 8, Sc. Géol., n° 49.
Frey, F. A., Haskin, M. A., Poetz, J., and Haskin, L. A. 1968. Rare-earth abundances in some basic rocks. *Journal of Geophysical Research*, **73**, 6085–6098.
Guibert, Ph. 1978. Contribution à l'étude du volcanisme de la chaîne des Virunga (Rép. du Zaïre); le volcan Mikeno. *Thèse Univ. Genève*, 152 p.

Green, D. H. 1973. Conditions of Melting of Basanite Magma from a Garnet Peridotite. *Earth and Planetary Science Letters*, 17, 456–465.
——, and Ringwood, A. E. 1967. The genesis of basaltic magmas. *Contr. Mineral. Petrol.*, 15, 103–190.
Hertogen, J., Vanlerberghe, L., and Namegabe, M. R. 1985. Geochemical evolution of the Nyiragongo volcano (Virunga, Western African Rift Zaïre). *Bulletin of Geological Society of Finland*, 14, 6.
Hopwood, A. T. and Lepersonne, J. 1953. Présence de formations d'âge miocène inférieur dans le fossé tectonique du lac Albert et de la basse Semliki (Congo Belge). *Ann. Soc. Géol. Belgique*, 77, 83–113.
Jones, P. W. 1976. Le magmatisme au stade initial de la fragmentation des plaques arabique, nubienne et somalienne. *Bull. Soc. Géol. Fr.*, (7), t. XVII, n° 4, 829–830.
Kampunzu, A. B. 1981. Le magmatisme du massif de Kahuzi (Kivu, Zaïre). Structure, pétrologie, signification et implication géodynamique. *Thèse ès-Sciences, Univ. Lubumbashi (Zaïre)*, 378 pp.
——, Caron, J. P.-H., Kanika, M., and Lubala, R. T. 1985. Evolution du rifting intracontinental et magmas tholéiitique, transitionnel et alcalin associés: cas du volcanisme phanérozoïque du rift Est-africain. 13th Coll. African Geol., St. Andrews – Scotland, *Occasional Publication* n° 3, CIFEG–Paris, 174–175.
——, Caron, J.P.-H., and Lubala, R. T. 1986. Rift-African volcanism, genesis of mantellic magmas and astheno-lithospheric dynamic. *Episodes* 9, 211–216.
——, Kanika, M., Caron, J.P.-H., Lubala, R. T., and Vellutini, P. J. 1984. Les basaltes transitionnels dans l'évolution des rifts continentaux: exemple de la Haute-Ruzizi dans le rift de l'Afrique Centrale (Kivu, Zaïre). *Geologische Rdsch.*, vol. 73, 3, 897–918.
——, Pottier, Y., and Vellutini, P. J. 1981. A propos des produits volcaniques de Cibinda, région de Bukava (Sud-Kivu, Zaïre). *Ann. Fac. Sc. Géol., Univ. Nat. Zaïre, Camps Lubumbashi*, 2, 21–30.
——, Vellutini, P. J., Caron, J.P.—H., Lubala, R. T., Kanika, M., and Rumvegeri, B. T. 1983. Le volcanisme et l'évolution structurale du Sud-Kivu (Zaïre). Un modèle de'interprétation géodynamique du volcanisme distensif intracontinental. *Bull. Cent. Rech. Explor., Prod. Elf-Aqitaine*, 7, 1, 257–271.
——, Kampunzu, A. B., Caron, J. P.-H., and Vellutini, P. J. 1982. Sur la nature et la signification possible des basaltes de la Lugulu au Sud-Kivu (Zaïre). *C.R. Ac. Sc. Paris*, série II, t. 294, pp. 325–328.
——, Kampunzu, A. B., Caron, J.P.-H., and Vellutini, P. J. 1984. Minéralogie et pétrologie des basaltes saturés tertiaires du Kahuzi-Biega (Rift du Kivu, Zaïre). *Ann. Soc. Géol. Belgique*, 107, 125–134.
MacDonald, G. A. and Katsura, T. 1964. Chemical composition of Hawaïan lavas. *Journal of Petrology*, 5, 82–133.
Meyer, A. 1954. Les basaltes du Kivu Méridional. *Notes volcanologiques du Congo Belge*, n° 2.
Mitchell, R. H. and Bell, K. 1976. Rare earth element geochemistry of potassic lavas from the Birunga and Toro–Ankole regions of Uganda, Africa. *Contrib. Mineral. Petrol.*, 58, 293–303.
Miyashiro 1978. Nature of alkalic volcanic rock series. *Contrib. Mineral. Petrol.*, 66, 91–104.
Norry, M. J. and Fitton, J. G. 1983. Compositional differences between oceanic and continental basic lavas and their significance. *In:* Hawkesworth, C. J. and Norry, M. J. (Eds), *Continental Basalts and Mantle Xenoliths*. 5–19.
——, Truckle, P. H., Lippard, S. J., Hawkesworth, C. J., Weaver, S. D., and Marriner, G. F. 1980. Isotopic and trace element evidence from lavas, bearing on mantle heterogeneity beneath Kenya. *Philosophical Transactions of the Royal Society, London*, A 297, 259–271.
Piccirillo, E. M., Justin-Visentin, E., Zanettin, B., Joron, J. L., and Treuil, M. 1979. Geodynamic evolution from plateau to rift: major and trace element geochemistry of the central eastern Ethiopian plateau volcanics. *N. Jb. Geol. Paläont. Abh.*, 158, 2, 139–179.
Pouclet, A. 1975. Histoire des grands lacs de l'Afrique Centrale. Mise au point des connaissances actuelles. *Rev. Geogr. Phys. Geol. Dyn.*, 2, 17, 5, 475–482.
——, 1976. Volcanologie du rift de l'Afrique Centrale; le Nyamulagira dans les Virunga. Essai de magmatologie du Rift. *Thèse d'Etat*, Orsay, Université de Paris. 610 pp.
——, 1979. Le magmatisme du rift de l'Afrique Centrale. Interprétation des données géochronologiques et relations structurales. 10ème *Colloque de Géologie Africaine*, Résumé, Montpellier, 140–141.
Roeder, P. L. and Emslie, R. F. 1970. Olivine–Liquid equilibrium. *Contr. Mineral. Petrol.*, 29, 275–289.
Simkin, T. and Smith, J. V. 1970. Minor-element distribution in olivine. *Journal of Geology*, 78, 304–325.
Stormer, J. C. 1973. Calcium zoning in olivine and its relationships to silica activity and pressure. *Geochim. Cosmochim. Acta*, 37, 1815–1821.
Thompson, R. N. 1982. Magmatism of the British Tertiary Volcanic Province. *Scottish Journal of Geology*, 18, 49–107.
Thoreau, J. and Chen, J. 1943. Les roches éruptives et métamorphiques du Kivu Central et Oriental. *Mém. Inst. Géol. Univ. Louvan*, IX, fasc. VIII, 18–26.
Thornton, C. P. and Tuttle, O. F. 1960. Chemistry of Igneous Rocks. I. Differenciation Index. *American Journal of Science*, 258, 664–684.
Vollmer, R. and Norry, M. J. 1983. Unusual isotopic variations in Nyiragongo nephelinites. *Nature*, 301, 141–143.

Wood, D. A., Joron, J. L. and Treuil, M. 1979. A re-appraisal of the use of trace elements to classify and discriminate between magma series erupted in different tectonic settings. *Earth and Planetary Science Letters,* **45**, 326–336.

Yoder, H. S. and Tilley, C. E. 1962. Origin of basalt magmas: an experimental study of natural and synthetic systems. *Journal of Petrology,* **3**, 342–532.

African transform continental margins: examples from Guinea, the Ivory Coast and Mozambique

J. Mascle, D. Mougenot, E. Blarez, M. Marinho, and P. Virlogeux

Laboratoire de Geodynamique sous-marine, BP 48, 06230 Villefranche-sur-Mer, France.

Marine Studies of selected sheared-type margins have been undertaken along both West and East African coastal regions in order to determine the geological structure and the evolution of this specific type of passive margin.

Geological and geophysical data have been obtained in the coastal waters off Guinea, Ivory Coast, and Ghana (Western Equatorial Africa) and off Mozambique (Eastern Africa).

The continental margin off the Guinea coastline can be divided into three distinct sectors. To the Northwest, the observed seismic stratigraphy and structures are typical of a rifted Central Atlantic margin of Jurassic age. To the south, several features such as steep scarps, basement ridges, volcanic piles, magnetic and gravity lineaments, etc., are interpreted in relation to a transcurrent motion. A southwestern marginal area contains lower tectonized sequences of Early Cretaceous age covered by post-Albian detrital sediments. The overall structure is interpreted as a Jurassic rifted margin reactivated by shearing during the Lower Cretaceous.

Off shore from the Ivory Coast and Ghana, two main marginal areas are also distinguished. To the east, south of Ghana, the upper margin is fractured by dissymetric and sedimented grabens, with the continental slope representing the unsedimented truncated African craton. South of the Ivory Coast, thick sedimentary basins, containing a Lower Cretaceous tectonized sequence developed between the continent and the Ivory Coast–Ghana Ridge, correspond to a composite feature consisting of deformed sedimentary wedges and basement structures.

Off the Mozambique coast, the strike-slip motion of Madagascar with respect to Africa, during the Upper Jurassic, has generated a series of *en echelon* structures. The Davie Ridge includes a series of sedimentary shear folds and basement structures. Later, in Cretaceous times, throughout the Cenozoic, and since Miocene times, the continental margin has been submitted to different tectonic stresses generating diversely-trending extensional features.

When shearing commences between two continental margins, it seems to generate either tensional features (e.g. Ivory Coast) or transpressive features (e.g. Mozambique) depending on the area. Later on, basement-involved structures and sedimentary wedges develop at the boundary between a thinned continental crust and a thick continental crust before connecting laterally with a typical oceanic fracture zone. The southern Guinea margin represents an unusual example where a rifted margin has been reactivated by transcurrent motion.

KEY WORDS   Shearing   Transform faults   Sedimentary basins   Seismic studies   Continental margins   Guinea   Ivory Coast   Mozambique

1. Introduction

Following the pioneering studies of Wilson (1965), Francheteau and Le Pichon (1972), Talwani and Eldholm (1973), Scrutton *et al.* (1974) on the concept of sheared (or transform) continental margins, Scrutton (1976) and Mascle (1976) proposed a clear distinction between rifted-type continental margins (resulting from tensional stresses during opening) and the transform type (where shearing stresses prevailed). Scrutton (1976, 1979, 1982) made several attempts to develop sections typical of sheared margins with a rapid transition from thin oceanic crust to thick continental crust being particularly emphasized. Mascle (1976, 1977) began to explore the different geological deformations generated during transform motion between two plates. In particular, the presence of a steep continental slope and of a lateral high, emplaced towards the oceanic basin, appears as common morphological and structural features at transform margins.

In order to improve our understanding of geological processes occurring at transform margins, the writers have undertaken comparative offshore studies of selected parts of transform margins around Africa. This paper is an attempt to illustrate and discuss, using new data, the various types of deformations which are generated along transform margins during their early evolution. We also show how transform-type continental margins (being tectonically and seismically active during part of their evolution) can help to understand better the early deformation occurring in orogenic belts.

A simplified sketch of Africa and its surrounding oceanic basins (Figure 1) demonstrates the dramatic contrast between areas of wide continental margins

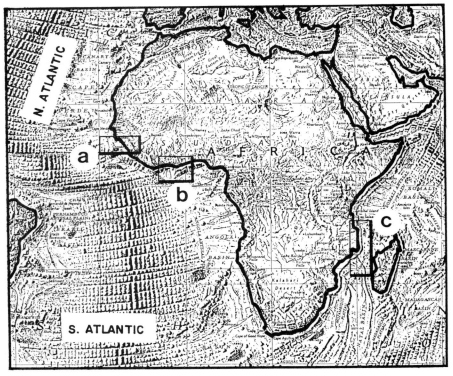

Figure 1. General physiographic sketch of Africa and surrounding oceanic basins (from the world ocean floor. National Geographic Society 1981). The Guinean, Ivory Coast–Ghana and Mozambique margins discussed in the text are respectively outlined by thick black blocks (a, b, c).

(facing deep oceanic basins) and portions with a very steep continental slope (for example, off-shore from the Ivory Coast or South Africa) showing an apparent lateral continuity with major oceanic fracture zones. Among the best known transform margins around Africa, are the South African margin, connected with the Agulhas fracture zone (Scrutton 1976), and the northern Gulf of Guinea margins related to the well known Equatorial fracture zone system (Delteil *et al.* 1974). Less apparent, and still debated, are the portions of the African margin facing the southwest Indian Ocean, between Durban and Northern Mozambique. Here, the steep slopes are most probably the result of a relative transcurrent motion between Africa and detached continental fragments such as the Falkland and Agulhas plateau, the Antarctic, and Madagascar (Segoufin and Patriat 1981; Lawner *et al.* 1985). We can also observe continental margin areas where a sharp change in direction delineates the marginal plateaux. This is the case for the Sao Paulo plateau east of Rio de Janeiro, or the Guinean plateau off western Guinea.

During the past three years our research work has been devoted to three of these areas (Figure 1), clearly related to early transform motions but now displaying a different geodynamic setting. They are respectively:

1. The Ivory Coast–Ghana margin, lying north of the Gulf of Guinea, an eastward structural extension of the Romanche fracture zone.

2. The northern Mozambique margin, which extends between Africa and Madagascar through the Davie Ridge.

3. The southwestern Guinean margin, which lies between Central and Equatorial Atlantic.

These three margins face oceanic basins of different ages. The Ivory Coast margin, for example, is a sector of a continental margin related to the evolution of the South Atlantic commencing during the Early Cretaceous. The Mozambique margin results from the motion between Africa and Madagascar in Late Jurassic–Early Cretaceous times. The Guinean margin faces westward towards the Jurassic Central Atlantic and southward towards the Equatorial Atlantic of probable Cretaceous age. A comparative study of the three areas should help to emphasize the common structural characteristics and differences of these transform margins.

2. The continental margin off Ivory Coast and Ghana

Along the Northern Gulf of Guinea (Figure 2) the trend of the continental margin is strongly oriented E–W. This direction has, for many years, been related to the relative motion between Africa and Brazil, witnessed by the equatorial fracture zones. Le Pichon and Hayes (1971), Francheteau and Le Pichon (1972) have discussed the relationship between the transform fault system and the respective tectonic fabric of the margins and the adjacent coastal basins. Arens *et al.* (1971), Delteil *et al.* (1974), Mascle (1977) have more precisely documented the structure of the margin west of the St Paul fracture zone and to the east of the Romanche fracture zone.

The Ghanaian continental slope lies exactly on the extension of the extinct Romanche fracture zone and represents the scar, within the upper continental

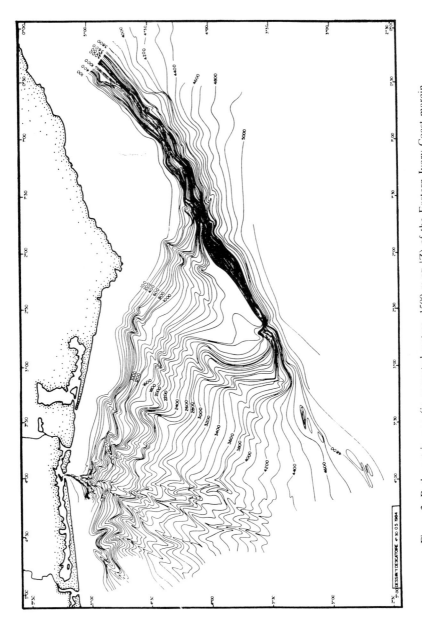

Figure 2. Bathymetric map (in corrected metres, 1500 m s^{-1}Z) of the Eastern Ivory Coast margin (from Blarez 1986). Note the prominent bathymetric SW–NE trend at the boundary between the margin and the deep oceanic basin. This trend represents the eastern extension, through the Ivory Coast Ridge and Ghanean slope, of the Romanche F–Z system.

crust, of the early transcurrent motion between Ghana and Northeast Brazil. The numerous seismic reflection lines and sampling stations (Figure 3) recently obtained in the margin allow us to distinguish several structural segments (Blarez 1986). Dredge hauls made along the Ghanian slope have produced pieces of different African cratonic geological units which indicate that the steep slope directly cuts across the outcropping African craton. In the absence of a detailed grid of multichannel seismic (MCS) lines, precise mapping of the Ghanian shelf is difficult.

However the different single channel lines available, as well as scattered MCS lines show tilted seismic sequences filling huge asymmetric grabens (with possibly some local syntectonic folding) cutting the shelf (Figure 4). Some bore holes suggest that the graben infilling consists of Lower Cretaceous terrigenous sediment, locally thick, resting on the dislocated African units and covered by thin, unconformable, post-Lower Cretaceous sediments. Towards the West, the Ghanian continental slope is extended by the Ivory Coast–Ghana Ridge (Figure 5).

This structural high has been, and still constitutes, a structural barrier for land-derived sediments, trapped between the borderland and the ridge, in a wide deep basin. This basin contains two 'litho-acoustic' sequences separated by a clear unconformity (Figure 6). The lower main part displays evidence of tectonic deformation, including features interpreted as shear folds, created during sedimentation. This sequence has been correlated with offshore commercial boreholes and dated as Early Cretaceous in age. The unconformity is between Late Albian to Early Cenomanian in age. The overlying sequence, locally intensively eroded (Figure 7), consists of Late Cretaceous to Cenozoic horizons giving good seismic reflections.

The ridge itself has been interpreted as a basement feature emplaced at the boundary between a northern thick continental crust and a southern oceanic crust (Figure 8). The marginal fracture ridge (Le Pichon and Hayes 1971) appears, in the light of recent data, to be a composite feature. We can observe all along its extension and particularly along its northern slope and on its top (Figure 6)

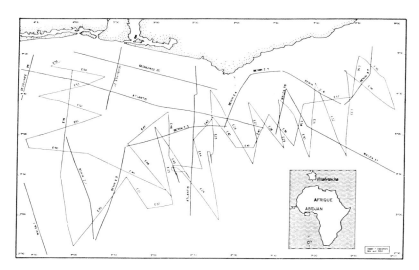

Figure 3. Seismic lines (single and multichannel) used for the structural compilation of the Ivory Coast margin (Blarez 1986).

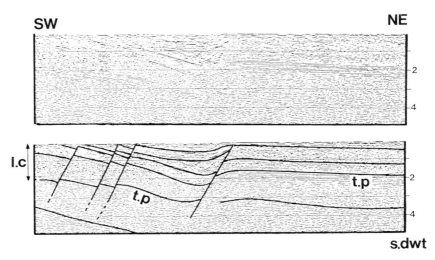

Figure 4. Multichannel seismic lines across the Ghanean continental shelf (from McGrew 1984) The proposed interpretation is based on a compilation of published bore holes (Blarez 1986) and on an interpretative sketch of the Ghanean shelf (Khan 1979).

deformed and broken seismic horizons presumed to be of sedimentary origin. Laterally, the top of the sequence correlates with the Albian–Cenomanian unconformity as defined in the deep basin. We therefore conclude that the ridge is not only a basement feature but the structural and bathymetric expression resulting from at least three intervening factors. A first cause of deformation (and maybe the most important one) could be due to the shearing mechanisms acting on a thick wedge of undercompacted sediments when both the African and Brazilian continental crusts were sliding against each other. A secondary source of deformation may be due to lateral readjustment (with a compressive component) at the end of the continent–continent contact. Finally, a third, more dubious, cause of deformation may be found in a re-heating and subsequent thermal uplift of the whole crust when a southern spreading centre was progressively moved along the area, during the Cretaceous, according to the different kinetic models of the South Atlantic (Rabinowitz and la Brechque 1979; Sibuet and Mascle 1978).

The Ivory Coast–Ghana ridge is progressively deepening toward the west and finally connects with an acoustic basement feature, well outlined by a strong magnetic anomaly and which represents the extinct Romanche fracture zone. This east–west ridge divides the deep abyssal plain into two crustal and sedimentary areas. North of it the basin, with more than 3s. (dwt) of sediments, represents oceanic extension of the deep Ivory Coast Basin. The southern area corresponds to a thinner sedimentary cover overlying a younger oceanic crust. We consider, on geological grounds, that the Ivory Coast–Ghana margin is a typical transform margin. The characteristics of this margin include a very sharp transition between a thick continental crust and a thin oceanic crust; and an apparent fracturing of the continental crust in elongated tensional features (asymmetric grabens). The effects of the fracturing are well expressed by the Ghanian shelf grabens which may also include a few syntectonic deformations such as the formation of an elongated structural high which may include a thick wedge of deformed sediments.

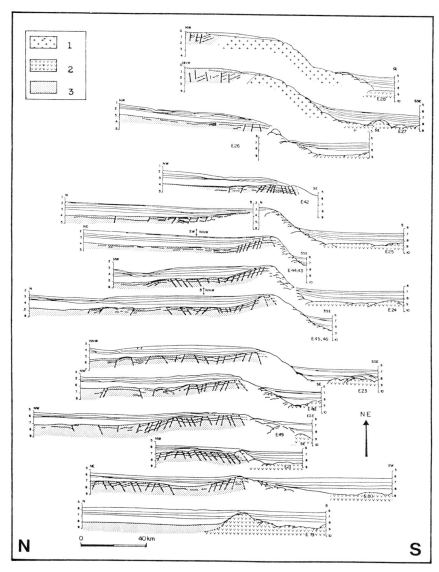

Figure 5. 14 tracing of seismic lines across the Ghanean shelf and slope and the Ivory Coast Ridge (see location of profiles on Figure 3; profiles E19 to E28).
1—African continental basement; 2—Oceanic basement; 3—Lower Cretaceous synrift deposits. Note that, north of the Ivory Coast Ridge the lower Cretaceous deposits are slightly deformed (shearing folds ?). Intensity of deformation appears to increase over the Ridge. The unconformity between deformed sediments (equivalent of synrift on a classic passive margin) and postrift (or post transcurrent phase) sediments is believed to be of upper Albian age.

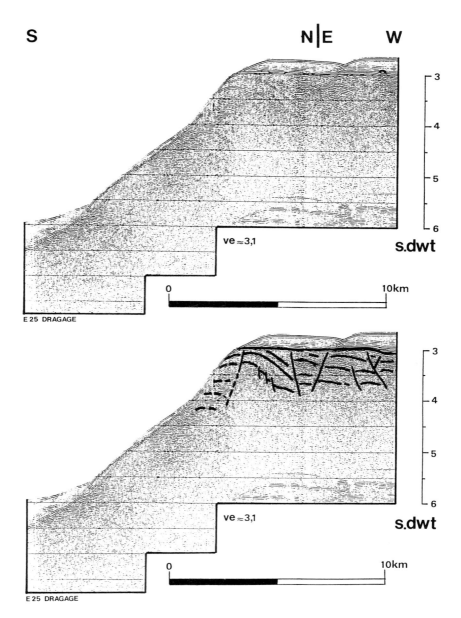

Figure 6. Detailed seismic section (single channel line; see location on Figure 3) across the Ivory Coast Ridge. Note deep, approximatively Northward dipping and faulted sedimentary sequences, unconformably covered by flat-lying post-Albian and Cenozoic deposits. Vertical scale in second double way travel time (Sdwtt).

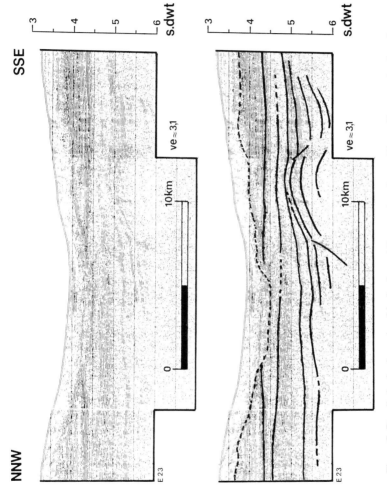

Figure 7. Detailed seismic section (single channel line; see location on Figure 3) across the deep Ivory Coast basin. Note the slightly folded deep sequence, attributed to Lower Cretaceous synsedimentary folding, unconformably covered by post-Albian sediments. Note also a strong erosional unconformity attributed to an Oligocene episode. Same scale as on Figure 6.

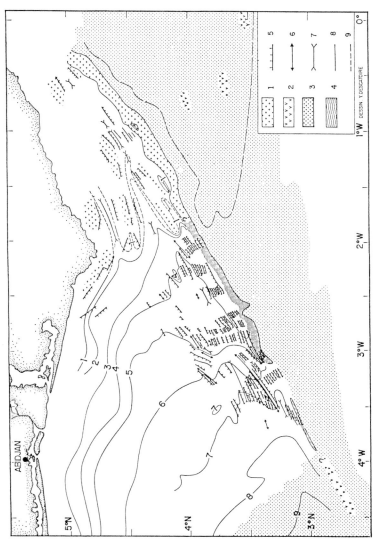

Figure 8. Structural sketch of the eastern Ivory Coast and western Ghanean margin based on the seismic interpretation of lines shown on Figure 3. (1) Subouthcropping African basement; (2) Oceanic basement ridges; (3) Oceanic basement; (4) Deformed sedimentary and basement wedge (Ivory Coast Ridge); (5) Main distensive faults; (6) Main 'anticline' axis; (7) Main 'synclines' and basin axis; (8) Isochrons (in second dwtt from the sea surface) of the Lower Cretaceous–Upper Cretaceous unconformity (Albian ?); (9) Western extension in the deep oceanic basin of Lower Cenomanian sedimentation.

Note that on the Ghanean shelf most of the observed structures correspond to elongated and dislocated basins (probably 'pull apart' basins) separated by a narrow system of faults. In contrast the deep Ivory Coast basin contains a series of folds, interpreted as synsedimentary shear folds developing in relation to the general dextral transcurrent motion between Africa and Brazil. The Ivory Coast Ridge may be interpreted as a complex deformed wedge of sediments together with a slight basement uplift (mechanical effect of transcurrent motion and thermal effect).

This ridge characterizes the lateral extension of the cratonic scar and shows evidences of both tensional (faults) and compressional deformation. The deformation (including probable shearing folds) display a marked decrease towards the continental borderland. The most important deformation phase appears to have been contemporaneous with the progressive shearing when a deep basin was developing between the two continental plates. The main deformation phase seems to end rather suddenly when the two continents were finally separated by the occurrence of oceanic crust. This leads to the prominent Mid-Cretaceous unconformity, and the presence, westward, of a prominent oceanic basement ridge dividing the abyssal plain into two areas of different ages.

We thus propose that tensional features are created during the early shearing motion when two thick cratonic crusts are in contact. The creation of a marginal ridge would then result directly from the progressive motion of the Brazilian continental crust against a thickly sedimented basin created between two portions of a transform margin. In the case of the Ivory Coast, the basin was totally surrounded during Early Cretaceous times by land-masses, and probably characterized by a huge sedimentation rate and a tremendous subsidence.

Finally, when the two continents parted, the constraints generated by the contact between two thick crusts vanished. This occurred in Mid-Cretaceous times and is well documented by the Late Albian/Early Cenomanian unconformity.

Later, the area as a whole was subjected to normal subsidence processes related to cooling of the lithosphere. It can, however, be speculated that due to the large offset of the Romanche Fracture zone, and the average rate of spreading of 2 cm y^{-1}, the southern spreading centre was progressively moving further along the margin. This progressive shift of a 'hot' centre may have produced a reheating of the area and a subsequent uplift.

3. The Northern Mozambique continental margin

The Mozambique Channel (Figure 9) between Eastern African and Madagascar corresponds to a wide, triangular-shaped sedimentary basin divided into two sectors by a N–S trending line of relief known as the Davie Ridge (Segoufin et al. 1978; Scrutton 1978). Plate tectonic reconstructions based on the magnetic anomaly pattern of the Somali Basin imply a southward motion of Madagascar relative to Africa (Segoufin and Patriat 1980; Rabinowitz et al. 1983). This motion occurred during the splitting of Eastern Gondwana, when Antarctica, Madagascar, and India also started to move southward during the Late Jurassic–Early Cretaceous. The drift of Madagascar result then in the creation of the Somali Basin and in the subsequent fabric of a transform-type margin along Northern Mozambique/ Tanzania. The Davie Ridge is inferred to be the morphostructural expression left by the transcurrent motion. Previous marine geophysical results have shown that the ridge is a block faulted feature (mainly along its western flank) superimposed on a sharp landward thickening of the crust (Recq 1982). The ridge is then interpreted as the result of superficial deformation occurring within the sedimentary layers and generated by shearing stresses (Heirtzler and Burroughs 1971).

New data have been recently collected along the Northern Mozambique margin (Figure 9; Mougenot et al. 1986a). They allow the precise definition of the geological structural pattern generated along this portion of this now fossilized transform margin. On morphological grounds (Vanney and Mougenot 1986) the area can be divided into three regions. A northern segment, between 9° and 13°

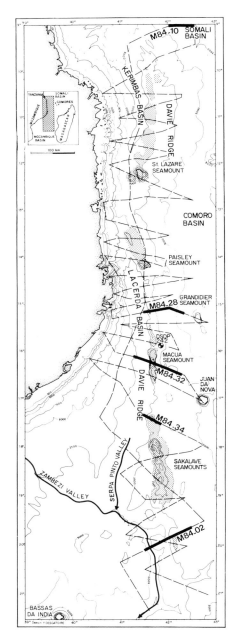

Figure 9. Simplified bathymetric map of the Northern Mozambique continental margin and of the Davie Ridge. The continental slope is disrupted downward by numerous basins : the Kerimbas and Lacerda basins wedged by a N–S trending line of submarine peaks; and the Davie Ridge. Bathymetry in metres (Vanney and Mougenot 1986). Dashed lines show the ship tracks of the MACAMO cruise. Seismic lines shown in Figures 11 to 13 are indicated by a continuous black line. The different seamounts of the Davie Ridge higher than 2,000 metres shown as dotted zones.

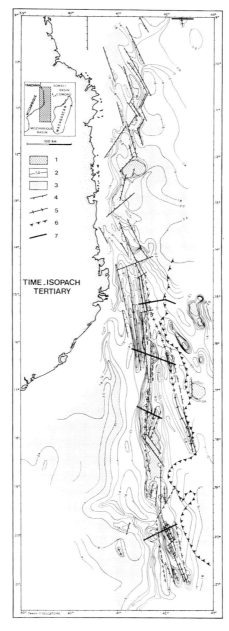

Figure 10. Structural sketch of the Davie Ridge. (1) Suboutcropping acoustic basement; (2) Isopachs of Cenozoic deposits (from sea-floor to reflector B); contour intervals are in seconds of two-way travel time; (3) Cenozoic deposits thicker than 1·6 S (dwtt); (4) Main faults (dashed: buried below Cenozoic); (5) Axis of main structural highs (basement ridge or anticline); (6) Western extension of possible upper Cretaceous volcanic flows; (7) Seismic lines illustrated respectively from north to south on Figures 11 to 13.

S, has a very steep upper continental slope and displays elongated deep basins as well as a N–S alignment of submarine domes forming the Northern Davie Ridge. From 13° to 17° S the narrow margin is bounded on its seaward side by a continuous line of prominent seamounts which constitutes the Central Davie Ridge. Finally, between 17° and 20° S, the Mozambique Channel is cut in a N–S direction by the Southern Davie Ridge.

According to Scrutton *et al.* (1981) and Rabinowitz *et al.* (1983), the Tanzania margin, to the north of the studied area, is well marked by a N–S gravity anomaly, indicating a strong crustal change. Below the thick sedimentary wedge, the margin also shows N–S trending basement features delineating a series of basins and bordering highs (Coffin *et al.* 1985). Our seismic data, only on the southern extremity of the Tanzania margin, also show a rather thick sedimentary wedge dislocated by Neogene distensive faults (Figure 10), interpreted in relation to a submerged branch of the East African rift system (Mougenot *et al.* 1986b). Apart from this recent fracturing, the lower acoustic sequence, believed to be of Cretaceous age, does not show any significant deformation. The only clear structure affecting the Cretaceous terranes is seen seaward (Figure 11).

The feature, which looks like a 20 km wide anticline, may represent a shear stress deformation ('positive flower structure') generated by the transcurrent motion of Madagascar along a fault, eastward of the Davie Ridge. Both the thick Cenozoic sedimentary cover and the Neogene tectonic activity may in fact prevent the observation of more conspicuous deformation at the level of Upper Jurassic–Lower Cretaceous terranes. It should be noted however that the neotectonic activity is concentrated along the N–S directions, which is in good agreement with theoretical trends generated by an early transcurrent motion along the Davie fracture zone (Scrutton *et al.* 1981).

Between 13° and 17° S, the Central Davie Ridge is made of narrow N–S trending basement features, facing the Comoro basin and bounding an elongated slope basin (the Lacerda basin) to the east. The seismic data (Figure 12) demonstrate that the features are tilted horsts, and dredge hauls have yielded volcanic breccias and sediments or metal-sediments like those of the African continental basins (Leclaire 1985). Seismic correlations with DSDP hole 242, indicate that the horsts have been emplaced before the Eocene (Simpson *et al.* 1974). We interpret the structural pattern as the effects of a Late Cretaceous phase of rifting with the creation of a subsequent new margin which resulted in the opening of the Mascarene basin to the east of Madagascar (northeastern motion of India; Masson 1984). As for the northern area, we believe that the N–S trending structures generated during the early transform evolution of the margin, have been reactivated by the new rift.

The Southern Davie Ridge offers a different configuration with the ridge made up of disconnected, NW–SE trending peaks. The relief consists of well layered, but deformed, seismic horizons clearly of a sedimentary nature. The strata are slightly deformed in wide anticlines (Figure 13). Each of the peaks comprises of such 'folds' bordering a fault system (Figure 10) locally intruded by magmatic injections (Sakalave Seamounts; Figure 9). The dating of the tectonized sequences (apparently never eroded) has been tentatively made with the help of bore hole data from Mozambique to the west (Lafourcade 1984) and DSDP hole 242 to the east. The correlations indicate a pre-Cenomanian age for the deformed units. Apparently the deformation phase was active in Early Cretaceous, contemporaneous with the progressive opening of the Somali basin.

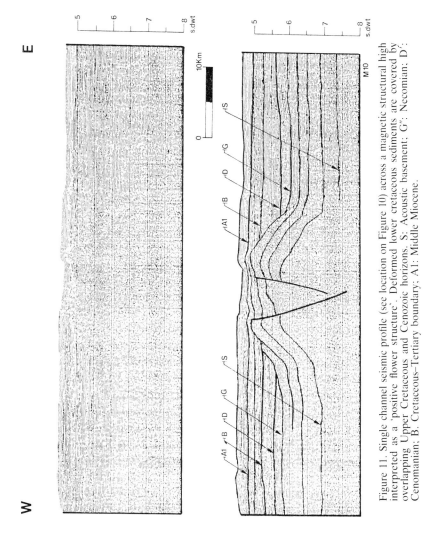

Figure 11. Single channel seismic profile (see location on Figure 10) across a magnetic structural high interpreted as a 'positive flower structure'. Deformed lower cretaceous sediments are covered by overlapping Upper Cretaceous and Cenozoic horizons. S: Acoustic basement; G'': Necomian; D': Cenomanian; B. Cretaceous–Tertiary boundary; A1: Middle Miocene.

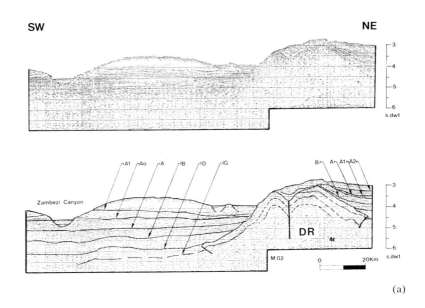

(a)

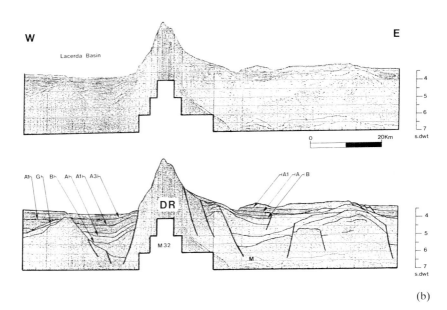

(b)

AFRICAN TRANSFORM CONTINENTAL MARGINS 553

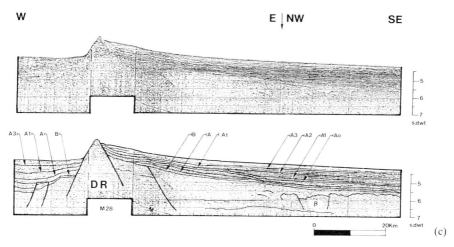

Figure 12. Single channel seismic lines (see location on Figure 10). The central Davie Ridge (D.R.) appears as a probable continental basement horst. The structure appears to have been remobilized as indicated by unconformities within the Tertiary cover. G: Neocomian; B: Cretaceous–Tertiary boundary; B: Volcanic flow; A: Eocene–Oligocene boundary; Ao: Oligocene–Miocene boundary; A1-2: Middle Miocene; A3: Late Miocene; M: Bottom multiple.

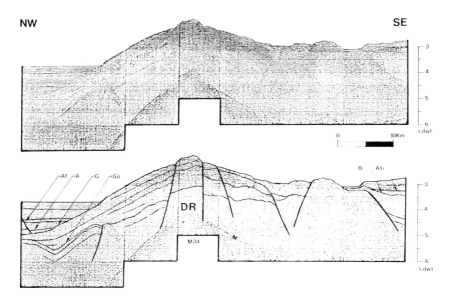

Figure 13. Single channel seismic profiles (see location on Figure 10). The southern Davie Ridge (D.R.) consists of a sedimentary wedge deformed in wide anticlines along a fault system and covered by overlying thick subhorizontal Late Cretaceous to Quanternary sediments. This structure, interpreted as a 'flower structure', may result from a transgressive component during the relative motion of Madagascar and Africa. Go: Latest Jurassic; G: Neocomian; D: Cenomanian; B: Cretacepis–Tertiary boundary; A: Eocene–Ogligocene boundary; Ao: Oligocene–Miocene boundary; A1-2: Middle Miocene; M: Bottom Multiple.

To summarize, the tectonic fabric of the Northern Mozambique margin (Figure 10) results from the superposition of at least three phases. The first phase was the creation of a transform margin, during Early Cretaceous times, as a consequence of the southward transcurrent motion of Madagascar relative to Africa (Figure 14). A second phase relates to the NE–SW opening of the Mascarene basin and to subsequent reactivation of N–S trends on the margin by distensive tectonics during Late Cretaceous times.. The third phase which began during the Mid-Miocene, is still active and is linked to the continued evolution of the East African Rift.

Such a complex polyphase tectonic history does not facilitate the recognition of typical features generated by early transform motion. Despite this, we speculate that the Davie Ridge represents the partial remnants of the structures generated by transcurrent motion. Before the Late Cretaceous remobilization, the Central Ridge resulted from independent small basement-block displacements during a convergent strike-slip motion. These features have been described as 'boudinage effects' by Sylvester and Smith (1976) along the San Andreas Fault. The Southern

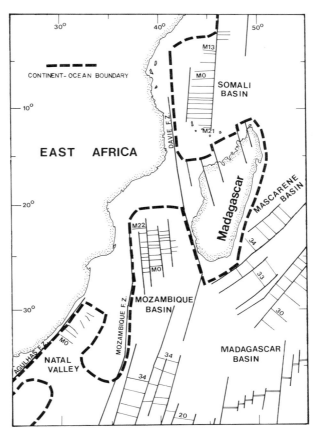

Figure 14. Highly simplified geotectonic sketch of the Eastern African margin and surrounding basins. Inferred ocean–continent boundary (dashed line), major fracture zones and associated oceanic magnetic anomalies are shown. Note the obliquity of oceanic trends between both Somali and Mozambique basins and the younger Mascarene basin.

Ridge illustrates the superficial manifestations as *en echelon* folds or 'positive flower structures' developed in a sedimentary basin in response to crustal shearing. Finally the whole ridge is superimposed on a sharp crustal discontinuity inherited from the transform motion and governing the later deformations.

4. The Southwestern Guinean continental margin

West of Guinea (Figure 15), the African continental margin shows a sudden change in its N–S general direction.

Although the margin is N–S trending off Gambia, the continental slope is oriented E–W off Conakry and includes a wide extension of the margin to the south known as the Guinean marginal plateau (Egloff 1972). The southern slope of this plateau shows a complex topography marked by a series of subcircular magnetic seamounts (Jones and Mgbatogu 1982). For many years the Southern Guinean margin has been interpreted as an extensional fracture zone forming an offset between two areas of the Central Atlantic (Le Pichon and Fox 1971; Jones and Mgbatogu 1982).

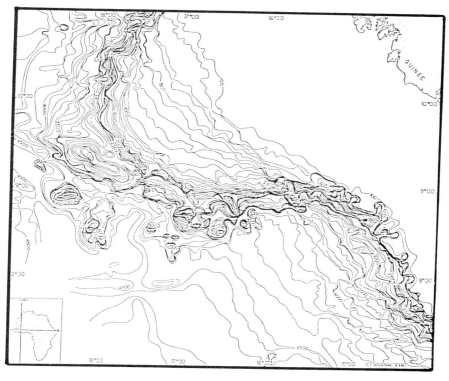

Figure 15. Detailed bathymetry (in metres) of the Guinean marginal plateau and adjacent slopes (from Marinho 1985). The map has been drawn using GEBCO data, Equamarge data, and a few unpublished lines. Note the three contrasting slope areas: to the west a wide dissected slope, to the south a narrow E–W trending slope showing numerous subcircular highs and between both areas a deep wide valley extending between the upper scarp and a series of NW–SE trending rises.

Recently Marinho (1985) and Mascle *et al.* (1986) have combined new data obtained from the whole of the marginal southwestern zone with previous geophysical results, notably those of Jones and Mgbatogu (1982). The results illustrate a dramatic contrast between the western and the southern areas of the Guinean margin. The western margin consists of a thick wedge of flat-lying Cretaceous to Cenozoic deposits and slightly faulted seismic horizons attributed to Jurassic terranes (Marinho 1985). It is believed that the Western Guinean margin represents a portion of the Jurassic Central Atlantic margin, corresponding to the Florida section along the Northeastern American margin (Olivet *et al.* 1974).

The seismic profiles across the southern Guinean slope show rather different characteristics. We observe below the plateau edge (Figure 16) a sequence of N–NE dipping seismic reflectors, displaced by faults and covered by a flat-lying (but eroded) series of well-layered reflectors. The lower sequence, which has been dredged along the slope edge (Mascle *et al.* 1986) is probably made up of Lower Cretaceous terrigenous sediments (including arkosic sandstones). Further correlation with two bore holes drilled off Guinea indicate that the unconformity is of Late Albian to Cenomanian age.

The seismic sections across the middle shelf (Figure 17) show a series of slope basins developing between the slope edge and a line of bordering highs, facing the deep oceanic basin (Sierra Leone basin). The slope basins include a lower seismic sequence, correlated with Lower Cretaceous terranes also unconformably covered by younger sediments. Eastwards, toward the continent, the slope basins and bordering highs are replaced by converging slope scarps where magnetic seamounts develop (Figure 18). These latter subcircular highs represent volcanoes as indicated by one dredge sample which recovered transitional-type basalt. Westwards, towards the Gambia abyssal plain, the slope laterally connects with a

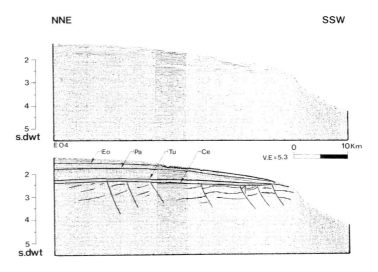

Figure 16. Interpretation of a seismic profile across the southern Guinea plateau. Eo: base of Eocene; Pa: base of Palaeocene; Tu: Turonian; Ce: Top of Cenomanian. Note the deep sequence of northward-dipping reflectors cut by faults and out cropping along the southern slope, where barren sandstones have been dredged up. Note also the southward thinning of the upper sedimentary sequence. Lower Eocene to Miocene sediments have been cored directly on top of the slope.

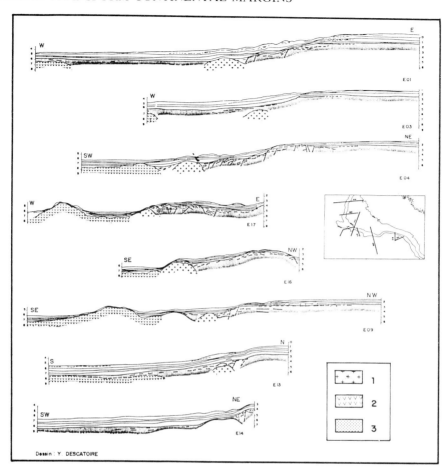

Figure 17. 8 tracing of seismic lines across the Guinea plateau and adjacent slopes (see location in insert). (1) Continental basement; (2) Oceanic basement; (3) Lower Cretaceous deposits.
E01 and E03 illustrate the northwestern margin, E04 and E05 the transition sector; E17, 16, 09, 13 and 14 the southern margin (Guinea Fracture zone) and the Southeastern section near Sierra Leone continental margin.

double scarp (the inner and outer scarp) delineating a sedimented, triangular shaped basin. Two main acoustic sequences separated by a strong unconformity are also present within the basin infill (Figure 19). A series of faults (including a few reverse faults) cut across the lower sequence attributed to Lower Cretaceous deposits.

In summary, the Guinean marginal plateau shows two contrasting features. To the west, the continental margin has many similarities with typical Central Atlantic margins and is therefore interpreted as the southern extension of the rifted Jurassic margins extending from Morocco to Guinea. To the south, we observe, in addition to a general east–west trend, many peculiarities (including pre-Cenomanian tectonized strata, elongated slope basins and bordering highs, important volcanic activity etc.) which indicate a different evolution for the margin. Mascle *et al.* (1986) have

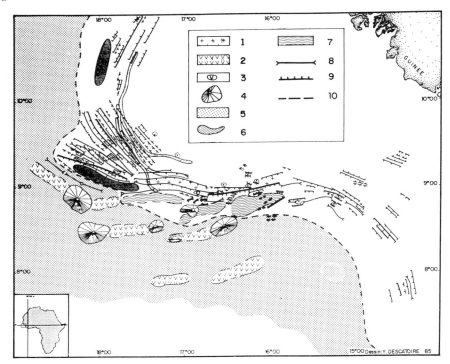

Figure 18. Structural sketch of the Guinean continental slope. (1) Continental basement; (2) Volcanic ridges; (3) Interbedded volcanism; (4) Volcanic dome; (5) Oceanic basement; (6) Lower slope basement ridges; (7) Sedimentary deformed wedge; (8) Main basin axis; (9) Main faults; (10) Extension of oceanic crust detected by seismic data.

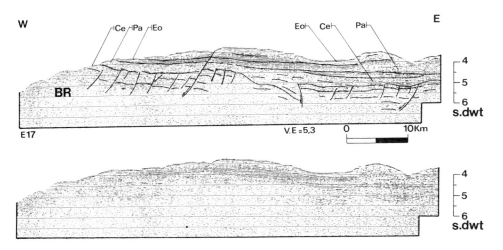

Figure 19. Seismic profile E17 across the lower margin between the western and southern Guinean slopes. Note the tectonically rotated lower sequence, correlated with the Lower Cretaceous and covered unconformably by post-Cenomanian deposits episodically affected by erosion. Faulting appears to have ceased before the Cenomanian unconformity.

proposed that the area was first a proximal southern margin of the Central Atlantic Jurassic margin, affected during the Early Cretaceous by transcurrent motion due to the progressive oblique opening of the Equatorial Atlantic.

If this hypothesis is correct, we can thus consider the Southern Guinean margin as a transform margin located at the junction between two oceanic areas of different ages. This unusual configuration may well explain both the general characteristics of the area as a transform margin and its peculiarities. Shearing stresses were generated by the oblique opening exerted on an earlier Jurassic fracture margin. As a consequence, the fracturing pattern is complex and is expressed as a series of closely spaced fracture zones injected along their trend, by volcanic and subvolcanic magmas.

5. Conclusions

Geological structures generated along three distinct transform margins have been studied. The Ivory Coast–Ghana margin is considered as a typical fossilized transform margin not subjected to later reactivation and resulting from a transcurrent motion between old and probably thick continental blocks of Africa and Brazil.

The transform type evolution of the Northern Mozambique margin ended before final disruption between the two cratonic plates. In fact, Madagascar and eastern Mozambique are still in contact as a continent to continent margin.

The southern Guinean plateau was generated in two episodes. During the first episode the area was evolving as the rifted margin of a Jurassic oceanic area (the central Atlantic); during the second period, a previously thinned and fractured margin was reactivated as a transform margin related to the opening of the Equatorial South Atlantic. Despite these different geodynamic settings, we believe that there are common structural characteristics which have emerged from this comparative study.

Considering the structures generated by the progressive shearing of two continental crusts, the most obvious features, we observe are:

1. A tectonic continental slope with a sharp transition between the cratonic crust and the oceanic crust. This slope may be expressed as a single scarp (off Ghana) or, if the motion is affecting an already fragile crust, by a complex system of closely spaced and disconnected scarps (off Southern Guinea). The resulting slope is not obvious off Mozambique because the transform motion was stopped before continental disruption.
2. The continent to continent motion also generates a system of elongated features in which thick coarse clastics are rapidly deposited with a high rate of sedimentation. Depending upon the prevailing conditions, tensional or compressional constraints can be dominant. The graben cutting across the Ghanean shelf, as well as the normal faults found along the Guinean slope, clearly indicate a prevailing distensive component there. The tensional constraints may have facilitated magmatic injections, as found off Guinea (where multifracturing may have helped this phenomenon) and off Ghana where volcanic rocks have been encountered in a few bore holes.
3. When the transform motion is progressive small rifted basins are created laterally between two fractures. The eastern Ivory Coast basin, the Mozambique Channel, and possibly the intermediate zone southwest of the Guinea

plateau, are examples of such rifted basins. Marine sediments were deposited during the initial opening, and parts of the sedimentary layers particularly near the transform fault, were progressively deformed during their deposition (Ivory Coast basin, southwestern Guinean margin) or just after sedimentary deposition, e.g. along the Davie Ridge. This leads to the formation of prominent positive features such as the Ivory Coast–Ghana Ridge or the Davie Ridge. Such ridges may partly consist of deformed sediments. They may be continuous elongated features (Ivory Coast Ridge) or disconnected features. One may observe a series of smaller ridges in lateral continuity with different continental scarps (Guinean margin) as multiple aligned features due to earlier tectonic lines of weakness.

4. Dependent on the prevailing constraints generated during the transcurrent motion, block faulting, 'en echelon' folds, 'flower structures', or a combination of the three deformations can be dominant. A tensional regime apparently prevailed along the Guinean and Ivory Coast area, while a probable compressive component is suspected along the Davie Ridge.

5. A common characteristic at a transform margin relates to the presence of a strong tectonic unconformity between two sedimentary series. The lower series can be considered as synchronous with the transform motion and records the tectonic effects of the motion, while the upper sedimentary series unconformably covers the syntectonic deposits. In the case of Ivory Coast, Ghana, or Guinea, the unconformity is tentatively explained as the result of relaxation of stress when the cratonic areas finally separated. Off Mozambique the unconformity is more simply explained by the termination of the relative drift between the African and Madagascan continental plates.

Acknowledgements. Contribution n.361 of the GEMC (Groupe d'Etude de la Marge Continentale) U.A. CNRS 718. We acknowledge IFREMER-CNEXO and TAAF (Territoires Australes et Antarctiques Francais) for providing shiptime during the cruises EQUAMARGE (off Guinea and Ivory Coast–Ghana) and MD.40 MACAMO (off Mozambique), respectively on board the R.V. 'Le Suroit' and 'Marion Dufresnes'. Financial support for data processing were obtained through CNRS (ATP GGO) and TAAF.

References

Arens, G., Delteil J. R., Valery, P., Damotte, B., Montadert, L., and Patriat, P. 1971. The continental margin of the Ivory Coast and Ghana. In: Delany F. M., (Ed.), *The Geology of the East Atlantic Continental Margin, 4, Africa,* Institute of Geological Sciences of Great Britain, London Rep. 70/16, 61–78.
Blarez, E., 1986. *La marge continentale de Côte d'Ivoire–Ghana. Structure et évolution d'une marge continentale transformante.* Thèse de doctorat de l'Université Paris VI, Paris. 188 pp.
Coffin, M. F., Rabinowitz, P. D., and Houtz, R. E. 1985. Crustal structure in the Western Somali Basin. In: (Coffin, M. F. (Ed.), *Evolution of the conjugate East African–Madagascar margins and Western Somali Basin.* Ph. D. Thesis, Columbia Univ., New York. 67 pp.
Delteil. J.R., Valery, P., Montadert, L., Fondeur, C., Patriat, P., and Mascle, J. 1974. Continental margin in the Northern part of the gulf of Guinea. In: Burke C. A. and Drake C. A. Eds, *Geology of Continental Margins,* Springer, New York. 297–311.
Egloff, J. 1972. Morphology of ocean basin seaward of Northwest Africa: Canary Islands to Monrovia, Liberia. *American Association of Petroleum Geologists Bulletin,* **54,** 694–706.
Francheteau, J. and Le Pichon, X. 1972. Marginal fracture zones as structural framework of continental margins in South Atlantic. *American Association of Petroleum Geologists Bulletin,* **56,** 991–1007.
Heirtzler, J. R. and Burroughs, R. M. 1971. Madagascar's palaeoposition: new data from the Mozambique Channel. *Science,* **174,** 488–490.

Jones, E. J. and Mgbatogu, C. C. S. 1982. The structure and evolution of the West African continental margin off Guinea, Guinea Bissau and Sierra Leone. In:Scrutton, R. A. and Talwani M. Eds, *The Ocean Floor*, John Wiley and Sons, England. 165–202.

Khan, M.H. 1979. Oil exploration in Ghana, West Africa. *Bulletin of O.N.G.C.*, **16**, 1–18.

Lafourcade, P. 1984. Etude géologique et géophysique de la marge continentale du Sud-Mozambique (17°S à 28°S). Thesis, Univ. Paris VI, Paris. 132 pp.

Lawner, L. A., Sclater, J. G., and Meinke, L. 1985. Mesozoic and Cenozoic reconstructions of the South Atlantic. *Tectonophysics*, **144**, 233–254.

Leclaire, L. 1985. La campagne RIDA/MD.39 dans le canal du Mozambique. *Courr. CNRS*, **60**, 8.

Le Pichon, X. and Fox, P. J. 1971. Marginal offsets, fracture zones and early opening of the North Atlantic. *Journal of Geophysical Research*, **76**, 6294–6308.

— and Hayes, D. E. 1971. Marginal offsets, fracture zones, and the early opening of the South Atlantic. *Journal of Geophysical Research*, **76**, 6283–6293.

MacGrew, H.J. 1984. Oil and gas developments in central and southern Africa in 1983. *American Association of Petroleum Geologists*, **68**, 1523–1599.

Marinho, M. O. 1985. *Le plateau marginal de Guinée. Transition entre Atlantique central et Atlantique équatorial*. Thesis. Univ. Paris VI, Paris. 183 pp.

Mascle, J. 1976. Atlantic-type continental margins: distinction of two basic structural types. *An. Acad. Bras. Sienc.*, **48**, 191–197.

— 1977. Le golfe de Guinée: un exemple d'évolution de marges atlantiques en cisaillement. *Mémoires de la Soc. Géol. France*, **55**, 128, 104.

— 1977. Marinho, M., and Wanneson, J. 1986. The structure of the Guinean continental margin: implications for the connection between central and south Atlantic. *Geologische Rundshau*, **75**, 57–70.

Masson. D.G. 1984. Evolution of the Mascarene Basin, Western Indian Ocean, and significance of the Amirante Arc. *Marine Geophysical Research*, **6**, 365–382.

Mougenot, D., Virlogeux, P., Vanney, J. R., and Malod, J. 1986a. La marge continentale au Nord du Mozambique : résultats préliminaires de la campagne MD40/MACAMO. *Bull. Soc. Géol. Fr.*, **2**, 419–422.

— Recq, M., Virlogeux, P., and Lepvrier, C. 1986b. A seaward extension of the East African Rift: the Kerimbas and Lacerda grabens (northern Mozambique continental margin). *Nature* (in press).

Olivet, J. L. *et al.* 1984. Cinématique de l'Atlantique nord et central. In: *Rapports scientifiques et techniques*. 54, 108 pp. Publ. CNEXO, Paris.

Rabinowitz, P. D. and la Breque, J. L. 1979. The mesozoic Atlantic ocean and evolution of its continental margin. *Journal of Geophysical Research*, **84**, 5973–6002.

—, Coffin, M. F., and Falvey, D. 1983. The separation of Madagascar and Africa. *Science*, **220**, 67–69.

Recq, M. 1982. Anomalies de propagation des ondes P à l'Est de la ride de Davie. *Tectonophysics*, **82**, 189–206.

Scrutton, R.A. 1976. Crustal structure at the continental margin of South Africa. *Geophys. J. R. Astro. Soc.*, **44**, 601–623.

— 1978 Davie fracture zone and the movement of Madagascar. *Earth and Planetary Science Letters*, **39**, 84–88.

— 1979. On sheared passive continental margins. *Tectonophysics*, **59**, 293–305.

— 1982. Crustal structure and development of sheared passive continental margins. In: Scrutton, R. A. (Ed.,). *Dynamics of Passive Margins*. Geodynamics series, **6**, American Geophysical Union New York. 133–140

— Duplessis, A., Barnaby, A.M., and Simpson, E.S.W. 1974. Contrasting structures and origins of the western and southeastern continental margins of southern Africa. In: Campbell, K. S. W., (Ed.), *Proceedings Third International Gondwana Symposium*, Canberra. 651–661.

— Heptonstall, W.B., and Peacock, J.H. 1981. Constraints on the motion of Madagascar with respect to Africa. *Marine Geology*, **43**, 1–20.

Segoufin, J., Leclaire, L., and Clocchiatti, M. 1978. Les structures du canal du Mozambique; le problème de la ride de Davie. *Ann. Soc. Géol. Nord*, **97**, 307–314.

— and Patriat, P. 1980. Existence d'anomalies mésozoïques dans le bassin de Somalie. Implications pour les relations Afrique–Antarctique–Madagascar. *C. R. Acad. Sc. Paris*, **291**, 85–88.

— and Patriat, P. 1981. Reconstructions de l'océan Indien occidental pour les époques des anomalies M 21, M 2 et 34. Paléoposition de Madagascar and Africa. *Bull. Soc. Géol. Fr.*, **23**, 6, 603–607.

Sibuet, J.C. and Mascle, J. 1978. Plate kinematic implications of equatorial fracture zone trends. *Journal of Geophysical Research*, **83**, 3401–3421.

Simpson, E.S.W., Schlich, R. *et al.* 1974. *Initial Reports of the D.S.D.P.*, **25**, U.S. Government Printing Office, Washington. 884 pp.

Sylvester A.G. and Smith, R.R. 1976. Tectonic transpression and basement controlled deformation in San Andreas fault zone, Salton Trough, California. *American Association of Petroleum Geologists*, **60**, 2081–2102.

Talwani, M. and Elpholm, O. 1973. The boundary between continental and oceanic basement at the margin of rifted continents. *Nature*, **241**, 325–330.

Vanney, J.R. and Mougenot, D. 1986. Carte bathymétrique de la 'Marge Nord-Mozambique et de la Chaîne Davie' at the scale 1:1.000.000 *Publ. Mission Recherche, Terres Australes et Antarctiques Françaises*, Paris.

Wilson, J.T. 1965. A new class of faults and their bearing on continental drift. *Nature*, **207**, 343–347.

Quaternary continental margin sedimentation off the southeast coast of South Africa

Frances Westall
Alfred Wegener Institut, Postfach 120161, D–2850 Bremerhaven, F.R.G.

A sedimentological study of Quarternary deposits in a deep-sea channel off the South African continental margin reveals that terrigenous and shelf (T and S) components from the sand fraction are concentrated on the upper continental slope and the channel floor. In comparison, sediments from the lower flanks of the channel contain relatively low amounts of these components. Based on these and seismic studies, four major controls on this distribution pattern have been established: (1) a marginal fracture ridge; (2) fluctuations in sea level; (3) dilution of the T and S components by biogenic carbonate; and (4) redistribution of the T and S components and dissolution of the carbonate by turbulent, corrosive bottom currents. Variability in the T and S component distribution patterns is related to changes in the influence of the last three controls due to climatic fluctuations. During interglacial periods, higher sea levels lead to the trapping of coarse-grained T and S components on the shelf; the shallow carbonate compensation depth (CCD) coupled with turbulent bottom currents leads to the dissolution of the (mainly planktonic) carbonate component on the floor of the Agulhas Passage; and redistribution of the T and S components by strong bottom currents is effected. Glacial periods are characterized by lower sea levels enabling T and S components to be transported directly offshelf into the passage; the deep CCD and weak bottom water turbulence leads to the dilution of the T and S component on the passage floor by carbonate; bottom currents are too weak to effectively redistribute this component.

KEY WORDS Terrigenous components   Shelf components   Sedimentary basin
          Continental margin   southern Africa

1. Introduction

The aim of this paper is to investigate the controls on terrigenous and shelf (T and S) component distribution in a deep-sea channel, the Agulhas Passage, off the southeast coast of South Africa. These components comprise detrital minerals as well as particles originating from the continental shelf, e.g. benthic foraminifera and authigenic glauconite.

Quaternary sediments off the east and southeast coasts of South Africa are being systematically investigated in a number of specific projects coordinated by the University of Cape Town (Westall 1984; Martin 1984; Dingle and Robson 1985). This paper reports part of the investigations carried out in the deep ocean basin area to the southeast of South Africa. Previous studies having a bearing on

0072–1050/87/TI0563–16$08.00
© 1987 by John Wiley & Sons, Ltd.

this project include a detailed investigation of shelf sediment distribution and transport (Flemming 1980, 1981), an investigation of Cenozoic sedimentation patterns on the Agulhas Plateau (Tucholke and Carpenter 1977), as well as general studies of carbonate and clay distribution in the surface sediments of the southwestern Indian Ocean (Kolla *et al.* 1976a, b).

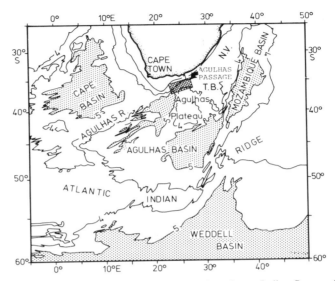

Figure 1. General bathymetry of the southeast Atlantic and southwest Indian Oceans showing location of the Agulhas Passage between the continental slope of South Africa and the Agulhas Plateau. N.V.—Natal Valley, T.B.—Transkei Basin.

The Agulhas Passage is a deep channel, 150 km long and 50 km wide, which has an average depth of 4800 m. It is situated between the continental slope south of Port Elizabeth and the Agulhas Plateau, and links the Transkei Basin to the north with the Agulhas Basin to the south (Figure 1). The passage has a relatively flat floor and is flanked on either side by steep slopes (Figure 2). The continental slope to the northwest of the passage forms part of a sheared continental margin (Scrutton and Du Plessis 1972) and is characterized by the presence of a marginal fracture ridge along the upper continental slope (Figure 3).

Sediment cover in the Agulhas Passage is variable. Sediment accumulations up to 2 s DT (double time) thickness forming large, current-moulded billows cover the continental slope. In contrast, thin sediments of 0·5–0·7 s DT thickness are found on the floor of the Agulhas Passage. Seismic sections across the area (Figure 3) suggest that the Agulhas Passage floor is both sediment starved as a result of sediment trapped upslope of the marginal fracture ridge (MFR), and also subject to strong bottom current scour, as evidenced by locally developed condensed sequences of sediment, erosional features, and current-moulded deposits. The northward thickening sedimentary sequence in the passage indicates that much of the sediment entering the area comes from the north (Natal Valley, Transkei Basin). This is consistent with the above observation that the MFR immediately adjacent to the passage acts as a dam to downslope sediment movement.

Sedimentological analyses of the Quaternary sediments from the Agulhas Passage and adjacent Transkei Basin reveal that these sediments range in type from

Figure 2. Bathymetry of the study area. Note the almost flat floor of the Agulhas Passage sandwiched between the steep continental slope and the Agulhas Plateau.

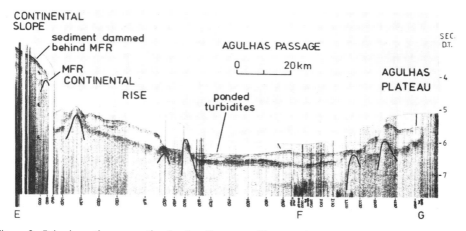

Figure 3. Seismic section across the Agulhas Passage to illustrate the damming effects of the marginal fracture ridge (MFR) on the downslope movement of sediment. Also illustrated are Quaternary turbidites ponded in the deepest part of the Agulhas Passage. The thick line drawn in on the section indicates basement. Location of the section shown in Figure 4.

pelagic foraminiferal oozes (on the Agulhas Plateau and upper continental slope), through hemipelagic sediments (lower flanks of the passage), down to terrigenous (turbiditic) deposits on the central passage floor (Westall 1984).

2. Methods

The sediment cover off the southeast continental margin of South Africa has been systematically mapped by single channel air gun surveys (Figure 4). Thirteen cores varying in length from 50–450 cm have been recovered from representative sediment facies in the study area (Figure 4). The core material was subjected

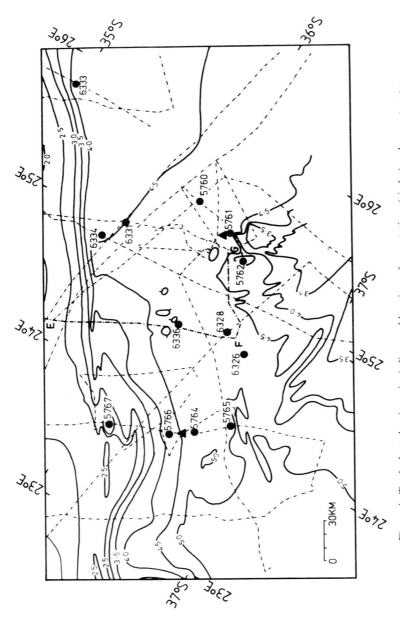

Figure 4. Track chart showing seismic lines and locations of core stations (circles) and current meter stations (triangles). Thickened track line labelled EFG is the track of the seismic record shown in Figure 3.

to standard sedimentological analyses. After being photographed and visually described, samples of about 20 g were taken every 5–10 cm for sedimentological purposes. Further cylindrical samples of about 3 cm³ were taken at 5–10 cm intervals for palaeomagnetic studies.

The analytical procedure was as follows: samples were dialysed for four days to remove the interstitial water and salts, then wet sieved at 63 μ to separate the sand and mud fractions. The pipette method was used to obtain the silt/clay ratio. Size analysis of the total sand fraction (including carbonate) was made with an automated settling tube (Flemming 1977; Flemming and Thum 1978). The wetting procedure of Thiede *et al.* (1976) was followed to enable the foraminiferal component to settle. A Coulter Counter was used to obtain the size distribution of the mud fraction (method after Shidler 1976).

Detrital mineral investigations have concentrated mainly on detailed binocular and scanning electron microscope (Cambridge S180) analysis of the sand fraction. The SEM (which was fitted with an EDAX, and later a KEVEX X-ray analyser) was used particularly to identify heavy minerals in the sediment. Point counts on 300–500 individual grains from each sample were made to determine the relative contributions of the biogenic components (planktonic and benthic foraminfera, radiolaria, and silicoflagellates), terrigenous components (quartz, feldspar, and various heavy minerals), and authigenic components (pyrite and glauconite). Representative silt fraction samples were also examined with the SEM to determine which components were present, but no quantitative analysis was attempted.

The carbonate content of the total sediment was measured using a 'Karbonat-Bombe' (Müller and Gastner 1971), which, for samples containing more than 10 per cent $CaCO_3$, has an accuracy of 2–5 per cent (results are unreliable for samples containing less than 10 per cent $CaCO_3$). Organic carbon determinations of the total sediment were made using the wet chemical method of Morgans (1956), which has an accuracy of ± 30 per cent at the 95 per cent confidence level for samples containing less than 1 per cent C_{org} (Birch, 1975). The composition of the clay minerals in the surface sediments, and also from a few subsurface samples, was determined using a Phillips X-ray diffractometer employing Ni-filtered CuKα radiation (method after Birch 1975). From the resulting diffractograms the proportions of the major clay minerals (illite, smectite, and kaolinite) were semiquantitatively estimated (after Johns *et al.* 1954).

A coarse stratigraphy for the cores was constructed from measurements of the magnetic inclination, declination, and intensity of the small cylindrical samples using a Digico Spinner Magnetometer.

The sedimentological data have been examined in two stages. Firstly, the distribution of the modern T and S components has been mapped from the core top data. From this and auxilliary information (e.g. carbonate content and seismic data) the factors controlling the T and S distribution in the passage have been determined. The second phase of the investigation involved the downcore analysis of these sedimentary characteristics in order to determine changes in the importance of the different controls with time, and, if possible, to relate them to climatic changes during the Quaternary.

3. Results

Detailed binocular and scanning electron microscope investigations of the sand fraction reveal that the terrigenous fraction consists mainly of quartz. Feldspar is a minor constituent, as are heavy minerals. A large variety of heavy mineral species has been identified, e.g. garnet, ilmenite, pyroxenes (including the relatively unstable pyroxene, pigeonite), amphiboles, zircon, chromite, and spinel. These detrital minerals are also in the silt fraction. Sand fraction concentrations in the sediments range from high values of 39–69 per cent on the upper flanks and on the passage floor to low values of 15–28 per on the lower flanks.

The distribution of the terrigenous components in the sand fraction of the surface sediments follows the same pattern as that of the sand fraction. The terrigenous component (quartz, feldspar, heavies) is concentrated on the upper continental slope (more than 40 per cent) as well as on the floor of the Agulhas Passage (more than 50 per cent) whilst, in comparison, the lower continental slope and the Agulhas Plateau are relatively starved of this component (20 per cent or less) (Figure 5).

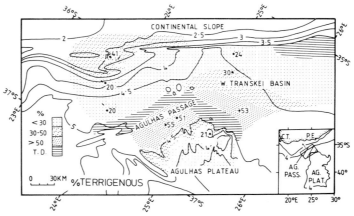

Figure 5. Distribution of the terrigenous component in the (non-turbiditic) surface sediments showing the greatest concentration of this component on the floor of the passage in the vicinity of the turbidite deposits (TD).

Another important terrigenous contribution to these sediments is clay which comprises between 18–51 per cent of the total sediment. The largest amount of clay is found at the base of the continental slope and on the Agulhas Plateau (36–51 per cent). XRD analysis reveals that the most common clay mineral is illite (60–70 per cent) with smectite, and kaolinite almost equally making up the rest of the clay fraction.

The shelf contribution to the sand fraction consists of two components: benthic foraminifera and glauconite (including terrigenous feldspar and quartz grains which have an overgrowth of authigenic glauconite). The latter component forms in the shallow marine environment of the continental shelf. The nearest source of glauconite for the Agulhas Passage is a small area on the continental shelf to the south of Port Elizabeth (Bremner 1978). In comparison to the terrigenous component, the shelf contribution to the sand fraction is of little volumetric importance. Despite this fact, the distribution of the shelf component shows some revealing patterns: benthic organisms generally form less than 2 per cent of the sand fraction, except on the passage floor where they contribute up to 4–5 per cent (Figure 6). Glauconite contents also tend to be low (2–4 per cent), but are similarly higher on the passage floor (up to 15 per cent or more in some horizons of the turbiditic sediments) than elsewhere in the Agulhas Passage.

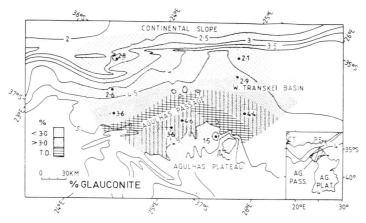

Figure 6. Distribution of glauconite in the sand fraction of the (non-turbiditic) surface sediments showing also the area where relatively large amounts of benthic foraminifera are found (vertical hatching). T.D.—turbidite deposits.

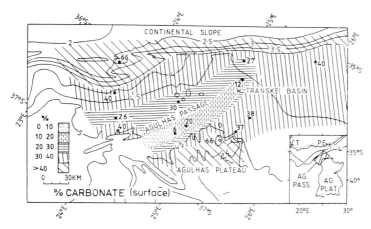

Figure 7. Distribution of carbonate content in the surface sediments. Lysocline occurs at a depth of 4000 m where there is a sharp increase in the rate of dissolution. The CCD occurs at a depth of about 4800 m where less than 10 per cent $CaCO_3$ is found in the sediment. T.D.—turbidite deposit.

The distribution of carbonate in the surface sediments shows a decreasing trend with increasing water depth (Figure 7). The lysocline occurs at a depth of 4000 m, whilst the carbonate compensation depth (CCD) occurs at about 4800 m. Very low carbonate contents are found in most cores from the passage floor and lower flanks of the continental slope (6–27 per cent). Exceptions are those cores in the axis of the channel (30–40 per cent in the turbidite sediments).

The general trends observed in the distribution of the T and S components in the surface sediments are also visible in the downcore sediments (Figure 8): cores from the very deepest part of the passage have higher concentrations of these components than those at the base of the continental slope or on the Agulhas Plateau. (Very little data is available from the upper continental slope because an unconformity at 25 cm depth in core 5767 (Figure 4) juxtaposes Middle Eocene sediment with Plio-Pleistocene sediment.) Cores from the passage floor exhibit

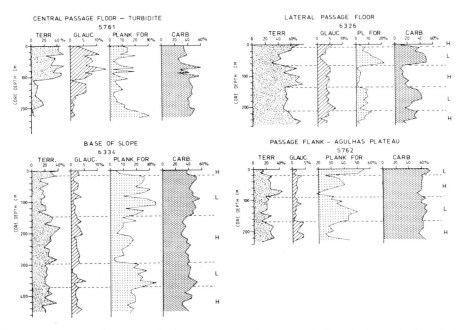

Figure 8. Downcore fluctuations in the terrigenous, glauconite (shelf) and planktonic foraminiferal components, as well as the total carbonate contents, of representative cores from each of the four main physiographical areas in the Agulhas Passage: passage flank, base of slope, passage floor–turbidite (central passage), and lateral passage floor (non-turbidite). These plots show a decrease in the amount of T and S components with increasing distance from the turbidite deposits: passage floor–base of slope–Agulhas Plateau. Also apparent is the negative correlation between T and S component concentration and the amount of planktonic foraminifera. Downcore cyclical compositional fluctuations have been labelled H (high T and S, low planktonic foraminifera) and L (low T and S, high planktonic foraminifera) horizons and have tentatively been correlated with interglacial and glacial periods respectively.

decreasing T and S component concentrations with increasing distance from the passage axis.

Marked downcore fluctuations in the concentrations of the terrigenous, shelf (glauconite), and planktonic foraminiferal contents are especially observable in cores from the floor of the Agulhas Passage (Figure 8). Cores from the Agulhas Plateau and lower flanks also exhibit these fluctuations, but they are not as strong. The terrigenous and shelf component fluctuations are positively correlated, but negatively correlated to those of the planktonic foraminifera. On the basis of these correlations, various subsurface horizons within the cores have been distinguished: those having high concentrations of T and S components and low concentrations of planktonic foraminifera (H horizons), and those having low T and S component concentrations coupled with high planktonic foraminifera (L horizons).

The vertical distribution of carbonate with depth in the cores shows that many of the cores have low surface carbonate values compared to subsurface values, in particular cores from the passage floor and lower flanks. Beneath this surface dissolution layer, however, carbonate contents are higher, although cores from the passage floor especially, and to a lesser extent those from the lower slope, contain subsurface dissolution horizons. The average downcore carbonate contents (calculated excluding data from the subsurface dissolution horizons) range from

34–49 per cent on the passage floor and base of slope, and between 56–63 per cent on the upper flanks (Figure 9). These data indicate that the subsurface lysocline remains at 4000 m but the higher carbonate values in cores from the floor of the Agulhas Passage suggest that the CCD must be deeper than 5000 m, except during the periods recorded by the subsurface dissolution horizons.

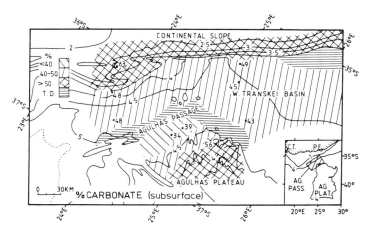

Figure 9. Distribution of average carbonate content in the subsurface sediment (excluding dissolution horizons). Lysocline is still at 4000 m, but relatively high carbonate contents on the passage floor show that the CCD is not present in the Agulhas Passage. T.D.—turbidite deposits.

4. Discussion

Clues to the origin of the terrigenous component are provided especially by the composition of the heavy mineral fraction. This composition suggests that the detrital minerals originate largely from the Karoo sediments, volcanics, and intrusive dolerites of Southern Africa, with some contribution from the metamorphosed basement granites, schists, and quartzites along the Natal coast and in Swaziland. The presence of unstable minerals such as pigeonite indicate rapid offshelf transport and deposition.

Of the clay minerals, illite is typically the weathering product of an arid continental climate and it is therefore to be expected that it makes up the bulk of the clay fraction in this area. The smectite probably originates from onland weathering of the Karoo basalts and perhaps also from *in situ* weathering of numerous basement peaks and outcrops in the Agulhas Passage area. The southerly transport of sediment from the Natal Valley explains the presence of kaolinite, a weathering product produced in the humid climate, in this area.

The distribution pattern of the T and S components in the Agulhas Passage suggests that there are four main controls on the influx and deposition of this component. The most important control is a structural feature — the marginal fracture ridge on the upper continental slope. Seismic reflection records show that sediment has been dammed on the upper continental slope behind the MFR (Figure 3). This has had the effect of preventing the direct downslope transport of T and S components from the immediately adjacent continental shelf, a fact which is born out by the distribution of T and S components in the surface sediments: instead of the gradual decrease in T and S components with increasing distance from the continental shelf that would be expected if there was direct downslope transport, we find that the base of the slope area is relatively starved of these components with respect to the upper slope and passage floor. This pattern suggests that the T and S components on the passage floor are transported into the area from elsewhere. Textural, compositional and structural data from the cores originating from the very deepest part of the Agulhas Passage and adjacent Transkei Basin (5761, 6336, 5765, Figure 4) show that these sediments are, in fact, turbidites. The textural data and the thickness of individual turbidite flows on the passage floor indicate that the source of the sediment is to the north of the passage — the texture of the turbidites in the passage coarsens northwards and individual turbidite flows thicken in the same direction.

Another major control of the introduction of T and S components into the passage is the height of the sea level. During the Quaternary sea levels have been influenced by climatic fluctuation (Vail et al. 1977). The lowering of sea levels during glacial periods led to the exposure of large areas of continental shelf. River and eroded shelf sediment was therefore transported directly into the deep ocean basins from the shelf edge. Conversely during interglacial periods sea levels are high and much of this material is trapped on the shelves. Evidence of a large influx of flat-surfaced, highly acoustically reflective turbiditic sediment into the Agulhas Passage in recent geologic history is found in the seismic records (Figure 3).

Dilution of the T and S component by biogenic carbonate (in this case, largely planktonic foraminifera) is a third control on the distribution of these sediments. The overall decrease in the carbonate content in the surface sediment (Figure 8) with increasing water depth is not completely paralleled by a corresponding increase in the T and S components with increasing water depth. From the observation that, on the upper continental slope, high T and S component concentrations are coupled with an overall high carbonate content, two facts may be deduced: firstly, that the bulk of the mud fraction must be of carbonate origin and secondly, that the coarse-grained T and S fraction must be transported by downslope gravity processes rather than by bottom current transport (the poor sorting of the upper continental slope sediments in comparison to the lower slope and passage floor sediments is supporting evidence for the lack of effective bottom current transport: Westall 1984).

Apart from the upper continental slope sediments, the negative correlation of carbonate component with T and S component strongly suggests that dilution by biogenic carbonate is an important influence on the distribution of this component. Observational and experimental evidence show that, in the world ocean, carbonate concentration generally decreases with increasing water depth and decreasing water temperature (Berger 1968, 1969, 1970, 1974, 1976). The data from the Agulhas Passage are no exception. The level at which there is a sharp increase in the rate of carbonate dissolution in the sediments, the lysocline, occurs at a depth of 4000 m in this region. The surface sediment data show that the modern, interglacial CCD occurs at a depth of 4800 m. Some subsurface sediment horizons on the floor of the Agulhas Passage have much higher carbonate concentrations than the surface sediments (Figure 8), thus indicating that, at times, the CCD occurred at depths greater than those found in the passage (i.e. greater than 5000 m). Binocular microscope analysis of the components confirm that in these horizons (L Horizons in Figure 8) and the T and S component in the passage floor sediments is diluted by a large input of planktonic carbonate.

There are a number of factors involved in the dissolution of carbonate (Volat et al. 1980) and therefore the dilution of the T and S component. One of the most important factors in the Agulhas Passage region is the nature of the bottom water mass. The modern bottom water in the study region is the Antarctic Bottom Water (AABW) which is very cold, having temperatures less than 1°C (Westall 1984; Camden-Smith et al. 1981), and undersaturated with respect to $CaCO_3$. Carbonate sediments coming into contact with the AABW are therefore readily dissolved. However, bottom water turbulence also plays an important role in the dissolution of carbonate, in the sense that the chemical diffusion layer at the sediment/water interface is broken up by turbulence, thus enhancing carbonate dissolution (Schink and Guinasso 1977a, b). A long-term current meter study (8–12 months) in the Agulhas Passage (Camden-Smith et al. 1981; Westall 1984) has revealed that turbulence is a characteristic of the bottom water in the Agulhas Passage. Peak velocity events with speeds ranging from 50–80 cm^{-1} have been recorded. The modern, shallow CCD is believed to be due to the presence of cold, corrosive, turbulent AABW in the Agulhas Passage (Westall 1984). Conversely, carbonate-rich horizons in the subsurface sediments indicate that, at certain periods during the past, the AABW was not present. This leads us to the hypothesis that the dissolution horizons (H Horizons in Figure 8) may generally represent interglacial periods whilst the intervening horizons (L Horizons) reflect glacial periods. The absence of AABW in the Agulhas Passage during glacial periods is very probable since this water mass has to pass through the southwest Indian Ridge in order to reach the Agulhas Basin/Passage and during glacial times, when the production of AABW was reduced, this would not have been possible (Volat et al. 1980).

Discussion of the influence of the AABW and carbonate dissolution on the dilution of the T and S component in the sediments leads us to the fourth factor controlling the distribution of these sediments, namely, their redistribution by strong bottom current activity. The distribution of T and S components in the surface sediments showing decreasing concentration with increasing distance from the turbidite deposits in the deepest areas of the passage (Figure 6) suggests that these deposits are the main source for the T and S components. Strong bottom currents are obviously effective in redistributing the turbidite sediments to the extent that T and S components are found upslope of the turbidites, even at a depth of 3900 m on the flanks of the Agulhas Plateau. The ability of the modern bottom currents to erode, transport and redeposit sediments on the floor and lower flanks of the Agulhas Passage has been verified in a sediment dynamics study (Westall 1984). This study concluded that, although the average bottom current velocities measured in the Agulhas Passage were not high enough to erode the sediments (6–20 cm^{-1}), intermittent peak velocities were (50–80 cm^{-1}), i.e. although sediment erosion and transport may not be continuous, it is intermittent and effective. Further evidence of this comes from the seismic records which show modern erosional channels and current-moulded bedforms in the surface of the turbidite deposits.

5. Model for Quaternary sedimentation in the Agulhas Passage

Based on the distribution of T, S and carbonate components in the surface sediments, and also in the subsurface sediments, a model of sedimentation has been developed to illustrate changes in the influence of the various factors controlling sedimentation in the Agulhas Passage during glacial and interglacial periods. These latter may be regarded as endmembers of a climatic continuum.

In constructing this model, the surface sediments are presumed to be of Holocene age and therefore representative of interglacial conditions. Preliminary oxygen isotope and nannoplankton studies (Fincham and Winter 1986) suggest that, at least on the floor of the Agulhas Passage, modern sediments have either not been deposited, or have been eroded and/or subject to dissolution. Data from the tops of cores for which there is a strong suspicion that much sediment has been removed (i.e. pre-Holocene sediment exposed) have therefore not been included in this exercise. A detailed O_2 isotope stratigraphy is at present being constructed (Winter personal communication) which will be used to check this presumption.

According to this model, during interglacial periods (H Horizons in Figure 8) large amounts of coarse T and S components are found on the passage floor due to the proximity of the turbidite source, dissolution of the carbonate by the cold, corrosive, and turbulent AABW and redistribution of the turbiditic sediments by this bottom water mass (Figure 10). Smaller amounts of T and S components are found on the lower flanks due to increased distance from source and dilution by carbonate because of deposition above the CCD.

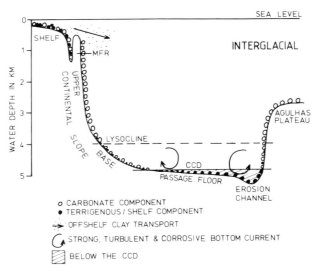

Figure 10. Model of terrigenous and shelf component distribution in the Agulhas Passage during the Quaternary (Part 1) — the *interglacial* period is characterized by a high sea level, shallow CCD, turbulent bottom water. The size of the particles drawn is representative of grain size, i.e. coarse-grained particles on the flanks, fine-grained at the base of the slope and medium-grained on the passage floor. The section is orientated NW–SE.

During glacial times (L Horizons in Figure 8) there is a change in the controls on the sedimentation pattern in the Agulhas Passage (Figure 11). Lower sea levels enhance the T and S component contribution leading to turbiditic sedimentation on the passage floor. The absence of the AABW leads to T and S component dilution by biogenic carbonate. During glacial periods, the AABW in this region

CONTINENTAL MARGIN SEDIMENTATION

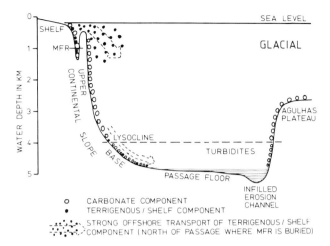

Figure 11. Model of terrigenous and shelf component distribution in the Agulhas Passage during the Quaternary (Part 2) — *glacial* period with low sea level, deep CCD and weak bottom currents. Other symbols as in Figure 10.

is replaced by a warmer, less turbulent, less corrosive, and weaker bottom water mass which cannot effectively redistribute the turbiditic sediments. Apart from the carbonate data, there is also evidence of this from the seismic records which show no indication of erosive bottom current activity in the pre-Holocene sediment cover (Westall 1984): e.g. ancient erosion channels have been infilled by turbiditic sediments and erosional surfaces have been covered by other deposits.

In conclusion, a stratigraphic model of glacial/interglacial alternations in the environment of sediment deposition in the Agulhas Passage has been constructed (Figure 12) showing that there is not much change in the glacial and interglacial sediments deposited on the upper continental slope, except that the lower rate of biogenic productivity in the surface waters during glacial episodes will mean a lower sedimentation rate for these sediments. Interglacial sediments at the base of the continental slope, however, contain a larger amount of T and S component than glacial sediments, but the sedimentation rate of the latter should be lower than during glacial times because of dissolution of the carbonate component. In the very deepest part of the Agulhas Passage interglacial sediments will be comparable to those of the rest of the passage floor, although the T and S component is more important. The glacial sediments will, however, consist entirely of turbidites deposited with a very high sedimentation rate. On the flanks of the Agulhas Passage only interglacial sediments contain T and S components. The sedimentation rate of these sediments is higher than during glacial times.

6. Conclusions

There are four main controls on the sedimentation pattern of the terrigenous and shelf components in the Agulhas Passage.

1. *Structural control.* The marginal fracture ridge on the upper continental slope effectively prevents the direct transport of T and S components by

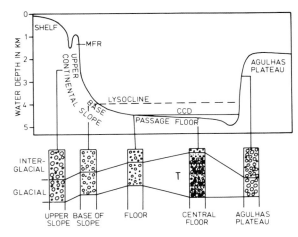

Figure 12. Model of terrigenous and shelf component distribution in the Agulhas Passage during the Quaternary (Summary). The expected fluctuations in T and S component distribution in time and space are illustrated here in a profile of sections across the passage. During interglacial times, coarse, mainly planktonic sediments with a small T and S component accumulate on the passage flanks; finer, hemipelagic sediments are found at the base of the continental slope; and medium-grained terrigenous sediments accumulate under the CCD on the passage floor. In glacial times there is not much change in the pattern of sedimentation on the passage flanks, except that the sedimentation rate (represented by the thickness of the deposits) is lower because of lower surface biogenic productivity, whilst on the flanks of the Agulhas Plateau there are no T and S components because of weak bottom currents. Sediments deposited at the base of the continental slope contain fewer T and S components for the same reason. On the deepest part of the Agulhas Passage floor turbidite deposits accumulate at a high sedimentation rate. In areas of the passage floor adjacent to these deposits the T and S components are diluted by biogenic carbonate because of deposition above the CCD, leading to a higher rate of sedimentation than during interglacial periods.

downslope gravity processes into the Agulhas Passage.

2. *Sea level*. Higher sea levels during interglacial periods trap especially the coarser T and S components on the continental shelf. Only the fine component is transported offshelf at offsets along the edge of the continental shelf by the strong surface current, the Agulhas Current. During lower glacial sea levels the T and S components are transported directly to the shelf edge, leading to a higher rate of terrigenous sedimentation in the passage.

3. *Lysocline and CCD*. These phenomena control the dilution of the T and S components by biogenic carbonate. The lysocline occurs at about 4000 m depth in both glacial and interglacial sediments, whilst the CCD is shallower (4800 m) during interglacial than glacial periods (greater than 5000 m). Passage floor glacial sediments are therefore diluted by biogenic carbonate whilst interglacial sediments are not.

4. *Bottom currents*. Bottom current turbulence contributes to the dissolution of carbonate and therefore to the dilution of the T and S components. The data indicate that interglacial bottom water (AABW) is more corrosive than glacial bottom water. Redistribution of the T and S components from their source in the deepest part of the Agulhas Passage (the turbidite deposits) is controlled by strong bottom current activity, but during glacial times, bottom current circulation is generally weak and redistribution by this means is not possible.

Acknowledgements. This research was supported by Council for Scientific and Industrial Research grants and scholarships from the University of Cape Town. The research was carried out at the University of Cape Town. Prof. D. Fütterer and Drs E. Klatt and H. Grobe are thanked for their helpful comments on the original manuscript. Frau G. Dansauer is thanked for her help with the photography.

References

Berger, W.H. 1968. Planktonic foraminifera: selective solution and palaeoclimatic interpretation. *Deep-Sea Research*, 151, 31–43.
—— 1969. Foraminiferal ooze: solution at depth. *Science*, 156, 383–395.
—— 1970. Planktonic foraminifera: selective solution and the lysocline. *Marine Geology*, 85, 111–138.
—— 1974. Deep sea sedimentation. In: Burk, C.A. and Drake, L.A. (Eds), *Geology of Continental Margins*. Springer, New York, 242–261.
—— 1976. Biogenous deep sea sediments: production, preservation and interpretation. In: Riley, J.P. and Chester, R. (Eds), *Chemical Oceanography*, 5, Academic Press, London, 261–388.
Birch, G.F. 1975. Sediments on the continental margin off the west coast of South Africa. *Joint GSO/UCT Marine Geoscience Bulletin*, 6. Department of Geology, University of Cape Town.
Bremner, J.M. 1978. Surficial sediments in Algoa Bay. *Joint GSO/UCT Marine Geoscience Technical Report*, 10, 66–74.
Camden-Smith, F., Perrins, L–A., Dingle, R.V. and Brundrit, G.B. 1981. A preliminary report on long-term bottom current measurements and sediment transport/erosion in the Agulhas Passage, Southwest Indian Ocean. *Marine Geology*, 39, M81–M88.
Dingle, R.V. and Robson, S. 1985. Slumps, canyons and related features on the continental margin off East London, SE Africa (SW Indian Ocean). *Marine Geology*, 67, 37–54.
Fincham, M.J. and Winter, A. 1986. Agulhas Current imprint on oxygen isotopes of Southwest Indian Ocean surface sediments. *Joint GSO/UCT Marine Geoscience Technical Report*, 16, 208–216.
Flemming, B.W. 1977. Depositional processes in Saldana Bay and Langebaan Lagoon. *Joint GSO/UCT Marine Geoscience Bulletin*, 8, Department of Geology, University of Cape Town.
—— 1980. Sand transport and bedform patterns on the continental shelf between Durban and Port Elizabeth (Southeast African continental margin). *Sedimentary Geology*, 26, 179–205.
—— 1981. Factors controlling sediment dispersal along the South-east African continental margin. *Marine Geology*, 42, 259–277.
—— and Thum, A.B. 1978. The Settling Tube — a hydraulic method for grain size analysis of sands. *Kieler Meeresforschung*, 4, 82–95.
Johns, W.D., Grim, R.E. and Bradley, W.F. 1954. Quantitative estimations of clay minerals by diffraction methods. *Journal of Sedimentary Petrology*, 24, 242–251.
Kolla, V., Bé, A. H. W. and Biscaye, P. E. 1976a. Calcium carbonate distribution in the surface sediments of the Indian Ocean. *Journal of Geophysical Research*, 81, 2605–2616.
——, Henderson, L. and Biscaye, P.E. 1976b. Clay mineralogy and sedimentation in the western Indian Ocean. *Deep-Sea Research*, 23, 949–961.
Martin, A.K. 1984. Plate tectonic status and sedimentary infill of the Natal Valley (S.W. Indian Ocean). *Joint GSO/UCT Marine Geoscience Bulletin*, 14, Department of Geology, University of Cape Town.
Morgans, J.F.S. 1956. Notes on the analysis of shallow water soft strata. *J. Anim. Ecol.*, 25, 367–387.
Müller, G. and Gastner, M. 1971. The 'Karbonat-Bombe', a simple device for the determination of the carbonate content in sediments, soils and other materials. *N. Jb. Miner. Mh.*, 10, 466–469.
Schink, D.R. and Guinasso, N.L. 1977a. Effects of bioturbation on sediment-seawater interaction. *Marine Geology*, 23, 133–154.
—— and —— 1977b. Modelling the influence of bioturbation and other processes on calcium carbonate dissolution on the sea floor. In: Anderson, N.R. and Malahoff, A. (Eds), *Fate of Fossil CO_2 in the Ocean*. Plenum, New York, 379–399.
Scrutton, R.A. 1976. Crustal structure at the continental margin south of South Africa. *Earth and Planetary Science Letters*, 19, 250–256.
—— and Du Plessis, A. 1972. Possible margin fracture ridge south of South Africa. *Nature*, 242, 180–182.
——, ——, Barnaby, A. and Simpson, E.S.W. 1974. Contrasting structures and origins of the western and southeastern continental margins of southern Africa. *Proceedings 3rd International Gondwana Symposium*, Canberra, 1973.
Shidler, G.L. 1976. A comparison of electronic particle counting and pipette techniques in routine mud analysis. *Journal of Sedimentary Petrology*, 46, 1017–1025.
Thiede, J., Chriss, T., Clauson, M. and Swift, S.A. 1976. *Settling Tubes for Size Analysis of Fine and Coarse Fractions of Oceanic Sediments*. Oregon State University, 87 pp.
Tucholke, B.E. and Carpenter, G.B. 1977. Sediment distribution and Cenozoic sedimentation patterns on the Agulhas Plateau. *Geological Society of America Bulletin*, 88, 1337–1346.

Vail, P.R., Mitchum, R.M. and Thompson, S. 1977. Seismic stratigraphy and global changes in sea level. In: Payton, C.E. (Ed.), *Seismic Stratigraphy — Applications to Hydrocarbon Exploration. American Association of Petroleum Geologists Memoir*, **26**, 49–212.

Volat, J-L., Pastouret, L. and Vergnaud-Grazzini, G. 1980. Dissolution and carbonate fluctuations in Pleistocene deep-sea cores — a review. *Marine Geology*, **34**, 1–28.

Westall, F. 1984. Current-controlled sedimentation in the Agulhas Passage, SW. Indian Ocean. *Joint GSO/UCT Marine Geoscience Bulletin*, **13**, Department of Geology, University of Cape Town.

Index

Accessory minerals 403–426
Acid dyke swarms 190, 201
Acritarchs 126–127, 134–137
Aerial photography: interpretation
  Mozambique 153
  Sudan 225–247
Age determinations
  Africa (general) 350
  Algeria 439
  Arabian-Nubian shield 163, 164
  Australia 351
  Cameroon 459
  Canary Islands (Fuerteventura) 444
  Corsica 362–372
  Guinea 441, 460
  Liberia 460
  Mali 58, 189–198, 441
  Morocco 439
  Niger 311
  Nigeria 311, 457–458, 460, 486–487
  South Africa 5, 6
  South America 350
  Sudan 163, 181
  Swaziland 5, 6
  West Africa 380
  Zaire 123–125

  K-Ar Nigeria 485–489, 491
  Rb-Sr Algeria 213
    Angola 88–92
    Mali 185–208
    Nigeria 272, 276–278
    Sudan 181–183
  Sm-Nd Sudan 163
Albitization Corsica 369
  Niger 317
  Nigeria 276, 280–281, 313–326, 403, 404, 413, 414, 417, 421, 422
  Saudi Arabia 317
Algeria
  Atlas domain 434–437, 440
  Hoggar 57–77, 213–223
Alkaline basalt 240, 451–452, 479, 481, 515–533
  granites
    Algeria 213–223
    Angola 303

Cameroon 300, 304, 458–459
Corsica 361–383
Egypt 304
Ethiopia 304
Libya 300
Malawi 306
Mali 196–202, 302, 306–307
Namibia 300, 302, 303
Niger 300, 304, 308, 378
Nigeria 297–311, 378, 389–390, 403–426, 457–459
Saudi Arabia 298–311
Sudan 166–167, 232–233, 239, 300, 304–311
lavas 457–468, 499, 517–518
provinces 297–311, 361–383
Alpine chain, north Africa 433–440
Amphiboles compositional variations 215, 314
Angola 85–102, 303
Anorogenic (see Magmatism alkaline)
Anorthosite 139–157, 374
Antarctica 351, 352, 353, 354
Arabian-Nubian shield 161–174, 175, 181, 225–247
Archaean
  climate 45–55
  cratons 140
  layered basic rocks 147
  Barberton granite/greenstone succession 5–22
  Cheshire Formation 45–55
  Diepgezet Group 9, 12
  Fig Tree Group 7
  In Hihaou granulite complex 57–77
  Malolotsha Group 9, 13
  Moodies Group 7
  Nyanzian system 25–43
  Onverwacht Group 7, 9
  Swaziland Supergroup 6
Atlantic ocean, African continental margins
  Ghana 539–547
  Guinea 555–559
  Ivory Coast 539–547
opening 433, 463
Australia 341, 351, 352, 353
Autocorrelation analysis 490

Banded iron Formation 7, 27
Basin oceanic (coastal)
  Agulhas 564
  Cameroon 459
  Cape 564
  Guinea 441–444
  Kerimbas 496
  Lacerda 496
  Madagascar 554
  Mauritania 441–443
  Mozambique 554, 564
  Senegal 441
  Somali 550, 554
 inland
  Chad 450–453
  Garoua 459
  Mamfe 429, 451, 468
  Niger 449–450
  Saharan 447–453
  Tanganyika 499
  Taoudenni 448–449
Batholith
  Algeria 213–223
  Corsica 362–363, 365, 367, 371, 378, 379
  Mali 185–208
  Sudan 178, 179
Bathymetry 540, 548, 555
Bayuda desert 231
Benue trough
  Abakaliki anticlinorium 430, 454–458, 469–491
  kink bands 476
  lower Benue valley 454, 468, 485
  middle Benue 455–459, 468
  strain ellipse (ammonites) 478–479
  structural domains 469
  upper Benue 455, 468
  Workum hills 470–491
Beryllium minerals 104, 256–257, 259, 260, 261, 262, 263, 264, 265, 266, 267
Biostratigraphy 127, 467
Biotite compositional variations 216
Bismuth minerals 263
Brazil 378
Burundi 106

Caesium minerals 256
Cainozoic (Cenozoic) 300, 303–304, 495–509, 537
Calc-silicates 153
Cameroon 300, 304, 311, 313–318, 321, 323–324, 328–330, 429
  line 380–381, 458–459
Canary islands 444–445
Cape Verde 445

Carbonates 47, 123, 471, 563–576
  compensation depth (CCD) 569, 572–573, 574
Carbonatites 299, 305, 361
Central Africa 103–119
Chad 450–453
Charnockite 141–143
Climatic fluctuations 563, 572
Coastal basins (for details see Basin)
Continental
  collision 185–208, 362–363
  crust 155, 361, 372, 377, 559
  drift 341–356
  fragmentation 371–373
  margins 429–430, 537–559, 563–565
  slope 539, 541, 543, 546, 547–549, 559
Corsica 361–383
Craton
  Central African 86
  Congo 104
  In Ouzzal 58
  Tanzanian 503
  West African 204
  Zimbabwe 139, 254
Cretaceous deformation 429, 467–491
Crustal contamination 367–368
Crystallinity index 482
Cyanobacteria 47, 52–53

Digital image processing 228
Diorite 232–233, 236
Dolerite dyke swarms 92, 123, 145–147, 154–156, 197, 460

Earthquake
  epicentre seismic activity 498, 500, 501, 511
East Africa 495–509
Egypt 161–167, 226, 231, 240–243, 304, 315, 321
Ethiopia 161–168, 304, 509
Europe 361–383

Fluid
  composition 318
  element mobility in 319–321, 396, 399, 401
  inclusions 40, 318–319
  origin 331–333
Foraminifera 563, 568, 570
Fuerteventura (Canary Islands) 444–445

Gabbro 139–157, 232–233, 236, 374, 377
Gemstones 253, 257, 262, 264, 266, 267
Geochemistry
  alkaline granites 213–223, 319–321,

INDEX 581

    327–333, 389–401
    alkaline lavas 392–394, 519–526
    ankaratrite 520–521
    anorthosite 149–151
    basanite 520–521
    calc-alkaline to alkaline granitoids 190–208, 213–223
    gabbro-norite 149–151
    greenstones 29
    hawaiite 520–521
    norite 149
    Pan-African basement 396–399
    pegmatites 278–287
    prospecting 279
    pyroxenite 149
    syenite 149, 481
    trachyandesite 392–294
    transitional basalts 481, 482
Geomorphological processes 234–236
Ghana 537, 539, 546, 559
Glaciation 341, 343, 346, 348, 349, 354, 355
Gneiss 139, 140–145, 155, 163, 236–237
    plagioclase 143–144
Gold
    Arabian-Nubian Shield 168–169
    Sudan 171, 175, 179, 226, 231–234, 240–243
    Zaire 104, 114
Gondwana 341–356, 429
Graben (see Rift)
Granites 140, 142, 143, 148, 153
    anorogenic 389–390, 394–401 (see Magmatism alkaline)
    Archaean 5, 8
    calc-alkaline to alkaline 190–208, 213–223
    hypersolvus 213–223, 362, 368–369, 377
    red, Angola 88–92
    subsolvus 201, 213–233, 361, 368–369, 371, 373, 377, 378, 379
Granulites 57–64, 139-143, 148, 155, 185, 187–189, 204–205
Greenstone belt
    Barberton, South Africa 5–22
    Belingwe, Zimbabwe 45, 47
    Migori, Kenya 25–43
Greisenization 313–317, 403–404, 415–416, 421, 423–424
Guinea 441, 460, 537–539
Gulf of Guinea 467, 444, 537, 539, 555–560

Hoggar (Algeria) 213–223
Hydrothermal alteration
    Barberton 8, 9–11, 14–15

Cameroon 311–324
Corsica 369
Mali 202
Mozambique 143–144, 156
Niger 311–324
Nigeria 280, 311–317, 369, 377, 380, 403–426
Saudi Arabia 311–324
Sudan 311–324
    fluids 311, 318, 369, 377, 381, 382–383

Iforas (Mali) 185–208
Illite crystallinity 455, 482–483
Image geological map 225–247
    processing 225–247
Indian ocean, African continental margin
    Agulhas passage, South Africa 564
    Mozambique 547–554
Inland basins (see Basin)
Intraplate domain 433
Island arc 175, 181–182, 185–187, 192–193, 205, 241
Isotopes
    ages 163, 164, 185–208, 364, 485–489, 491
    carbon 45–52
    oxygen 45–51, 574
    ratios 365–368, 369, 270, 379
    rehomogenization 369
    strontium isotopic ratios 191–205, 365–370, 379
    sulphur 332–334
    systematics 328–331, 332–333
Ivory Coast 444, 537–547, 559–560

Kaolinization 143–144
Karoo (=Karroo) 145, 147
Kasai, Zaire 121–137
Kenya 25–43, 161–165, 168, 171, 496, 502–503
Kerguelen Islands 369
Kerogen 45, 48–52
Kibaran
    cycle, Angola
        Cahama-Otchinjau Formation 85–100
        Chela Group 85–100
        Leba-Tchamalindi Formation 85–100
        Luana Group 97–100
        Malombe Group 97–100
        Oendolongo (s.s) Group/Formation 85–100
    deformation 107, 115
    granites 88–92
    pegmatites 253–254, 259–264, 277
    structural evolution 107, 115
Kimberlites 303, 460

Landsat imagery 225–247
Layered intrusions 151
Liberia 460
Libya 300
Lithium minerals 104, 110, 254–256, 260, 261, 262, 263, 264, 265, 267
Lopolith 156, 157
Lufilian arc 140
Lysocline 569, 571, 576

Madagascar 253, 254, 256, 266–267, 537, 539, 547–555
Magma
  evolution 185–208
  fractionation 369–370, 374–375, 382
  mixing 392
  source 299, 327–331
Magmatism
  alkaline
    Angola 303
    Cameroon 300, 304, 459
    Canary Islands (Fuerteventura) 444
    Corsica 361–383
    Egypt 167, 304
    Ethiopia 304
    Guinea 441
    Libya 300
    Malawi 306
    Namibia 300, 302, 303
    Niger 300, 304, 308, 378, 450
    Nigeria 261, 300–311, 361–383, 389–390, 400–401, 403–404, 455–460, 479–481
    Sudan 167, 232–233, 236, 239, 240, 300, 304–311
    Tanzania 499
    Western Mediterranean 362–363, 365, 367, 371, 378, 379
    Zaire 107–108, 115, 517–533
  calc-alkaline
    Iforas batholith, Mali 191–196
    northern Red Sea Hills 179, 181, 182
    western Mediterranean 371
  cordilleran-type 191–193
  gabbro-anorthosite Mozambique 157
  pegmatites, Nigeria 280–281, 281–291
  tholeiitic
    Chad basin 451
    East African rift 495, 508
    Fuerteventura (Canary Islands) 444
    Guinea 441
    Maio Island (Cape Verde) 445–446
    Western Rift, Zaire 517
  transitional basalts, dyke swarms Western Mediterranean 371

transitional calc-alkaline to alkaline
  granites Algeria 213–223
  granites Mali 185–208
  basalts, dyke swarms Western Mediterranean 371
Malawi 140–144, 305–306
Mali 58, 181, 185–208, 302, 306–309, 378, 380, 442
Mediterranean 361–383
Mesozoic 232–233, 235–236, 239–240, 300–303, 362–383, 471–491, 537–560
Metamorphism 236, 237
  amphibolite-granulite facies 163–166
  anchizonal to epizonal 454, 482
  conditions of (PT paths) 71–76
  contact aureoles 430, 483
  granulite facies 57–77, 140–148, 176, 204–204
  greenschist facies 5–22, 27, 31, 107, 180, 185–190, 430
  mineral assemblages 58–64
  retrogressive 58, 145, 153
Metasedimentary rocks 6–13, 127, 163, 176, 182, 187–189, 232–233, 237
Metasomatism 8, 154, 403–426
  acid 313–326, 403–424
  late stage 313–326
  potash 313–326, 403, 414, 416
  silica 9–11, 15, 313–327
  sodic 276, 280–281, 313–327, 403–404, 413–422
Metavolcanic rocks 8, 13, 27–28, 237
Microclinization 321–326, 403, 404, 416, 423
Mineral chemistry
  biotite 74
  boulangerite 39
  bournonite 39
  clinopyroxene 73, 529–531
  cordierite 65, 71
  freibergite 39
  garnet 65, 68–69
  gudmundite 37
  olivine 529
  orthopyroxene 64, 65–67
  plagioclase 72, 532
  sapphirine 65, 70
  titanomagnetite 532
Mineralization 104, 108–119, 167, 168–171
  exploration 225–247
  greenstones 15–22, 33–40
  lode 403–426
  pegmatite 271–288, 253–272
  structural controls 33
  styles 323

INDEX 583

Cameroon 321–327
Kenya 25–26, 31–40
Mozambique 139
Niger 321–327
Nigeria 321–327, 403–426, 467
Saudi Arabia 321–327
Sudan 167–171, 175, 240–243, 321–327
Western Mediterranean 369
Monzodiorite 195–196
Monzonite 195–196
Morocco 434–441
Mozambique 139–157, 266, 537, 539, 547–555, 559–560
  belt 139, 140, 154, 163, 171, 253–257, 303
MSS imagery 226, 229, 234, 243, 244, 496
Mylonite 17, 177, 179–180

Namaqualand 263
Namibia 255, 256, 258, 264–266, 300, 302–303
Net-veining 374
Niger 280–304, 308–310, 313–317, 323–324, 329, 380–381, 449–450
  delta 467–468
Nigeria 271–288, 297–340, 369–383, 389–401, 411–426, 433–466, 457–460
Nile valley 163, 165, 168, 171, 176
Niobium minerals 258, 259, 260, 261, 263, 264, 265, 266, 267, 276, 280, 284–286, 403–426
Norite 139–157
North Africa 429, 433–440
Nubian
  desert 225–247
  Formation 232–233, 235–236, 239
  Province 304, 312, 317, 429 (see also Arabian-Nubian shield)

Oceanic
  basins 495, 539, 546 (for details see Basin)
  bottom currents 573, 574, 576
  closure 185–208
  crust 495, 509, 538, 541, 542
  sediments origin of 571, 572
Olivine compositional variations 374–375
Ophiolites 22, 165, 167, 177, 179, 182, 372
Orogeny belts 538
  Alpine fold belt 371, 372, 377, 429, 434–440
  Damaran 253, 254, 264
  Eburnean 86, 185, 187, 188, 189
  Hercynian (Variscan) belt 361–363, 379

Irumide belt 139, 140, 154
Kibaran 103–119, 154 (see also separate entries)
Lomamian 107
Pan African (see also separate entries)
Trans-Saharan belt 185–208
Zambezi belt 139, 140, 154
Ores and economic mineral deposits (see also Gold) 168–172, 272, 276, 285, 321–327, 403–426
  albite (cleavelandite) 256, 258
  amblygonite-montebrasite 254, 260–261
  apatite 263
  base metals 26, 33–43, 171–172, 241, 403–427, 467–491
  beryl 256–257, 259–263
  cassiterite 104, 258, 260, 261, 262, 264, 265, 267, 272, 276, 285–287, 369,403–426,
  chromite 168
  chrysoberyl 256–257
  chromite 168
  columbite-tantalite 104, 110, 258, 260, 403–426
  corundum 148
  eucryptite 256, 262
  euxinite 263
  fergusonite 263
  ferrocolumbite 258
  gadolinite 263
  graphite 142
  haematite 413, 417, 421, 422
  hafnian-zircon 262
  ilmenite 143, 409, 412, 416, 417, 420, 421, 424
  iron 7, 27
  lithiophilite-triphylite 254, 263
  manganotantalite 258
  monazite 257, 267
  nigerite 258
  origin 287
  oxides 35, 168, 314–317, 321–327, 403–426
  petalite 256
  pollucite 256, 263
  pyrochlore 259, 417, 420
  spodumene 255, 259, 263
  stannite 258
  stratiform 171–172
  sulphide 26, 33–43, 241, 314–317, 321–327, 403–426
  syngenetic 109
  tapiolite 258
  titanomagnetite 532
  uranium minerals 104, 109, 139, 147, 241–242, 403–443

wodginite 258
wolframite 262, 264, 404, 418
xenotime 257, 266, 267

Palaeomagnetism 341–356
Palaeontology 125, 478
Palaeostress 495–509
Palaeozoic 231, 236, 272, 300–302, 341–356, 362, 377–381
   stratigraphic record 345–351
Palynology 121–137
Pan-African granites
   Algeria 213–223
   Arabian-Nubian shield 171
   Mali 181, 185–208
   Sudan 178, 179
   West Africa 378, 379
   orogeny 361, 367, 377, 378, 379, 380
   pegmatites 271–276 et seq.
   stress regimes 185–208
Pegmatites
   Madagascar 266–269
   Mozambique 253–272, 266
   Namibia 253–254, 263, 264–266
   Nigeria 271–288
   Rwanda 260
   South Africa 263–264
   Sudan 168, 232–233
   Uganda 259
   Zaire 110, 260–261
   Zimbabwe 261–263
Permo-Trias 362–383
Phanerozoic 305–340, 361–383, 467–491, 537–560
Photogeological mapping 226
Placer deposits 240–242
Plate
   collision 175–183, 185–211, 433–440
   motion 341–356
   subduction 181–182, 185, 195, 206
   sutures 175–183, 185–186
Polar wandering 341–356
Proterozoic Bushimay Supergroup 121–137
   upper 139–157, 161–163, 175–183, 231–236, 272–274
   middle Buganda-Toro system 105
   lower Kidal assemblage 190–191
Pyroxene compositional variations 216, 374–375, 530–531
Pyroxenite 147

Quaternary 431, 563–577

Rare earths
   alkaline granites 168, 221–222, 320–321

alkaline lavas 522, 527
basic rocks 150–152
halogen complexes 399
minerals 257, 260, 266, 267, 403–426
pegmatite 110
RBV imagery 226, 228, 234, 496
Reconnaissance geological mapping 225–247
Red Sea Hills 165, 166, 168, 171, 175–183, 231
Remote sensing 225–247, 430, 495–496
Rifts (grabens)
   geometry 496–501
   geomorphology 430, 495, 496, 500
   model 506–510
   palaeostress orientations 501–506
   Benue (see separate entries)
   Bukavu 515–533
   Bongor 451
   Central African 451–453
   Chad 453
   Djourab 451–453
   Doba-Birao 451–453
   East African 430, 495–510, 550–554
   Edward 496
   Ethiopian 496–497
   Gongola 451
   Gregory 495–509
   Kafra 451–452
   Kivu 496, 515–533
   Madagascar 550
   Malawi 495
   Messen'ya 451
   Mobutu 496, 498
   N'Géledji 452
   Rukwa 499–501
   Tanganyika 496–508
   Tefidet 450–452
   Ténéré 452–453
   Termit 451–453
   Western 430, 495–510, 515–533
   Yola 451
Ring complexes
   alkaline 167, 170, 177–178, 202, 304–308 et seq., 350–351, 361, 365–377, 378, 381, 390, 403, 450, 457–458
   calc-alkaline 166, 177, 178
Rwanda 103, 253, 255, 260

Satellite imagery 224–247, 496
Saudi Arabia 161, 163, 165, 166, 168, 171, 298, 304, 306, 308, 313, 315, 317, 323, 324, 328
Sea level changes 563, 572, 574–576
Sedimentary

INDEX

basins 343, 345, 346, 347, 348, 349, 355, 467, 468, 496, 499, 537, 542, 547, 560, 572
minerals 567, 568, 571
rocks detrital 345, 346
Sedimentation
 continental 232–233, 235–236, 239–240
 glacial 572, 573, 574, 575, 576
 interglacial 572, 573, 574, 575, 576
 marine 344–349
 shelf 431, 563–577
 structural control on 430, 485, 575
 terrigenous components 431, 563–577
Sediments
 sampling 565, 567
 volcaniclastic 9, 10, 16, 26–27, 165–166, 172, 185, 189, 190, 191–192, 195
Seismic profiles 496, 498, 537, 541–545, 548–553, 556–558, 564–567, 571, 572
Shaba province, Zaire 103, 121–137
Sierra Leone coastal region 443
Sinai 161
SIR imagery 228, 229, 234
Somalia 161, 163, 165, 168, 171
South Africa 5–22, 263–264, 342, 346, 347, 348, 350
 coastal waters 563–577
 tectonics 501
South America 343, 344, 345, 346, 347, 348, 349, 350–351, 354, 355, 539, 541, 542, 559
Spinifex textures 10
Stromatolites 45–55, 123, 125
Structures
 cleavage 469–476, 482
 controls on mineralization 31
 deformation 14–22, 107, 203, 467–491
 evolution 5–22, 104, 107, 115, 122–126, 181–183, 189, 434–462, 469–479, 495–509, 526–527
 fault motions 437–441, 499, 501, 509
 fracture zones 235, 429, 443
 kink bands 476
 lineaments 153, 297–304, 429–430, 460–463, 498–499
 lithospheric faults 361, 363, 441
 setting 58, 14–22, 139–157, 163–167, 176–179, 186–190, 231–233, 272, 300–304, 327, 362–365, 468–469, 526–529
 shear zones
  African continental margins 537–560
  Algeria 58–59
  Central Africa 462
  Mali 58, 187–190

Mozambique 140
Nigeria 274, 301, 363, 372–373, 382, 469, 474–478
North Africa 461
Saudi Arabia 175–183
Sudan 175–183, 240
West Africa 452, 454, 457, 461, 463
slickenslides 499, 501, 509
thrusts 18-20
transform faults 443, 495–509, 537–559
Sudan 161–171, 175–183, 225–247, 300, 304, 313, 315, 317, 318, 323, 324
Swaziland 5, 6
Syenite 148, 153, 479, 480

Tantalum minerals 253, 258, 260, 261, 262, 263, 264, 265, 266, 267, 272, 284–286, 403–425
Tanzania 103, 260, 499, 503
Tectonism 161, 165, 175–183, 185–208, 363, 467–491, 495–509
Troughs (see Benue; Rifts)
Terranes 166–172, 236–239
Tertiary
 alkaline lavas 515–533
 Kerri-Kerri Formation 455
Tete province 139–157
Thorium minerals 403–426
Tillites 343, 346, 348, 349
Tin mineralization 104, 109–118, 168, 253, 258, 260, 261, 262, 265, 267, 271–275, 282–286, 369, 403–426
Tonalite 193–195
Trace elements
 alkaline granites 319–321, 327–333, 392–401
 alkaline lavas 519–528
 cassiterite 278, 286
 greisens 319–321, 404–426
 hydrothermal rocks 319–321, 404–426
 pegmatites 278–282
 transitional granitoids 213–223
Trachytic dyke swarms 459
Transgressions 349, 355
Tuareg shield 214
Tungsten mineralization 104, 109–118, 168, 404, 418
Tunisia 434–441
Turbidites 572, 573, 574, 575

Uganda 103, 161, 163, 171, 253, 254, 255, 259, 260, 305
Ultramafic intrusions 109

Virunga volcanic chain 500, 509
Volcanic rocks
 alkaline basalts 517–533

andesites   195
calc-alkaline   362–363, 371
mafic-ultramafic   8, 13, 27
phonolite   458, 499
pillow lavas   27, 123
tholeiitic basalts   452, 459, 517
trachyte   458

Ethiopia   509
Kenya   26
Nigeria   392–394, 479–481
South Africa   8, 13
Zaire   123–124, 501, 515–533

West Africa   204, 351–354, 361, 377–383, 451–463, 537–547, 555–560
Wilson cycle   185, 186

Yemen   161, 165, 166
Yttrium minerals   257, 266

Zaire   103, 110, 121–137, 253–255, 260–261, 515–533
Zambia   103, 154
Zimbabwe   45–54, 154, 254, 255, 256, 261–263
Zircon zoned   408–426